Levodopa-Induced Dyskinesia in Parkinson's Disease

Susan H. Fox • Jonathan M. Brotchie
Editors

Levodopa-Induced Dyskinesia in Parkinson's Disease

Springer

Editors
Susan H. Fox
Toronto Western Hospital
Toronto, ON
Canada

Jonathan M. Brotchie
Toronto Western Hospital
Toronto, ON
Canada

ISBN 978-1-4471-6502-6 ISBN 978-1-4471-6503-3 (eBook)
DOI 10.1007/978-1-4471-6503-3
Springer London Heidelberg New York Dordrecht

Library of Congress Control Number: 2014951222

Printed on acid-free paper

Springer is part of Springer Science+Business Media (www.springer.com)

Preface

Parkinson's disease (PD) is one of the most common neurodegenerative diseases, with the incidence rising along with the aging population. Treating the motor symptoms of PD with the dopamine precursor, levodopa, is extremely effective and, to this day, one of the successes of modern clinical pharmacology. Unfortunately, the almost unavoidable development of involuntary movements, levodopa-induced dyskinesia (LID), reduces the functional benefit of levodopa and impacts on PD patients' quality of life. While the degree of functional impact varies such that not all patients have, or will develop, LID that is bothersome, globally the problem and its impact on individuals and the socioeconomic well-being of society is one that will only increase. Indeed, LID limits the utility of many PD drugs, including some that have only recently reached the market, that would otherwise improve the motor symptoms.

In this book, we have brought together the leading experts in the field to provide a comprehensive text on the current state of the art with respect to understanding LID. The book covers all aspects of LID from clinical phenomenology; current treatment and management practices; clinical trial design; epidemiology; and finally, a comprehensive review of preclinical and clinical studies into the pathophysiology of LID, including plasticity, where we review a wide range of neurotransmitter systems. Of course, all of these also provide insights into the neural mechanisms underlying nonmotor basal ganglia functions that may extend into a better understanding of behavioral and cognitive issues that also affect PD patients as well as having potential to be applied to hyperkinetic movement disorders outside PD.

We hope this text provides an A to Z of LID and gives both neuroscientists and practicing clinicians an insight into the management, both current and future, of LID.

Toronto, ON, Canada Susan H. Fox
Toronto, ON, Canada Jonathan M. Brotchie

Contents

Contributors

Matthieu Bastide Department of Neurodegenerative Diseases, Université de Bordeaux, Institut des Maladies Neurodégénératives, UMR 5293, Bordeaux, France

CNRS, Institut des Maladies Neurodégénératives, UMR 5293, Bordeaux, France

Erwan Bezard Motac Neuroscience Ltd, Manchester, UK

Department of Neurodegenerative Diseases, Université de Bordeaux, Institut des Maladies Neurodégénératives, UMR 5293, Bordeaux, France

CNRS, Institut des Maladies Neurodégénératives, UMR 5293, Bordeaux, France

Christopher Bishop, PhD Behavioral Neuroscience Program, Department of Psychology, Binghamton University, Binghamton, NY, USA

Isabelle Boileau-Boire Movement Disorder Clinic, University of Toronto, Toronto Western Hospital, Toronto, ON, Canada

Tanuja Bordia Center for Health Sciences, SRI International, Menlo Park, CA, USA

Jonathan M. Brotchie Toronto Western Research Institute, Toronto Western Hospital, Toronto, ON, Canada

Paolo Calabresi Laboratorio di Neurofisiologia, Fondazione Santa Lucia, Rome, Italy

Clinica Neurologica, Facoltà di Medicina e Chirurgia, Università degli Studi di Perugia, Ospedale S. Maria della Misericordia, Perugia, Italy

Manolo Carta Department of Biomedical Sciences, Section of Physiology, Cagliari University, Cittadella Universitaria, Monserrato, Italy

M. Angela Cenci Basal Ganglia Pathophysiology Unit, Department of Experimental Medical Science, Lund University, Lund, Sweden

Kallol Ray Chaudhuri, DSc, MD, FRCP Department of Neurology, National Parkinson Foundation International Centre of Excellence, King's College Hospital, King's Health Partners, London, UK

Miguel Coelho Clinical Pharmacology Unit, de Medicina Molecular, Faculty of Medicine, University of Lisbon, Lisbon, Portugal

Department of Neurosciences, Neurology Service, Hospital de Santa Maria, Lisbon, Portugal

Carlo Colosimo, MD Departments of Neurology and Psychiatry, Sapienza University of Rome, Rome, Italy

Joaquim J. Ferreira Clinical Pharmacology Unit and Laboratory of Clinical Pharmacology and Therapeutics, Instituto de Medicina Molecular, Faculty of Medicine, University of Lisbon, Lisbon, Portugal

Department of Neurosciences, Neurology Service, Hospital de Santa Maria, Lisbon, Portugal

Camino Fidalgo Department of Biomedical Sciences, Section of Physiology, University of Cagliari, Cittadella Universitaria, Monserrato, Italy

Susan H. Fox Movement Disorder Clinic, University of Toronto, Toronto Western Hospital, Toronto, ON, Canada

Veronica Francardo Basal Ganglia Pathophysiology Unit, Department of Experimental Medical Science, Lund University, Lund, Sweden

Andrea Giuffrida Department of Pharmacology, Biomedical Research Development, University of Texas Health Science Center, San Antonio, TX, USA

Biomedical Research Development, University of Texas Health Science Center at San Antonio, San Antonio, TX, USA

Christopher G. Goetz, MD Department of Neurological Sciences, Rush University Medical Center, Chicago, IL, USA

Michael P. Hill, PhD Toronto Western Research Institute, Toronto, ON, Canada

Philippe Huot Division of Neurology, Department of Pharmacology, Faculty of Medicine, University of Montreal and Centre Hospitalier de l'Université de Montréal, Montreal, QC, Canada

Beom S. Jeon, MD, PhD Department of Neurology, Seoul National University Hospital, College of Medicine, Seoul National University, Seoul, South Korea

Tom H. Johnston, PhD Division of Brain, Imaging and Behaviour – Systems Neuroscience, Toronto Western Research Institute, Toronto, ON, Canada

Vincent A. Jourdain, MSc Neuroscience Research Unit, Centre de recherche du CHU de Québec, Quebec, QC, Canada

Faculty of PharmacyLaval University, Quebec, QC, Canada

Regina Katzenschlager Department of Neurology, Karl Landsteiner Institute of Neuroimmunological and Neurodegenerative Disorders, Danube Hospital/Donauspital, Vienna, Austria

Wai Kin D. Ko Department of Neurodegenerative Diseases, Université de Bordeaux, Institut des Maladies Neurodégénératives, UMR 5293, Bordeaux, France

CNRS, Institut des Maladies Neurodégénératives, UMR 5293, Bordeaux, France

Jee-Young Lee, MD Department of Neurology, Seoul National University-Seoul Metropolitan Government Boramae Medical Center, Seoul, South Korea

Department of Neurology, College of Medicine, Seoul National University, Seoul, South Korea

Alex Martinez Department of Biology, Trinity University, San Antonio, TX, USA

Department of Pharmacology, Biomedical Research Development, University of Texas Health Science Center, San Antonio, TX, USA

Marcelo Merello, MD, PhD Raul Carrea Institute for Neurological Research (FLENI), Universidad Católica Argentina, CONICET, Buenos Aires, Argentina

Raul Carrea Institute for Neurological Research, Buenos Aires, Argentina

Nicolas Morin, B.Pharm, MSc Neuroscience Research Unit, Centre de recherche du CHU de Québec, Quebec, QC, Canada

Faculty of Pharmacy, Laval University, Quebec, QC, Canada

Renato P. Munhoz, MD, MSc Division of Neurology, Department of Medicine, Toronto Western Hospital, University of Toronto, Toronto, ON, Canada

Jose A. Obeso, MD, PhD Movement Disorders Laboratory, Neuroscience Department, Center for Applied Medical Research (CIMA), University of Navarra, Pamplona, Spain

Michael S. Okun, MD The Center for Movement Disorders and Neurorestoration, College of Medicine, University of Florida, Gainesville, FL, USA

Department of Neurology, University of Florida College of Medicine, Gainesville, FL, USA

Corinne Y. Ostock Behavioral Neuroscience Program, Department of Psychology, Binghamton University, Binghamton, NY, USA

Therese Di Paolo, PhD Neuroscience Research Unit, Centre de recherche du CHU de Québec, Quebec, QC, Canada

Faculty of Pharmacy, Laval University, Quebec, QC, Canada

Xiomara Perez Center for Health Sciences, SRI International, Menlo Park, CA, USA

Barbara Picconi Laboratory of Neurophysiology, Fondazione Santa Lucia, I.R.C.C.S., Rome, Italy

Maryka Quik Center for Health Sciences, SRI International, Menlo Park, CA, USA

Paula Ravenscroft, PhD Division of Brain, Imaging and Behaviour – Systems Neuroscience, Toronto Western Research Institute, Toronto, ON, Canada

Michael Samuel, MD, FRCP Department of Neurology, National Parkinson Foundation International Centre of Excellence, King's College Hospital, King's Health Partners, London, UK

East Kent Hospitals NHS Foundation Trust, William Harvey Hospital, Ashford, Kent, UK

Ines Trigo Damas, PhD Movement Disorders Laboratory, Neuroscience Department, Center for Applied Medical Research (CIMA), University of Navarra, Pamplona, Spain

Elisabetta Tronci Department of Biomedical Sciences, Section of Physiology, University of Cagliari, Cittadella Universitaria, Monserrato, Italy

Danhui Zhang Center for Health Sciences, SRI International, Menlo Park, CA, USA

Panagiotis Zis, MSc, MRCPsych, PhD Department of Neurology, National Parkinson Foundation International Centre of Excellence, King's College Hospital, King's Health Partners, London, UK

Department of Neurology, Evangelismos General Hospital, Athens, Greece

Chapter 1
Phenomenology of Levodopa-Induced Dyskinesia

Panagiotis Zis, Kallol Ray Chaudhuri, and Michael Samuel

Abstract Levodopa has been effective against the motor features of Parkinson's disease for several decades. However, it is observed that long-term treatment with levodopa can be complicated by the development of various types of response fluctuations as well as dyskinesias. The latter, once established, tend to remain persistent although they can be reduced by some pharmacological and neurosurgical manipulations. These situations can lead to a significant source of disability, and their treatment options require significant expertise and costs. Therefore, efforts are made to minimize or prevent the appearance of long-term dyskinesia and fluctuations. In this chapter, we will consider the phenotypes of levodopa-induced dyskinesias.

Keywords Levodopa-induced dyskinesia • Phenomenology • Levodopa • Dyskinesia • On-off

---

Electronic supplementary material Supplementary material is available in the online version of this chapter at 10.1007/978-1-4471-6503-3_1. Videos can also be accessed at http://www.springerimages.com/videos/978-1-4471-6502-6.

P. Zis, MSc, MRCPsych, PhD
Department of Neurology, National Parkinson Foundation International Centre of Excellence, King's College Hospital, King's Health Partners, London, UK

Department of Neurology, Evangelismos General Hospital, Athens, Greece

K.R. Chaudhuri, DSc, MD, FRCP
Department of Neurology, National Parkinson Foundation International Centre of Excellence, King's College Hospital, King's Health Partners, London, UK

M. Samuel, MD, FRCP (✉)
Department of Neurology, National Parkinson Foundation International Centre of Excellence, King's College Hospital, King's Health Partners, London, UK

East Kent Hospitals NHS Foundation Trust, William Harvey Hospital, Ashford, Kent, UK
e-mail: m.samuel@nhs.net

S.H. Fox, J.M. Brotchie (eds.), *Levodopa-Induced Dyskinesia in Parkinson's Disease*, DOI 10.1007/978-1-4471-6503-3_1

Introduction

Since the description of "paralysis agitans" by James Parkinson, in 1817, it took about 150 years for an effective symptomatic therapy to be found for some of the motor symptoms [1]. In the 1950s, it was believed that the surgical interventions of pallidotomy (which is distinct from today's pallidotomy) and thalamotomy were more effective compared to the medical treatment that was available at the time [2]. However, studies in the 1950s [3, 4] led, in 1960, to the discovery that catecholamines are abnormally metabolized in Parkinson's disease (PD), thus making dopamine precursors an appealing treatment option [5–8].

Since the 1960s [9], numerous trials with either oral or intravenous levodopa (L-3, 4-dihydroxyphenylalanine) in PD have been reported, with mostly favorable outcomes. These findings suggested that levodopa is effective against rigidity and especially akinesia and that it acts probably through transformation into dopamine in the brain. Moreover, combination with monoamine oxidase inhibitors or surgery was considered also effective. The early and short-term reported side effects of the use of levodopa included nausea, loss of appetite, and hypotension [1].

Years later, however, it was observed that long-term treatment with levodopa is often complicated by the development of various types of response fluctuations as well as dyskinesia [10, 11]. The latter, once established, tend to remain persistent although can be reduced by some pharmacological manipulations. When such fluctuations occur, there sometimes arises the difficulty of clinically balancing "OFF" symptoms (tremor, akinesia, rigidity) with "ON" dyskinesia and fluctuations. These situations are difficult to treat and sometimes require the introduction of invasive or surgical therapies for PD [12]. These situations lead to a significant source of disability, and treatment options require significant expertise and costs. Therefore, efforts are made to minimize or prevent the appearance of long-term dyskinesia and fluctuations [13]. In this chapter, we will consider the phenotypes of levodopa-induced motor complications as both fluctuations and levodopa-induced dyskinesia tend to occur simultaneously. Before doing so, we provide some definitions. An "OFF" period applies to the motor state where the patient is experiencing tremor, akinesia, or rigidity either because medication has not been taken, or because its effect has worn "OFF," or because medication was taken but it did not have an effect. On the contrary, an "ON" period applies to the motor state of the patient when these symptoms, including mobility, have been reduced/abated by treatment [14].

Response Fluctuations in Levodopa Treatment

Treatment with levodopa offers motor and some nonmotor symptom control [15]. Especially in early initiation of levodopa treatment, a sustained response can be achieved in relatively small doses. However, with disease progression this "honeymoon" period is reduced and patients experience episodes of recurring "OFF" parkinsonian symptoms, which can occur at any time during the day and particularly during the night due to the drug's metabolism (short half-life) [16]. Apart from the motor fluctuations, patients can also suffer from oscillations of nonmotor symptoms, such as anxiety, drenching sweats, slowness of thinking, fatigue, and irritability [17]. The motor fluctuations during levodopa treatment, which are the main focus of this chapter, can broadly be classified into three different patterns:

Wearing "OFF"

This is the earliest and most common sign of the loss of symptom control and usually occurs after a few years on levodopa. The pathophysiological mechanism is believed to be related to the progressive loss of presynaptic dopaminergic cell terminals, leading to subtherapeutic levels of levodopa concentration between doses. This pattern refers to the observation that the duration of the beneficial effect of each levodopa dose becomes progressively shorter [18]. "End-of-dose" deterioration is generally predictable and usually improves by taking another dose of levodopa, increasing the previous dose of levodopa, or the addition of levodopa-enhancing agents, such as monoamine oxidase inhibitors (MAOIs) and catechol-O-methyltransferase (COMT) inhibitors. Examples of these are the patients' reports of early morning tremor, nocturnal immobility, or nighttime "OFF" dystonia, improving rapidly after the first morning dose [19].

Delayed "ON"

This is another pattern of response fluctuation. One causal factor may be impaired or delayed absorption of oral levodopa, either in the proximal jejunum or across the blood-brain barrier, where neutral amino acids compete with levodopa uptake [20]. Another potential causal factor is the erratic gastric emptying commonly seen in PD [21]. A further mechanism involving lack of synaptic "buffering" secondary to progressive cell loss may also be conceptualized. Delayed "ON" refers to the observation that the interval between the intake of a levodopa dose and induction of the subsequent "ON" effect is delayed or is absent (no "ON" despite taking medication) until the patient experiences the benefit from subsequent therapy.

Random "ON-OFF"

Transitions from "OFF" to "ON," and vice versa, are usually predictable in early stage disease. Random "ON-OFF" is a pattern of response oscillations that can be seen with disease progression and therefore most commonly presents in the advanced stages of PD. These fluctuations can be fast. They seem unrelated to the timing of the last levodopa oral dose. Although the exact mechanism remains unknown, it is hypothesized that pharmacodynamics and neuroplastic changes in striatal medium spiny neurons and the basal ganglia circuitry may contribute to their generation [19].

Definitions of Dyskinesia and Its Typical Forms

The word dyskinesia is a Greek word (δυσκινησία), literally meaning "bad movement." In medical definitions, "dyskinesia" generally refers to any nonvoluntary movement, other than tremor [22]. In general, the spectrum of parkinsonian levodopa-induced dyskinesia includes a variety of movements, such as choreic, dystonic, athetoid, and ballistic movements.

The word chorea derives from the Greek word χορός (chorus), meaning "dance." Choreic movements are irregular, spasmodic, and involuntary movements that are not rhythmic, but appear to flow from one muscle to the next [23]. It refers to mainly distal muscles and is best seen in the fingers, the toes, the wrists, and/or the ankles. Chorea can also be seen in the cranial facial muscles and neck. In non-parkinsonian patients, facial choreic movements can of course be related to other pathologies, including those induced by neuroleptics (tardive dyskinesia), affecting the tongue, lips, cheeks, eyebrows, and forehead. Facial choreic movements in Parkinson's disease do not seem as common as limb dyskinesia.

Dystonia is a Greek word (δυστονία), meaning "bad tone." Dystonia is defined as a movement disorder that is sustained or intermittent and is characterized by involuntary, patterned, and often repetitive contractions of opposing muscles, causing twisting movements or abnormal postures, or both. Dystonia is often initiated or worsened by voluntary action and associated with overflow muscle activation, the latter referring to an involuntary movement of one body part when another moves voluntarily. In non-parkinsonian patients, dystonia is classified along 2 axes: clinical characteristics (which include age at onset, body distribution, temporal pattern, and associated features) and etiology (which includes nervous system pathology and inheritance) [24]. By analogy, dystonia in PD represents similar patterns of involuntary movements, but the etiology is usually either wearing "OFF" of antiparkinsonian medication or a subtype of "ON" dyskinesia, called diphasic dyskinesia. Such dystonic symptoms can result in significant pain and functional disabilities [25].

Ballism derives from the Greek verb βάλλω, meaning "throw." Movements that are performed with maximal velocity and acceleration can be considered as ballistic. These actions are electrophysiologically characterized by high firing rates, brief contraction times, and high rates of force development [26]. Clinically the move-

Table 1.1 Types and characteristics of levodopa-induced dyskinesia

	Most commonly affected site	Most common type of movements	Associated pain
LID type			
OFF-period dyskinesia	Lower limbs	Dystonic	+++
ON period	Extremities and neck	Choreic or choreoathetoid	–
Peak-dose dyskinesia			
Square-wave dyskinesia			
Diphasic	Extremities and neck	Choreic, choreoathetoid, dystonic, or ballistic	±

ments affect mainly proximal muscles, i.e., sudden abrupt violent movement of the whole arm or leg [27].

Athetoid movements – the typical movements of athetosis (also a Greek word, αθέτωσις) – are slow, involuntary, writhing movements of the fingers, hands, toes, and feet and in some cases can affect the arms, the legs, the neck, and the tongue.

Phenotypically, levodopa-induced dyskinesia (LID) can look like any of the above and typically develop with disease progression and with repeated dopamine replacement therapy in Parkinson's disease [25]. Most often, LID is characterized by idiosyncratic mixtures of dystonia and chorea. Although they are termed "levodopa induced," they also occur with dopamine agonists [28, 29] but their prevalence after starting agonist use seems less than with levodopa. Hence, allowing for other potential side effects, patient preference, and other individual patient factors, there is sometimes a preference for agonist use in early disease in an attempt to minimize the induction and delay the onset of drug-induced fluctuations and dyskinesia. Additionally, when fluctuations and dyskinesia start to appear, another strategy is the introduction of agonist and reduction of levodopa to attempt to delay and/or reduce the existing dyskinesia in PD. Dyskinesia can also be induced by the addition of COMT inhibitors to standard levodopa preparations, since COMT inhibitors delay breakdown of levodopa and therefore boost the action of levodopa, despite the stability of the oral dose of levodopa that was taken. In this chapter, therefore, when describing phenomenology, the term "levodopa-induced dyskinesia" will be taken to mean dopaminergic drug-induced dyskinesia in PD, which is related to all dopaminergic therapy, whether it be levodopa, agonist, or COMT inhibitors.

Timing of Levodopa-Induced Dyskinesia

Several patterns of expression of LID can be described, especially based on the timing of their appearance in relation to the "ON-OFF" cycle of the patient. Table 1.1 summarizes the different types of LID and the associated characteristics.

"ON"-Period Dyskinesia

"ON"-period dyskinesia is the most common type of LID as they present in the 70–80 % of the patients who experience dyskinesia [22]. This type of LID occurs when the dopaminergic stimulation in the brain is maximal or increased and subsequently the patient is in an "ON" motor state. Usually the movements are choreiform or choreoathetoid, but occasionally ballistic [19]. The clinical manifestation of "ON"-period dyskinesia includes the appearance of restlessness and continuous jerky, involuntary movements most frequently affecting the extremities and the trunk. Although, typically they affect the limbs, both upper and lower (Video 1.1), and the head (Video 1.2), respiratory muscles can also be affected.

The first manifestation of LID usually occurs ipsilateral to the most affected side of PD. According to the onset and duration, "ON"-period dyskinesia can be classified into two different subtypes: "peak-dose" and "square-wave" dyskinesia.

Peak-Dose Dyskinesia

Peak-dose dyskinesia, initially termed the "improvement-dyskinesia-improvement" (I-D-I) response [30], is the most common subtype, occurring at the time of best "ON" response to levodopa, when the dopaminergic stimulation is ongoing. Before and after the period of dyskinesia, the patient experiences some "ON" time without dyskinesia. For example, after a dose of medication, a patient may experience 1 h of good "ON" without dyskinesia, followed by 20 min of "ON" with dyskinesia, followed by 45 min of good "ON" without dyskinesia, before wearing "OFF" occurs if a further dose of medication is not taken. Although peak-dose dyskinesias are predominantly choreic, they can also be dystonic as painful muscle cramping.

Square-Wave Dyskinesia

Although peak-dose dyskinesia may initially be reduced or disappear with lowering of the levodopa dose, the time occupied by the dyskinesia can expand to fill the entire "ON" period of benefit. That is, the patient does not have an "ON" period without dyskinesia. As a result, the relief of akinesia (lack of movement) alternates with movement "overshoot" (excessive movement) that they represent [31].

"OFF"-Period Dyskinesia ("OFF" Dystonia)

"OFF"-period dyskinesia occurs when the dopaminergic stimulation in the brain is low and subsequently the patient is "OFF" or transitioning from "ON" to "OFF." This most commonly occurs during the night (nocturnal dyskinesia) or before the first levodopa dose in the morning (early morning dyskinesia) or just after taking the

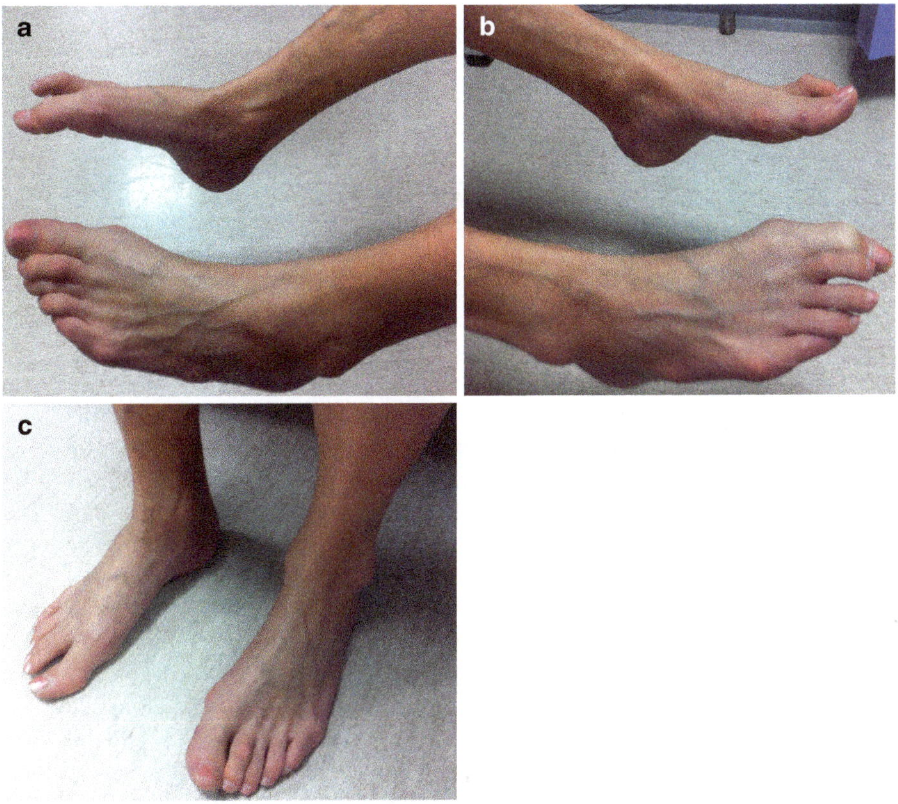

Fig. 1.1 "OFF" right foot dystonia. The right foot shows plantar flexion and toes curling down (**a**). In this case, it was painful. During the same time point, the left foot has a normal position (**b**), showing asymmetry affecting the more severely affected side. In this case, the right foot has not developed contracture (**c**), as can be seen when the patients stand, the abnormal right foot posture is corrected

levodopa dose. The latter is presumed to be because the brain dopamine concentration has not yet reached a therapeutic level despite the recent intake of the levodopa. Alternatively, it may occur just before the next levodopa dose, as the drug effect of the previous dose wears "OFF." The phenotype of OFF-period dyskinesia is predominantly dystonic ("OFF" dystonia) [31]. The clinical presentation often affects the lower limbs, usually ipsilaterally to the side most affected by the disease [19]. A characteristic manifestation includes foot inversion and painful flexion of the toes (Fig. 1.1). Long term, dystonia can be sustained enough to cause contracture.

Although "OFF" dystonia can be seen in patients with PD not yet treated [22] and, thus, one could argue that "OFF"-period dyskinesia should not considered to be levodopa induced, some studies have showed that it can also be a drug-induced phenomenon, as it can resolve when levodopa is stopped or the dosage is appreciably reduced [32–34]. However, it can also be combated by taking more dopaminergic medication.

Diphasic Dyskinesia

Diphasic (or biphasic) dyskinesia, also termed "dyskinesia-improvement-dyskinesia" (D-I-D) response [30], is more difficult to treat and fortunately less commonly observed than I-D-I. Typically D-I-D occurs at two different time points of a single dose cycle: once at the beginning (for a few minutes) and once at the end (also for a few minutes). These dyskinetic episodes are separated by a "best ON" period in which less or no dyskinesia occurs. For example, the first effect of taking levodopa may be noted by a patient after 30 min after drug ingestion. After this, a patient may enter a "start of dose" moderate/severe dyskinetic period for 20 min, followed by a "best ON" period with less dyskinesia for 40–60 min, followed by an "end-of-dose" moderate/severe dyskinetic period of 15 min, before the patient would become "OFF" if a further dose of medication is not taken. The duration of each component of this cycle can vary, but the cycle itself can be stereotyped with each dose of levodopa. At the time points of dyskinesia, the levodopa level is presumed to be changing – rising at the beginning of dose period and falling at the end of dose. The patient enters from the "OFF" state to the "ON" state, or vice versa, but having a period of severe dyskinesia sometime during the transition phases [19].

Diphasic dyskinesia can affect arms and legs but usually affect the lower limbs more [35]. Phenomenologically they involve repetitive rapidly alternating dystonic flexion/extension foot movements or leg kicking in a stereotyped pattern, often associated with high-stepping and bizarre gaits [36]. However, occasionally diphasic dyskinesia can be ballistic or a mixture of dystonia and ballism [37]. In this case, diphasic dyskinesia is not only disabling per se, but also painful. A notable feature of this type of LID is that while the lower limbs are moving involuntarily, the upper half of the body can exhibit parkinsonian signs, e.g., tremor [38].

Other Forms of Levodopa-Induced Dyskinesia

Although levodopa-induced dyskinesia generally affects the limbs, trunk, head, and neck, less common dyskinesias associated with levodopa include respiratory [39] and ocular muscle dyskinesia [40].

Respiratory Dyskinesia

Respiratory muscle dyskinesia has been described in non-parkinsonian patients, e.g., tardive dyskinesia induced by neuroleptic drugs [41]. Historically, respiratory difficulties associated with levodopa therapy were observed in a small number of patients with postencephalitic parkinsonism who developed "respiratory crises" with gasping, panting, breath holding, and irregular respiratory depth and rate [42]. In PD, respiratory dyskinesia can be a rare, but disabling complication of treatment with levodopa and presents as an irregular, tachypneic pattern of respiration alternating with brief periods of apnea, in a pattern consistent with a central origin [43]. The underlying pathophysiological mechanisms are not well understood, but a possible explanation for an irregular and fast breathing is the loss of control over diaphragmatic and intercostal muscles [44]. It is not clear why the frequency of levodopa-induced respiratory muscle dyskinesia is low compared with other voluntary muscles. In the limbs, "ON" dyskinesia appears first in the limbs which are more affected by the parkinsonism and so are more bradykinetic when "OFF," supporting the view that the propensity to develop levodopa-induced dyskinesia is related to the degree of nigral–striatal denervation. Bradykinesia of the voluntary respiratory muscles is not typically seen in PD, and so it can be hypothesized that this is why levodopa-induced dyskinesia of the voluntary muscles is not seen frequently in PD. Although respiratory disturbance has been reported without coexisting limb dyskinesia [45], it has also been associated with laryngeal dystonia and/or orofacial dyskinesia [39].

Ocular Dyskinesia

Although infrequent, levodopa-induced dyskinesia can involve the extraocular muscles. Ocular dyskinesia involves repeated and stereotyped movements of the eyes. These eye movements occur simultaneously with choreoathetoid limb movements [46, 47], during the peak effect of levodopa [47]. They can involve both axes, vertical and horizontal. It has been suggested that the main direction of gaze deviation can be towards the side most affected by PD [40]. The spectrum of clinical manifestation varies from being phasic and brief to dystonic and sustained for several seconds [40]. Ocular dyskinesia in PD patients differs from oculogyric crises in non-PD patients in several aspects. Oculogyric crises are usually tonic, and sustained in duration, and can last up to few hours. Moreover, in an oculogyric crisis, voluntary control over the eyes is short and visual fixation is difficult during an oculogyric crisis. In PD levodopa-induced ocular dyskinesia, eye movements are usually brief and stereotyped with some ability to control fixation [40].

Impact of Duration of Treatment

Duration of levodopa therapy has been considered a significant risk factor for developing dyskinesia. In the first years on levodopa, the percentage of patients suffering from dyskinesia is low, varying from less than 3 % after 6 months of treatment [48] to approximately 33 % of the patients after 20 months of treatment [49]. After 3 years on levodopa, the rates increase further and vary from 26 % [50] to 54 % [51]. After 5 years on levodopa, approximately one half of the patients have been reported to develop dyskinesia induced by levodopa [28, 52, 53].

In long-term follow-up studies, the incidence of dyskinesia increases further. After 10 years on levodopa, 52–78 % of the patients will develop dyskinesia [54–56], and after 15 years of treatment, more than nine out of ten patients will suffer from dyskinesia [57]. Thus, it is largely accepted that dyskinesia is likely to occur when on levodopa for many years [58], although one cannot determine if this is an exclusive effect of treatment duration or a combined effect of neurodegeneration, disease duration, and dose. Further, these studies need to be interpreted with some caution in their application to clinical practice, as it is clear that the frequency of levodopa-induced dyskinesia (after being on treatment for a specific period of time) differs from study to study. This variability might be caused by the different methodology used, the different clinical settings, and the different studied populations [25]. For example, while younger patients seem to have a slower progression of the disease, they present with earlier onset and higher rate of levodopa-induced dyskinesia [58, 59].

Impact of Levodopa Dose

The ability of levodopa to induce dyskinesia and alleviate extrapyramidal symptoms has generally been considered as a continuous dose-dependent pharmacological spectrum [60]. Moreover, the fact that in modern series, levodopa-induced dyskinesia appears to occur later than in earlier series [61] suggests that the higher levodopa doses used in the past lead to the appearance of earlier dyskinesia. In a large retrospective study, one risk for dyskinesia was a higher initial levodopa dose [62]. In a cross-sectional study, higher daily levodopa dose was found to be associated with dyskinesia, after adjusting for other risk factors [63]. In the DATATOP study, patients with levodopa-induced dyskinesia were taking higher levodopa doses at the time of appearance of dyskinesia, compared to patients without dyskinesia [49]. However, the cumulative dose of levodopa did not differ significantly, as was also observed in another smaller, retrospective study [64].

The exact impact of levodopa dose on the appearance of dyskinesia is difficult to be calculated, as PD patients might receive different doses of levodopa on different time points throughout the day. Moreover, their drug regimen changes as the disease progresses. However, it is common clinical practice to keep the dose and duration of levodopa to that which provides a good "ON" period, in an effort to keep both dose and duration of levodopa exposure to be adequate for motor control for the individual's requirements and yet minimize the long-term risk of developing dyskinesia.

Impact of Levodopa Type

Levodopa is currently available as an oral preparation as standard-release, controlled-release, and dispersible tablets. However, levodopa can also be administered intravenously [65] or intrajejunally [66]. Controlled-release formulations are thought to reduce the fluctuations in plasma levodopa levels and have been suggested to be superior to immediate-release formulations with decreased "OFF" time and reduced levodopa dosing frequency in clinical trials [67–69]. However, other studies suggested that controlled-release formulations may not be superior to standard-release preparations [70] but in some studies are associated with increased dyskinesia [70–74]. Hence there is no uniform conclusion in selecting one form over another when attempting to treat dyskinesia. Intravenous levodopa has been directly compared to oral levodopa. In general, intravenous administration shows more prolonged and stable clinical response [65] with minimal side effects [75], including improvement in dyskinesia [76]. Practical considerations have hindered clinical use and intravenously administered levodopa is only used in research studies.

In contrast, the recent introduction of continuous intrajejunal infusion in some countries has been a novel strategy for the treatment of motor complications in patients with advanced PD. This offers more stable plasma levodopa concentration compared to oral therapy [77–79]. Intrajejunal levodopa not only improves motor fluctuations in PD but also reduces dyskinesia severity and duration [80–83]. Therefore, although intrajejunal infusion is an expensive and technically demanding option, it may be promising for some patients who have already developed dyskinesia and fluctuations in advanced PD. As it is not employed in early disease, one cannot determine if early use would prevent the development of dyskinesia. The variability of the response to levodopa treatment when a different route of administration is selected highlights the fact that several pharmacokinetic factors, including the erratic absorption and short elimination half-life of levodopa, in combination with a pharmacodynamic post-receptor response play a significant role in the motor fluctuations and levodopa-induced dyskinesia [84].

Finally, a new capsule formulation of carbidopa-levodopa, the IPX066, has been studied lately versus immediate-release carbidopa-levodopa in patients with Parkinson's disease and motor fluctuations. This extended-release formulation has been shown to potentially be a useful treatment for patients with Parkinson's disease who have motor fluctuations, with potential benefits including decreased off-time and reduced levodopa dosing frequency [67].

Conclusion

Levodopa-induced dyskinesia includes a wide spectrum of movements such as choreic, dystonic, athetotic, and ballistic movements. The pattern is usually described based on the timing of their appearance in relation to "ON-OFF" cycle of the patient. Various factors seem to be associated with levodopa-induced dyskinesia, which include the duration on levodopa treatment, its dosage, and its type. Dyskinesia has been associated with decreased quality of life [13, 61, 85]. However, some PD patients express a significantly greater preference for the "ON" state (with dyskinesia) rather than the "OFF" state (without dyskinesia) [86]. Therefore, treatment choices for advanced PD, including the management of dyskinesia, are not only determined by the clinical expression but also by the patient's preference [87].

Acknowledgment The authors are grateful to Dr Davide Martino MD, PhD, King's College Hospital, King's Health Partners, London, for critical review of this chapter.

References

1. Barbeau A. L-dopa therapy in Parkinson's disease: a critical review of nine years' experience. Can Med Assoc J. 1969;101(13):59–68.
2. Schwab RS, England AC. Medical and surgical therapy of Parkinson's disease. In: Fields WS, Charles C, editors. Pathogenesis and treatment of parkinsonism. Springfield: Thomas Publisher; 1958. p. 203.
3. Carlsson A, Lindqvist M, Magnusson T, Waldeck B. On the presence of 3-hydroxytyramine in brain. Science. 1958;127(3296):471.
4. Carlsson A. The occurrence, distribution and physiological role of catecholamines in the nervous system. Pharmacol Rev. 1959;11(2 Part 2):490–3.
5. Ehringer H, Hornykiewicz O. Distribution of noradrenaline and dopamine (3-hydroxytyramine) in the human brain and their behavior in diseases of the extrapyramidal system. Klin Wochenschr. 1960;38:1236–9.
6. Barbeau A. Preliminary observations on abnormal catecholamine metabolism in basal ganglia diseases. Neurology. 1960;10:446–51.
7. Barbeau A. Dopamine and basal ganglia diseases. Arch Neurol. 1961;4:97–102.
8. Barbeau A, Murphy GF, Sourkes T. Excretion of dopamine in diseases of basal ganglia. Science. 1961;133(3465):1706–7.
9. Degkwitz R, Frowein R, Kulenkampff C, Mohs U. On the effects of L-dopa in man and their modification by reserpine, chlorpromazine, iproniazid and vitamin B6. Klin Wochenschr. 1960;38:120–3.
10. Shaw KM, Lees AJ, Stern GM. The impact of treatment with levodopa on Parkinson's disease. Q J Med. 1980;49(195):283–93.
11. Friedman A. Levodopa-induced dyskinesia: clinical observations. J Neurol. 1985;232(1):29–31.
12. Volkmann J, Albanese A, Antonini A, Chaudhuri KR, Clarke CE, de Bie RM, Deuschl G, Eggert K, Houeto JL, Kulisevsky J, Nyholm D, Odin P, Ostergaard K, Poewe W, Pollak P, Rabey JM, Rascol O, Ruzicka E, Samuel M, Speelman H, Sydow O, Valldeoriola F, van der Linden C, Oertel W. Selecting deep brain stimulation or infusion therapies in advanced Parkinson's disease: an evidence-based review. J Neurol. 2013;260(11):2701–14.
13. Thanvi B, Lo N, Robinson T. Levodopa-induced dyskinesia in Parkinson's disease: clinical features, pathogenesis, prevention and treatment. Postgrad Med J. 2007;83(980):384–8.
14. Nutt JG. On-off phenomenon: relation to levodopa pharmacokinetics and pharmacodynamics. Ann Neurol. 1987;22(4):535–40.
15. Fasano A, Ricciardi L, Lena F, Bentivoglio AR, Modugno N. Intrajejunal levodopa infusion in advanced Parkinson's disease: long-term effects on motor and non-motor symptoms and impact on patient's and caregiver's quality of life. Eur Rev Med Pharmacol Sci. 2012;16(1):79–89.
16. Müller T. Motor complications, levodopa metabolism and progression of Parkinson's disease. Expert Opin Drug Metab Toxicol. 2011;7(7):847–55.
17. Witjas T, Kaphan E, Azulay JP, Blin O, Ceccaldi M, Pouget J, Poncet M, Chérif AA. Nonmotor fluctuations in Parkinson's disease: frequent and disabling. Neurology. 2002;59(3):408–13.
18. van Laar T. Levodopa-induced response fluctuations in patients with Parkinson's disease: strategies for management. CNS Drugs. 2003;17(7):475–89.
19. Hametner E, Seppi K, Poewe W. The clinical spectrum of levodopa-induced motor complications. J Neurol. 2010;257 Suppl 2:S268–75.
20. Leenders KL, Poewe WH, Palmer AJ, Brenton DP, Frackowiak RS. Inhibition of L-[18F]fluorodopa uptake into human brain by amino acids demonstrated by positron emission tomography. Ann Neurol. 1986;20(2):258–62.

21. Jost WH. Gastrointestinal dysfunction in Parkinson's disease. J Neurol Sci. 2010;289(1–2):69–73.
22. Zesiewicz TA, Sullivan KL, Hauser RA. Levodopa-induced dyskinesia in Parkinson's disease: epidemiology, etiology, and treatment. Curr Neurol Neurosci Rep. 2007;7(4):302–10.
23. Jankovic JJ, Fahn S. Dystonic disorders. In: Jankovic JJ, Tolosa E, editors. Parkinson's disease and movement disorders. 4th ed. Philadelphia: Lippincott Williams & Wilkins; 2002. p. 331–57.
24. Albanese A, Bhatia K, Bressman SB, Delong MR, Fahn S, Fung VS, Hallett M, Jankovic J, Jinnah HA, Klein C, Lang AE, Mink JW, Teller JK. Phenomenology and classification of dystonia: a consensus update. Mov Disord. 2013;28(7):863–73.
25. Fabbrini G, Brotchie JM, Grandas F, Nomoto M, Goetz CG. Levodopa-induced dyskinesia. Mov Disord. 2007;22(10):1379–89.
26. Zehr EP, Sale DG. Ballistic movement: muscle activation and neuromuscular adaptation. Can J Appl Physiol. 1994;19(4):363–78.
27. Jankovic J. Motor fluctuations and dyskinesia in Parkinson's disease: clinical manifestations. Mov Disord. 2005;20 Suppl 11:S11–6.
28. Rascol O, Brooks DJ, Korczyn AD, De Deyn PP, Clarke CE, Lang AE. A five-year study of the incidence of dyskinesia in patients with early Parkinson's disease who were treated with ropinirole or levodopa. 056 Study Group. N Engl J Med. 2000;342(20):1484–91.
29. Constantinescu R, Romer M, McDermott MP, Kamp C, Kieburtz K, CALM-PD Investigators of the Parkinson Study Group. Impact of pramipexole on the onset of levodopa-related dyskinesia. Mov Disord. 2007;22(9):1317–9.
30. Muenter MD, Sharpless NS, Tyce GM, Darley FL. Patterns of dystonia ("I-D-I" and "D-I-D-") in response to l-dopa therapy for Parkinson's disease. Mayo Clin Proc. 1977;52(3):163–74.
31. Quinn NP. Classification of fluctuations in patients with Parkinson's disease. Neurology. 1998;51(2 Suppl 2):S25–9.
32. Ilson J, Fahn S, Côté L. Painful dystonic spasms in Parkinson's disease. Adv Neurol. 1984;40:395–8.
33. Melamed E. Early-morning dystonia. A late side effect of long-term levodopa therapy in Parkinson's disease. Arch Neurol. 1979;36(5):308–10.
34. Poewe WH, Lees AJ, Stern GM. Dystonia in Parkinson's disease: clinical and pharmacological features. Ann Neurol. 1988;23(1):73–8.
35. Alegre M, López-Azcárate J, Alonso-Frech F, Rodríguez-Oroz MC, Valencia M, Guridi J, Artieda J, Obeso JA. Subthalamic activity during diphasic dyskinesia in Parkinson's disease. Mov Disord. 2012;27(9):1178–81.
36. Obeso JA, Grandas F, Vaamonde J, Luquin MR, Artieda J, Lera G, Rodriguez ME, Martinez-Lage JM. Motor complications associated with chronic levodopa therapy in Parkinson's disease. Neurology. 1989;39:11–9.
37. Marconi R, Lefebvre-Caparros D, Bonnet AM, Vidailhet M, Dubois B, Agid Y. Levodopa-induced dyskinesia in Parkinson's disease phenomenology and pathophysiology. Mov Disord. 1994;9:2–12.
38. Luquin MR, Scipioni O, Vaamonde J, Gershanik O, Obeso JA. Levodopa-induced dyskinesia in Parkinson's disease: clinical and pharmacological classification. Mov Disord. 1992;7:117–24.
39. Jankovic J, Nour F. Respiratory dyskinesia in Parkinson's disease. Neurology. 1986;36(2):303–4.
40. Grötzsch H, Sztajzel R, Burkhard PR. Levodopa-induced ocular dyskinesia in Parkinson's disease. Eur J Neurol. 2007;14(10):1124–8.
41. Rich MW, Radwany SM. Respiratory dyskinesia. An underrecognized phenomenon. Chest. 1994;105(6):1826–32.
42. Sacks OW, Kohl M, Schwartz W, Messeloff C. Side-effects of L-dopa in postencephalic parkinsonism. Lancet. 1970;1(7654):1006.
43. Rice JE, Antic R, Thompson PD. Disordered respiration as a levodopa-induced dyskinesia in Parkinson's disease. Mov Disord. 2002;17(3):524–7.

44. De Letter M, Santens P, De Bodt M, Van Maele G, Van Borsel J, Boon P. The effect of levodopa on respiration and word intelligibility in people with advanced Parkinson's disease. Clin Neurol Neurosurg. 2007;109(6):495–500.
45. De Keyser J, Vincken W. L-dopa-induced respiratory disturbance in Parkinson's disease suppressed by tiapride. Neurology. 1985;35(2):235–7.
46. LeWitt PA. Conjugate eye deviations as dyskinesia induced by levodopa in Parkinson's disease. Mov Disord. 1998;13(4):731–4.
47. Linazasoro G, Van Blercom N, Lasa A, Indakoetxea B, Ruiz J. Levodopa-induced ocular dyskinesia in Parkinson's disease. Mov Disord. 2002;17(1):186–7.
48. Fahn S, Oakes D, Shoulson I, Kieburtz K, Rudolph A, Lang A, Olanow CW, Tanner C, Marek K, Parkinson Study Group. Levodopa and the progression of Parkinson's disease. N Engl J Med. 2004;351(24):2498–508.
49. Parkinson Study Group. Impact of deprenyl and tocopherol treatment on Parkinson's disease in DATATOP patients requiring levodopa. Ann Neurol. 1996;39(1):37–45.
50. Oertel WH, Wolters E, Sampaio C, Gimenez-Roldan S, Bergamasco B, Dujardin M, Grosset DG, Arnold G, Leenders KL, Hundemer HP, Lledó A, Wood A, Frewer P, Schwarz J. Pergolide versus levodopa monotherapy in early Parkinson's disease patients: The PELMOPET study. Mov Disord. 2006;21(3):343–53.
51. Parkinson Study Group. Pramipexole vs levodopa as initial treatment for Parkinson disease: a randomized controlled trial. Parkinson Study Group. JAMA. 2000;284(15):1931–8.
52. Ahlskog JE, Muenter MD. Frequency of levodopa-related dyskinesia and motor fluctuations as estimated from the cumulative literature. Mov Disord. 2001;16(3):448–58.
53. Holloway RG, Shoulson I, Fahn S, Kieburtz K, Lang A, Marek K, McDermott M, Seibyl J, Weiner W, Musch B, Kamp C, Welsh M, Shinaman A, Pahwa R, Barclay L, Hubble J, LeWitt P, Miyasaki J, Suchowersky O, Stacy M, Russell DS, Ford B, Hammerstad J, Riley D, Standaert D, Wooten F, Factor S, Jankovic J, Atassi F, Kurlan R, Panisset M, Rajput A, Rodnitzky R, Shults C, Petsinger G, Waters C, Pfeiffer R, Biglan K, Borchert L, Montgomery A, Sutherland L, Weeks C, DeAngelis M, Sime E, Wood S, Pantella C, Harrigan M, Fussell B, Dillon S, Alexander-Brown B, Rainey P, Tennis M, Rost-Ruffner E, Brown D, Evans S, Berry D, Hall J, Shirley T, Dobson J, Fontaine D, Pfeiffer B, Brocht A, Bennett S, Daigneault S, Hodgeman K, O'Connell C, Ross T, Richard K, Watts A, Parkinson Study Group. Pramipexole vs levodopa as initial treatment for Parkinson disease: a 4-year randomized controlled trial. Arch Neurol. 2004;61(7):1044–53.
54. López IC, Ruiz PJ, Del Pozo SV, Bernardos VS. Motor complications in Parkinson's disease: ten year follow-up study. Mov Disord. 2010;25(16):2735–9.
55. Hauser RA, Rascol O, Korczyn AD, Jon Stoessl A, Watts RL, Poewe W, De Deyn PP, Lang AE. Ten-year follow-up of Parkinson's disease patients randomized to initial therapy with ropinirole or levodopa. Mov Disord. 2007;22(16):2409–17.
56. Rajput AH, Fenton ME, Birdi S, Macaulay R, George D, Rozdilsky B, Ang LC, Senthilselvan A, Hornykiewicz O. Clinical-pathological study of levodopa complications. Mov Disord. 2002;17(2):289–96.
57. Hely MA, Morris JG, Reid WG, Trafficante R. Sydney Multicenter Study of Parkinson's disease: non-L-dopa-responsive problems dominate at 15 years. Mov Disord. 2005;20(2):190–9.
58. Manson A, Stirpe P, Schrag A. Levodopa-induced-dyskinesia clinical features, incidence, risk factors, management and impact on quality of life. J Parkinsons Dis. 2012;2(3):189–98.
59. Jankovic J. Parkinson's disease therapy: treatment of early and late disease. Chin Med J (Engl). 2001;114(3):227–34.
60. Mouradian MM, Heuser IJ, Baronti F, Fabbrini G, Juncos JL, Chase TN. Pathogenesis of dyskinesia in Parkinson's disease. Ann Neurol. 1989;25(5):523–6.
61. Del Sorbo F, Albanese A. Levodopa-induced dyskinesia and their management. J Neurol. 2008;255 Suppl 4:32–41.
62. Van Gerpen JA, Kumar N, Bower JH, Weigand S, Ahlskog JE. Levodopa-associated dyskinesia risk among Parkinson disease patients in Olmsted County, Minnesota, 1976–1990. Arch Neurol. 2006;63(2):205–9.

63. Hashim HZ, Norlinah MI, Nafisah WY, Tan HJ, Raymond AA, Tamil AM. Risk factors and predictors of levodopa-induced dyskinesia among multiethnic Malaysians with Parkinson's disease. Int J Neurosci. 2014;124:187–91.
64. Blanchet PJ, Allard P, Grégoire L, Tardif F, Bédard PJ. Risk factors for peak dose dyskinesia in 100 levodopa-treated parkinsonian patients. Can J Neurol Sci. 1996;23(3):189–93.
65. Quinn N, Marsden CD, Parkes JD. Complicated response fluctuations in Parkinson's disease: response to intravenous infusion of levodopa. Lancet. 1982;2(8295):412–5.
66. Kurlan R, Rubin AJ, Miller C, Rivera-Calimlim L, Clarke A, Shoulson I. Duodenal delivery of levodopa for on-off fluctuations in parkinsonism: preliminary observations. Ann Neurol. 1986;20(2):262–5.
67. Hauser RA, Hsu A, Kell S, Espay AJ, Sethi K, Stacy M, Ondo W, O'Connell M, Gupta S, IPX066 ADVANCE-PD investigators. Extended-release carbidopa-levodopa (IPX066) compared with immediate-release carbidopa-levodopa in patients with Parkinson's disease and motor fluctuations: a phase 3 randomised, double-blind trial. Lancet Neurol. 2013;12(4):346–56.
68. Ahlskog JE, Muenter MD, McManis PG, Bell GN, Bailey PA. Controlled-release Sinemet (CR-4): a double-blind crossover study in patients with fluctuating Parkinson's disease. Mayo Clin Proc. 1988;63(9):876–86.
69. Hutton JT, Morris JL, Bush DF, Smith ME, Liss CL, Reines S. Multicenter controlled study of Sinemet CR vs Sinemet (25/100) in advanced Parkinson's disease. Neurology. 1989;39(11 Suppl 2):67–72.
70. Koller WC, Hutton JT, Tolosa E, Capilldeo R. Immediate-release and controlled-release carbidopa/levodopa in PD: a 5-year randomized multicenter study. Carbidopa/Levodopa Study Group. Neurology. 1999;53(5):1012–9.
71. Cedarbaum JM, Hoey M, McDowell FH. A double-blind crossover comparison of Sinemet CR4 and standard Sinemet 25/100 in patients with Parkinson's disease and fluctuating motor performance. J Neurol Neurosurg Psychiatry. 1989;52(2):207–12.
72. Sage JI, Mark MH. Comparison of controlled-release Sinemet (CR4) and standard Sinemet (25 mg/100 mg) in advanced Parkinson's disease: a double-blind, crossover study. Clin Neuropharmacol. 1988;11(2):174–9.
73. Deleu D, Jacques M, Michotte Y, Ebinger G. Controlled-release carbidopa/levodopa (CR) in parkinsonian patients with response fluctuations on standard levodopa treatment: clinical and pharmacokinetic observations. Neurology. 1989;39(11 Suppl 2):88–92.
74. Stocchi F, Quinn NP, Barbato L, Patsalos PN, O'Connel MT, Ruggieri S, Marsden CD. Comparison between a fast and a slow release preparation of levodopa and a combination of the two: a clinical and pharmacokinetic study. Clin Neuropharmacol. 1994;17(1):38–44.
75. Quinn N, Parkes JD, Marsden CD. Control of on/off phenomenon by continuous intravenous infusion of levodopa. Neurology. 1984;34(9):1131–6.
76. Schuh LA, Bennett Jr JP. Suppression of dyskinesia in advanced Parkinson's disease. I. Continuous intravenous levodopa shifts dose response for production of dyskinesia but not for relief of parkinsonism in patients with advanced Parkinson's disease. Neurology. 1993;43(8):1545–50.
77. Nyholm D, Askmark H, Gomes-Trolin C, Knutson T, Lennernäs H, Nyström C, Aquilonius SM. Optimizing levodopa pharmacokinetics: intestinal infusion versus oral sustained-release tablets. Clin Neuropharmacol. 2003;26(3):156–63.
78. Stocchi F, Bonamartini A, Vacca L, Ruggieri S. Motor fluctuations in levodopa treatment: clinical pharmacology. Eur Neurol. 1996;36 Suppl 1:38–42.
79. Antonini A, Tolosa E. Apomorphine and levodopa infusion therapies for advanced Parkinson's disease: selection criteria and patient management. Expert Rev Neurother. 2009;9(6):859–67.
80. Nyholm D, Nilsson Remahl AI, Dizdar N, Constantinescu R, Holmberg B, Jansson R, Aquilonius SM, Askmark H. Duodenal levodopa infusion monotherapy vs oral polypharmacy in advanced Parkinson disease. Neurology. 2005;64(2):216–23.
81. Antonini A, Isaias IU, Canesi M, Zibetti M, Mancini F, Manfredi L, Dal Fante M, Lopiano L, Pezzoli G. Duodenal levodopa infusion for advanced Parkinson's disease: 12-month treatment outcome. Mov Disord. 2007;22(8):1145–9.

82. Eggert K, Schrader C, Hahn M, Stamelou M, Rüssmann A, Dengler R, Oertel W, Odin P. Continuous jejunal levodopa infusion in patients with advanced parkinson disease: practical aspects and outcome of motor and non-motor complications. Clin Neuropharmacol. 2008;31(3):151–66.
83. Fernandez HH, Vanagunas A, Odin P, Espay AJ, Hauser RA, Standaert DG, Chatamra K, Benesh J, Pritchett Y, Hass SL, Lenz RA. Levodopa-carbidopa intestinal gel in advanced Parkinson's disease open-label study: interim results. Parkinsonism Relat Disord. 2013;19(3):339–45.
84. Rascol O. Extended-release carbidopa-levodopa in Parkinson's disease. Lancet Neurol. 2013;12(4):325–6.
85. Encarnacion EV, Hauser RA. Levodopa-induced dyskinesia in Parkinson's disease: etiology, impact on quality of life, and treatments. Eur Neurol. 2008;60(2):57–66.
86. Khlebtovsky A, Rigbi A, Melamed E, Ziv I, Steiner I, Gad A, Djaldetti R. Patient and caregiver perceptions of the social impact of advanced Parkinson's disease and dyskinesia. J Neural Transm. 2012;119(11):1367–71.
87. Clarke CE, Worth P, Grosset D, Stewart D. Systematic review of apomorphine infusion, levodopa infusion and deep brain stimulation in advanced Parkinson's disease. Parkinsonism Relat Disord. 2009;15(10):728–41.

Chapter 2
Dyskinesia Rating Scales in Parkinson's Disease

Christopher G. Goetz and Carlo Colosimo

Abstract An increasing scientific focus in Parkinson's disease research involves the identification of new treatments to improve or prevent dyskinesia. One of the major roadblocks to progress has been the clear delineation of a single best scale to assess the burden of dyskinesia. The Movement Disorder Society has developed a program to evaluate existing dyskinesia scales with prespecified criteria that must be met to receive the designation of Recommended. Among the many available dyskinesia scales, a small number meet the minimal criteria, and investigators and clinicians can select among these scales the one that best fits the need of the assessments. If patient perceptions are the primary focus, the Lang-Fahn and PD-DYS-26 are likely to be selected, with the latter having the stronger clinimetric profile. If only objective assessments are desired, the AIMS can be chosen for impairment and the RDRS can be chosen for disability. The UDysRS has a very strong clinimetric profile and combines both patient perceptions and objective assessments of disability and impairment, providing the most comprehensive measurement tool for the overall burden of dyskinesia.

Keywords Dyskinesia • Parkinson's disease • Rating scales

Introduction

Drug-induced dyskinesia is common in Parkinson's disease (PD) and is frequently the source of social and physical disability [1]. In cross-sectional studies, dyskinesia is reported in 20–60 % of chronically treated PD patients, and recognized risk factors include young age at disease onset, long disease duration, and high total dosage of levodopa and other dopaminergic drugs [2–4]. Whereas dyskinesia is

C.G. Goetz, MD (✉)
Department of Neurological Sciences, Rush University Medical Center,
Suite 755, 1725 W. Harrison Street, Chicago, IL 60612, USA
e-mail: cgoetz@rush.edu

C. Colosimo, MD
Department of Neurology and Psychiatry, Sapienza University of Rome, Rome, Italy
e-mail: Carlo.Colosimo@uniroma1.it

S.H. Fox, J.M. Brotchie (eds.), *Levodopa-Induced Dyskinesia in Parkinson's Disease*, DOI 10.1007/978-1-4471-6503-3_2

often a bilateral problem, it usually first appears on the body side most affected by parkinsonian symptoms [3]. The cranial region, specifically facial muscles and neck as well as the trunk, is also frequently involved [4]. As discussed elsewhere in this book, different forms of dyskinesia have a proclivity to affect different body parts, for example, peak-dose dyskinesia primarily affect the head, neck, and upper extremities, end-of-dose dyskinesia affects primarily the trunk and legs, and finally dystonia usually affects one or both feet. Regardless of clinical form, dyskinesia can impair health-related quality of life [1]. As such, continued efforts to improve dyskinesia are ongoing and require an accurate rating of dyskinesia using clinical measures that are valid, reproducible, and able to detect clinically meaningful change.

Several rating scales have been used in clinical practice since the 1970s for the assessment of dyskinesia in PD. Some are part of global scales that measure overall motor disability in PD, others are adaptations of scales developed from other forms of dyskinesia such as tardive dyskinesia, and some have been specifically developed for dyskinesia assessment in PD [5, 6]. In the last decade, although new pharmacological and surgical treatments for advanced PD have been developed and tested, efforts have been limited by the lack of a single, widely accepted clinical rating instrument for dyskinesia [7].

The development of a single clinical rating instrument that captures all features of dyskinesia has been challenging for several reasons [6]. First, there are multiple issues related to dyskinesia, namely, anatomical distribution of dyskinesia, intensity of movements, disability or impact on activities of daily living, time spent with dyskinesia, and patient perceptions. Second, the actual phenomenology of dyskinesia, whether choreic or dystonic, is important to register and rate. Finally, discrimination from other motor features of parkinsonism, specifically tremor, is important to capture. All these features influence the items that need to be included in a given scale and the anchors related to those items.

Because of the impact of dyskinesia on patients with PD, the Movement Disorder Society (MDS) organized a systematic review of the clinimetric properties of the scales used to measure dyskinesia in PD [8]. This review fits into a larger program to evaluate rating scales for all elements of PD and therefore followed a standard process, described below. The overall findings of this program have been consolidated into a recent textbook [9], and this chapter is largely based on the chapter in that book devoted to dyskinesia [10].

The MDS Rating Scale Evaluation Programs

The organization of the systematic review of rating scales for PD, including the evaluation of dyskinesia rating scales, follows an established methodology. This process includes scale identification, selection, and appraisal strategies [8], using terminology and definitions developed for the Appendix of Ancillary Scales to complement the MDS-sponsored revision of the Unified Parkinson's Disease Rating Scale (MDS-UPDRS) [11, 12].

Table 2.1 Descriptive characteristics of the scales [10]

Scale	Time to complete (minutes)	Patient historical rating	Clinical examination	Administration burden[a]
AIMS	15	No	Yes	+
UPDRS	20[b]	Yes	Yes	+
MDS-UPDRS	20[b]	Yes	Yes	+
Obeso (CAPIT)	2	Yes	Yes	+
Rush Dyskinesia Rating Scale	5	No	Yes	+
CDRS	10	No	Yes	+
Lang-Fahn	5	Yes	No	+
PDYS-26	10	Yes	No	+
UDysRS	15	Yes	Yes	+

AIMS Abnormal Involuntary Movement Scale, *UPDRS* Unified Parkinson's Disease Rating Scale, *MDS-UPDRS* revised version of the *UPDRS*, *CDRS* Clinical Dyskinesia Rating Scale, *PDYS-26* Parkinson Disease Dyskinesia Scale, *UDysRS* Unified Dyskinesia Rating Scale

[a]Administration burden was rated as: + (easy, e.g., summing up of the items), ± (moderate, e.g., visual analogue scale (VAS) or simple formula), – (difficult, e.g., VAS in combination with formula, or complex formula, ? (no information found on rating method)

[b]Time necessary to complete all the scales

The official definitions for the critiques are: a scale is considered Recommended if it has been applied to PD populations, if there are data on its use in studies beyond the group that developed the scale, and if it has been studied clinimetrically and found to be valid, reliable, and sensitive to change. A scale is considered Suggested if it has been applied to PD populations, but only one of the other criteria applies. A scale is Listed if it meets only one of the three criteria defined for Recommended scales. Because of the paucity of demonstrated treatments for dyskinesia, the clinimetric criterion for rating dyskinesia scales does not categorically require responsiveness to be established. In the event that a scale fulfills the requirements of reliability and validity, the criterion is considered to be met, although the absence of responsiveness is noted as a weakness of the given scale. Details of the scale identification process for the entire PD rating scale program and for dyskinesia rating scales are outlined in full detail elsewhere [9, 10].

Dyskinesia Rating Scales

Nine published rating scales for dyskinesia in PD are identified and described below with the designations based on the three key criteria described above (Tables 2.1 and 2.2): Abnormal Involuntary Movement Scale (AIMS) [13], The Unified Parkinson's Disease Rating Scale (UPDRS) Part IV [14], its recent revision as part of the MDS-UPDRS [12], the Obeso Dyskinesia Rating Scale [15, 16],

Table 2.2 Key evaluation points of the scales

Scale	Applied in PD	Applied beyond original authors	Successful clinimetric testing	Qualification
AIMS	X	X	X[a]	Recommended
UPDRS	X	X	0[b]	Suggested
MDS-UPDRS	X	X	0[b]	Suggested
Obeso (CAPIT)	X	X	0	Suggested
Rush Dyskinesia Rating Scale	X	X	X	Recommended
CDRS	X	0	X	Suggested
Lang-Fahn	X	X	X	Recommended[c]
PDYS-26	X	X	X	Recommended[c]
UDysRS	X	X	X	Recommended[c]

For an explanation of the qualification groups, see text

AIMS Abnormal Involuntary Movement Scale, *UPDRS* Unified Parkinson's Disease Rating Scale, *MDS-UPDRS* revised version of the *UPDRS*, *CDRS* Clinical Dyskinesia Rating Scale, *PDYS-26* Parkinson Disease Dyskinesia Scale, *UDysRS* Unified Dyskinesia Rating Scale

[a]AIMS has several modified versions and it is not entirely clear whether clinimetric analyses are uniform across all versions

[b]Clinimetric testing not performed specifically on the Part IV

[c]In the original report [8], this scale was designated as Suggested, but based on newer published data, the authors have reclassified as Recommended on the premise that this designation will be officially authorized when an updated report is generated

the Rush Dyskinesia Rating Scale [17], the Clinical Dyskinesia Rating Scale [18], the Lang-Fahn Activities of Daily Living Dyskinesia Scale [19], the Parkinson Disease Dyskinesia Scale (PDYS-26) [20], and the Unified Dyskinesia Rating Scale (UDysRS) [21]. Home diaries for patients' self-assessment of dyskinesia have been developed [22], but these rating instruments are primarily focused on motor fluctuations. Nonetheless, a brief discussion of their use in dyskinesia is provided at the end of this chapter.

AIMS: Abnormal Involuntary Movement Scale

Scale Description

The AIMS is a clinician-rated instrument to assess the severity of abnormal movements in seven body areas [13], each area is scored from 0 to 4 (absent, minimal, mild, moderate, severe). Three additional items rate the global severity of abnormal movements, disability derived from the abnormal movements, and patient's awareness of the abnormal movement. The scale includes specific instructions to standardize the evaluation and requires the examiner to observe the patient sitting quietly at rest and again while carrying out selected motor tasks. The highest severity of the abnormal movements is rated. If movements only occur upon activation procedures such as opening and closing of the mouth, finger tapping, standing, and sitting, but are not seen spontaneously, the severity rating is ranked as one level lower than if the same intensity is seen spontaneously. The scale does not provide word anchors to explain the designations of absent, minimal, mild, moderate, and sever, so that these final designations may be biased by the rater's experience. The scale was originally developed for rating tardive dyskinesia in psychiatric patients but has been used for rating of Huntington's disease- and PD-related dyskinesia. Modifications that exclude the one-point reduction caveat for movements seen only with activation have also been used. In some cases, the questions related to dental hygiene and the wearing of dentures are excluded, since these questions are often considered more important in tardive dyskinesia than in dyskinesia related to Parkinson's disease.

The AIMS is quick to administer and takes only about 10 minutes to complete. It can therefore be repeated several times during the day for time-based assessments. It focuses on anatomy and does not give any indication whether the dyskinesia is dystonic or choreic in character. There is no patient input into the ratings, so the disability ratings rely on the clinician's judgment rather than patient perceptions. Because it is a scale developed for tardive dyskinesia, it emphasizes face and neck movements that may not be the primary focus of dyskinesia in PD.

Key Evaluation Issues and Recommendation Status

AIMS meets the criterion for use in PD and use by multiple authors studying dyskinesia effects by drugs and surgery [23, 24]. Clinimetric data rely mainly on inter- and intra-rater coefficients, and in non-PD patients, the scale showed high inter-rater and test-retest reliability for tardive dyskinesia [25, 26]. Only the original version of the scale has been clinimetrically assessed, whereas none of the modified versions have undergone validation testing. Further, the clinimetric properties of the scale have been only partly tested in PD. In one study the mean correlation coefficient (R) between two raters for total score was 0.81 ($p<0.01$) [20]. Internal consistency, concurrent validity, discrimination validity, and content validity have not been examined in PD patients. A moderate correlation was found between a modified AIMS version and involuntary movement amplitude derived from accelerometers in dyskinetic PD subjects [27]. Therefore, whereas sufficient testing has been performed to meet the clinimetric criterion, the data are not strong for specific reference to PD dyskinesia. As a final designation, the AIMS meets the criteria of a Recommended scale, but with limitations that include limited clinimetric data in PD patients, poor documentation of phenomenological subtypes of dyskinesia, and no information on the impact of dyskinesia on the patient's quality of life or perceptions of health.

The Unified Parkinson's Disease Rating Scale (UPDRS)

Scale Description

The UPDRS was developed by incorporating elements from previous scales, including the Columbia, Webster, King's College, Northwestern University Disability, UCLA, and New York University Parkinson's Disease Rating Scales to provide a comprehensive assessment of disability and impairment in PD [14]. The development of this scale involved multiple trial versions, and the final published scale is officially known as the UPDRS version 3.0 [14]. The UPDRS has been the most widely used rating scale for PD, in routine clinical practice and clinical trials, although it is increasingly being replaced by the newer version, called the Movement Disorder Society Unified Parkinson's Disease Rating Scale (MDS-UPDRS) [11, 12] . Part IV of the UPDRS focuses on motor complications and includes rating items to assess dyskinesia over the past week. Four issues are rated: historical information on dyskinesia duration (dividing the waking day into four quartiles), an overall assessment of intensity, the amount of painful dyskinesia, and the presence or absence of early morning dystonia.

The main strengths of the Part IV of the scale are the very short time required and the clear anchors with defined time frame for assessment. Whereas the UPDRS as a full scale demonstrates good inter- and intra-rater reliability, the individual or collective items covering dyskinesia have not been independently studied from a clinimetric perspective. Even with this weakness, the original UPDRS, specifically the items covering the disability due to dyskinesia and the amount of the waking day when dyskinesia is present, has been the primary outcome measure in clinical trials for antidyskinetic agents [23].

Key Evaluation Issues and Recommendation Status

The UPDRS meets the criteria of having been used to assess PD and having been used by multiple authors. A teaching videotape is available, standardizing the practical application of the scale, serving as an important tool to enhance inter-rater reliability [28]. Because the items related to dyskinesia have not been specifically studied from a clinimetric perspective, this criterion is not met, and the UPDRS therefore is designated as a Suggested scale.

The MDS-Revised Unified Parkinson's Disease Rating Scale (MDS-UPDRS)

Scale Description

Because of scientific advances and relevant limitations identified in the original UPDRS scale [29], the MDS developed a revision of the UPDRS, termed the MDS-UPDRS. After development, this scale underwent extensive clinimetric testing and validation before official presentation [12]. Although developed in English, official translations that meet clinimetric criteria for equivalency with the English version are now available in French, German, Italian, Spanish, Slovak, and Estonian with active programs in process for several additional languages. In regard to dyskinesia, Part IV has a series of questions to measure time spent with dyskinesia, severity of movements, and impact on the patient's life. Although part of a larger scale, the dyskinesia questions comprise a single statistical factor within clinimetric profile of the MDS-UPDRS.

Key Evaluation Issues and Recommendation Status

The MDS-UPDRS has been adopted in several clinical trials of PD by authors unrelated to the original authors and has been designated by the NIH Common Data Elements as the Recommended scale to assess overall PD impairment and specifically motor impairment (www.commondataelements.ninds.nih.gov). However, the dyskinesia items have not been specifically used as a primary outcome in a clinical trial of dyskinesia, and extensive clinimetric data are not available yet on this subsection of the overall scale. Therefore, the MDS-UPDRS dyskinesia items meet only two of the three criteria and are a Suggested scale for the evaluation of dyskinesia. Even with further clinimetric testing, because the questions on dyskinesia are few and general in their nature, this subsection of the MDS-UPDRS is likely to serve primarily as a screening assessment rather than a primary outcome measure for definitive studies.

Obeso Dyskinesia Rating Scale

Scale Description

This scale was one of the first assessment tools specifically developed to measure dyskinesia in PD. It consists of several questions that combine the patient's historical perceptions regarding dyskinesia and the examiner's objective rating of dyskinesia severity. Disability is assessed using two categories of information: severity (graded 0–5) and duration (graded 0–5). These scores are combined to provide a single score based on the mean of the two subscores. The intensity score combines two clinical issues, namely, patient awareness of abnormal movements and the actual observed intensity of such movements. The duration score, similar to the UPDRS Part IV duration item, divides the waking day into four quartiles.

The main strength of the scale is that it very easy to apply, and instructions to the rater are clear, except for the absence of the time frame of reference.

Key Evaluation Issues and Recommendation Status

Following development, this scale was included in the widely used Core Assessment Program for Intracerebral Transplantations (CAPIT) protocol for evaluation of patients undergoing neurosurgical interventions for PD [15, 16]. Subsequently, however, the scale was not widely used and has never been explored from a clinimetric point of view. As such, the Obeso Dyskinesia Rating Scale meets two of the three criteria (used in PD and by multiple authors, but no clinimetric validation), and it is designated as a Suggested scale for dyskinesia in PD.

Rush Dyskinesia Rating Scale [RDRS]

Scale Description

The Rush Dyskinesia Rating Scale [17] focuses on observation-based ratings of dyskinesia-related disability during prescribed tasks of daily living. In the original scale, the rater observes the patient walking, drinking from a cup, and putting on and buttoning a coat. A revision included a fourth task, namely, communication while a patient is sitting and speaking or reading. The greatest degree to which dyskinesia interferes with function is rated on a 0–4 scale that includes descriptors (0, absent; 1, minimal severity, no interference with voluntary motor acts; 2, dyskinesia may impair voluntary movements but patient is normally capable of undertaking most motor acts; 3, intense interference with movement control and daily life activities are greatly limited; 4, violent dyskinesia, incompatible with any normal motor task). In addition, the rater indicates which types of dyskinesia (chorea, dystonia, other) are present and which single type is most disabling.

In the original version, after all activities were observed, the single highest rating of disability was entered as the score, and in modifications, each activity was rated separated with the final score being the sum of all activity ratings. A videotape accompanies the original publication and demonstrates different degrees of severity, examples of chorea without dystonia, dystonia without chorea, and mixed dyskinesia [17]. The main strengths of the scale are the focus on functional disability of dyskinesia and its very rapid application, allowing for repeated ratings over time, especially for acute pharmacological studies. In several studies, the RDRS has been combined with the AIMS as two complementary measures utilized together, one focusing on disability and one on impairment. Because the scale is based on pre-specified tasks, if a patient has maximal dyskinesia during a specific activity not included in the protocol, there is no way to capture this disability with the RDRS.

Key Evaluation Issues and Recommendation Status

This scale is specifically designed for PD and has been applied in many clinical trials beyond the original authors [23, 30]. The scale has been demonstrated to have high intra-rater and inter-rater reliability. The scale also has been rated highly for its ease of application, appropriateness of tasks for reflecting disability, and overall utility. Clinimetric testing revealed relatively high inter- and intra-rater reliability. It therefore meets the three criteria and is designated as a Recommended scale, but likely needs to be used along with another scale, like the AIMS, in order to provide a fuller overall view of dyskinesia. Given this limitation, efforts to combine elements of the AIMS and RDRS led to the development of the Unified Dyskinesia Rating Scale (see below).

Clinical Dyskinesia Rating Scale (CDRS)

Scale Description

CDRS independently evaluates severity of hyperkinesia and dystonic postures in PD patients and is scored for each body region (face, neck, trunk, right and left upper extremities, right and left lower extremities) [18]. Scores range from 0 (none observed) to 4 (extreme), with use of 0.5-scoring intervals permitted for 6 items. The maximum total score for each subscale (dyskinesia and dystonia) is 28. Ratings are based on patient observation at rest and during activation. The scale does not provide an estimate of disability or functional compromise from dyskinesia. The scale has been validated in raters with and without experience in clinical trials for dyskinesia. Because it is short to administer, the scale has primarily been used to measure severity during acute levodopa challenge testing, applicable during both "on" and "off" conditions. No specific instructions on its use are described.

Key Evaluation Issues and Recommendation Status

This scale is designed specifically for PD but has not been used outside its developers. Clinimetrically, inter-rater reliability in PD patients was explored for different groups of raters (neurologists, neurosurgeons, and nurses specialized in PD), several of them being inexperienced with formal dyskinesia rating programs. The CDRS proved to have excellent inter-rater reliability for hyperkinesia ($W=0.88$) and moderate reliability for dystonia ($W=0.44$). Overall test-retest reliability was satisfactory (Kendall's tau = 0.74). Dystonia ratings had less concordance (with some Kendall tau coefficients as low as 0.31). The scale proved to be valid across all disease stages, but the scale's sensitivity to change (over time or to treatment) has not been demonstrated. Based on these collective findings, the CDRS meets two of the three criteria (use in PD and clinimetric testing), but it has not been used outside of its originators, resulting in a designation of a Suggested scale.

Lang-Fahn Activities of Daily Living Dyskinesia Scale

Scale Description

The Lang-Fahn Activities of Daily Living Dyskinesia Scale focuses on patient perceptions of disability related to dyskinesia. An ordinal scale (0–4, 0 representing no dyskinesia and 4 representing inability to perform the task) assesses the patient's report on five activities of daily living potentially impacted by dyskinesia at their maximum severity over the past few days (handwriting or drawing, cutting food and handling utensils, dressing, hygiene, and walking). The scale is completed by the physician based on historical information provided by the patient. Patients are asked to recall their function over the last few days and respond based on the worst interference by dyskinesia. Further information regarding the dyskinesia pattern such as diphasic, peak dose, or dystonia is not captured by the scale. Like other scales based on patients' declaration, it does not take into account that many patients will defer activities until dyskinesia resolves and, therefore, might state that dyskinesia is not disabling for a given task since the activity was deferred.

Key Evaluation Issues and Recommendation Status

This scale is specific for PD but has not been used widely in publications unrelated to the original authors (Parkinson Study Group) [19, 30, 31]. One study attempted validation of this scale [19], and documented a moderate correlation with patient diary information completed 1 week prior to the visit [19] and some correlation with the Clinician's Global Impression and the Patient's Global Impression. Core issues like test-retest reliability have not been clearly established. In a recent study, however, the scale was responsive to change in dyskinesia during a placebo-controlled assessment of amantadine [32]. Whereas the original MDS report designated this scale as Suggested [8], when the evaluation group updates its official report, the new information on responsivity warrants a likely redesignation of the Lang-Fahn Scale as Recommended for the assessment of patient-perceived disability from dyskinesia.

Parkinson Disease Dyskinesia Scale (PDYS-26)

Scale Description

The Parkinson Disease Dyskinesia Scale (PDYS-26) is a 26-item, patient-based measure "for quantifying the impact of dyskinesia on activities of daily living" in PD using a 0–4 option for each activity (0 = no interference, 4 = Activity impossible) [20]. Items include basic, instrumental, and social daily activities. The question for each item is about interference with the given activity over the past week by involuntary movements when they were at their most intense state. The scale is focused on choreic dyskinesia, and in the instructions, dyskinesia is equated to "involuntary movements," with specific warnings to exclude both tremor and dystonia. The PDYS-26 can be rapidly completed with clear and concise instructions to patients, although some questions have redundancy.

Key Evaluation Issues and Recommendation Status

This scale is specifically designed to assess PD dyskinesia and has been used in a number of studies conducted by investigators outside the developing team [32–34]. The scale was developed with strong clinimetric testing following Item Response Theory (Rasch analysis) principles and methodology. Later, it was validated by means of the Rasch analysis, again, and also applying Classical Test Theory methods. PDYS-26 has satisfactory acceptability, with no floor or ceiling effect and appropriate distribution of scores [20]. The internal consistency was very high (alpha = 0.97), perhaps related to redundancy, and the item homogeneity coefficient resulted satisfactory (0.59). The test-retest reliability was excellent (for the total score, ICC = 0.92). Concerning the convergent construct validity, PDYS-26 showed strong correlation with the UPDRS dyskinesia items and the sum of these items. Correlation was moderate/high with items of the Rush Dyskinesia Rating Scale ($R = 0.36$–0.78) and variable with the components of the AIMS ($R = 0.20$–0.84). Factor Analysis identified a single factor, explaining 58 % of the variance. It also demonstrated responsivity to amantadine treatment in comparison to placebo treatment [32]. In the original assessment document [8], this scale was designated as Suggested, but based on these findings, PDYS-26 fulfills the criteria to be Recommended and will likely be officially designated as such in a future update as a measure for assessing patient's perception of functional impact from dyskinesia in PD.

The Unified Dyskinesia Rating Scale (UDysRS)

Scale Description

The Unified Dyskinesia Rating Scale (UDysRS) is the newest rating scale developed specifically for the assessment of dyskinesia in PD [21]. The UDysRS contains both self-evaluation questions (by the patient alone or with their caregivers) and items that are assessed directly by the physician to rate abnormal movements objectively during prescribed activities of daily living. The time frame for the patient rating of dyskinesia refers to the prior week (including the day of which the examination is performed). The UDysRS consists of two primary sections (Historical and Objective). The Historical section is divided in two parts, the first focusing on "on dyskinesia" and the second focusing on "off dystonia." The Objective section is a rater-based assessment and is derived from a combination of the principles of the AIMS and the RDRS. Impairment or intensity of dyskinesia is rated in several body regions and disability is rated during four motor tasks: communication, drinking from a cup, dressing, and ambulation. All questions are scored on a scale from 0 to 4 (0 = normal, to 4 = severe) with a total possible score of 104. The scale provides specific instructions for all questions and comes with a teaching DVD including a training exercise [35]. This training program, based on various examples of dyskinesia and anchored in scores generated by experts, is aimed at increasing homogeneity of ratings among and within raters and centers. There is an official Spanish translation that has completed clinimetric validation. Other language translation programs are in process or planned

Key Evaluation Issues and Recommendation Status

This scale is designed specifically for PD-associated dyskinesia and, although new, is already part of numerous clinical trials of anti-dyskinesia agents [35 and see www.ClinicalTrials.gov. Internal consistency, factor structure, intra-rater and inter-rater reliabilities, temporal stability, and reproducibility of the scale have been established with a well-defined clinimetric program [21, 35, 36]. Further, in a trial comparing different scales in their ability to detect responsivity to amantadine, the UDysRS total score showed the highest effect size ($eta^2=0.138$) for detecting treatment-related change, with all other scales (Lang-Fahn, PDYS-26, Rush Dyskinesia Rating Scale, and AIMS) having effect sizes <0.1 [32]. In the original report [8], this scale was marked as Suggested, but, based on the new clinimetric data, the UDysRS fulfills the criteria to be designated as Recommended and will likely be officially designated as such for the overall assessment of dyskinesia, including both patient-based and rater-based ratings, in the official update planned for the future.

Home Diaries

In the MDS-sponsored evaluation of dyskinesia scales, home diaries were not officially critiqued, as these tools are primarily focused on the identification of motor fluctuations [22, 37]. Daily diaries are completed by patients at home and generally cover 30 min increments throughout a 24 h day. The most commonly used diary in clinical research is the Parkinson's Disease Diary, developed by Hauser and colleagues, and this tool includes five categories: sleep, OFF, ON without dyskinesia, ON with non-troublesome dyskinesia (dyskinesia that do not impede activities), and ON with troublesome dyskinesia that interfere with function [22]. Another diary model is the CPSIT-PD (Core Assessment Program for Surgical Interventional Therapies in Parkinson's Disease) ON-OFF Diary, and this tool has slightly different options: OFF, partial OFF, ON, and ON with dyskinesia of any type or severity [37]. Clinimetric studies of these two tools have been performed, but the assessments cover all aspects of motor fluctuations without specific focus on dyskinesia validation.

Future Perspectives

Given that very few effective treatments are available for treating or prevention of development of dyskinesia [38], there is significant scientific and pharmaceutical interest in the development of new antidyskinetic agents. One of the major roadblocks to progress has been the clear delineation of a single best scale to assess the burden of dyskinesia. The criteria utilized to assess dyskinesia scales have identified a small number of scales that meet the minimal criteria for Recommended and investigators and clinicians need to select among these scales the one that best fits the need of the assessments. If patient perceptions are the primary focus, the Lang-Fahn and PD-DYS-26 are likely to be selected, with the latter having the stronger clinimetric profile. If only objective assessments are desired, the AIMS can be chosen for impairment and the RDRS can be chosen for disability. The UDysRS has a very strong clinimetric profile and combines both patient perceptions and objective assessments of disability and impairment, providing the most comprehensive measurement tool for the overall burden of dyskinesia. Scales involving objective assessments are currently based on ratings in the physician's office, but, with the development of smart phones and electronic camera devices, it is reasonable to consider adaptations for home assessments. Further, a number of mechanical devices that record movement are available that may allow monitoring of dyskinesia in the home environment and thereby allow a longitudinal documentation of dyskinesia within the patient's day-to-day lifestyle. These technologies, however, need to discriminate dyskinesia from volitional movement and tremor in PD in order to be valid measurements of dyskinesia. The anatomical variability of dyskinesia, its multiple phenomenological forms, and the mechanical burden of wearing detection devices limit the practicality of such instruments today, but they can be reasonably envisioned for the future. Such devices may be particularly useful for detecting the true temporal pattern of dyskinesia, because a selective reduction of awareness of dyskinesia during the patient's "on" state is well described and likely falsely minimizes the true amount of daily time spent with dyskinetic movements [39].

Acknowledgments The authors would like to thank the following additional members besides themselves of the original MDS task force on scales to assess dyskinesia in Parkinson's disease: Pablo Martínez-Martín, MD (Madrid, Spain), Giovanni Fabbrini, MD (Rome, Italy), Robert A. Hauser, MD (Tampa, FL, USA), Marcelo Merello, MD, PhD (Buenos Aires, Argentina), Janis Miyasaki, MD (Toronto, Canada), Werner Poewe, MD (Innsbruck, Austria), Cristina Sampaio, MD (Lisbon, Portugal), Olivier Rascol, MD (Toulouse, France), Glenn T. Stebbins, PhD (Chicago, IL, USA), and Anette Schrag, MD, PhD (London, UK). Dr. Anette Schrag, as chair of the 2013 MDS Committee on Rating Scales, reviewed the contents of this chapter.

References

1. Damier P. Drug-induced dyskinesias. Curr Opin Neurol. 2009;22:394–9.
2. Fabbrini G, Brotchie JM, Grandas F, Nomoto M, Goetz CG. Levodopa-induced dyskinesias. Mov Disord. 2007;22:1379–89.
3. Mones RJ, Elizan TS, Siegel GJ. Analysis of L-dopa induced dyskinesias in 51 patients with parkinsonism. J Neurol Neurosurg Psychiatry. 1971;34:668–73.
4. Fabbrini G, Defazio G, Colosimo C, Suppa A, Bloise M, Berardelli A. Onset and spread of dyskinesias and motor symptoms in Parkinson's disease. Mov Disord. 2009;24:2091–6.
5. Martinez-Martin P, Cubo E. Scales to measure parkinsonism. Handb Clin Neurol. 2007;83:289–327.

6. Goetz CG. Rating scales for dyskinesias in Parkinson's disease. Mov Disord. 1999;14 Suppl 1:48–53.
7. Colosimo C, Fabbrini G, Berardelli A. Drug insight: new drugs in development for Parkinson's disease. Nat Clin Pract Neurol. 2006;2:600–10.
8. Colosimo C, Martínez-Martín P, Fabbrini G, et al. Task force report on scales to assess dyskinesia in Parkinson's disease: critique and recommendations. Mov Disord. 2010;25:1131–42.
9. Sampaio C, Goetz CG, Schrag A. Rating scales in Parkinson's disease. New York: Oxford University Press; 2012.
10. Colosimo C. Dyskinesia rating scales in Parkinson's disease. In: Sampaio C, Goetz CG, Schrag A, editors. Rating scales in Parkinson's disease. New York: Oxford University Press; 2012. p. 84–98.
11. Goetz CG, Fahn S, Martinez-Martin P, Poewe W, Sampaio C, Stebbins GT, Stern MB, Tilley BC, Dodel R, Dubois B, Holloway R, Jankovic J, Kulisevsky J, Lang AE, Lees A, Leurgans S, Lewitt PA, Nyenhuis D, Olanow CC, Rascol O, Schrag A, Teresi JA, Van Hilten JJ, Lapelle N. Movement Disorder Society-sponsored revision of the Unified Parkinson's Disease Rating Scale (MDS-UPDRS): process, format, and clinimetric testing plan. Mov Disord. 2007;22:41–7.
12. Goetz CG, Tilley BC, Shaftman SR, et al. Movement Disorder Society-sponsored revision of the Unified Parkinson's Disease Rating Scale (MDS-UPDRS): scale presentation and clinimetric testing results. Mov Disord. 2008;23:2129–70.
13. Guy W. Abnormal Involuntary Movement Scale. ECDEU assessment manual for psychopharmacology. Washington, DC: US Government Printing Office; 1976. p. 534–7.
14. Fahn S, Elton RL. Members of the UPDRS Development Committee. Unified Parkinson's Disease Rating Scale. In: Fahn S, Marsden CD, Calne DB, Goldstein M, editors. Recent development in Parkinson's disease, vol. 2. Florham Park: MacMillan Healthcare information; 1987. p. 153–63.
15. Langston JW, Widner H, Goetz CG, Brooks DB, Fahn S, Freeman T, Watts R. Core assessment program for intracerebral transplantation (CAPIT). In: Lindvall O, Bjorkland A, Widner H, editors. Intracerebral transplantation in movement disorders. Amsterdam: Elsevier; 1991. p. 227–41.
16. Langston JW, Widner H, Goetz CG, Brooks D, Fahn S, Freeman T. Core assessment program for intracerebral transplantations (CAPIT). Mov Disord. 1992;7:2–13.
17. Goetz CG, Stebbins GT, Shale HM, et al. Utility of an objective dyskinesia rating scale for Parkinson's disease: inter- and intrarater reliability assessment. Mov Disord. 1994;9:390–4.
18. Hagell P, Widner H. Clinical rating of dyskinesias in Parkinson's disease: utility and reliability of a new rating scale. Mov Disord. 1999;14:448–55.
19. Parkinson Study Group. Evaluation of dyskinesias in a pilot, randomized, placebo-controlled trial of remacemide in advanced Parkinson's disease. Arch Neurol. 2001;58:1660–8.
20. Katzenschlager R, Schrag A, Evans A, et al. Quantifying the impact of dyskinesias in PD. The PDYS-26: a patient-based outcome measure. Neurology. 2007;69:555–63.
21. Goetz CG, Nutt JG, Stebbins GT. The unified dyskinesias rating scale: presentation and clinimetric profile. Mov Disord. 2008;23:2398–403.
22. Hauser RA, Deckers F, Lehert P. Parkinson's disease home diary: further validation and implications for clinical trials. Mov Disord. 2004;19:1409–13.
23. Goetz CG, Damier P, Hicking C, et al. Sarizotan as a treatment for dyskinesias in Parkinson's disease: a double-blind placebo-controlled trial. Mov Disord. 2007;22:179–86.
24. Rodrigues JP, Walters SE, Watson P, Stell R, Mastaglia FL. Globus pallidus stimulation improves both motor and nonmotor aspects of quality of life in advanced Parkinson's disease. Mov Disord. 2007;22:1866–70.
25. Whall AL, Engle V, Edward A, Bobel L, Haberland C. Development of a screening program for tardive dyskinesias: feasibility issue. Nurs Res. 1983;32:151–6.
26. Sweet RA, DeSensi EG, Zubenko GS. Reliability and applicability of movement disorders rating scales in the elderly. J Neuropsychiatry Clin Neurosci. 1993;5:56–60.
27. Hoff JI, van den Plas AA, Wagemans EAH, van Hilten JJ. Accelerometric assessment of levodopa-induced dyskinesias in Parkinson's disease. Mov Disord. 2001;16:58–61.

28. Goetz CG, Stebbins GT, Chmura TA, et al. Teaching tape for the motor section of the Unified Parkinson's Disease Rating Scale. Mov Disord. 1995;10:263–6.
29. Movement Disorder Society Task Force on Rating Scales for Parkinson's Disease. The Unified Parkinson's Disease Rating Scale (UPDRS): status and recommendations. Mov Disord. 2003;18:738–50.
30. Fahn S, Oakes D, Shoulson I, et al. Levodopa and the progression of Parkinson's disease. N Engl J Med. 2004;351:2498–508.
31. Holloway RG, Shoulson I, Fahn S, et al. Pramipexole vs levodopa as initial treatment for Parkinson disease: a 4-year randomized controlled trial. Arch Neurol. 2004;61:1044–53.
32. Goetz CG, Stebbins GT, Chung KA, Hauser RA, Miyasaki JM, Nicholas AP, Poewe W, Seppi K, Rascol O, Stacy MA, Nutt JA, Tanner CM, Urkowitz A, Jaglin JA, Ge S. Which dyskinesia scale best detects treatment response? Multiple dyskinesia measures used in a double-blind placebo controlled trial. Mov Disord. 2013;28:341–6.
33. Lewitt PA, Hauser RA, Lu M, Nicholas AP, Weiner W, Coppard N, Leinonen M, Savola JM. Randomized clinical trial of fipamezole for dyskinesia in Parkinson's disease (Fjord study). Neurology. 2012;79:163–9.
34. Carroll CB, Bain PG, Teare L. Cannabis for dyskinesia in Parkinson's disease: a randomized double-blind crossover study. Neurology. 2004;63:1245–50.
35. Goetz CG, Nutt JG, Stebbins GT, Chmura TA. Teaching program for the Unified Dyskinesia Rating Scale. Mov Disord. 2009;24:1296–8.
36. Goetz CG, Stebbins GT, Theeuwes A, Stocchi F, Ferreira JJ, van de Witte S, Bronzova J. Temporal stability of the Unified Dyskinesia Rating Scale. Mov Disord. 2011;26:2556–9.
37. Defer G, Widner H, Marie RM, Remy P, Levivier M. Core assessment program for surgical interventional therapies in Parkinson's disease (CAPSIT-PD). Mov Disord. 1999;14:572–84.
38. Fox SH, Katzenschlager R, Lim S-Y, Rvina B, Seppi K, Coelho M, Poewe W, Rascol O, Goetz CG, Sampaio C. The movement disorder society evidence-based medicine review update: treatments for the motor symptoms of Parkinson's disease. Mov Disord. 2011;26(S3):S2–41.
39. Amanzio M, Monteverdi S, Giordano A, Soliveri P, Filippi P, Geminiani G. Impaired awareness of movement disorders in Parkinson's disease. Brain Cogn. 2010;72:337–46.

Chapter 3
Epidemiology of Levodopa-Induced Dyskinesia

Miguel Coelho and Joaquim J. Ferreira

Abstract Several factors may explain the heterogeneity found in reports of the prevalence and incidence of levodopa-induced dyskinesia (LID), namely, the type of design of the studies, the length of follow-up, and the instruments used to measure LID. In order to show these figures as clear as possible, this chapter reviews the epidemiology of LID found in three main design studies: community-based observational studies, hospital-based observational studies, and clinical trials.

In a landmark paper by Ahlskog and Muenter, the authors reviewed the studies from the pre-levodopa and modern era and found similar numbers for the prevalence and incidence of LID between observational studies and clinical trials. Nevertheless, a distinction between community-based and hospital-based studies was not done. The frequency of dyskinesia was higher and earlier in pre-levodopa compared to modern era studies. In the modern era, the frequency of LID was about 40 % by 5 years and 90 % by 9–15 years.

However, some recent studies found slightly smaller percentages of LID in community-based studies. In contrast, recent hospital-based cohorts found LID in more than 90 % of the patients by 10–15 years, although they were troublesome in only 12 %. The available data show that the figures for LID are similar worldwide and no major difference exists between populations with different ethnic background.

Clinical trials have proved to be a strong tool to compare the incidence of LID between different drugs or drug regimens. However, their strict inclusion and exclusion criteria may limit the external validation of their findings. Most often, RCTs have compared the frequency of LID in the experimental drug arm with that in the levodopa arm. The ELLDOPA trial found that patients receiving 600 mg/day of

M. Coelho (✉)
Clinical Pharmacology Unit, Instituto de Medicina Molecular,
Faculty of Medicine, University of Lisbon, Lisbon, Portugal

Department of Neurosciences, Neurology Service, Hospital de Santa Maria, Lisbon, Portugal
e-mail: migcoelho2002@yahoo.es

J.J. Ferreira
Clinical Pharmacology Unit, Laboratory of Clinical Pharmacology and Therapeutics,
Instituto de Medicina Molecular, Faculty of Medicine, University of Lisbon, Lisbon, Portugal

Department of Neurosciences, Neurology Service, Hospital de Santa Maria, Lisbon, Portugal

S.H. Fox, J.M. Brotchie (eds.), *Levodopa-Induced Dyskinesia in Parkinson's Disease*, DOI 10.1007/978-1-4471-6503-3_3

levodopa had a significantly higher frequency of LID (16.5 %) compared to placebo (3.3 %) and that dose of levodopa >300 mg/day carried a significantly higher risk of developing LID. Several trials compared the incidence of LID between dopamine agonists (DAs) and levodopa in early PD patients. As a rule, the incidence of LID was lower and delayed in patients taking DA, even if levodopa could be added later to the DA arm in open label. This difference was substantially higher before patients had started supplemental levodopa. As an example, the frequency of LID was 20 % in the ropinirole group compared to 45 % in the levodopa group at 5 years and 52.4 % compared to 77.8 % at 10 years, respectively. The pramipexole trials reported a frequency of LID of 24.5 % in the pramipexole group compared to 54 % in the levodopa group at 4 years, while at 6 years the frequency was 20.4 % with pramipexole compared to 36.8 % with levodopa. Interestingly, the 14-year follow-up of a UK trial found that LIDs were present in 56 % of the patients in the bromocriptine arm compared to 58 % in the levodopa arm, and additionally the frequency of moderate/severe LID was also similar in both arms (bromocriptine 35 % vs levodopa 39 %). The addition of entacapone to levodopa/carbidopa (LCE) was associated with a shorter time to the onset of and a higher frequency of LID (42 %) compared to the levodopa/carbidopa group (32 %). Patients taking selegiline were also found to have twice the hazard ratio to develop dyskinesia than patients on placebo.

Risk factors for LID have been identified. Several studies have shown that patients with a younger age at PD onset have a higher risk of LID, but this difference may lose significance in the long term. Females also appear to be at a higher risk for LID but this might be biased by a lower weight, yielding a higher concentration of levodopa. Longer disease duration, longer PD duration before the initiation of levodopa, and more advanced disease have been associated with a higher frequency of LID. On the other hand, levodopa carries a higher risk of LID compared to DA, whereas the dose and the duration of levodopa treatment are also associated with a higher frequency of LID.

Keywords Levodopa • Dyskinesia • Frequency • Prevalence • Incidence • Motor complications

Introduction

Levodopa remains the "gold-standard" antiparkinsonian drug to treat the motor symptoms of Parkinson's disease (PD). However, its long-term use is associated with the development of levodopa-induced motor complications (MC), namely, motor fluctuations and dyskinesia.

The emergence of MC in the natural history of PD is of paramount importance since it is a milestone in terms of greater disability, poorer quality of life (QoL), and complex drugs regimens. In fact, motor complications are the major indication for advanced therapies in the management of PD, such as functional surgery and apomorphine and levodopa/carbidopa intestinal gel pumps.

As such, knowledge of the epidemiology of levodopa-induced dyskinesia (LID) is very important in order to plan individual patient management and research for future strategies to treat and prevent the emergence of LID.

Several factors may account for discrepancies found in reports of the prevalence and incidence of LID. Although LID manifests as different phenomenology, most observational and interventional studies assess peak-dose dyskinesia, which is an important bias when estimating the epidemiology of LID. Figures based on observational studies or clinical trials introduce another type of bias, as these studies tend to recruit different samples of patients, and the same is true whether observational studies are community or hospital based. The epidemiology of LID is also influenced by the instruments used to capture the event, for example, whether they are patient centered or investigator centered. Similar to other studies, prospective and retrospective reports tend to find different values for LID. Moreover, some authors report the occurrence of MC as a whole, without differentiating motor fluctuations from LID. Finally, many reports do not distinguish between troublesome and non-troublesome dyskinesia, which have distinct clinical implications.

This chapter covers the epidemiology of LID and aims to present the information in such a way that the reader more easily realizes the influence of biases in the reporting of figures for LID.

Observational Studies

Community-Based Studies

A landmark paper by Ahlskog and Muenter [1] reviewed the epidemiology of LID reported in studies from the pre-levodopa era and from the post-levodopa or modern era. Papers from 1966 until September 2000 were included, and the authors searched series where patients had been treated with levodopa monotherapy. They excluded series with patients previously treated with dopamine agonists (DAs), but they allowed patients using concomitant amantadine, selegiline, or anticholinergics. The authors also excluded studies that specifically addressed MC in young-onset PD (<40 years). Series were stratified into pre-levodopa era and modern era, since in general, patients from the pre-levodopa era had a longer duration of parkinsonism at the onset of levodopa therapy; this stratification would help to understand whether the increase in frequency of MC with duration of levodopa therapy primarily reflects disease duration or levodopa treatment duration. Series were also grouped according to the duration of levodopa treatment. Comparing the frequency of LID between prospective clinical trials and observational studies yielded similar figures so that the overall frequency of LID can apply to both observational studies and clinical trials. However, a distinction between community- and hospital-based studies was not available in the review. Median duration of parkinsonism before onset of levodopa therapy was 6–10 years in pre-levodopa era studies compared to 2–3 years in modern era. The frequency of

Table 3.1 Levodopa-induced dyskinesia reported in community-based observational studies

Observational studies					
Community-based					
Cross-sectional		Longitudinal			
Schrag and Quinn 00 [4]	Wickremaratchi 11 [5]	Dotchin 11 [3]	Evans 11 [2]	Ahlskog and Muenter 01 [1]	Van Gerpen 06 [6] (retrospective)
19 %	23 %	25 % by 8.5 yrs of LD therapy	35 % by 8 yrs of FU	40 % by 5 yrs of LD therapy	30 % by 5 yrs of LD therapy and 59 % by 10 yrs of LD therapy

FU follow-up, *LD* Levodopa, *LID* Levodopa-induced dyskinesia, *yrs* years

dyskinesia was higher and much earlier in pre-levodopa era studies. By 6 months, about 50 % of patients developed dyskinesia in pre-levodopa era studies, compared to <10 % by 12 months in modern era studies. In pre-levodopa era, the frequency of LID (50 %) was maintained throughout the 2.5–3.5 years interval, with very few data beyond that time point. In contrast, the frequency of LID increased slowly in modern era studies, reaching about 33 % by 4–6 years and 90 % by 9–15 years. This difference in the frequency of LID may be accounted for by the difference in parkinsonism duration before the initiation of levodopa therapy, suggesting that greater depletion of dopaminergic striatal terminals may increase the likelihood of LID (Table 3.1).

A community-based prospective cohort of incident PD cases followed up for 8 years [2] enrolled 132 PD patients, with mean age at onset of 70.1 years. Using survival analysis, around 35 % of the patients (exact figure not given) developed LID diagnosed according to UPDRS part IV, with an estimated time to dyskinesia onset of 6.6 years. The only independent predictor for dyskinesia was the daily dose of levodopa equivalents (mean 429 mg), although duration of levodopa treatment predicted time to onset of LID. A smaller prospective study in rural Tanzania [3] identified 32 patients with PD who were followed up for 3 years. Median age at PD onset was 69 years and median duration of disease at study entry was 5.1 years. All patients were treated with levodopa monotherapy. Sixteen patients were available for evaluation at final follow-up. At the last visit, median age of these 16 patients was 81.5 years, median disease duration 8.5 years, median Hoehn and Yahr (HY) stage 3, and patients were taking a small mean daily levodopa dose of 384.4 mg. Four (25 %) patients developed LID, of peak-dose and end-of-dose subtype.

Schrag and Quinn [4] assessed the prevalence of MC in an unselected community-based sample of 124 patients with PD. This was a cross-sectional study and most information was retrieved retrospectively. Peak-dose, diphasic, and off-period dyskinesia were assessed. Patients had a mean age of 71.3 years and a mean disease duration of 6.8 years and were taking levodopa for 5.2 years with a mean daily dose of 423.3 mg. Eighty-seven (70 %) patients were taking levodopa, one-third was taking selegiline and about 15 % were taking DA or anticholinergics. Overall, twenty-four patients (19 %) had experienced LID (peak-dose in 25 %, diphasic in 6 %, and off-period dystonia in 10 %). The frequency of LID was 28 % among patients taking levodopa and it was 40 % in those taking >300 mg/day. The rate of MC increased with disease severity and duration and with dose and duration of levodopa treatment. The frequency of LID was 4 % in patients with ≤5 years of PD duration, 12 % after disease duration of 6–9 years, and 53 % in those with ≥10 years of disease. This rate increased to 7, 18 and 57 %, respectively, in patients taking levodopa. When considering only patients with good response to levodopa, the rates were 6, 20, and 67 %, respectively. The occurrence of dyskinesia also increased with higher stages of HY, from 29 % in HY stage 2–2.5, 43 % in HY 3, and 60 % in HY 4–5. The frequency of dyskinesia according to the duration of levodopa therapy was 13 % in patients taking levodopa for ≤5 years, 36 % after 6–9 years of levodopa therapy, and 100 % in those with ≥10 years of levodopa. Mean time to onset of dyskinesia was 6.7 years from symptoms onset and 5.7 years from levodopa start. There was a moderate positive correlation between emergence of MC and the time elapsed between disease onset and start of levodopa. However, the authors did not find a correlation between disease severity and onset of levodopa when controlled for disease duration, suggesting that a more aggressive disease was not the cause for an earlier start of levodopa and thus for an earlier emergence of MC. Younger patients at PD onset were found to have earlier dyskinesia, whereas current older patients developed dyskinesia later even after controlling for levodopa dose. In a logistic regression analysis using dyskinesia as the dependent variable, treatment duration and duration of disease significantly predicted the occurrence of dyskinesia; however, treatment duration did so more strongly than duration of disease. Nevertheless, the sensitivity of the model to predict dyskinesia was low at 75 %. In another cross-sectional study, Wickremaratchi et al. [5] reported on the motor phenotype of PD according to age at onset. They conducted a community-based study supplemented by regional referrals to enrich the sample with early onset (<45 years) PD cases (EOPD). Three hundred fifty-eight patients were included (70 EOPD), of whom 125 were community based. Mean age at onset was 56 years, mean current age was 65 years, and the mean disease duration was 9 years. Levodopa mean duration of treatment was 5 years and the mean daily dose was 428 mg. Community-based cases were older at onset (64 vs 51 years), older at time of assessment (72 vs 61 years), and had a shorter disease duration (7 vs 9 years) than referral cases. Most EOPDs were referrals but they did not significantly differ from EOPS cases from the community. Eighty (23 %) patients developed LID, a figure similar to that found by Schrag and Quinn [4]. LID strongly related to age at disease onset, with an increased risk for those with an onset <55 years (35 %) compared to those with disease onset ≥55 years (11 %); multivariate analysis showed that age at onset, duration of disease, duration of levodopa treatment, and levodopa dose all independently predicted occurrence of LID. For those with onset <55 years, duration of disease and levodopa dosage were stronger predictors for LID than duration of levodopa treatment. Younger patients also spent more time of the waking day in a dyskinetic state. EOPD cases reported more dystonia than later-onset patients, either pretreatment dystonia or levodopa-induced off-period and peak-dose dystonia.

In a retrospective study, Van Gerpen et al. [6] identified 126 incident PD cases diagnosed in Olmsted County, Minnesota, between 1976 and 1990 and followed up for a median of 11 years. Only peak-dose and diphasic LID were retrieved. The median age at PD onset (69 years) was older than usually reported and the median age at levodopa start was 72 years, while the median daily dose of levodopa for the first year was 450 mg. Reflecting prescription patterns at the time, all patients were initially treated with levodopa, and adjunct medication was limited. Using Kaplan-Meier estimates, the risk of dyskinesia was 30 % by 5 treatment years and 59 % by 10 treatment years, while the risk for dyskinesia severe enough to cause drug adjustment was 17 % and 43 % by 5 treatment and 10 treatment years, respectively. After 10 years of levodopa, the risk of dyskinesia not adequately controlled by medication adjustment was 12 %. Severe and uncontrolled LID occurred in only 1 patient after 10 years of treatment. Median time to dyskinesia onset was 8 years. In multivariate analysis, older age decreased the risk of dyskinesia, whereas higher levodopa dose at disease onset increased that risk.

Hospital-Based Studies

As a rule, hospital-based cohorts tend to report higher frequencies of LID, which is usually explained due to referral bias, as more complex patients or with younger onset are referred to dedicated clinics (Table 3.2). The first prospective, longitudinal cohort of new onset Chinese PD patients included 171 cases that were follow-up for a mean of 11 years [7]. Mean age at disease onset was 62.2 years, mean HY at 1.9, while mean disease duration at the end of follow-up was 11.4 years. Fifty (29.2 %) died during the study. First occurrence of dyskinesia was recorded although the instrument to capture the event was not reported. Almost all (99 %) patients were exposed to levodopa, of whom 63.9 % developed LID. Chinese in addition to Thai or mixed origin patients were also enrolled in a multicenter study in Thailand [8]. One hundred fifty-four patients were asked retrospectively about the occurrence of MC. Young-onset patients were excluded. Current mean age was 68.1 years while mean age at onset was 61.2 years. PD duration was ≥10 years in 13.6 % of patients. Levodopa treatment was reported in 98.1 % of patients (mean daily dose 409 mg), and 85 % were classified as excellent responders. Dyskinesia was present in 23.7 % of patients and 10.5 % complained of morning dyskinesia, presumably morning dystonia. A univariable analysis identified younger age at onset, longer disease duration, higher severity, longer treatment duration, and higher levodopa dose as predictors of MC.

The Sydney Multicenter Study was initially conducted as a double-blind RCT comparing the efficacy and safety of low-dose levodopa with low-dose bromocriptine in 149 levodopa-naive PD patients [9]. Results on the surviving patients were reported after a mean follow-up of 15 and 20 years [10, 11]. The mean age at PD onset was 56 years for the 52 surviving patients at the 15-year assessment; all but 3 were on levodopa with a mean daily dose of 734 mg [10]. LID had been experienced by 49 (94 %) patients, and they were disabling according to UPDRS part IV in 6 (12 %) [10]. The mean duration of treatment to LID onset was 5.3 years, with a significant difference between levodopa (4.2 years) and bromocriptine (6.9 years) arms. Thirty patients were still surviving at 20 years, and LIDs were present in all those taking >300 mg of levodopa but were disabling in only 6 [11].

Table 3.2 Levodopa-induced dyskinesia reported in hospital-based observational studies

Observational studies														
Hospital-based														
Cross-sectional							Longitudinal							
Papapetropou. 04 [21]	Fabbrini 09 [16]	Muller 07 [24] (survey to physicians)	Hashim 13 [18]	Juri-Clav. 07 [17]	Hassin-Baer 11 [19] (retrospective)	Coelho 10 [20]	Sato 06 [22] (retrospective)	Kulkantrakorn 06 [8] (retrospective)	Cabo López 10 [14]	Auyeung 12 [7]	McColl 02 [12]	Mazzella 05 [23] (retro-prospective)	Garcia-Ruiz 12 [15]	Hely 05 [10]
27.4 %	33.5 %	34 %	44 %	47.2 %	57.4 %	62 %	8.5 %, 35.1 % and 62.8 % by 5, 10 and 15 yrs of PD	23.7 % by 7 yrs of PD	38 % by 5 yrs of PD and 71 % by 10 yrs of PD	63.9 % by 11.4 yrs of PD	72 % by 2.3 yrs of LD therapy	88 % by 9.9 yrs of LD therapy	91 % by 10 yrs of PD	94 % by 15 yrs of PD

LD Levodopa, *LID* Levodopa-induced dyskinesia, *PD* Parkinson's disease, *yrs* years

Using an interesting design, patients were videotaped during 3 challenge tests after 200/50 mg of levodopa/carbidopa and scored regarding parkinsonism and dyskinesia in order to assess how the motor response to levodopa evolutes in the long term (6 and 10 years) [12, 13]. Thirty-four patients were recruited from the time of treatment initiation and followed up for a mean of 8 years [12]. All patients except one were started with levodopa and mean daily dose at study end was 606 mg. Mean age was 64 years at study onset, and 16 reached final assessment at 94 months [12]. LID emerged in 72 % of the patients after mean treatment duration of 28 months; their severity increased over time, and they were significantly more severe in fluctuators than in non-fluctuators and also in those with higher amplitude of motor response to levodopa [12].

Cabo López et al. [14] conducted a long-term prospective study of "de novo" PD patients in a naturalistic environment of a clinic-based practice. Sixty-four patients recruited between 1992 and 2002 were evaluated at baseline and 5 years, of whom 38 were available for assessment at 10 years. Reasons for lost to follow-up were not reported, and neither the clinical characteristics nor the drug use of those patients who withdrew from study was recorded as well. Patients were asked about MC at each 6-month follow-up visit. Mean age at PD onset was 61.3 years and mean baseline motor UPDRS score was 17.6 points. Patients were grouped according to initial therapy (levodopa vs non-levodopa) in a nonrandomized fashion. At baseline, 43 patients were initially treated with levodopa and these had a worse score in motor UPDRS compared to those initially not treated with levodopa. At 5 years, all 64 patients were on levodopa and 38 % reported LID (levodopa 51.2 % vs non-levodopa 9.5 %), and at 10 years LID occurred in 71 % (levodopa 84 % vs non-levodopa 46.2 %), whereas 25 % of the patients did not develop MC at 5 years and only 1 of 38 was free of MC at 10 years. Mean daily levodopa dose at last assessment was 550 mg. In a similar study, 45 early PD patients (mean age at diagnosis 58.5 years) were followed for 10 years with assessments every 6 months [15]. Initial antiparkinsonian therapy was chosen by the treating neurologist, and levodopa was the initial drug in 25 patients. There were no clinical differences between patients started on levodopa or on other antiparkinsonian drugs. At 5 years, all but 1 patient were on levodopa, and at 10 years LIDs were present in 91 % of the patients. Factors that predicted a higher frequency of LID were younger age at onset (OR 0.90; 95 % CI 0.82–0.98), being female (OR 12.87; 95 % CI 1.91–167.9), and levodopa as initial therapy (OR 8.31; 95 % CI 1.42–82.77).

Several cross-sectional studies have assessed the frequency of LID. Fabbrini et al. [16] found that 33.5 % of 307 PD patients recruited during a 12-month period had LID. Patients had a mean age of 69.1 years, mean disease duration of 8.5 years and had dyskinesia for a mean of 4.6 years. Thirty-nine patients with LID were treated with levodopa monotherapy, 64 were treated with levodopa and DA, and the mean daily dose of levodopa was 625 mg. All patients had peak-dose dyskinesia and 27 % had additionally diphasic dyskinesia, and in most their severity was mild to moderate. Patients with LID had a younger age at PD onset, a longer duration of disease and more severe symptoms, a longer duration of treatment with levodopa, and more motor fluctuations than patients without LID. In 124 Chilean PD patients, LIDs were identified in 47.2 % of the patients [17]. Mean age for the whole sample was

66.1 years and patients had a mean PD duration of 8.1 years; levodopa was taken by 92.7 % of the patients with a mean daily dose of 713 mg. Female patients and a higher levodopa dose were associated with a higher prevalence of LID. A Malaysian study found a similar frequency of LID (44 %) in 95 multiethnic Malaysian PD patients (64.3 % Chinese, 31 % Malays, and 3.7 % Indians and other ethnic groups) [18], assessing LID by UPDRS part 4. Mean age at PD onset was 58.5 years and mean disease duration was 6 years, while median duration of levodopa therapy was 3 years and its mean daily dose was 550 mg. Peak-dose LIDs were present in 59.5 % of the cases and diphasic in 26.2 %. In 54.8 % they involved >1 body region, in 23.8 % affected the face and neck, the lower limbs were involved in 14.3 %, the whole body in 11.9 %, and the trunk and the upper limbs were affected in 4.8 and 2.4 % of the patients, respectively. LID had a median duration of 52.5 min. while peak-dose dyskinesia started 60 min. after levodopa intake. The independent predictors of LID were the longer duration of levodopa therapy, younger onset age, and a higher total daily levodopa. The strongest predictor of LID was the duration of levodopa therapy, as a 1-year increase in the duration of levodopa therapy predicted a 42.4 % ($p=0.003$, 95 % of CI 11.3–17.9 %) increase in the odds of developing dyskinesia. The total daily levodopa dose and onset age ranked, respectively, the second and third strongest risk factors for LID. A slightly higher frequency of LID was found in 155 Israeli PD patients treated with levodopa and with good clinical data in medical files [19]. Mean age at PD onset was 61.1 years but duration of PD was not reported. Time from PD diagnosis to levodopa onset was 1.5 years, whereas time from PD diagnosis to development of LID was 5.3 years. Eighty-nine (57.4 %) had developed LID, and these were predicted by a younger PD age onset and a longer latency between diagnosis and initiation of levodopa, while female patients had a shorter time to LID onset. The clinical characteristics of late-stage Parkinson's disease have been reported by Coelho et al. [20] in a study of 50 PD patients in stage 4 or 5 of HY during on-state. Mean disease duration was 18 years, mean age of patients was 74.1 years, and they were taking a mean levodopa daily dose of 785 mg. LID were present in 31 (62 %) patients and they were troublesome in 13. They were rated as severe completely disabling by 8 % of the patients and in a similar percentage they occurred for >75 % of the waking day, according to UPDRS part 4. Peak-dose dyskinesia affected 30 % of the cases, diphasic dyskinesia 18 %, and in 14 % of the patients both peak-dose and diphasic LID were reported. Papapetropoulos et al. [21] compared MC in sporadic and familial PD and found an overall frequency of LID of 27.4 %. The frequency of LID increased with disease duration and severity in both groups. Only in those patients with ≤5 years of PD duration did familiar PD patients report significantly more dyskinesia (30.4 %) than sporadic PD patients (4.5 %). Importantly, familial cases with a genetic diagnosis of parkin or alpha-synuclein mutations were excluded. Mean age at onset (61 years), disease duration (7 years), levodopa treatment duration (6 years), daily dose (440 mg), and mean time from onset to levodopa start (1.2 years) were similar in both groups. Time from disease onset to development of LID was similar between familiar and sporadic PD, whereas familiar PD patients started LID significantly earlier than sporadic PD patients after onset of levodopa (3.4 vs 4.1 years, respectively). Peak-dose dyskinesia was the predominant LID, while 3 patients reported diphasic dyskinesia.

The Juntendo Parkinson Study Group [22] included 1,768 Japanese PD patients in a retrospective study to assess long-term prognosis in PD. Frequency and time to onset of MC were assessed in a sub-cohort of 1,183 individuals, whose mean age at PD onset was 59.3 years, mean disease duration was 6.4 years, and mean time to levodopa start was 2.3 years. After 5, 10, and 15 years of disease, the frequency of LID was 8.5, 35.1, and 62.8 %, respectively. Regression models showed that females and patients with an onset ≤50 years had a significantly shorter time to the emergence of MC, while no association was found between the initial symptoms (tremor onset, bradykinesia onset, or gait onset) or the initial type of treatment and the time to develop LID. A higher frequency of LID (88 %) was found in another retrospective study conducted in 116 patients on levodopa monotherapy [23]. The authors reviewed the medical records of patients treated between 1965 and 1992 and followed regularly until death for a mean of 7 years. The sample included patients diagnosed before the advent of levodopa, thus mean time between PD diagnosis and start of levodopa was considerably long (6.7 years) compared to modern series. Mean age at PD onset was 58.7 years, mean PD duration was 16.1 years, while mean duration of levodopa treatment was 9.9 years (mean daily dose 992.4 mg). Mean time of onset of LID after levodopa was 1.4 years. None of the independent variables predicted the occurrence of LID, although younger age at onset was associated with a higher risk of dyskinesia but without reaching statistical significance. The heterogeneity of this sample may explain the lack of association found between risk factors and the development of LID.

An interesting survey was conducted by Müller et al. [24], where 380 PD specialists from 7 countries worldwide were inquired about LID in their patients. Physicians estimated that 34 % of their patients experienced LID, although differences were present between countries. Physicians additionally completed retrospectively the patient record forms of their last 5 patients with LID. The mean age of 1,900 patients was 69 years and the mean duration of PD was 8.9 years; 94 % of patients were receiving levodopa at the time of dyskinesia onset (monotherapy 38 %; combination with DA 56 %), while 5 % were treated with DA monotherapy. Most patients (60 %) were receiving levodopa for 1–6 years when dyskinesia were first diagnosed, although the occurrence of LID was unrelated to the duration of levodopa therapy. LID were moderate-to-completely disabling in 57 % of the patients, and 18 % of them suffered LID for more than 50 % of the waking day.

Clinical Trials

Randomized clinical trials (RCTs) are another source of information for extracting numbers concerning the frequency of LID. Strict inclusion and exclusion criteria may limit the external validation of the findings of an RCT, whereas type of design, comparators, choice of outcomes and measurement tools, duration of the trial, and sample size call for attention when interpreting results regarding the frequency of LID. Nonetheless, RCTs are powerful research tools and ideal experimental paradigms to assess and compare the frequency of LID under specific drug treatments. Usually, RCTs assessing the emergence of LID have compared the frequency with which LID develops with the experimental drug with that due to levodopa. We are reporting only trials not included in the review by Ahlskog and Muenter et al. [1] (Table 3.3).

Levodopa

The ELLDOPA trial [25] assessed the effect of levodopa on the rate of progression of parkinsonism in 361 patients with early PD. The primary outcome was the change in the total score of UPDRS, while dyskinesia was depicted as adverse events. Patients were allocated to placebo or to three different doses of levodopa (150, 300, or 600 mg daily). After 40 weeks, patients receiving levodopa at 600 mg had a significantly higher frequency of LID (16.5 %) compared to placebo (3.3 %), whereas the frequency of dyskinesia was similar to that of placebo in those patients receiving 150 or 300 mg of levodopa. These results suggest that doses of levodopa >300 mg/day are associated with a significantly higher risk of developing LID.

In an open-label RCT, Caraceni et al. [26] compared the frequency of MC in 473 early PD patients allocated to either levodopa, DA, or selegiline monotherapy, with the possibility of later addition of levodopa if necessary. After a mean follow-up of 3 years, the frequency of LID was significantly higher in levodopa-treated patients (27.1 %) compared to DA-treated patients (14.8 %) but not to selegiline-treated patients (20.7 %). Patients on levodopa developed LID after a mean of 26 months, compared to 18 and 21 months in those on DA and selegiline, respectively. LID in the selegiline and DA arms became more frequent when levodopa was added.

Dopamine Agonists

RCTs comparing the frequency of LID between DA and levodopa differ in the methods of dyskinesia assessment, number of treatment arms, allowance for levodopa rescue, disability and QoL assessment, and duration of follow-up. Nevertheless, they have all shown a consistent lower frequency of or a delay to the onset of LID in early PD patients randomized to DA monotherapy compared with those randomized to levodopa monotherapy, even if levodopa could be added later to the DA arm in open label [27–34]. The magnitude of this difference was substantially higher when comparing patients not requiring supplemental levodopa or before they started rescue levodopa. Additionally, the majority of these RCTs have found that the DA had a weaker effect on motor disability than levodopa [27–35]. Although DAs were associated with a lower frequency of LID, most RCTs found a similar frequency of disabling dyskinesia in the long term (4–10 years) as well as equivalent scores in QoL scales between patients initially randomized to DA or levodopa [27–34].

Table 3.3 Levodopa-induced dyskinesia reported in clinical trials (selection)

Clinical trials											
Levodopa		Pergolide	Ropinirole			Pramipexole		Bromocriptine	Entacapone		Selegiline
PSG 04 (ELLDOPA) [25]	Caraceni 01 [26]	Oertel 06 [27]	Rascol 00 [28]	Rascol 06 [29]	Hauser 07 [30]	PSG 04 [31]	PSG 09 9 [32]	Katzenschlager 08 [35]	Hauser 09 (FIRST-STEP) [37]	Stocchi 10 (STRIDE-PD) [36]	Shoulson 02 [38]
LID at 40 weeks: LD 600 mg 16.5 % vs PL 3.3 %	LID at 3 yrs: LD 27.1 % vs DA 14.8 % vs SL 20.7 %	LID at 3 yrs: PER 16.3 % vs LD 32.9 %	LID at 5 yrs: RP 20 % vs LD 45 %	LID before LD supplementation: RP 5 % vs LD 36 %	LID at 10 yrs: RP 52.4 % vs LD 77.8 %	LID at 4 yrs: PR 24.% vs LD 54 %	LID at 6 yrs: PR 20.4 % vs LD 36.8 %	LID at 14 yrs: BR 56 % vs LD 58 %	LID at 39 weeks: LCE 5.3 % vs LD 7.4 %	LID at 134 weeks: LCE 42 % vs LD 32 %	LID at 2 yrs: SL 33.8 % vs PL 19.4 %

BR bromocriptine, *DA* dopamine agonists, *LCE* Levodopa/carbidopa/entacapone, *LD* Levodopa, *LID* Levodopa-induced dyskinesia *PER* pergolide, *PL* placebo, *PR* pramipexole, *RP* ropinirole, *SL* selegiline, *yrs* years

The PELMOPET study [27] stands out as a good model regarding the comparative frequency of LID between DA and levodopa because rescue therapy with levodopa was not allowed and no open-label extension was conducted, albeit it poorly reflects routine clinical practice. In this strict monotherapy trial for 3 years, the incidence of LID and the time to onset of LID were significantly lower and longer with pergolide than with levodopa. If emergence of LID was defined by the first UPDRS IVa positive score, LID in the pergolide group was 16.3 % compared to 32.9 % in the levodopa group, whereas it was 8 % in the pergolide arm compared to 26 % in the levodopa arm if only a positive answer to question 32 of UPDRS IVa was considered. These values are in the range of those reported in the ropinirole trial before levodopa supplementation (ropinirole 5 % vs levodopa 36 %), where a positive score on UPDRS IVa item 32 was the primary outcome [28, 29]. The total frequency of LID was 20 % in the ropinirole group compared to 45 % in the levodopa group at 5 years, whereas it was 52.4 % compared to 77.8 % at 10 years, respectively [28, 30]. Hauser et al. [30] found a median time to develop LID of 8.6 years in patients from the original ropinirole group compared to 7 years in patients from the original levodopa group. The ropinirole trial identified younger age at baseline, worse motor disability at baseline, and higher levodopa dose at dyskinesia onset as the best predictors of time to development of dyskinesia [28, 29]. The pramipexole trials reported similar results, with a frequency of LID at 4 years of 24.5 % in the pramipexole arm compared to 54 % in the levodopa arm, while in the final visit at 6 years LID was reported by 20.4 % of the patients initially randomized to pramipexole and by 36.8 % of those originally randomized to levodopa [31, 32]. Katzenschlager et al. [35] reported the 14-year follow-up results of the open-label randomized Parkinson's Disease Research Group of the United Kingdom (PDRG-UK) trial comparing three initial treatments (levodopa, levodopa + selegiline, and bromocriptine) in PD, extending the previous results of the 10-year follow-up reported by Lees et al. [34]. The levodopa+selegiline arm was stopped prematurely due to higher mortality in an interim analysis. Of the original 782 patients, data was available for 109, whose mean age at disease onset was 57 years and mean disease duration was 19 years. All but 4 patients were on levodopa at final follow-up. LIDs were present in 58 % of the patients in the levodopa arm compared to 56 % in the bromocriptine arm. Similarly, the difference in the prevalence of moderate/severe LID according to investigators rating was also nonsignificant between the treatment arms (levodopa 39 % vs bromocriptine 35 %).

COMT Inhibitors (Entacapone)

The STRIDE-PD study [36] compared the time to onset and frequency of LID between levodopa/carbidopa (LC) and levodopa/carbidopa/entacapone (LCE) in early PD patients requiring initiation of levodopa. The results showed that initiation of levodopa with LCE was associated with a significant shorter time to onset of LID and also the frequency of LID was higher in the LCE group (42 %)

compared to the LC group (32 %). Either group had a frequency of LID in the range of that reported for the levodopa treatment arms in the DA trials. A subanalysis showed that this significant difference was lost if patients on DA were excluded (LCE 34 % vs LC 35.5 %). The occurrence of dyskinesia was more frequent in patients younger than 65 years, women, patients with a lower body weight (<75 kg), a disease duration <2 years, and in patients receiving >400 mg/day of levodopa. There was a trend for better motor function in patients taking LCE. A prior study (FIRST-STEP) [37], evaluating the efficacy, safety, and tolerability of LCE compared to LC in early PD, found that the frequency of LID was not statistically different between LCE (5.3 %) and LC (7.4 %), although the methods to depict dyskinesia were the same in both STRIDE-PD and FIRST-STEP trials. In contrast to the STRIDE-PD, the primary outcome of the FIRST-STEP was the difference in the total score of UPDRS parts II and III, its duration was much shorter (39 weeks), and patients on DA were excluded.

MAO-B Inhibitors (Selegiline and Rasagiline)

From the initial 800 PD patients enrolled in the parent Deprenyl and Tocopherol Antioxidative Therapy of Parkinsonism (DATATOP) clinical trial, a second independent randomization was carried out for 368 patients who had required levodopa and were on selegiline [38]. In a double-blind design, patients were allocated either to withdraw selegiline and change to a placebo of selegiline or to continue on selegiline. The primary outcome was the development of wearing off, dyskinesia, or on-off motor fluctuations during a mean follow-up of 2 years. At baseline, patients had a mean age of 67 years and a mean in-study exposure to selegiline of 4.4 years; mean HY stage was 2.1 and mean SE was 86, whereas mean UPDRS motor score was 20. LIDs were present in 37 % (placebo) and 40 % (selegiline) of the patients at baseline, and patients randomized to placebo had a slightly higher dose and exposure to levodopa. During the follow-up, patients on selegiline had twice the hazard ratio to develop dyskinesia than patients on placebo, with a frequency of LID of 33.8 and 19.4 %, respectively. Selegiline patients taking at baseline levodopa for >3 years had a frequency of LID of 45 % compared to 25 % in those taking levodopa for less than 3 years. The difference in the frequency of LID lost significance when only patients with no LID at the second randomization were analyzed.

We found no figures for the incidence of LID in the trials of rasagiline.

Risk Factors

Risk factors for LID are covered more extensively in Chap. 4 in this book. Here, we briefly touch upon the most relevant ones; many of which have already been mentioned along previous sections.

Characteristics of the Patient

Age

Patients with younger age at disease onset have been associated with a higher risk to develop LID [5, 15, 18, 39–43]. In a recent systematic review (SR), Wickremaratchi et al. [39] found that patients with a younger age at onset had a higher risk of LID as well as an increased rate of dystonia at onset and during treatment, either as peak-dose or off-period motor symptom. Interestingly, this SR identified very few prospective population-based studies assessing the effect of PD onset age on the clinical features of PD and that the definition of young-onset PD (YOPD), usually defined as those with onset <40 years, varied widely between reports. Even in patients with a disease onset after 40 years, the incidence of LID drops with increasing age of PD onset, as found by Kumar et al. [41] in a review of the medical records of 91 patients who had been treated with levodopa for >5 years. The incidence of LID declined markedly with disease onset after 60 years, and it was only 14–16 % in those patients with PD onset after 70 years. These results were confirmed in 3 recent studies [5, 43, 15]. In 358 community-based and regional PD patients, the OR for development of LID in those patients with onset <55 years was 3.8 (95 % CI 1.8–8.0), while it was 2.1 (95 % CI 1.0–4.8) in those with onset <45 years [5]. Importantly, mutations in parkin or PINK1 genes did not explain this difference in the frequency of LID [5]. The studies by Ku et al. and Garcia-Ruiz et al. suggest that the difference in the incidence of LID according to PD age onset loses significance in the long term (10 years) [43, 15]. In the study by Schrag and Quinn [4], age or age at disease onset was not associated with a higher prevalence of LID; however, younger patients and patients with younger age at PD onset developed LID earlier than older patients or those older at disease onset.

Gender

The risk of LID has been found to be higher in female patients in some reports [44, 45, 17, 15] but not in others [18, 4]. A higher prevalence of LID in females may be confounded by a lower weight in females, yielding a significantly higher plasma concentration of levodopa [46, 47]. Hassin-Baer et al. [19] found that female patients, although not associated with a higher risk of LID, had a shorter time to LID onset.

Characteristics of the Disease

Regarding features of PD, the characteristics that more frequently have been associated with a higher risk of LID are longer disease duration, longer PD duration before initiation of levodopa therapy, and more advanced disease [1, 4, 48, 42, 18], albeit the OR of each of these risk factors varies between reports.

Characteristics of the Drug Treatment

Several clinical trials have shown that levodopa carries a higher risk of LID compared to DA [27–35]. The duration of levodopa treatment has also been associated with a higher prevalence of LID [1, 4, 48, 18], and one study has found that the effect of the duration of levodopa treatment was even stronger than that of PD duration [4]. Higher levodopa dose has also been associated with a higher frequency of LID [1, 4, 48, 42, 18, 17].

Conclusions

Evidence shows that the prevalence and incidence of LID vary considerably depending on the instrument used to capture the event, the type of PD patients recruited, the design of the study, the length of follow-up, and the drugs used to treat parkinsonism. No strong data support a striking difference in the frequency of LID among different ethnic populations. A great percentage of PD patients will develop LID during the disease course, especially if attending a dedicated PD clinic. Levodopa substantially increases the risk of LID, although this difference and its clinical implication might dissipate in the long term.

References

1. Ahlskog JE, Muenter MD. Frequency of levodopa-related dyskinesias and motor fluctuations as estimated from the cumulative literature. Mov Disord. 2001;16(3):448–58.
2. Evans JR, Mason SL, Williams-Gray CH, et al. The natural history of treated Parkinson's disease in an incident, community based cohort. J Neurol Neurosurg Psychiatry. 2011;82(10):1112–8.
3. Dotchin C, Jusabani A, Walker R. Three year follow up of levodopa plus carbidopa treatment in a prevalent cohort of patients with Parkinson's disease in Hai, Tanzania. J Neurol. 2011;258(9):1649–56.
4. Schrag A, Quinn N. Dyskinesias and motor fluctuations in Parkinson's disease. A community-based study. Brain. 2000;123(Pt 11):2297–305.
5. Wickremaratchi MM, Knipe MD, Sastry BS, et al. The motor phenotype of Parkinson's disease in relation to age at onset. Mov Disord. 2011;26(3):457–63.
6. Van Gerpen JA, Kumar N, Bower JH, Weigand S, Ahlskog JE. Levodopa-associated dyskinesia risk among Parkinson disease patients in Olmsted County, Minnesota, 1976–1990. Arch Neurol. 2006;63(2):205–9.
7. Auyeung M, Tsoi TH, Mok V, et al. Ten year survival and outcomes in a prospective cohort of new onset Chinese Parkinson's disease patients. J Neurol Neurosurg Psychiatry. 2012;83(6):607–11.
8. Kulkantrakorn K, Tiamkao S, Pongchaiyakul C, Pulkes T. Levodopa induced motor complications in Thai Parkinson's disease patients. J Med Assoc Thai. 2006;89(5):632–7.
9. Hely MA, Morris JG, Traficante R, Reid WG, O'Sullivan DJ, Williamson PM. The Sydney multicentre study of Parkinson's disease: progression and mortality at 10 years. J Neurol Neurosurg Psychiatry. 1999;67(3):300–7.

10. Hely MA, Morris JG, Reid WG, Trafficante R. Sydney Multicenter Study of Parkinson's disease: non-L-dopa-responsive problems dominate at 15 years. Mov Disord. 2005;20(2):190–9.
11. Hely MA, Reid WG, Adena MA, Halliday GM, Morris JG. The Sydney multicenter study of Parkinson's disease: the inevitability of dementia at 20 years. Mov Disord. 2008;23(6):837–44.
12. McColl CD, Reardon KA, Shiff M, Kempster PA. Motor response to levodopa and the evolution of motor fluctuations in the first decade of treatment of Parkinson's disease. Mov Disord. 2002;17(6):1227–34.
13. Reardon KA, Shiff M, Kempster PA. Evolution of motor fluctuations in Parkinson's disease: a longitudinal study over 6 years. Mov Disord. 1999;14(4):605–11.
14. Lopez IC, Ruiz PJ, Del Pozo SV, Bernardos VS. Motor complications in Parkinson's disease: ten year follow-up study. Mov Disord. 2010;25(16):2735–9.
15. Garcia-Ruiz PJ, Del Val J, Fernandez IM, Herranz A. What factors influence motor complications in Parkinson disease?: a 10-year prospective study. Clin Neuropharmacol. 2012;35(1):1–5.
16. Fabbrini G, Defazio G, Colosimo C, Suppa A, Bloise M, Berardelli A. Onset and spread of dyskinesias and motor symptoms in Parkinson's disease. Mov Disord. 2009;24(14):2091–6.
17. Juri-Claveria C, Aguirre MC, Viviani GP, Chana-Cuevas P. Risk factors associated with the development of motor complications in Parkinson's disease. A study in a Chilean population. Rev Neurol. 2007;45(2):77–80.
18. Hashim HZ, Norlinah MI, Nafisah WY, Tan HJ, Raymond AA, Tamil AM. Risk factors and predictors of levodopa-induced dyskinesia among multiethnic Malaysians with Parkinson's disease. Int J Neurosci. 2014;124:187–91.
19. Hassin-Baer S, Molchadski I, Cohen OS, et al. Gender effect on time to levodopa-induced dyskinesias. J Neurol Nov. 2011;258(11):2048–53.
20. Coelho M, Marti MJ, Tolosa E, et al. Late-stage Parkinson's disease: the Barcelona and Lisbon cohort. J Neurol. 2010;257(9):1524–32.
21. Papapetropoulos S, Argyriou AA, Ellul J, Chroni E. Comparison of motor fluctuations and L-dopa-induced dyskinesias in patients with familial and sporadic Parkinson's disease. Eur J Neurol. 2004;11(2):115–9.
22. Sato K, Hatano T, Yamashiro K, et al. Prognosis of Parkinson's disease: time to stage III, IV, V, and to motor fluctuations. Mov Disord. 2006;21(9):1384–95.
23. Mazzella L, Yahr MD, Marinelli L, Huang N, Moshier E, Di Rocco A. Dyskinesias predict the onset of motor response fluctuations in patients with Parkinson's disease on L-dopa monotherapy. Parkinsonism Relat Disord. 2005;11(3):151–5.
24. Muller T, Woitalla D, Russ H, Hock K, Haeger DA. Prevalence and treatment strategies of dyskinesia in patients with Parkinson's disease. J Neural Transm. 2007;114(8):1023–6.
25. Fahn S, Oakes D, Shoulson I, et al. Levodopa and the progression of Parkinson's disease. N Engl J Med. 2004;351(24):2498–508.
26. Caraceni T, Musicco M. Levodopa or dopamine agonists, or deprenyl as initial treatment for Parkinson's disease. A randomized multicenter study. Parkinsonism Relat Disord. 2001;7(2):107–14.
27. Oertel WH, Wolters E, Sampaio C, et al. Pergolide versus levodopa monotherapy in early Parkinson's disease patients: the PELMOPET study. Mov Disord. 2006;21(3):343–53.
28. Rascol O, Brooks DJ, Korczyn AD, De Deyn PP, Clarke CE, Lang AE. A five-year study of the incidence of dyskinesia in patients with early Parkinson's disease who were treated with ropinirole or levodopa. 056 Study Group. N Engl J Med. 2000;342(20):1484–91.
29. Rascol O, Brooks DJ, Korczyn AD, et al. Development of dyskinesias in a 5-year trial of ropinirole and L-dopa. Mov Disord. 2006;21(11):1844–50.
30. Hauser RA, Rascol O, Korczyn AD, et al. Ten-year follow-up of Parkinson's disease patients randomized to initial therapy with ropinirole or levodopa. Mov Disord. 2007;22(16):2409–17.
31. Holloway RG, Shoulson I, Fahn S, et al. Pramipexole vs levodopa as initial treatment for Parkinson disease: a 4-year randomized controlled trial. Arch Neurol. 2004;61(7):1044–53.
32. Parkinson Study Group CALM Cohort Investigators. Long-term effect of initiating pramipexole vs levodopa in early Parkinson disease. Arch Neurol. 2009;66(5):563–70.

33. Bracco F, Battaglia A, Chouza C, et al. The long-acting dopamine receptor agonist cabergoline in early Parkinson's disease: final results of a 5-year, double-blind, levodopa-controlled study. CNS Drugs. 2004;18(11):733–46.
34. Lees AJ, Katzenschlager R, Head J, Ben-Shlomo Y. Ten-year follow-up of three different initial treatments in de-novo PD: a randomized trial. Neurology. 2001;57(9):1687–94.
35. Katzenschlager R, Head J, Schrag A, Ben-Shlomo Y, Evans A, Lees AJ. Fourteen-year final report of the randomized PDRG-UK trial comparing three initial treatments in PD. Neurology. 2008;71(7):474–80.
36. Stocchi F, Rascol O, Kieburtz K, et al. Initiating levodopa/carbidopa therapy with and without entacapone in early Parkinson disease: the STRIDE-PD study. Ann Neurol. 2010;68(1):18–27.
37. Hauser RA, Panisset M, Abbruzzese G, Mancione L, Dronamraju N, Kakarieka A. Double-blind trial of levodopa/carbidopa/entacapone versus levodopa/carbidopa in early Parkinson's disease. Mov Disord. 2009;24(4):541–50.
38. Shoulson I, Oakes D, Fahn S, et al. Impact of sustained deprenyl (selegiline) in levodopa-treated Parkinson's disease: a randomized placebo-controlled extension of the deprenyl and tocopherol antioxidative therapy of parkinsonism trial. Ann Neurol. 2002;51(5):604–12.
39. Wickremaratchi MM, Ben-Shlomo Y, Morris HR. The effect of onset age on the clinical features of Parkinson's disease. Eur J Neurol. 2009;16(4):450–6.
40. Kostic V, Przedborski S, Flaster E, Sternic N. Early development of levodopa-induced dyskinesias and response fluctuations in young-onset Parkinson's disease. Neurology. 1991;41((2(Pt 1))):202–5.
41. Kumar N, Van Gerpen JA, Bower JH, Ahlskog JE. Levodopa-dyskinesia incidence by age of Parkinson's disease onset. Mov Disord. 2005;20(3):342–4.
42. Kum WF, Gao J, Durairajan SS, et al. Risk factors in development of motor complications in Chinese patients with idiopathic Parkinson's disease. J Clin Neurosci. 2009;16(8):1034–7.
43. Ku S, Glass GA. Age of Parkinson's disease onset as a predictor for the development of dyskinesia. Mov Disord. 2010;25(9):1177–82.
44. Lyons KE, Hubble JP, Troster AI, Pahwa R, Koller WC. Gender differences in Parkinson's disease. Clin Neuropharmacol. 1998;21(2):118–21.
45. Zappia M, Annesi G, Nicoletti G, et al. Sex differences in clinical and genetic determinants of levodopa peak-dose dyskinesias in Parkinson disease: an exploratory study. Arch Neurol. 2005;62(4):601–5.
46. Zappia M, Crescibene L, Arabia G, et al. Body weight influences pharmacokinetics of levodopa in Parkinson's disease. Clin Neuropharmacol. 2002;25(2):79–82.
47. Bachmann CG, Zapf A, Brunner E, Trenkwalder C. Dopaminergic treatment is associated with decreased body weight in patients with Parkinson's disease and dyskinesias. Eur J Neurol. 2009;16(8):895–901.
48. Benbir G, Ozekmekci S, Apaydin H, Delil S, Erginoz E. A hospital-based study: risk factors in development of motor complications in 555 Parkinson's patients on levodopa therapy. Clin Neurol Neurosurg. 2006;108(8):726–32.

Chapter 4
Risk Factors for Levodopa-Induced Dyskinesia

Jee-Young Lee and Beom S. Jeon

Abstract Levodopa-induced dyskinesia (LID) is a disabling motor complication of chronic dopaminergic therapy in patients with Parkinson's disease (PD). The most important risk factor for LID is levodopa therapy: longer treatment duration and high daily levodopa dose. Longer duration of PD, more severe disease, and younger age at PD onset are also significant risk factors. However, there is a profound inter-individual difference in the susceptibility to LID; thus, discoveries from recent genetic association studies are worth reviewing although they are limited because of small sample size, lack of replication, and poor pathophysiological backgrounds. Three major suggestions from the studies are as follows: first, a *DRD2* gene haplotype with the functional consequence of a low generalized activity on the presynaptic D2 receptor may be associated with high risk of dyskinesia; second, a *DRD3* variant with low receptor-binding affinity may be associated with the diphasic form of dyskinesia; and third, a *COMT* low-activity allele may be associated with earlier onset of dyskinesia. Further investigations on other genes regarding dopaminergic and nondopaminergic modulators in the basal ganglia would enhance our understanding of LID susceptibility as well as reveal the possible mechanism of it.

Keywords Parkinson's disease • Levodopa-induced dyskinesia • Levodopa • Genetic susceptibility • Risk factors

---

J.-Y. Lee, MD
Department of Neurology, Seoul National University-Seoul Metropolitan Government Boramae Medical Center, Boramae 5, Dongjak-gu, Seoul 156-707, South Korea

Department of Neurology, College of Medicine, Seoul National University, 28 Yongon-dong, Chongno-gu, Seoul 110-744, South Korea

B.S. Jeon, MD, PhD (✉)
Department of Neurology, Seoul National University Hospital, College of Medicine, Seoul National University, 28 Yongon-dong, Chongno-gu, Seoul 110-744, South Korea
e-mail: brain@snu.ac.kr

S.H. Fox, J.M. Brotchie (eds.), *Levodopa-Induced Dyskinesia in Parkinson's Disease*, DOI 10.1007/978-1-4471-6503-3_4

Text

Since the discovery of levodopa as a treatment of Parkinson's disease (PD) in the 1960s, it has been a gold standard treatment of PD for half a century. However, after the introduction of levodopa therapy, patients have encountered disabling and occasionally medically refractory dyskinesia due to the levodopa medications. For the past three decades, studies have been accumulating regarding the risk factors and predictors of levodopa-induced dyskinesia (LID). Knowledge from those studies has become a basis for treatment guidelines in PD geared toward minimizing LID. However, LID is still a potential disabling and inevitable problem in patients with PD; thus, exploration of individual susceptibility to LID is an emerging issue from both a clinical and research perspective.

In this chapter, we will review the clinical risk factors and predictors of LID in the first part and secondly, will discuss the genetic risk factors for LID covering an update from recent genetic association studies.

Clinical Risk Factors

Duration of Levodopa Treatment

The strongest risk factor for LID is levodopa treatment. Epidemiologic studies have suggested that LID occurs more frequently with a longer levodopa treatment duration. In a community-based survey of 124 patients with PD, the prevalence of dyskinesia among those on levodopa therapy increased as the disease duration increased with a prevalence of 7 % in those with a disease duration of 5 years or less, 18 % for a duration of 6–9 years, and 57 % for a duration of 10 years or more [1]; however, the levodopa treatment duration was more significantly associated with dyskinesia risk than with PD duration in this cohort [1]. In another community-based retrospective study, the estimated rate of dyskinesia was 30 % at 5 years of treatment and 59 % at 10 years [2]. Prospective studies also showed that the LID frequency increases with treatment duration similar to those in the retrospective study reports. However, the frequency of LID varied among reported studies, probably because of differences in the definition of dyskinesia and the methods of detection. In the DATATOP study which included 352 de novo PD patients who had required levodopa treatment, dyskinesia appeared in about 20 % of the patients with a follow-up of 20.5 months [3]. With difference to community-based cohort studies, levodopa dosages were actively adjusted for optimal motor control in this interventional study, which might cause relatively high levodopa dosages. In addition, dyskinesia was detected by investigators, not by patients; thus, assessments might be much sensitive. In another large interventional study, the CR First study, where a regular and slow releasing form of levodopa was compared, dyskinesia appeared in only 22 % of patients after 5 years of treatment [4]. The quite low frequency in this study might be explained by less inclusion of young-onset patients and by detection of dyskinesia being based on patients' diary defined as presence of 10 % or more "ON with dyskinesia" during waking hours.

Duration of PD and Hoehn and Yahr Stage

In a large study reviewing 74 publications spanning 1966 through 2000, Ahlskog and Muenter have summarized that levodopa therapy can cause dyskinesia in 50 % of patients by the 5th to 6th month of therapy in patients diagnosed during the pre-levodopa era, whereas the median dyskinesia frequency in patients diagnosed during the post-levodopa era was approximately 40 % by 4th to 6th year of therapy [5]. Early onset of LID in patients during the pre-levodopa era was probably due to the patients having longer durations of PD when they started the levodopa therapy. In a retrospective comparison of three PD groups that included the Hoehn and Yahr stage (HY) 1 ($n=17$), HY 2 ($n=13$), and HY 3 ($n=10$), when they started the levodopa therapy, the time to onset of dyskinesia was significantly shorter in the HY 3 group [6] although the sample size was too small. Thus, more severe PD and a longer duration of the disease are also important risk factors. In line with this, LID appears more prominently in the worst parkinsonian side [7].

Levodopa Dosage

Levodopa usually does not induce dyskinesia in a normal individual, but a very high dose can produce LID as evidenced by studies on normal monkeys [8, 9]. In hemiparkinsonian rats, the levodopa dosage seems to be critically involved in dyskinesia via loss of synaptic depotentiation [10]. There are many studies reporting that higher doses of levodopa therapy may cause LID earlier and more frequently [1, 11, 12]. In a prospective trial comparing different doses of levodopa in which patients were randomly allocated 150, 300, and 600 mg/day and followed-up for 40 weeks of treatment and 2 weeks of washout (ELLDOPA study) [13], the frequency of LID was significantly high in the highest dose group (3.2, 2.3, and 16.5 %, respectively) [13]. Another large prospective trial (STRIDE-PD) where early PD patients requiring initiation of levodopa therapy were randomly allocated to levodopa + carbidopa or levodopa + carbidopa + entacapone treatments and followed-up for 134 weeks again supported the fact that levodopa dosage is closely related to the risk of developing LID [14]. In this study, 12.1 % of the patients receiving less than 400 mg/day of levodopa had dyskinesia, whereas 55.8 % of those receiving more than 600 mg/day developed dyskinesia at the end of the study [14].

Studies suggested that females are more likely to develop dyskinesia; [15] however, when given the same dose of levodopa plus decarboxylase inhibitor, the area under the levodopa plasma concentration time curve corrected for kilogram body weight (AUC_w) is significantly higher in women with a reduced oral clearance [16]. Thus, with the same dose of levodopa, females are exposed to a higher AUC_w compared to males. Furthermore, body weight is reported to affect the pharmacokinetics of levodopa which may influence the onset of dyskinesia [17, 18]. A subanalysis of a prospective trial comparing ropinirole versus levodopa as an initial therapy (REAL-PET) revealed that the levodopa dose per kilogram body weight is the most significant in the development of LID, whereas female gender, absolute levodopa dose, body weight, disease duration, and motor unified PD rating scores were not significant by multiple logistic regression analysis [19]. In a retrospective Japanese PD study to investigate the efficacy of low-dose levodopa treatment in 92 patients, the mean levodopa dosage was 186.4 ± 75.3 mg/day and the mean duration of follow-up was 6.2 ± 3.3 years [20]. There was a very low rate of dyskinesia in the overall patients (6.5 %) as well as in those ($n = 19$) with a treatment duration of 10 years or more (21.1 %) [20]. Therefore, treatment with a high levodopa dose relative to a patient's body weight is a significant risk factor of LID.

Early Use of Dopamine Receptor Agonist

Large prospective trials of dopamine agonists (ropinirole, pramipexole, cabergoline, and pergolide) versus levodopa as the initial monotherapy [21–24] have consistently shown a lower frequency of dyskinesia in the dopamine agonist-treated group at the cost of having reduced or marginally reduced symptomatic improvement and tolerability disadvantages over levodopa. Possibly, the initial dopamine agonist treatment prior to the levodopa therapy or concomitant administration of a dopamine agonist to levodopa can reduce the dose of levodopa, or the initial dopamine agonist monotherapy can delay the initiation of levodopa therapy. Open-label follow-up of the CALM-PD cohort for up to 6 years of initial randomized cohorts showed that the incidence of dyskinesia was still low in the initial pramipexole-treated group compared with the initial levodopa-treated group [25]. A 10-year follow-up of the initial 5-year treatment cohort of ropinirole versus levodopa also showed that the time to onset of dyskinesia was significantly longer in the initial ropinirole-treated group [26]. However, initial reduction of dyskinesia frequency was not sustained in a longest follow (14 years) cohort of bromocriptine versus levodopa trial [27]. There was no significant difference in the prevalence of disabling dyskinesia between the two groups in the long term. Thus, early dopamine agonist use probably delays only the time to onset of LID.

Young Age at Onset

Many epidemiologic studies have shown that young age at onset of PD is an important risk factor for LID [1, 2, 11, 14, 15, 19, 28, 29]. The 5-year risk of developing LID in a population-based cohort was 50 % in patients with onset between 40 and 59 years of age, whereas it was 16 % in those with an onset over 70 years of age [28]. In vivo functional imaging studies using positron emission tomography have shown that "dopamine turnover" to dopamine synthesis and storage rate is inversely correlated with age at the onset of PD [30]. The "dopamine turnover" rate is more accentuated by levodopa treatment than by dopamine agonist use [31], which can be implicated in the pathogenesis of LID. Young-onset PD patients might have more compensatory mechanisms to dopaminergic cell loss in the basal ganglia; such changes may make younger patients more vulnerable to developing LID [32]. In a recent voxel-based grey matter volume and cortical thickness analysis, young-onset dyskinetic PD patients tended to have more nigral abnormality, whereas late-onset dyskinetic patients had more cortical abnormalities [33]. In terms of preventing or delaying LID in younger PD subjects, high-dose dopamine agonist therapy in young-onset PD patients may increase the risk of impulse control disorders (ICD) [34]; thus, strategy for the dopaminergic therapy should take into consideration an individual's condition to minimize potential treatment-related complications including LID and ICD while maintaining functional abilities.

Genetic Risk Factors

As discussed above, long-term high-dose levodopa therapy is a strong risk factor for LID in PD. However, some patients do not develop LID despite a long period of levodopa therapy, whereas other patients develop LID soon after the initiation of therapy. Both the profound interindividual difference and the high prevalence of LID in young-onset PD patients suggest that there may be a genetic susceptibility. In the post-levodopa era, identification of genetic risk factors for LID has become an issue to avoid this disabling motor complication.

Chronic pulsatile administration of levodopa leads to plastic changes in the presynaptic and postsynaptic neurotransmitter systems in the basal ganglia and related circuits (as discussed further in subsequent chapters). Genetic polymorphisms regarding the presynaptic and postsynaptic structures involved in plastic changes are potential substrates for genetic susceptibility. There may be affinity or expression level differences in postsynaptic dopamine receptors, or differences in presynaptic receptors and transporter protein-binding availabilities or enzymatic activity differences involved in levodopa metabolism. In addition to dopaminergic transmission, non-dopaminergic modulators in the basal ganglia circuit also play important roles in the establishment of LID [35]. Thus, genetic polymorphisms of these molecules or receptors are good targets for genetic association studies on LID in PD.

Although a considerable number of studies have managed to identify the genetic associations with LID, consistently replicated results are very limited because small sample sizes (probably from difficulties in recruiting patients and samples) and population stratification might hamper the identification of true genetic associations regarding LID. Differences in the definition of dyskinetic subjects versus non-dyskinetic controls among the studies are also an unsolved problem. No studies have yet prospectively evaluated the genetic association with LID because whether a patient has dyskinesia from a retrospective evaluation is not accurate in PD.

Despite all these issues, the results of genetic association studies to date can provide valuable insight into the pathophysiology of LID and new knowledge for the genetic research of PD.

Dopamine Receptor and Dopamine Transporter Genes

Genetic association studies on dopamine receptors and transporter genes are summarized in Table 4.1. There have been many studies conducted on *DRD2* variants, but for other subtypes, studies are rare. The *TaqI* polymorphism was initially thought to be a *DRD2* variant but later revealed to be located downstream of it, called the ankyrin repeat domain *ANKK1*. *ANKK1* alters the expression level of NF-κB-regulated genes which regulate *DRD2* expression [36, 37]. The *Taq*IA variant produces amino acid changes (Glu713Lys). The A1(T) allele has been reported to be associated with low-dopamine D2 receptor availability [38–40] and with the increased striatal activity of L-amino acid decarboxylase [41]. Thus, it has been suggested that the *Taq*IA variant mainly alters "presynaptic" D2 receptors, which exert negative control on dopamine release. There was a report of a significant association with this variant and motor fluctuation in PD [42], but two independent studies on LID have reported that there was no association. However, a recent comprehensive study on the haplotype of *DRD2* combined with *ANKK1* gene variants reported that LID is associated with the TTCTA haplotype (including A1 of *TaqI*A) in PD [43]. In this study, Rieck et al. analyzed 6 variants in the *DRD2*/*ANKK1* gene among 199 Brazilian patients (*n* for dyskinesia = 83): the -141C insertion/deletion (Ins/Del -rs1799732), rs2283265, rs1076560, C957T (rs6277), *Taq*IA (rs1800497), and rs2734849. They found that a haplotype (TTCTA) derived from rs2283265, rs1076560, C957T, *Taq*IA, and rs2734849 among the *DRD2*/*ANKK1* gene region was associated with LID. The former two variants are putatively involved in alternative splicing, thus might decrease the expression of the *DRD2* short splice variant when compared with the *DRD2* long variant, and the product of the short splice variant is mainly located in the presynaptic fiber as an autoreceptor [44]. The latter three variants were associated with reduced dopamine D2 receptor availability or reduced affinity [38–41, 45, 46]. This result is very interesting. It was first approached by haplogroup analysis regarding LID in PD, and this haplotype is linked to functional consequences because a generalized reduced expression of *DRD2* mainly at the presynaptic level results in a lower negative feedback signal and a loss of control over dopamine release [43] despite the high dopamine content in the synaptic cleft.

Table 4.1 Summary of genetic association studies on dopamine receptors and dopamine transporter genes with levodopa-induced dyskinesia in Parkinson's disease

Genes	Gene variants	Dyskinesia type	Results	References
DRD1	3 variants (5′ UTR, 3′ UTR, 5q35.1 codon421)	Peak dose	No association with any of the genetic variants	[48]
DRD2	STR, (CA)n	Peak dose	Low risk of dyskinesia: carrying 13 or 14 alleles	[48]
		Peak dose	Low risk in men: carrying 13 or 14 alleles	[47]
		Early vs. late dyskinesia	High risk of dyskinesia: carrying 14/15 genotype	[49]
	*Taq*IA, IB, ID, Val96Ala, Pro310Ser, Ser311Cys, −141Cins/del	Not specified	No association with any of these genetic variants	[61]
	*Taq*IA	Diphasic, peak dose	No association with either dyskinesias	[29]
	Haplotype (−141Cins/del, rs2283265, rs1076560, C957T, *Taq*IA, rs2734849) at DRD2/ANKK1 region	Not specified	High risk of dyskinesia: TTCTA haplotype	[43]
DRD3	Ser9Gly(rs6280)	Chorea, dystonia, motor fluctuation	No association with any motor complications	[62]
		Diphasic, peak dose	High risk of diphasic dyskinesia: AA genotype	[29]
			No association with peak-dose dyskinesia	
		Not specified	No association	[61]
DRD4	48-bp VNTR, 12-bp VNTR, 13-bp deletion	Not specified	No association	[61]
DRD5	T978C	Motor fluctuation, only	No association, no study on dyskinesia	[64]
SLC6A3	40-bp VNTR	Not specified	High risk of dyskinesia: 9-copy allele	[61]
		Not specified	No association	[67]

SLC6A3 dopamine transporter gene, *DRD1* dopamine receptor D1 gene, *DRD2* dopamine receptor D2 gene, *DRD3* dopamine receptor D3 gene, *DRD4* dopamine receptor D4 gene, *DRD5* dopamine receptor D5 gene, *STR* short tandem repeat, *VNTR* variable number tandem repeat, *UTR* untranslated region

Three other studies investigated the short tandem repeat (CA_n) of the *DRD2* gene for an association with LID in PD. However, the results are controversial because two studies showed a protective effect for the 13 or 14 allele, whereas one study showed an increasing risk for carrying the 14/15 genotype [47–49]. The functional significance of this variant has not been clarified.

The D1 receptor is involved in the expression of LID via selective loss of long-term depotentiation with close relationship to glutamate receptor ionotropic N-methyl-D-aspartate (NMDA) subunit 2b signaling [50, 51]. The D1 direct pathway is suggested to be hyperactive for both peak dose and diphasic forms of LID [35]. However, studies on the *DRD1* variant with LID in PD are very rare. There is one study reporting a *DRD1* gene variant with LID [48] in which Oliveri et al. investigated three independent variants, but there was no significant association. The *DRD1* variants including rs4532 have been recently reported to be pharmacogenomic markers of treatment response to antipsychotics in schizophrenia patients [52]; thus, further studies are worth doing to analyze this variant for LID in PD.

Three studies have investigated the *DRD3* Ser9Gly variant, in which the "Ser" allele represents a relatively low receptor-binding affinity [53]. The function of D3 is not yet well known, but an initial suggestion of its function is that the postsynaptic D3 receptor may act as an inhibitory control over behavioral or locomotor activity [54]. The D3 receptor distributes mainly in the mesolimbic region in normal conditions, but in the presence of LID, its expression is increased in the putamen and globus pallidus shown by studies involving 6-hydroxydopamine-lesioned rats [55] and a nonhuman primate 1-methyl to 4-phenyl to 1,2,3,6-tetrahydropyridine model expressing LID [56]. This overexpression of D3 in the motor part of the basal ganglia circuit is correlated with the severity of LID [56, 57]. Interestingly, the expression of D3 mainly occurs in D1-expressing neurons [57, 58] in these models. Brain-derived neurotrophic factor (BDNF) can trigger the expression of D3 [59]. An interesting suggestion for the implication of D3 in the pathogenesis of LID is the alternative splicing of the *DRD3* gene. Because the D3 receptor has the highest affinity to dopamine among the dopamine receptors, in a homeostatic process, alternative splicing would occur with chronic levodopa treatment. The alternative splicing variant of DRD3 called D3nf can dimerize with D3, trafficking it into the intracytoplasmic space removing the D3 receptors from the synaptic membrane [60]. This process results in loss of normal inhibitory control of postsynaptic D3 receptors on D1, thus enhancing the D1 direct pathway.

The first study on *DRD3* Ser9Gly showed no association with dyskinesia (not specified as either peak dose or diphasic or both) [61]. Another study analyzed associations of *DRD3* Ser9Gly with various kinds of motor complications including dyskinesia, separately analyzed as choreic forms and dystonic forms, and motor fluctuations. But there was no significant association [62]. An interesting finding in this study was that for dystonic forms of dyskinesia, the levodopa load was not a significant factor unlike choreic dyskinesia and motor fluctuation. The authors suggested that intrinsic mechanisms outweigh the contribution of pharmacological issues in the development of dystonia in PD [62]. Dystonic forms of dyskinesia may appear as a severe form of peak-dose dyskinesia or as a feature of diphasic dyskinesia [7], and intrinsic and genetic susceptibilities may be more important for this kind of dyskinesia rather than the usual peak-dose choreiform dyskinesia [63]. In a third study regarding *DRD3*, peak-dose and diphasic dyskinesia were separately analyzed because these two forms of dyskinesia are quite different in clinical features, and distinct pathophysiological substrates are expected to be involved [7, 29, 35, 63]. It was revealed that *DRD3* Ser allele is significantly associated with the risk of diphasic dyskinesia but not with peak-dose dyskinesia [29]. The results of that study have yet to be replicated by other groups, but the possible role of the D3 receptor in diphasic dyskinesia is worth further investigation in the future.

There are few studies regarding the association of the *DRD4* variant with LID in PD. Kaiser et al. reported no significant association [61]. There is no published literature regarding the *DRD5* variant for LID, but one study reported it had no association with motor fluctuation in PD [64].

Dopamine transporter (DAT) is located in the presynaptic nerve terminals for the reuptake of synaptic dopamine. In early PD, it is reported that dopamine transporter expression is reduced to compensate for dopamine loss [65]. This compensatory change was more exaggerated in young-onset patients [30], which is related to an increase in the "dopamine turnover" rate. In a functional imaging study using positron emission tomography (PET), this downregulation of dopamine transporters was more exaggerated in patients with LID than in patients without LID [66]. From this background, DAT gene (=*SLC6A3*) variants may be associated with LID in PD. There are few studies regarding the *SLC6A3* variant. One study reported that the 40-base pair variable number tandem repeat (VNTR) 9-copy allele was associated with an increased risk of LID in PD [61]. However, another study using [^{123}I]-N-omega-fluoropropyl-2beta-carbomethoxy-3beta-(4-iodophenyl)nortropane ([^{123}I]-FP-CIT) single-photon emission computed tomography (SPECT) and VNTR genotyping revealed that there was no difference in [^{123}I]-FP-CIT bindings as well as in the frequencies of LID between the carriers of the 9-copy allele and 10-copy homozygotes [67]. Thus, further studies are needed to confirm the role of the *SLC6A3* variants in LID.

COMT and MAO

The catechol-*O*-methyltransferase (COMT) is an enzyme that degrades levodopa into 3-*O*-methyldopa (3-OMD) which is opposite to the process of decarboxylation in which levodopa enzymatically becomes dopamine. COMT is present in both the periphery and brain; thus, theoretically high COMT activity needs higher levodopa doses to produce dopamine at a similar level. The most studied polymorphism of the *COMT* gene is the Val158Met substitution (rs4680) of the membrane-bound COMT isoform found predominately in the brain. The Met substitution produces a low-activity allele (usually designated as L in the literature) which leads to a higher bioavailability of levodopa [68]. There are many studies on COMT variants and the association with interindividual pharmacokinetic differences of levodopa therapy, daily dosages of levodopa, and response sensitivity [69–73], as well as cognitive functioning in PD [74, 75]. With dyskinesia risk, there have been several cross-sectional studies, but the results are not convincing as to any consistent association (Table 4.2) [71–73, 76, 77]. However, one recent study with a longitudinal follow-up of 5 years has shown that carriers of the low-activity allele had an increase in the risk of dyskinesia 2 times or more compared to the high-activity allele homozygotes [78]. The proposed dopamine metabolism change between the Met and Val carriers was also shown by a [^{18}F]DOPA PET study in PD patients [79]. Thus, further studies are warranted to replicate this genetic association with a well-designed protocol to adjust for potential clinical risk factors of LID in the future.

Table 4.2 Summary of genetic association studies on COMT and MAOB with levodopa-induced dyskinesias in Parkinson's disease

Genes	Gene variants	Dyskinesia type	Results	References
COMT	Val158Met (rs4680)	Not specified	High risk of dyskinesia: Val allele (longitudinal study)	[78]
		Not specified	No association	[85]
		Not specified	No association	[72]
		Not specified	No association	[77]
		Not specified	No association	[71]
		Not specified	High risk of dyskinesia: no significant after Bonferroni correction	[76]
	Haplotype (rs6269:A>G; rs4633C>T; rs4818:C>G; rs4680:A>G)	Not specified	High-activity haplotype(G_C_G_G): high prescribed dose	[73]
			No significant influence on LID	
MAOA	T941G (rs6223)	Not specified	No association	[85]
MAOB	A644G (rs1799836)	Not specified	No association	[77]

MAOA monoamine oxidase A gene, *MAOB* monoamine oxidase B gene, *COMT* catechol-O-ethyltransferase

In the brain, dopamine is metabolized by intraneuronal monoamine oxidase A (MAOA) and by glial and astrocyte MAOA and monoamine oxidase B (MAOB) [80]. The A644G (rs1799836) variant of *MAOB* in intron 13 is associated with a change in the enzyme activity [81]. The MAOB activity is higher for carriers of allele G compared to allele A [82, 83]. In PD, carriers of the *MAOB* A allele have been shown to be more efficiently treated with low daily doses of levodopa [69]. However, regarding the risk of LID, the *MAOB* variant was not shown to have a significant effect (Table 4.2) [77]. A synonymous substitution of T to G in exon 8 of the *MAOA* gene is thought to promote *MAOA* mRNA expression [84], but it also had no significant association with LID in PD [85].

Non-dopaminergic Modulators

The glutamate *N*-methyl-D-aspartate receptor subunit 2B (NMDA) gene (*GRIN2B*) is a possible candidate gene for LID in PD because of the role of NMDA subunit 2b-mediated signaling in the generation of dyskinesia and the antidyskinetic effect of NMDA antagonists [50, 51]. Lee et al. investigated three gene variants, C366G (rs7301328), C2664T(rs1806201), and T-200G(rs1019385), and found no significant association with dyskinesia, either diphasic or peak-dose type (Table 4.3) [29]. Because these three variants do not have functional consequences, it is still possible that a functional variant of *GRIN2B*, which is yet to be identified, could be associated with either diphasic or peak-dose dyskinesia. However, a marginal association was found between the C2664T variant with peak-dose dyskinesia in that study cohort; thus, this variant is worth investigating again in future studies. This variant is reportedly associated with age at onset in Huntington's disease and with the differences in responsiveness to antipsychotics in schizophrenia [29, 86, 87].

The brain-derived neurotrophic factor (BDNF) is known to stimulate dopamine release and has a role in striatal plasticity by modulating receptors in the basal ganglia. Specifically, BDNF acts on D1 and D3 receptor expression and serotonergic fiber sprouting which was shown to be involved in LID by rodent models [59, 88–90]. Because the Val66Met variant of *BDNF* results in reduced activity-dependent secretion of BDNF [91, 92], this variant may be associated with the risk of LID. There were two studies that investigated the association of the Val66Met variant with LID in PD (Table 4.3) [85, 93]. One study showed that patients with Met/Met and Met/Val alleles had an earlier onset of dyskinesia [93], whereas the other study with pathologically confirmed PD patients showed that the Val66Met variant was not significantly associated with dyskinesia onset [85]. In these two studies, other clinical risk variables of LID were not completely controlled for and there might be some discrepancies in the detection of dyskinesia due to the retrospective nature of the studies; thus, further studies with a prospective follow-up are warranted in the future.

Serotonergic innervation to the striatum is proposed to be a potential contributor to LID [94–96] and serotonergic modulation may attenuate LID in PD [35, 63]. The promoter region of the serotonin transporter gene variant (5-HTTLPR) is linked to the transcriptional activity of this gene [97] because the S allele for low transcriptional activity results in greater facilitation of serotonergic transmission. In one study, the frequency of the S allele was higher in the peak-dose dyskinesia group but without any statistical significance (Table 4.3) [29]. In that study, the frequency of the minor allele S was too low; thus, a small sample size might be responsible for the negative results. For other serotonergic gene variants, there is a paucity of data.

Endogenous opioids are modulators in the basal ganglia pathway, and altered opioid transmission was reported in PD patients with LID by an in vivo functional imaging study using [^{11}C]diprenorphine PET [98]. Because the A188G variant of the opioid mu receptor gene (*OPRM1*) is linked to high binding affinity for β-endorphin [99], Strong et al. investigated this variant regarding LID risk (Table 4.3). They found that the G allele was associated with an increased risk of earlier onset of dyskinesia, but the statistical significance was only marginal ($p=0.05$) [49]. The exact role of many kinds of endogenous opioids and receptor subtypes have yet to be clarified; thus, further research is required in the future (see Chap. 12).

Lastly, the angiotensin-converting enzyme gene (*ACE*) and the apolipoprotein E gene (*APOE*) have been studied because of their relationship with a younger onset age of PD and emerging knowledge about the role of the rennin-angiotensin system on dopamine-mediated neuroinflammation and oxidative stress in the basal ganglia [100]. However, studies have yet to show any significant associations with LID in PD (Table 4.3) [101–103].

Table 4.3 Summary of genetic association studies on nondopaminergic modulators with levodopa-induced dyskinesia in Parkinson's disease

Genes	Gene variants	Dyskinesia type	Results	References
GRIN2B	C366G, C2664T, T-200G	Diphasic, peak dose	No association with either dyskinesia	[29]
BDNF	Val66Met	Not specified	High risk of dyskinesia: Met allele	[93]
		Not specified	No association	[85]
SLC6A4	5-HTTLPR	Diphasic, peak dose	No association with either dyskinesia	[29]
OPRM1	A118G	Early vs. late dyskinesia	Association with early dyskinesia: GG or GA genotype ($p=0.05$)	[49]
APOE	ε2, ε3, ε4 allele	Not specified	No association	[103]
ACE	Insertion/deletion (I/D) polymorphism in intron 16 of a 287-base pair Alu repeat sequence	Not specified	No association	[101]
		Not specified	No association	[102]

BDNF brain-derived neurotrophic factor gene, *GRIN2B* N-methyl-D-aspartate receptor subunit 2B gene, *SLC6A4* serotonin transporter gene, *5-HTTLPR* promoter region of serotonin transporter gene variant, *APOE* apolipoprotein gene, *ACE* angiotensin-converting enzyme gene, *OPRM1* opioid mu receptor gene

Conclusions

Currently there is no ultimate way to prevent LID in PD therapy. However, strategies to reduce levodopa dosages and the early use of dopamine agonists particularly in young-onset PD patients can help to reduce LID as well as to delay onset of it. Recent association studies for LID suggest that genetic predispositions to increasing synaptic dopamine content and dopamine turnover rate make an individual prone to develop LID. Further works on molecular machineries to modulate synaptic dopamine content and metabolism could enhance understanding the mechanism of LID and future development of preventive therapy for it.

References

1. Schrag A, Quinn N. Dyskinesia and motor fluctuations in Parkinson's disease. A community-based study. Brain. 2000;123:2297–305.
2. Van Gerpen JA, Kumar N, Bower JH, Weigand S, Ahlskog JE. Levodopa-associated dyskinesia risk among Parkinson disease patients in Olmsted County, Minnesota, 1976–1990. Arch Neurol. 2006;63(2):205–9.
3. Parkinson Study Group. Impact of deprenyl and tocopherol treatment on Parkinson's disease in DATATOP patients requiring levodopa. Ann Neurol. 1996;39:37–45.

4. Block G, Liss C, Reines S, Irr J, Nibbelink D. Comparison of immediate-release and controlled release carbidopa/levodopa in Parkinson's disease. A multicenter 5-year study. Eur Neurol. 1997;37:23–7.
5. Ahlskog JE, Muenter MD. Frequency of levodopa-related dyskinesia and motor fluctuations as estimated from the cumulative literature. Mov Disord. 2001;16(3):448–58.
6. Kostić VS, Marinković J, Svetel M, Stefanova E, Przedborski S. The effect of stage of Parkinson's disease at the onset of levodopa therapy on development of motor complications. Eur J Neurol. 2002;9(1):9–14.
7. Fahn S. The spectrum of levodopa-induced dyskinesia. Ann Neurol. 2000;47 Suppl 1:S2–11.
8. Pearce RKB, Heikkila M, Lindern IB, Jenner P. L-dopa induces dyskinesia in normal monkeys: behavioral and pharmacokinetic observations. Psychopharmacology (Berl). 2001;156:402–9.
9. Togasaki DM, Tan L, Protell P, Di Monte DA, Quik M, Langston JW. Levodopa induces dyskinesia in normal squirrel monkeys. Ann Neurol. 2001;50:254–7.
10. Picconi B, Paillé V, Ghiglieri V, Bagetta V, Barone I, Lindgren HS, Bernardi G, Angela Cenci M, Calabresi P. l-DOPA dosage is critically involved in dyskinesia via loss of synaptic depotentiation. Neurobiol Dis. 2008;29(2):327–35.
11. Benbir G, Ozekmekçi S, Apaydin H, Delil S, Erginöz E. A hospital-based study: risk factors in development of motor complications in 555 Parkinson's patients on levodopa therapy. Clin Neurol Neurosurg. 2006;108(8):726–32.
12. Hauser RA, McDermott MP, Messing S. Factors associated with the development of motor fluctuations and dyskinesia in Parkinson disease. Arch Neurol. 2006;63(12):1756–60.
13. Fahn S. Parkinson's disease, the effect of levodopa, and the ELLDOPA trial. Arch Neurol. 1999;56:529–35.
14. Warren Olanow C, Kieburtz K, Rascol O, Poewe W, Schapira AH, Emre M, Nissinen H, Leinonen M, Stocchi F, Stalevo Reduction in Dyskinesia Evaluation in Parkinson's Disease (STRIDE-PD) Investigators. Factors predictive of the development of Levodopa-induced dyskinesia and wearing-off in Parkinson's disease. Mov Disord. 2013;28(8):1064–71.
15. Hassin-Baer S, Molchadski I, Cohen OS, Nitzan Z, Efrati L, Tunkel O, Kozlova E, Korczyn AD. Gender effect on time to levodopa-induced dyskinesia. J Neurol. 2011;258(11):2048–53.
16. Martinelli P, Contin M, Scaglione C, Riva R, Albani F, Baruzzi A. Levodopa pharmacokinetics and dyskinesia: are there sex-related differences? Neurol Sci. 2003;24(3):192–3.
17. Arabia G, Zappia M, Bosco D, Crescibene L, Bagalà A, Bastone L, Caracciolo M, Scornaienghi M, Quattrone A. Body weight, levodopa pharmacokinetics and dyskinesia in Parkinson's disease. Neurol Sci. 2002;23 Suppl 2:S53–4.
18. Zappia M, Crescibene L, Arabia G, Nicoletti G, Bagalà A, Bastone L, Caracciolo M, Bonavita S, Di Costanzo A, Scornaienchi M, Gambardella A, Quattrone A. Body weight influences pharmacokinetics of levodopa in Parkinson's disease. Clin Neuropharmacol. 2002;25(2):79–82.
19. Sharma JC, Ross IN, Rascol O, Brooks D. Relationship between weight, levodopa and dyskinesia: the significance of levodopa dose per kilogram body weight. Eur J Neurol. 2008;15(5):493–6.
20. Kitagawa M, Tashiro K. Low-dose levodopa therapy in Japanese patients with Parkinson's disease: a retrospective study. Intern Med. 2005;44(9):939–43.
21. Rascol O, Brooks DJ, Korczyn AD, De Deyn PP, Clarke CE, Lang AE. A five-year study of the incidence of dyskinesia in patients with early Parkinson's disease who were treated with ropinirole or levodopa. 056 Study Group. N Engl J Med. 2000;342(20):1484–91.
22. Parkinson Study Group. Pramipexole vs levodopa as initial treatment for Parkinson disease: a randomized controlled trial. Parkinson Study Group. JAMA. 2000;284(15):1931–8.
23. Bracco F, Battaglia A, Chouza C, Dupont E, Gershanik O, Marti Masso JF, Montastruc JL, PKDS009 Study Group. The long-acting dopamine receptor agonist cabergoline in early Parkinson's disease: final results of a 5-year, double-blind, levodopa-controlled study. CNS Drugs. 2004;18(11):733–46.

24. Oertel WH, Wolters E, Sampaio C, Gimenez-Roldan S, Bergamasco B, Dujardin M, Grosset DG, Arnold G, Leenders KL, Hundemer HP, Lledó A, Wood A, Frewer P, Schwarz J. Pergolide versus levodopa monotherapy in early Parkinson's disease patients: the PELMOPET study. Mov Disord. 2006;21(3):343–53.
25. Parkinson Study Group CALM Cohort Investigators. Long-term effect of initiating pramipexole vs levodopa in early Parkinson disease. Arch Neurol. 2009;66(5):563–70.
26. Hauser RA, Rascol O, Korczyn AD, Jon Stoessl A, Watts RL, Poewe W, De Deyn PP, Lang AE. Ten-year follow-up of Parkinson's disease patients randomized to initial therapy with ropinirole or levodopa. Mov Disord. 2007;22(16):2409–17.
27. Katzenschlager R, Head J, Schrag A, Ben-Shlomo Y, Evans A, Lees AJ, Parkinson's Disease Research Group of the United Kingdom. Fourteen-year final report of the randomized PDRG-UK trial comparing three treatments in PD. Neurology. 2008;71(7):474–80.
28. Kumar N, Van Gerpen JA, Bower JH, Ahlskog JE. Levodopa-dyskinesia incidence by age of Parkinson's disease onset. Mov Disord. 2005;20(3):342–4.
29. Lee JY, Cho J, Lee EK, Park SS, Jeon BS. Differential genetic susceptibility in diphasic and peak-dose dyskinesia in Parkinson's disease. Mov Disord. 2011;26(1):73–9.
30. Sossi V, de la Fuente-Fernández R, Schulzer M, Adams J, Stoessl J. Age-related differences in levodopa dynamics in Parkinson's: implications for motor complications. Brain. 2006;129:1050–8.
31. Storch A, Wolz M, Beuthien-Baumann B, Löhle M, Herting B, Schwanebeck U, Oehme L, van den Hoff J, Perick M, Grählert X, Kotzerke J, Reichmann H. Effects of dopaminergic treatment on striatal dopamine turnover in de novo Parkinson disease. Neurology. 2013;80(19):1754–61.
32. Sossi V, de la Fuente-Fernández R, Schulzer M, Troiano AR, Ruth TJ, Stoessl AJ. Dopamine transporter relation to dopamine turnover in Parkinson's disease: a positron emission tomography study. Ann Neurol. 2007;62(5):468–74.
33. Cerasa A, Salsone M, Morelli M, Pugliese P, Arabia G, Gioia CM, Novellino F, Quattrone A. Age at onset influences neurodegenerative processes underlying PD with levodopa-induced dyskinesia. Parkinsonism Relat Disord. 2013;19(10):883–8.
34. Lee JY, Lee EK, Park SS, Lim JY, Kim HJ, Kim JS, Jeon BS. Association of DRD3 and GRIN2B with impulse control and related behaviors in Parkinson's disease. Mov Disord. 2009;24(12):1803–10.
35. Brotchie JM. Nondopaminergic mechanisms in levodopa-induced dyskinesia. Mov Disord. 2005;20:919–31.
36. Huang W, Payne TJ, Ma JZ, Beuten J, Dupont RT, Inohara N, Li MD. Significant association of ANKK1 and detection of a functional polymorphism with nicotine dependence in an African–American sample. Neuropsychopharmacology. 2009;34(2):319–30.
37. Bontempi S, Fiorentini C, Busi C, Guerra N, Spano P, Missale C. Identification and characterization of two nuclear factor-kB sites in the regulatory region of the dopamine D2 receptor. Endocrinology. 2007;148(5):2563–70.
38. Thompson J, Thomas N, Singleton A, Piggott M, Lloyd S, Perry EK, Morris CM, Perry RH, Ferrier IN, Court JA. D2 dopamine receptor gene (*DRD2*) Taq1 A polymorphism: reduced dopamine D2 receptor binding in the human striatum associated with the A1 allele. Pharmacogenetics. 1997;7(6):479–84.
39. Pohjalainen T, Rinne JO, Någren K, Lehikoinen P, Anttila K, Syvälahti EK, Hietala J. The A1 allele of the human D2 dopamine receptor gene predicts low D2 receptor availability in healthy volunteers. Mol Psychiatry. 1998;3(3):256–60.
40. Ritchie T, Noble EP. Association of seven polymorphisms of the D2 dopamine receptor gene with brain receptor-binding characteristics. Neurochem Res. 2003;28(1):73–82.
41. Laakso A, Pohjalainen T, Bergman J, Kajander J, Haaparanta M, Solin O, Syvälahti E, Hietala J. The A1 allele of the human D2 dopamine receptor gene is associated with increased activity of striatal l-amino acid decarboxylase in healthy subjects. Pharmacogenet Genomics. 2005;15(6):387–91.
42. Wang J, Liu ZL, Chen B. Association study of dopamine D2, D3 receptor gene polymorphisms with motor fluctuations in PD. Neurology. 2001;56(12):1757–9.

43. Rieck M, Schumacher-Schuh AF, Altmann V, Francisconi CL, Fagundes PT, Monte TL, Callegari-Jacques SM, Rieder CR, Hutz MH. DRD2 haplotype is associated with dyskinesia induced by levodopa therapy in Parkinson's disease patients. Pharmacogenomics. 2012;13(15):1701–10.
44. Zhang Y, Bertolino A, Fazio L, Blasi G, Rampino A, Romano R, Lee ML, Xiao T, Papp A, Wang D, Sadée W. Polymorphisms in human dopamine D2 receptor gene affect gene expression, splicing, and neuronal activity during working memory. Proc Natl Acad Sci U S A. 2007;104(51):20552–7.
45. Hirvonen MM, Laakso A, Nagren K, Rinne JO, Pohjalainen T, Hietala J. C957T polymorphism of the dopamine D2 receptor (DRD2) gene affects striatal DRD2 availability in vivo. Mol Psychiatry. 2004;9(12):1060–1.
46. Hirvonen MM, Laakso A, Nagren K, Rinne JO, Pohjalainen T, Hietala J. C957T polymorphism of dopamine D2 receptor gene affects striatal DRD2 in vivo availability by changing the receptor affinity. Synapse. 2009;63(10):907–12.
47. Zappia M, Annesi G, Nicoletti G, Arabia G, Annesi F, Messina D, Pugliese P, Spadafora P, Tarantino P, Carrideo S, Civitelli D, De Marco EV, Cirò-Candiano IC, Gambardella A, Quattrone A. Sex differences in clinical and genetic determinants of levodopa peak-dose dyskinesia in Parkinson disease: an exploratory study. Arch Neurol. 2005;62(4):601–5.
48. Oliveri RL, Annesi G, Zappia M, Civitelli D, Montesanti R, Branca D, Nicoletti G, Spadafora P, Pasqua AA, Cittadella R, Andreoli V, Gambardella A, Aguglia U, Quattrone A. Dopamine D2 receptor gene polymorphism and the risk of levodopa-induced dyskinesia in PD. Neurology. 1999;53(7):1425–30.
49. Strong JA, Dalvi A, Revilla FJ, Sahay A, Samaha FJ, Welge JA, Gong J, Gartner M, Yue X, Yu L. Genotype and smoking history affect risk of levodopa-induced dyskinesia in Parkinson's disease. Mov Disord. 2006;21(5):654–9.
50. Calabresi P, Di Filippo M, Ghiglieri V, Picconi B. Molecular mechanisms underlying levodopa-induced dyskinesia. Mov Disord. 2008;23 Suppl 3:S570–9.
51. Picconi B, Centonze D, Håkansson K, Bernardi G, Greengard P, Fisone G, Cenci MA, Calabresi P. Loss of bidirectional striatal synaptic plasticity in levodopa induced dyskinesia. Nat Neurosci. 2003;6:501–6.
52. Ota VK, Spíndola LN, Gadelha A, dos Santos Filho AF, Santoro ML, Christofolini DM, Bellucco FT, Ribeiro-dos-Santos ÂK, Santos S, Mari Jde J, Melaragno MI, Bressan RA, Smith Mde A, Belangero SI. DRD1 rs4532 polymorphism: a potential pharmacogenomic marker for treatment response to antipsychotic drugs. Schizophr Res. 2012;142(1–3):206–8.
53. Lundstrom K, Turpin MP. Proposed schizophrenia-related gene polymorphism: expression of the Ser9Gly mutant human dopamine D3 receptor with the Semliki Forest virus system. Biochem Biophys Res Commun. 1996;225:1068–72.
54. Waters N, Svensson K, Haadsma-Svensson SR, Smith MW, Carlsson A. The dopamine D3 receptor: a postsynaptic receptor inhibitory on rat locomotor activity. J Neural Transm. 1993;94:11–9.
55. Bordet R, Ridray S, Carboni S, Diaz J, Sokoloff P, Schwartz JC. Induction of dopamine D3 receptor expression as a mechanism of behavioral sensitization to levodopa. Proc Natl Acad Sci U S A. 1997;94:3363–7.
56. Bézard E, Ferry S, Mach U, Stark H, Leriche L, Boraud T, Gross C, Sokoloff P. Attenuation of levodopa-induced dyskinesia by normalizing dopamine D3 receptor function. Nat Med. 2003;9:762–7.
57. Guigoni C, Aubert I, Li Q, Gurevich VV, Benovic JL, Ferry S, Mach U, Stark H, Leriche L, Håkansson K, Bioulac BH, Gross CE, Sokoloff P, Fisone G, Gurevich EV, Bloch B, Bezard E. Pathogenesis of levodopa-induced dyskinesia: focus on D1 and D3 dopamine receptors. Parkinsonism Relat Disord. 2005;11:S25–9.
58. Bordet R, Ridray S, Schwartz JC, Sokoloff P. Involvement of the direct striatonigral pathway in levodopa-induced sensitization in 6-hydroxydopaine-lesioned rats. Eur J Neurosci. 2000;12:2117–23.
59. Guillin O, Diaz J, Carroll P, Griffon N, Schwartz JC, Solokoff P. BDNF controls dopamine D3 receptor expression and triggers behavioral sensitization. Nature. 2001;411:86–9.

60. Richtand NM. Behavioral sensitization, alternative splicing, and D3 dopamine receptor-mediated inhibitory function. Neuropsychopharmacology. 2006;31(11):2368–745.
61. Kaiser R, Hofer A, Grapengiesser A, Gasser T, Kupsch A, Roots I, Brockmöller J. L -dopa-induced adverse effects in PD and dopamine transporter gene polymorphism. Neurology. 2003;60(11):1750–5.
62. Paus S, Gadow F, Knapp M, Klein C, Klockgether T, Wüllner U. Motor complications in patients from the German competence network on Parkinson's disease and the *DRD3* Ser9Gly polymorphism. Mov Disord. 2009;24:1080–4.
63. Fabbrini G, Brotchie JM, Grandas F, Nomoto M, Goetz CG. Levodopa-induced dyskinesia. Mov Disord. 2007;22(10):1379–89.
64. Wang J, Liu ZL, Chen B. Dopamine D5 receptor gene polymorphism and the risk of levodopa-induced motor fluctuations in patients with Parkinson's disease. Neurosci Lett. 2001;308(1):21–4.
65. Lee CS, Samii A, Sossi V, Ruth TJ, Schulzer M, Holden JE, Wudel J, Pal PK, de la Fuente-Fernandez R, Calne DB, Stoessl AJ. In vivo positron emission tomographic evidence for compensatory changes in presynaptic dopaminergic nerve terminals in Parkinson's disease. Ann Neurol. 2000;47(4):493–503.
66. Troiano AR, de la Fuente-Fernandez R, Sossi V, Schulzer M, Mak E, Ruth TJ, Stoessl AJ. PET demonstrates reduced dopamine transporter expression in PD with dyskinesia. Neurology. 2009;72(14):1211–6.
67. Contin M, Martinelli P, Mochi M, Albani F, Riva R, Scaglione C, Dondi M, Fanti S, Pettinato C, Baruzzi A. Dopamine transporter gene polymorphism, spect imaging, and levodopa response in patients with Parkinson disease. Clin Neuropharmacol. 2004;27(3):111–5.
68. D'souza UM, Craig IW. Functional polymorphisms in dopamine and serotonin pathway genes. Hum Mutat. 2006;27:1–13.
69. Białecka M, Droździk M, Kłodowska-Duda G, Honczarenko K, Gawrońska-Szklarz B, Opala G, Stankiewicz J. The effect of monoamine oxidase B (MAOB) and catechol-O-methyltransferase (COMT) polymorphisms on levodopa therapy in patients with sporadic Parkinson's disease. Acta Neurol Scand. 2004;110(4):260–6.
70. Lee MS, Lyoo CH, Ulmanen I, Syvänen AC, Rinne JO. Genotypes of catechol-O-methyltransferase and response to levodopa treatment in patients with Parkinson's disease. Neurosci Lett. 2001;298(2):131–4.
71. Lee MS, Kim HS, Cho EK, Lim JH, Rinne JO. COMT genotype and effectiveness of entacapone in patients with fluctuating Parkinson's disease. Neurology. 2002;58(4):564–7.
72. Contin M, Martinelli P, Mochi M, Riva R, Albani F, Baruzzi A. Genetic polymorphism of catechol-O-methyltransferase and levodopa pharmacokinetic-pharmacodynamic pattern in patients with Parkinson's disease. Mov Disord. 2005;20(6):734–9.
73. Białecka M, Kurzawski M, Klodowska-Duda G, Opala G, Tan EK, Drozdzik M. The association of functional catechol-O-methyltransferase haplotypes with risk of Parkinson's disease, levodopa treatment response, and complications. Pharmacogenet Genomics. 2008;18(9):815–21.
74. Foltynie T, Goldbery TE, Lewis SG, Blackwell AD, Kolachana BS, Weinberger DR, et al. Planning ability in Parkinson's disease is influenced by the COMT val158met polymorphism. Mov Disord. 2004;19:885–91.
75. Williams-Gray CH, Hampshire A, Barker R, Owen AM. Attentional control in Parkinson's disease is dependent on COMT val158met genotype. Brain. 2008;131:397–408.
76. Watanabe M, Harada S, Nakamura T, Ohkoshi N, Yoshizawa K, Hayashi A, Shoji S. Association between catechol-O-methyltransferase gene polymorphisms and wearing-off and dyskinesia in Parkinson's disease. Neuropsychobiology. 2003;48(4):190–3.
77. Torkaman-Boutorabi A, Shahidi GA, Choopani S, Rezvani M, Pourkosary K, Golkar M, Zarrindast MR. The catechol-O-methyltransferase and monoamine oxidase B polymorphisms and levodopa therapy in the Iranian patients with sporadic Parkinson's disease. Acta Neurobiol Exp (Wars). 2012;72(3):272–82.
78. de Lau LM, Verbaan D, Marinus J, Heutink P, van Hilten JJ. Catechol-O-methyltransferase Val158Met and the risk of dyskinesia in Parkinson's disease. Mov Disord. 2012;27(1):131–5.

79. Wu K, O'Keeffe D, Politis M, O'Keeffe GC, Robbins TW, Bose SK, Brooks DJ, Piccini P, Barker RA. The catechol-O-methyltransferase Val(158)Met polymorphism modulates fronto-cortical dopamine turnover in early Parkinson's disease: a PET study. Brain. 2012;135:2449–57.
80. Moussa BH, Youdim MB, Edmondson D, Tipton KF. The therapeutic potential of monoamine oxidase inhibitors. Nat Rev Neurosci. 2006;7:295–309.
81. Kurth JH, Kurth MC, Poduslo SE, Schwankhaus JD. Association of a monoamine oxidase B allele with Parkinson's disease. Ann Neurol. 1993;33:368–72.
82. Garpenstrand J, Ekblom K, Forslund G, Rylander L, Oreland L. Platelet monoamine oxidase activity is related to MAOB intron 13 genotype. J Neural Transm. 2000;107:523–30.
83. Costa-Mallen P, Kelada SN, Costa LG, Checkoway H. Characterization of the in vitro transcriptional activity of polymorphic alleles of the human monoamine oxidase-B gene. Neurosci Lett. 2005;383:171–5.
84. Pinsonneault JK, Papp AC, Sadee W. Allelic mRNA expression of X-linked monoamine oxidase a (MAOA) in human brain: dissection of epigenetic and genetic factors. Hum Mol Genet. 2006;15:2636–49.
85. Cheshire P, Bertram K, Ling H, O'Sullivan SS, Halliday G, McLean C, Bras J, Foltynie T, Storey E, Williams DR. Influence of Single Nucleotide Polymorphisms in COMT, MAO-A and BDNF Genes on Dyskinesia and Levodopa Use in Parkinson's Disease. Neurodegener Dis. 2014;13:24–8.
86. Arning L, Saft C, Wieczorek S, Andrich J, Kraus PH, Epplen JT. NR2A and NR2B receptor gene variations modify age at onset in Huntington disease in a sex-specific manner. Hum Genet. 2007;122:175–82.
87. Hong CJ, Yu YW, Lin CH, Cheng CY, Tsai SJ. Association analysis for NMDA receptor subunit 2B (GRIN2B) genetic variants and psychopathology and clozapine response in schizophrenia. Psychiatr Genet. 2001;11:219–22.
88. Do T, Kerr B, Kuzhikandathil EV. Brain-derived neurotrophic factor regulates the expression of D1 dopamine receptors. J Neurochem. 2007;100:416–28.
89. Mamounas LA, Altar CA, Blue ME, Kaplan DR, Tessarollo L, Lyons WE. BDNF promotes the regenerative sprouting, but not survival, of injured serotonergic axons in the adult rat brain. J Neurosci. 2000;20:771–82.
90. Mattson MP, Maudsley S, Martin B. BDNF and 5-HT: a dynamic duo in age-related neuronal plasticity and neurodegenerative disorders. Trends Neurosci. 2004;27:589–94.
91. Chen ZY, Patel PD, Sant G, Meng CX, Teng KK, Hempstead BL, Lee FS. Variant brain-derived neurotrophic factor (BDNF) (Met66) alters the intracellular trafficking and activity-dependent secretion of wild-type BDNF in neurosecretory cells and cortical neurons. J Neurosci. 2004;24:4401–11.
92. Egan MF, Kojima M, Callicott JH, Goldberg TE, Kolachana BS, Bertolino A, Zaitsev E, Gold B, Goldman D, Dean M, Lu B, Weinberger DR. The BDNF val66met polymorphism affects activity-dependent secretion of BDNF and human memory and hippocampal function. Cell. 2003;112:257–69.
93. Foltynie T, Cheeran B, Williams-Gray CH, Edwards MJ, Schneider SA, Weinberger D, Rothwell JC, Barker RA, Bhatia KP. BDNF val66met influences time to onset of levodopa induced dyskinesia in Parkinson's disease. J Neurol Neurosurg Psychiatry. 2009;80(2):141–4.
94. Carta M, Carlsson T, Kirik D, Björklund A. Dopamine released from 5-HT terminals is the cause of L-DOPA-induced dyskinesia in parkinsonian rats. Brain. 2007;130:1819–33.
95. Rylander D, Parent M, O'Sullivan SS, Dovero S, Lees AJ, Bezard E, Descarries L, Cenci MA. Maladaptive plasticity of serotonin axon terminals in levodopa-induced dyskinesia. Ann Neurol. 2010;68(5):619–28.
96. Carlsson T, Carta M, Muñoz A, Mattsson B, Winkler C, Kirik D, Björklund A. Impact of grafted serotonin and dopamine neurons on development of L-DOPA-induced dyskinesia in parkinsonian rats is determined by the extent of dopamine neuron degeneration. Brain. 2009;132:319–35.

97. Heins A, Jones DW, Mazzanti C, Goldman D, Ragan P, Hommer D, Linnoila M, Weinberger DR. A relationship between serotonin transporter genotype and in vivo protein expression and alcohol neurotoxicity. Biol Psychiatry. 2000;47:643–9.
98. Piccini P, Weeks RA, Brooks DJ. Alterations in opioid receptor binding in Parkinson's disease patients with levodopa-induced dyskinesia. Ann Neurol. 1997;42(5):720–6.
99. Bond C, LaForge KS, Tian M, Melia D, Zhang S, Borg L, Gong J, Schluger J, Strong JA, Leal SM, Tischfield JA, Kreek MJ, Yu L. Single-nucleotide polymorphism in the human mu opioid receptor gene alters beta-endorphin binding and activity: possible implications for opiate addiction. Proc Natl Acad Sci U S A. 1998;95:9608–13.
100. Labandeira-Garcia JL, Rodriguez-Pallares J, Dominguez-Meijide A, Valenzuela R, Villar-Cheda B, Rodríguez-Perez AI. Dopamine-Angiotensin interactions in the basal ganglia and their relevance for Parkinson's disease. Mov Disord. 2013;28(10):1337–42.
101. Lin JJ, Yueh KC, Lin SZ, Harn HJ, Liu JT. Genetic polymorphism of the angiotensin converting enzyme and L-dopa-induced adverse effects in Parkinson's disease. J Neurol Sci. 2007;252(2):130–4.
102. Pascale E, Purcaro C, Passarelli E, Guglielmi R, Vestri AR, Passarelli F, Meco G. Genetic polymorphism of Angiotensin-Converting Enzyme is not associated with the development of Parkinson's disease and of L-dopa-induced adverse effects. J Neurol Sci. 2009;276(1–2):18–21.
103. Molchadski I, Korczyn AD, Cohen OS, Katzav A, Nitzan Z, Chapman J, Hassin-Baer S. The role of apolipoprotein E polymorphisms in levodopa-induced dyskinesia. Acta Neurol Scand. 2011;123(2):117–21.

Chapter 5
Pharmacological Treatment Options for Levodopa-Induced Dyskinesia

Regina Katzenschlager

Abstract The current pharmacological treatment options for dyskinesia in Parkison's disease are based on three principles: adjusting the ongoing treatment, adding specific antidyskinetic drugs, or administering dopaminergic drugs continuously via pump systems.

In patients with a wide therapeutic window, an overall dose reduction may be sufficient to improve dyskinesia without motor worsening. Once fluctuations become troublesome, individual adjustments are required, which may include reducing levodopa while increasing longer-acting classes of drugs, such as dopamine agonists and MAO-B and COMT inhibitors. When motor complications have become complex, or if tolerability is impaired, these classes of drugs may need to be reduced or discontinued and levodopa at short intervals may be the only oral option.

Substances shown to have specific antidyskinetic effects, without worsening motor function, include the antiglutamatergic drug amantadine and the antipsychotic clozapine (which carries a risk of leukopenia and requires regular blood checks).

In patients with motor complications refractory to all adjustments of oral and transdermal treatments and where deep brain stimulation is either contraindicated or not desired, continuous delivery of levodopa into the jejunum (via a percutaneous gastric tube) or of the dopamine agonist apomorphine subcutaneously may greatly improve motor fluctuations and dyskinesia.

Keywords Parkinson's disease • Dyskinesia • Treatment • Intrajejunal levodopa • Apomorphine infusion

R. Katzenschlager (✉)
Department of Neurology, Karl Landsteiner Institute for Neuroimmunological and Neurodegenerative Disorders, Danube Hospital/Donauspital, Vienna, Austria
e-mail: regina.katzenschlager@wienkav.at

S.H. Fox, J.M. Brotchie (eds.), *Levodopa-Induced Dyskinesia in Parkinson's Disease*, DOI 10.1007/978-1-4471-6503-3_5

Introduction

The treatment of motor fluctuations is often relatively straightforward as long as OFF periods are limited to end-of-dose effects and gastrointestinal absorption remains reliable. Once dyskinesia has developed, however, management becomes more complex. While mild dyskinesia may go unnoticed by patients and may not necessarily require changes in a patient's medication, its emergence always indicates that a stage of the illness has been reached where the therapeutic window is narrowing, and physicians must be vigilant and intervene as soon as dyskinesia worsens. The goal of any therapeutic intervention for dyskinesia needs to be discussed with the patient at each stage. Depending on a patient's personal and professional circumstances, even slight dyskinesia may be socially extremely undesirable and may require immediate action. At the other end of the spectrum, there are those patients who prefer being ON and able to move, despite dyskinesia, to their OFF states. In the vast majority of patients, however, dyskinesia is associated with impaired quality of life and treatment is an important goal.

Peak-Dose Dyskinesia

Dose reductions can often relieve dyskinesia as long as wearing-off is mild or not apparent at all – indicating that the patient still has a wide therapeutic window. However, with the progression of the neurodegenerative process, the dopaminergic doses required for adequate motor control increase and often lead to worsening of dyskinesia. The approach to a patient with motor fluctuations and dyskinesia therefore always requires an individualized adjustment of the medication, aiming at the lowest possible dose that will control parkinsonian motor problems without inducing troublesome dyskinesia.

For some patients with motor complications that have become refractory to all adjustments of oral or transdermal treatments, deep brain stimulation (DBS) – or, in some cases, lesional surgery – is an efficacious option (see Chap. 6). Before recommending surgery, all available and suitable options for the pharmacological management of dyskinesia should be considered for each patient. In patients not eligible for, or reluctant to undergo DBS, pharmacological alternatives include dopaminergic drugs delivered via pump systems, where available.

The current pharmacological treatment options for peak-dose dyskinesia are based on several principles:

1. Adjusting intervals and doses of the available and suitable oral or transdermal dopaminergic treatments
2. Adding oral drugs with direct antidyskinetic effects
3. Administering antiparkinsonian drugs continuously via pumps

Biphasic Dyskinesia

Similar principles apply to the management of biphasic dyskinesia, although this has rarely been studied as a separate entity in treatment studies. In addition, it has been recommended that increasing levodopa to levels exceeding the upper threshold of biphasic dyskinesia may be useful. However, this approach may worsen peak-dose dyskinesia. Overall, clinical practice suggests that biphasic dyskinesia is often more difficult to manage than dyskinesia limited to ON periods.

OFF Period Dystonia

In contrast, OFF-period dystonia, which is typically painful, is often also considered a form of dyskinesia but occurs during the times of lowest levodopa plasma levels, typically in the morning or during marked OFFs. The approach here generally includes all measures to improve motor fluctuations. In addition, specific rapidly acting rescue drugs can be very helpful. These are listed in Table 5.1.

Adaptations to Current Antiparkinsonian Drug Treatment

Adjusting Levodopa, Dopamine Agonist, and Enzyme Inhibitor Administration

Reductions in the overall dopaminergic dose will lead to an improvement of dyskinesia. This can be achieved by reducing individual doses or by extending the intervals between drug intakes. If the patient still has a wide therapeutic window (meaning an ability to reduce the dopamergic dose without worsening PD motor symptoms) or if the initial dose was higher than currently needed by the patient, this may be the only intervention required.

The aim is to determine, for each patient, which drug or drug combinations and which doses provide sufficient motor control while inducing as little dyskinesia as possible. Levodopa has the highest propensity to worsen dyskinesia. If doses can be lowered and some levodopa can be replaced with a dopamine agonist, dyskinesia often improves. This has been demonstrated for numerous oral and transdermal dopamine agonists: Examples include rotigotine and pramipexole, which were shown to increase daily ON time without troublesome dyskinesia by 2.8 and 2.7 h, respectively, compared with 1.4 h on placebo ($p<0.001$) [1], ropinirole (although the significance of the difference from placebo was not stated in that study) [2], and several ergot dopamine agonists. The controlled-release formulation of ropinirole also reduces ON time with troublesome dyskinesia [3]. There is no evidence of clinically relevant differences among the dopamine agonists in this indication [4, 5].

However, due to lower antiparkinsonian potency of all oral and transdermal dopamine agonists, it is usually not possible to administer these agents as monotherapy in patients who have already developed motor complications.

Reductions of levodopa may result in worse control of the parkinsonian motor symptoms, leading to increased daily OFF duration. This can sometimes be counteracted by shortening the intervals between levodopa intakes.

If these steps are insufficient to control dyskinesia, reducing or discontinuing MAO-B inhibitors and COMT inhibitors should be considered. With shorter-acting drugs such as levodopa, entacapone, or standard-release dopamine agonists, it is sometimes possible to focus individual dose reductions on those times of the day when dyskinesia is usually present or more marked in an individual patient (e.g., the second half of the day for many patients).

Prolonged-Release, Transdermal, and High-Dose Dopamine Agonist Treatment

Achieving a more continuous delivery of a dopaminergic drug may improve not only OFF duration but also dyskinesia. However, for the oral dopamine agonists, there is no evidence that the new controlled-release formulations are associated with less dyskinesia compared with the respective immediate-release formulations. Similarly, transdermal rotigotine does not appear to offer a superior benefit with respect to managing dyskinesia compared with shorter-acting oral dopamine agonists [1, 6]. Patients who have developed motor complications will continue to require oral levodopa, which has a very short half-life. It is likely that this outweighs any benefits derived from the more stable plasma curves achieved by these dopamine agonists, whose antiparkinsonian efficacy is not sufficient to replace levodopa.

The improvement in dyskinesia observed when oral levodopa is switched to continuous intrajejunal administration is in keeping with the concept that it is not the pharmacological agent as such that drives dyskinesia but rather the manner in which it is administered. In the past, attempts were made to achieve more continuous drug delivery by using very high doses of oral dopamine agonists, thereby enabling a reduction in oral levodopa. In some studies, maximum licensed daily doses were considerably exceeded [7–11]. This approach is currently not generally encouraged, mainly due to the side effect profile of dopamine agonists. Many undesirable effects have been found to be dose dependent, such as somnolence, unplanned episodes of sleep, and hallucinations [12]. The specific impact of dopamine agonists on risk assessment and impulse control, which is currently being investigated in more depth, raises concerns with respect to using higher than licensed doses.

Management of Increasingly Complex Motor Complications and Disease Progression

Eventually, some patients will reach a stage when they fluctuate between OFF and ON periods throughout the day, with all ON periods associated with troublesome peak-dose dyskinesia, or with additional biphasic dyskinesia. The predictability of OFFs and dyskinesia may thus get lost. In the case of such complex motor complications, where managing OFFs can only be achieved at the expense of increasing dyskinesia, it may become necessary to reduce or discontinue even the dopamine agonist and to administer frequent doses of levodopa throughout the day to prevent OFFs.

Finding the right combination of doses and intervals that provide good motor control for as much of the day as possible, without inducing dyskinesia, may require frequent follow-up visits and adjustments. The more advanced the patient is, the less likely he or she manages to maintain a state of good motor control without troublesome dyskinesias throughout the day, with any of the oral or transdermal pharmacological treatments currently available.

Oral Treatments with (Potentially) Specific Antidyskinetic Effects

There is currently a limited choice of orally administered agents which have been proven to relieve dyskinesia without worsening motor disability. While research efforts are directed at detecting and investigating such agents, only two are currently available, amantadine and clozapine, with limited evidence for some other agents.

Glutamatergic Antagonists

Amantadine

A pathogenic role of striatal N-methyl-D-aspartate (NMDA) receptor changes in the formation of dyskinesia has been suggested (see Chap. 13). Amantadine is a nonselective NMDA receptor antagonist and has been reported to reduce levodopa-induced dyskinesia, both in animal models and in patients with PD.

The effect of amantadine (300 mg/day) on dyskinesia was assessed in a 3-week randomized, placebo-controlled trial in 18 patients [13]. ON time with dyskinesia and the Abnormal Involuntary Movement Scale (AIMS) after a levodopa challenge test improved significantly on amantadine, as did OFF time duration. In a placebo-controlled crossover trial (2 weeks on each treatment, 1 week washout period), dyskinesia severity was assessed by self-scoring using visual analogue scales and was reduced by approximately 50 % on amantadine compared with placebo. Similarly, Unified Parkinson's Disease Rating Scale (UPDRS), part IV/items 32 and 33 (duration and disability due to dyskinesia), also significantly improved with amantadine [14].

A randomized, placebo-controlled trial found a dyskinesia reduction by 45 % with amantadine (300 mg/day), on video recordings after levodopa challenge tests, using a modified Rush Dyskinesia Rating Scale (RDS) [15]. UPDRS part IV (motor complications) was also significantly superior compared to placebo. As open-label follow-up indicated a recurrence of dyskinesia after a mean of 4.9 months, compared with 1.3 months on placebo, the authors concluded that the antidyskinetic effect of amantadine may be only transitory.

In a subsequent study, 32 patients who had been on stable amantadine for the treatment of dyskinesia for at least 1 year were randomized to amantadine (at the same dose) or placebo for 3 weeks [16]. The mean dose of amantadine was around 300 mg/day. A significant increase in dyskinesia, as measured on UPDRS items 32 and 33, occurred in patients switched to placebo, from 3.06 (95 % CI, 2.1–4.03) at baseline to 4.28 (95 % CI, 3.1–5.4). In contrast, there was no change on continuing amantadine. While the difference between the arms did not reach significance (likely due to insufficient power), only the placebo patients experienced a significant and clinically relevant increase in ON time with troublesome dyskinesia (from 1.7 to 3.5 h/day).

A crossover study evaluated amantadine 300 mg/day (for 27 days) compared to placebo for dyskinesia in a Japanese population [17]. The drug was titrated at weekly intervals so the maximal dose was only taken for 1 week. Adjusted odds ratio for an improvement on RDS with amantadine (assessed on home videos) was 6.7 [95 % CI, 1.4–31.5]. UPDRS IVa (dyskinesia items 32–35) also improved significantly on amantadine.

A small randomized, placebo-controlled study in 18 patients found no significant change in dyskinesia severity but a significant improvement in daily duration of dyskinesia and impact on daily acitivities [8].

Safety

Amantadine may induce or worsen neuropsychiatric adverse effects, including hallucinations, confusion, agitation, and psychosis. Leg edema may occur, as may skin changes (livedo reticularis). Prolongation of QT intervals has been observed. There have been case reports of reversible corneal edema, and clinicians need to be vigilant about patients reporting sudden visual changes. Myoclonus may also occur in patients with impaired renal function.

Summary

The evidence shows that amantadine is efficacious as an oral antidyskinetic drug and that the antidyskinetic effect is sustained for at least 1 year. The average dose in the studies demonstrating dyskinesia reduction was 300 mg/day. The potential of amantadine to worsen neuropsychiatric problems should be taken into account when considering its use, particularly in elderly patients. Similarly, a combination with drugs that have an effect on the QT interval should be avoided. The evidence from randomized-controlled studies enabled the Evidence Based Medicine (EBM) Task Force of the International Parkinson and Movement Disorder Society (MDS) to classify amantadine as efficacious in the treatment of dyskinesia and as being clinically useful in this indication. Its safety was considered to be acceptable, without the need for specialized monitoring [5].

Memantine

Very limited evidence exists for a potential antidyskinetic effect of memantine. In a small double-blind crossover study in 12 patients, no improvement was found in the dyskinesia elicited by levodopa challenge tests [18], in contrast to uncontrolled observations of some effect on dyskinesia in PD patients with dementia [19].

Neuroleptics

Clozapine

Clozapine is an atyical neuroleptic which has been investigated as a treatment for dyskinesia. Its exact mechanism of action is unclear but may relate to the rate of binding to striatal dopamine D2 receptors, which is of shorter duration than with other neuroleptics. Other mechanisms such as its serotonergic properties may also have a role. Uncontrolled studies of clozapine for dyskinesia suggested a reduction by around 50 % on high doses [20]. A randomized, placebo-controlled 10-week study of clozapine (average dose around 40 mg/day) was conducted in 50 patients with disabling dyskinesia [21]. Clozapine was associated with a significant reduction in ON time with dyskinesia (−1.7 h) compared with placebo (−0.74 h), without changes in OFF time duration. Video scoring of dyskinesia after levodopa challenge tests showed significantly less dyskinesia at rest than at baseline. However, during activation using mental calculation tasks, no significant improvement in severity was observed.

In clinical practice, many patients experience worsening of dyskinesia during emotional changes or with physical activities. Therefore, the change in dyskinesia measured at rest in this study remains of somewhat unclear clinical significance.

Safety

Clozapine use requires mandatory regular blood testing because of the risk of agranulocytosis. There is evidence to show that the risk of this potentially life-threatening adverse effect decreases over time. In a US study, the frequency in the second 6 months of treatment was found to be 0.70/1,000 patient-years and after the first year, 0.39/1,000 patient-years [22]. Sedation, sialorrhea, and orthostatic hypotension are not uncommon. Myocarditis, cardiomyopathy, and pericarditis have been observed. Additional risks associated with all neuroleptics include QT prolongation, metabolic changes such as diabetes, hyperlipidemia and weight gain, and cerebrovascular and cardiovascular morbidity.

Summary

The EBM Review by the MDS in its most recent update [23] classified clozapine as being efficacious in the treatment of dyskinesia and clinically useful, but had previously stated concerns with respect to its clinical usefulness, based on safety issues – particularly, but not limited to, the risk of leukopenia – and on the slightly inconsistent results in the outcome measures of the single available randomized study. In practical terms, if all other noninvasive measures have failed and if a patient is not eligible for device-aided treatments (pumps or surgery), clozapine can be considered for dyskinesia management. The usual initial dose is 12.5 mg and this should be increased slowly to a maximum of around 75 mg/day (higher doses have been used in studies), administered as one dose at nightime or in two doses.

Quetiapine

As an atypical neuroleptic with some similarities to clozapine, quetiapine has also been investigated for the treatment of dyskinesia.

A small retrospective chart review reported on 22 patients who had received quetiapine for dyskinesia at varying doses (mean 239 mg/day in the 11 patients available for follow-up) and for varying periods of time (22 months in the 15 patients classified as responders) [24]. The retrospective nature and the lack of a validated scale limit any conclusions from this observation.

In a small randomized crossover study, nine patients received 25 mg of quetiapine at nighttime or placebo for 2 weeks, with a 1-week washout period [25]. Patient diaries showed no significant difference in any parameter nor did subjective dyskinesia assessments or blinded video ratings of levodopa challenge tests. The subsequent uncontrolled administration of 50 mg/day was associated with a slight dyskinesia reduction on some scales, but drowsiness and daytime sleep episodes occurred frequently on this dose.

Overall, the role of quetiapine in the management of dyskinesia has not been determined with certainty. However, its propensity to induce sedation and the lack of convincing improvements in the reported small studies make this drug unlikely to gain a role in the majority of patients.

Anticonvulsants

Levetiracetam

Levetiracetam had shown an antidyskinetic effect in MPTP-lesioned primates and had been suggested to improve dyskinesia in PD patients in several small, open, uncontrolled studies.

In a randomized, placebo-controlled crossover trial (2-week treatment period), power was lost due to low recruitment (38 subjects). ON time with dyskinesia was reduced by 75 min, but statistical comparisons were not clearly defined. The change in UPDRS item 32 (dyskinesia duration) was significant, but RDS after a levodopa challenge was not [26]. In contrast, two randomized, placebo-controlled studies did not find significant improvements with levetiracetam: In an 11-week study (mean dose 1,800 mg) in 34 patients, none of the endpoints, including modified AIMS, UPDRS IV, ON time with and without dyskinesia, and a challenge test, were significantly changed [27]; a small (16 patients) 6-week study did not find any improvements on 1,500–2,000 mg/day [28]. The MDS EBM Review classified levetiracetam as having insufficient evidence for a role in the treatment of dyskinesia [23].

Other Anticonvulsants

Zonisamide was investigated for the management of motor fluctuations [29]. Post hoc analysis suggested an improvement in disability due to dyskinesia (UPDRS item 33). Zonisamide is currently licensed only in Japan for the treatment of PD.

Licensed Drugs Shown to Be Ineffective in Clinical Studies

Based on preclinical data or preliminary findings in humans, many agents with various suggested mechanisms have been investigated in patients with dyskinesia. Among those licensed for other indications, agents lacking efficacy for levodopa-induced dyskinesia in PD include naltrexone [30], cannabis [31], perampanel [32], and riluzole [33].

The atypical neuroleptic, olanzapine has been investigated for dyskinesia in PD. A randomized, placebo-controlled study had to be terminated early due to marked motor worsening in the olanzapine arm [34], similar to studies of olanzapine for psychosis in PD patients [35], and additional evidence of detrimental motor effects in PD patients exists. Olanzapine is therefore not efficacious and should not be used for the management of dyskinesia.

Licensed Drugs with Uncertain Efficacy on Dyskinesia

Very limited and conflicting evidence exists for the serotonergic agent, buspirone, and for the beta adrenergic antagonist, propanolol [36].

Device-Aided Treatments Providing Continuous Drug Delivery

Switching from oral levodopa to an intrajejunal application of the same drug has been shown to improve motor complications, including dyskinesia.

Evidence from large but uncontrolled studies suggests that a similar effect may be achieved with the dopamine agonist apomorphine when administered by continuous subcutaneous infusion.

Levodopa/Carbidopa Intestinal Gel Infusion (LCIG)

Levodopa/carbidopa intestinal gel (LCIG) is a carbomethylcellulose aqueous gel administered via a portable infusion pump into the jejunum, where it is released continuously and absorbed. The pump contains a cassette with the gel suspension and is attached to a permanent endoscopic gastrostomy (PEG) tube. The insertion of the tube requires a minor surgical intervention. Gastric emptying, which becomes less reliable as PD progresses, is bypassed. This delivery method leads to relatively stable levodopa plasma levels [37], and numerous uncontrolled studies have shown an improvement in motor fluctuations and in dyskinesia.

Until recently, only limited evidence from small randomized studies had confirmed this effect on motor complications [38, 39]. A randomized study compared nasoduodenal levodopa as monotherapy (oral levodopa was allowed overnight) with optimized conventional therapy (which included subcutaneous apomorphine injections or infusion in eight patients) in a crossover design with two 3-week treatment phases and no washout phase [39]. Per protocol, 19 of 24 subjects completed the study. Daily levodopa doses ranged between 456 and 3,556 mg; doses in the conventional arm were not stated. A non-validated video rating scale was devised, ranging from −3 (severe OFF) to +3 (severe dyskinesia). The primary efficacy variable was the percentage of ratings between −1 and +1, accepting mild parkinsonism or mild dyskinesia. The median percentage in this range was 81 % on conventional therapy and 100 % on infusion ($p<0.01$). Total UPDRS (but not part 3 subscore) was significantly improved with LCIG. There were no differences in adverse effects.

A relatively large, placebo-controlled, randomized, double-blind study of intrajejunal levodopa infusion has now been completed [40]: This compared LCIG with oral levodopa in 71 patients over 12 weeks, using a double-dummy design. Patients were randomized to LCIG infusion plus placebo capsules or encapsulated levodopa tablets plus placebo gel infusion. Sixty-six patients completed the study. The primary endpoint, OFF time reduction, was significantly greater with LCIG (4.04 versus 2.14 h on immediate-release levodopa). ON time without troublesome dyskinesia was significantly different and was increased by 4.1 h on LCIG versus 2.2 on immediate-release levodopa ($p=0.0059$). Quality of life was also significantly in favor of LCIG. The most common adverse events were complications related to the device insertion (89 %), abdominal pain (42 %), procedural pain (32 %), and nausea (25 %). Reasons for discontinuation were peritonitis, psychosis, and post-procedural discharge, in one patient each.

Several uncontrolled studies had reported improvements in dyskinesia compared with baseline [41–44]. One retrospective multicenter review found at least some dyskinesia improvement in 94.7 % of 75 patients assessed for efficacy (out of 91 patients started on LCIG) [41]. Interim results of an ongoing prospective, uncontrolled multicenter study in 192 patients reported a significant OFF time reduction from baseline and a decrease in ON time with troublesome dyskinesia from 1.5 to 0.9 h/day at week 12, and this was sustained (1.0 h/day) at week 54; the difference was significant at each time point [42].

Safety

Most adverse effects of LCIG occur in relation to the device itself and include technical and surgical complications such as infections, including peritonitis. In the prospective multicenter study of 192 patients, seven were reported to have developed peritonitis, all within 2 weeks of PEG tube insertion [42]. Disconnection, clogging, coiling, or kinking of the tube are not uncommon. Complications of device insertion have been reported in 21–57 % of patients [42]. Mechanical interference of certain foods such as asparagus may occur [45]. Recurrent pancreatits due to duodenal ulceration following traction of the tube has been described [46].

A retrospective review reported on 91 patients with advanced PD who had received LCIG as last-line treatment [41]. At least one technical problem occurred in 62.6 % of these patients; peritonitis, in 4.3 %. Even though 65 % of the patients had visual hallucinations at entry, severe psychosis occurred in only 2.2 %.

With respect to neuropsychiatric problems, it is likely that LCIG has more beneficial than negative effects. There are reports of improvements including complete resolution of impulse control disorders (ICDs) including gambling [47], which are closely linked to dopamine agonist use, and of dopamine dysregulation syndrome [48]. This involves craving for higher doses, dose increases, and associated behavioral changes. However, newly onset neuropsychiatric problems including dopamine dysregulation syndrome have been described [49]. Confusion and hallucinations occur infrequently.

Cases of new onset of polyneuropathy have been reported in patients on LCIG, and this is currently being investigated further. Evidence has emerged that PD as such is associated with an increased risk of axonal, predominantly sensory polyneuropathy, with duration of levodopa exposure contributing to the risk [50]. However, in a large prospective study, five patients out of 192 developed polyneuropathy, classified as serious in four, and three had acute to subacute onset. The mechanism has not yet been determined but may but may involve raised homocysteine levels resulting from increased COMT enzyme activity associated with high dose levodopa treatment [51]. It has not yet been established whether vitamin B12 supplemetation is indicated when levels are normal, but regular screening for vitamins B12 and B6 and homocysteine levels, as well as baseline neurography, is advisable. Malabsorption has been suspected, and this may also underlie weight loss which is occasionally observed [52].

The majority of hospital admissions due to a complication related to LCIG have occurred during the first year of treatment. In order to detect and manage any problems early, patients and carers must be instructed carefully and emergency contacts must be provided.

Summary

LCIG translates the principle of dopaminergic drug delivery into clinical practice and its clinical use as an infusion has demonstrated that the same drug that induces dyskinesia when taken orally is capable of relieving dyskinesia when administered continuously. Apart from the beneficial effect of LCIG on motor fluctuations, considerable and clinically relevant improvements in dyskinesia have been reported [40], and dyskinesia reduction has now also been demonstrated in a placebo-controlled, randomized study. Safety issues are mostly related to the device and to infections, including peritonitis. The treatment requires a dedicated multidisciplinary team with good collaboration between neurology and gatroenterology services.

Apomorphine Infusion

Apomorphine is a non-ergot dopamine agonist which must be administered parenterally due to its low bioavailability. When injected subcutaneously, motor symptom relief is equivalent to levodopa, with a considerably faster onset (5–15 min) and shorter duration (mean 40 min) of effect. A randomized study demonstrated that 95 % of OFF periods were successfully treated by apomorphine injections compared to 23 % on placebo, but ON time with troublesome dyskinesia also increased [53].

Several uncontrolled studies showed marked reductions in daily OFF time from baseline when apomorphine is administered via continuous subcutaneous infusion during the waking hours, but randomized comparisons with other treatments or placebo have not yet been performed. The largest, retrospective, study was multicenter and reported a reduction in daily OFF time by 4.3 h [54].

Some uncontrolled studies have also reported reductions in dyskinesia severity compared to baseline, by 34 % up to 83 % (the latter was found in a small study of patients on apomorphine monotherapy). One study used blinded rating of dyskinesia severity following levodopa and apomorphine challenge tests before and 6 months after initiating apomorphine infusion and showed a reduction in dyskinesia severity by 34–44 % on video ratings [55]. The maximum dyskinesia improvement has been observed to occur after several months; mean daily doses in studies reporting effects on dyskinesia have been in the range of 100 mg [56].

Dyskinesia reduction may be more marked in patients who manage to substantially reduce their oral dopaminergic therapy. "Apomorphine monotherapy" has been defined as infusion only during the waking day with discontinuation of oral drugs, except in the morning and at night [55–58]. While a proportion of patients do achieve actual apomorphine infusion monotherapy (45 out of 63 patients in one retrospective study), others do not tolerate complete discontinuation of oral drugs during daytime. However, the overall reduction of short-acting agents appears to be the determinant factor for dyskinesia reduction [55, 59], and a forced switch to strict monotherapy probably is not required. The observed improvement is in keeping with the current concept of dyskinesia formation and believed to be due to the replacement of pulsatile with continuous drug delivery.

In practical terms, infusion treatment is usually initiated on an inpatient basis although this is not an absolute requirement if frequent visits are possible and increases are done very slowly. Domperidone is a peripheral dopamine receptor blocker which counteracts adverse effects such as nausea and is usually used as a premedication for 1–3 days before starting apomorphine. Domperidone has been linked to QT prolongations, and attempts to decrease or discontinue it should be made within weeks if possible. Alternatively, trimethobenzamide may be used if domperidone is not available. Oral dopamine agonists and subsequently other oral antiparkinsonian drugs are gradually withdrawn over weeks to months, while flow rates of apomorphine are increased. While the standard daily duration of infusion is around 14–16 h, some patients with severe nocturnal OFFs benefit from 24-h administration, with lower flow rates during the night. The pump is usually worn on a belt around the patient's waist. The needle is inserted into the abdominal skin into rotating injection sites. During the initial inpatient stay, patients and carers are instructed in handling the pump, including hygiene measures.

Safety

Potential side effects of apomorphine infusion include dopaminergic effects which may occur on any dopamine agonist, including nausea, orthostatic hypotension, leg edema, or somnolence.

Skin nodule formation is very common but is usually mild. Histologically, this was found to be aseptic panniculitis. Rarely, medically relevant skin problems such as abscesses or ulcerations occur, which may require surgical treatment. Occasionally, widespread nodules may impair reliable and stable absorption of apomorphine.

Hemolytic anemia is probably rare (below 1 %) [54] but regular screening is required. Recommended intervals for full blood counts are usually 3–6 months. Coombs test has been described to turn positive in 6–12.5 % of patients although this may be reversible. Hemolytic anemia requires discontinuation of apomorphine and treatment in collaboration with hematology specialists.

Neuropsychiatric adverse effects may occur. As with other dopaminergic drugs, a subgroup of vulnerable patients develop ICDs. No comparative studies exist to show whether ICDs are more common than with other dopamine agonists. As pump treatments are generally used in patients with advanced disease, neuropsychiatric problems typically associated with long disease duration and high dopaminergic doses may occur: These include punding, a behavioral disorder with repetitive, prolonged activities resembling normal recreational or domestic activities (such as cleaning, using a computer), and dopamine dysregulation syndrome, which possibly occurs at a similar rate as on high doses of levodopa although comparative data are lacking and improvement in some neuropsychiatric features has also been observed in non-randomsied studies [60, 61]. Confusion or hallucinations are not uncommon but typically occur in cognitively impaired patients. It is currently unknown whether this adverse effect is more common than with oral dopamine agonists [62, 63].

Technical problems rarely result in cessation of the pump and improvement in some neuropsychiatric features has also been observed in non-randomsied studies [60, 61]. Issues occasionally seen include clotting of connections, arrest of the pump, and disconnection of the syringe. Even if a technical problem cannot be dealt with immediately, e.g., using contact numbers provided to patients and carers, there is little or no risk involved for the patients if they are clearly instructed which oral/transdermal medication to use for the interim.

Summary

Apomorphine infusion has been shown in many uncontrolled studies to markedly improve OFF time, and this has often been observed to be accompanied by dyskinesia reduction, particularly when oral drugs could be reduced. While these observations are supported by long-standing clinical use of the treatment, data from randomized studies are still being awaited. Safety issues include those typical of dopamine agonists as well as skin changes and, rarely, hemolytic anemia, requiring regular follow-up (Table 5.1).

Table 5.1 Clinical features and their impact on the decision for one infusion therapy over the other

	Apomorphine pump	LCIG
Slight hallucinations	+/−	+
Hallucinations > slight	−	+/−
Impulse control disorders	−	+
Mild cognitive impairment to mild dementia	+/−	+
Moderate dementia	−	+/−
L-dopa-unresponsive postural and gait problems, falls	+/−	+/−
Orthostatic hypotension	−	+/−

Choice of Infusion Therapy: Shared Features and Differences

- *The principal indication for apomorphine and intrajejunal infusion is identical*: motor complications which have become refractory to all adaptations of oral and transdermal treatment options, determined by the patient's perception and the impact of motor complications on each patient's quality of life.
- Continuous delivery of dopaminergic drugs by pumps has the potential to ameliorate motor complications, and both LCIG and apomorphine may be considered for the treatment of dyskinesia. A randomized study of LCIG has been completed, making the evidence for LCIG more robust than for apomorphine infusion. The bulk of the evidence from uncontrolled studies suggests that OFF time reduction is likely similar with LCIG and apomorphine infusion. The effect on dyskinesia may potentially be less predictable with apomorphine, particularly if oral medication cannot be reduced sufficiently. However, studies to definitely show the effect of apomorphine infusion on dyskinesia are still lacking and no data from randomized studies comparing the two infusions are available.
- All device-aided treatments that are available and for which a patient is suitable should be discussed with the patient. The choice must be made on an individual basis. If a patient meets contraindications for DBS or decides against surgery, the factors listed in Table 5.1 may be helpful in deciding whether both pumps are options for a given patient (if available) or a specific recommendation for one over the other should be made.
- It has been suggested that device-aided treatments are currently often considered and discussed with patients too late in the course of the disease [64].
- Contraindications for pump treatments are less strict than for DBS, and importantly, there is no age limit. Both LCIG and apomorphine infusion may be tried in patients with mild cognitive impairment and LCIG may be considered on an individual basis in patients with up to a moderate degree of dementia. Either treatment in cognitively impaired patients depends on adequate caregiver support.

- In contrast to DBS, gait and balance problems unresponsive to levodopa are not contraindications but the patient and caregiver should be made aware that these problems will likely not improve on infusion therapy.
- Interdisciplinary teams including a nurse with special interest in PD are best suited for the management of patients with any of the device-aided treatments. For LCIG, additional good collaboration with a gastroenterology service must be in place (Table 5.2).

Table 5.2 Practical recommendations

1. Peak-dose dyskinesia
Reduce individual levodopa dose, at the risk of increasing OFF time. The latter can be compensated for by increasing the number of daily doses of levodopa, or by increasing a dopamine agonist, or both
Discontinue or reduce MAO-B inhibitors, or COMT inhibitors, or both, at the risk of worsening of wearing-off
Add amantadine – most studies have used oral 200–400 mg/day
DBS in patients without contraindications
Add atypical antipsychotic clozapine, at doses of 12.5–75 mg/day (up to 200 mg/day). The use of clozapine is limited by potential serious adverse events (agranulocytosis, myocarditis), but it has stronger evidence for an antidyskinetic effect than quetiapine
Apomorphine continuous subcutaneous infusion
Intrajejunal levodopa infusion
2. Biphasic dyskinesia
This type of dyskinesia can be very difficult to treat and has not been specifically investigated in high-quality studies
Apart from surgical treatment (DBS/lesional), the following pharmacological approaches may be tried:
The strategies described for peak-dose dyskinesia can be considered
Increasing the size and frequency of levodopa doses or a dopamine agonist, at the risk of increasing peak-dose dyskinesia
Larger, less frequent doses may give more predictable responses
Apomorphine and intrajejunal levodopa infusion can be tried
3. Off-period and early-morning dystonia
All strategies used to improve wearing-off can be applied, including long-acting and transdermal dopamine agonists and rasagiline
Dispersible formulations of levodopa
Subcutaneous injections of apomorphine
Apomorphine and intrajejunal levodopa infusion as 24 h application may be appropriate
Additional doses of levodopa (with or without a COMT inhibitor) or dopamine agonist upon awakening during the night may be helpful
Botulinum toxin can be employed in severe OFF-period and early-morning dystonia in selected cases

Adapted from Ferreira et al. [4]

References

1. Poewe WH, Rascol O, Quinn N, Tolosa E, Oertel WH, Martignoni E, Rupp M, Boroojerdi B, SP 515 Investigators. Efficacy of pramipexole and transdermal rotigotine in advanced Parkinson's disease: a double-blind, double-dummy, randomised controlled trial. Lancet Neurol. 2007;6:513–20.
2. Barone P, Lamb J, Ellis A, Clarke Z. Sumanirole versus placebo or ropinirole for the adjunctive treatment of patients with advanced Parkinson's disease. Mov Disord. 2007;22:483–9.
3. Pahwa R, Stacy MA, Factor SA, Lyons KE, Stocchi F, Hersh BP, Elmer LW, Truong DD, Earl NL, EASE-PD Adjunct Study Investigators. Ropinirole 24-hour prolonged release: randomized, controlled study in advanced Parkinson disease. Neurology. 2007;68:1108–15.
4. Ferreira JJ, Katzenschlager R, Bloem BR, Bonuccelli U, Burn D, Deuschl G, Dietrichs E, Fabbrini G, Friedman A, Kanovsky P, Kostic V, Nieuwboer A, Odin P, Poewe W, Rascol O, Sampaio C, Schüpbach M, Tolosa E, Trenkwalder C, Schapira A, Berardelli A, Oertel WH. Summary of the recommendations of the EFNS/MDS-ES review on therapeutic management of Parkinson's disease. Eur J Neurol. 2013;20:5–15.
5. Fox SH, Katzenschlager R, Lim SY, Ravina B, Seppi K, Coelho M, Poewe W, Rascol O, Goetz CG, Sampaio C. The Movement Disorder Society Evidence-Based Medicine Review Update: Treatments for the motor symptoms of Parkinson's disease. Mov Disord. 2011;26 Suppl 3:S2–41.
6. Stocchi F, Giorgi L, Hunter B, Schapira AH. PREPARED: comparison of prolonged and immediate release ropinirole in advanced Parkinson's disease. Mov Disord. 2011;26:1259–65.
7. Cristina S, Zangaglia R, Mancini F, Martignoni E, Nappi G, Pacchetti C. High-dose ropinirole in advanced Parkinson's disease with severe dyskinesias. Clin Neuropharmacol. 2003;26:146–50.
8. da Silva-Júnior FP, Braga-Neto P, Sueli Monte F, de Bruin VM. Amantadine reduces the duration of levodopa-induced dyskinesia: a randomized, double-blind, placebo-controlled study. Parkinsonism Relat Disord. 2005;11:449–52.
9. Facca A, Sanchez–Ramos J. High-dose pergolide monotherapy in the treatment of severe levodopa-induced dyskinesias. Mov Disord. 1996;11:327–9.
10. Odin P, Oehlwein C, Storch A, Polzer U, Werner G, Renner R, Shing M, Ludolph A, Schüler P. Efficacy and safety of high-dose cabergoline in Parkinson's disease. Acta Neurol Scand. 2006;113:18–24.
11. Schwarz J, Scheidtmann K, Trenkwalder C. Improvement of motor fluctuations in patients with Parkinson's disease following treatment with high doses of pergolide and cessation of levodopa. Eur Neurol. 1997;37:236–8.
12. Storch A, Trenkwalder C, Oehlwein C, Winkelmann J, Polzer U, Hundemer HP, Schwarz J. High-dose treatment with pergolide in Parkinson's disease patients with motor fluctuations and dyskinesias. Parkinsonism Relat Disord. 2005;11:393–8.
13. Verhagen Metman L, Del Dotto P, van den Munckhof P, Fang J, Mouradian MM, Chase TN. Amantadine as treatment for dyskinesias and motor fluctuations in Parkinson's disease. Neurology. 1998;50:1323–6.
14. Luginger E, Wenning GK, Bösch S, Poewe W. Beneficial effects of amantadine on Levodopa-induced dyskinesias in Parkinson's disease. Mov Disord. 2000;15:873–8.
15. Thomas A, Iacono D, Luciano AL, et al. Duration of amantadine benefit on dyskinesia of severe Parkinson's disease. J Neurol Neurosurg Psychiatry. 2004;75:141–3.
16. Wolf E, Seppi K, Katzenschlager R, Hochschorner G, Ransmayr G, Schwingenschuh P, Ott E, Kloiber I, Haubenberger D, Auff E, Poewe W. Long-term antidyskinetic efficacy of amantadine in Parkinson's disease. Mov Disord. 2010;25:1357–63.

17. Sawada H, Oeda T, Kuno S, Nomoto M, Yamamoto K, Yamamoto M, Hisanaga K, Kawamura T, Amantadine Study Group. Amantadine for dyskinesias in Parkinson's disease: a randomized controlled trial. PLoS One. 2010;5(12):e15298.
18. Merello M, Nouzeilles MI, Cammarota A, Leiguarda R. Effect of memantine (NMDA antagonist) on Parkinson's disease: a double-blind crossover randomized study. Clin Neuropharmacol. 1999;22:273–6.
19. Varanese S, Howard J, Di Rocco A. NMDA antagonist memantine improves levodopa-induced dyskinesias and "on-off" phenomena in Parkinson's disease. Mov Disord. 2010;25:508–10.
20. Bennett Jr JP, Landow ER, Schuh LA, et al. Suppression of dyskinesias in advanced Parkinson's disease. II. Increasing daily clozapine doses suppress dyskinesias and improve parkinsonism symptoms. Neurology. 1993;43:1551–5.
21. Durif F, Debilly B, Galitzky M, Morand D, Viallet F, Borg M, Thobois S, Broussolle E, Rascol O. Clozapine improves dyskinesias in Parkinson disease: a double-blind, placebo-controlled study. Neurology. 2004;62(3):381–8.
22. Schulte P. Risk of clozapine-associated agranulocytosis and mandatory white blood cell monitoring. Ann Pharmacother. 2006;40:683–8.
23. Fox MDS Website 2013: http://www.movementdisorders.org/UserFiles/file/update-on-treatments-for-motor-symptoms-of-PD.pdf?new.
24. Baron MS, Dalton WB. Quetiapine as treatment for dopaminergic-induced dyskinesias in Parkinson's disease. Mov Disord. 2003;18:1208–9.
25. Katzenschlager R, Manson A, Evans A, Watt H, Lees AJ. Quetiapine for drug-induced dyskinesias in Parkinson's disease: a double-blind cross-over study. J Neurol Neurosurg Psychiatry. 2004;75:295–7.
26. Stathis P, Konitsiotis S, Tagaris G, Peterson D, VALID-PD Study Group. Levetiracetam for the management of levodopa-induced dyskinesias in Parkinson's disease. Mov Disord. 2011;26:264–70.
27. Wolz M, et al. Levetiracetam for levodopa-induced dyskinesia in Parkinson's disease: a randomized, double-blind, placebo-controlled trial. J Neural Transm. 2010;117:1279–86.
28. Wong KK, Alty JE, Goy AG, Raghav S, Reutens DC, Kempster PA. A randomized, double-blind, placebo-controlled trial of levetiracetam for dyskinesia in Parkinson's disease. Mov Disord. 2011;26:1552–5.
29. Murata M, Hasegawa K, Kanazawa I. Zonisamide improves motor function in Parkinson disease: a randomized, double-blind study. Neurology. 2007;68:45–50.
30. Manson AJ, Katzenschlager R, Hobart J, Lees AJ. High dose naltrexone for dyskinesias induced by levodopa. J Neurol Neurosurg Psychiatry. 2001;70:554–6.
31. Carroll CB, Bain PG, Teare L, Liu X, Joint C, Wroath C, Parkin SG, Fox P, Wright D, Hobart J, Zajicek JP. Cannabis for dyskinesia in Parkinson disease: a randomized double-blind crossover study. Neurology. 2004;63:1245–50.
32. Lees A, Fahn S, Eggert KM, Jankovic J, Lang A, Micheli F, Mouradian MM, Oertel WH, Olanow CW, Poewe W, Rascol O, Tolosa E, Squillacote D, Kumar D. Perampanel, an AMPA antagonist, found to have no benefit in reducing "off" time in Parkinson's disease. Mov Disord. 2012;27:284–8.
33. Braz CA, Borges V, Ferraz HB. Effect of riluzole on dyskinesia and duration of the on state in Parkinson disease patients: a double-blind, placebo-controlled pilot study. Clin Neuropharmacol. 2004;27:25–9.
34. Manson AJ, Schrag A, Lees AJ. Low-dose olanzapine for levodopa induced dyskinesias. Neurology. 2000;55:795–9.
35. Goetz CG, Blasucci LM, Leurgans S, Pappert EJ. Olanzapine and clozapine: comparative effects on motor function in hallucinating PD patients. Neurology. 2000;55:789–94.
36. Guridi J, González-Redondo R, Obeso JA. Clinical features, pathophysiology, and treatment of levodopa-induced dyskinesias in Parkinson's disease. Parkinsons Dis. 2012;18:943159.
37. Nyholm D, Askmark H, Gomes-Trolin C, Knutson T, Lennernäs H, Nyström C, Aquilonius SM. Optimizing levodopa pharmacokinetics: intestinal infusion versus oral sustained-release tablets. Clin Neuropharmacol. 2003;26:156–63.

38. Kurth MC, Tetrud JW, Tanner CM, Irwin I, Stebbins GT, Goetz CG, Langston JW. Double-blind, placebo-controlled, crossover study of duodenal infusion of levodopa/carbidopa in Parkinson's disease patients with 'on-off' fluctuations. Neurology. 1993;43:1698–703.
39. Nyholm D, Nilsson Remahl AI, Dizdar N, Constantinescu R, Holmberg B, Jansson R, Aquilonius SM, Askmark H. Duodenal levodopa infusion monotherapy vs oral polypharmacy in advanced Parkinson disease. Neurology. 2005;64:216–23.
40. Olanow CW, Kieburtz K, Odin P, Espay AJ, Standaert DG, Fernandez HH, Vanagunas A, Othman AA, Widnell KL, Robieson WZ, Pritchett Y, Chatamra K, Benesh J, Lenz RA. Antonini A; for the LCIG Horizon Study Group. Continuous intrajejunal infusion of levodopa-carbidopa intestinal gel for patients with advanced Parkinson's disease: a randomised, controlled, double-blind, double-dummy study. Lancet Neurol. 2013. Lancet Neurol. 2014;13:141–9.
41. Devos D, French DUODOPA Study Group. Patient profile, indications, efficacy and safety of duodenal levodopa infusion in advanced Parkinson's disease. Mov Disord. 2009;24: 993–1000.
42. Fernandez HH, Vanagunas A, Odin P, Espay AJ, Hauser RA, Standaert DG, Chatamra K, Benesh J, Pritchett Y, Hass SL, Lenz RA. Levodopa-carbidopa intestinal gel in advanced Parkinson's disease open-label study: interim results. Parkinsonism Relat Disord. 2013; 19:339–45.
43. Syed N, Murphy J, Zimmerman T, Mark M, Sage J. Ten years' experience with enteral levodopa infusions for motor fluctuations in Parkinson's disease. Mov Disord. 1998; 13:336–8.
44. Zibetti M, Merola A, Artusi CA, Rizzi L, Angrisano S, Reggio D, De Angelis C, Rizzone M, Lopiano L. Levodopa/carbidopa intestinal gel infusion in advanced Parkinson's disease: a 7-year experience. Eur J Neurol. 2014;21(2):312–8.
45. Schrader C, Böselt S, Wedemeyer J, Dressler D, Weismüller TJ. Asparagus and jejunal-through-PEG: an unhappy encounter in intrajejunal levodopa infusion therapy. Parkinsonism Relat Disord. 2011;17:67–9.
46. Jitkritsadakul O, Jagota P, Petchrutchatachart S, Sansopha L, Rerknimitr R, Bhidayasiri R. Recurrent pancreatitis as a rare complication of duodenal levodopa infusion treatment. Mov Disord. 2013;28:1308–10.
47. Catalán MJ, de Pablo-Fernández E, Villanueva C, Fernández-Diez S, Lapeña-Montero T, García-Ramos R, López-Valdés E. Levodopa infusion improves impulsivity and dopamine dysregulation syndrome in Parkinson's disease. Mov Disord. 2013;28(14):2007–10.
48. Cannas A, Solla P, Marrosu MG, Marrosu F. Dopamine dysregulation syndrome in Parkinson's disease patients on duodenal levodopa infusion. Mov Disord. 2013;28:840–1.
49. Raudino F, Garavaglia P, Pianezzola C, Riboldazzi G, Leva S, Guidotti M, Bono G. Long-term experience with continuous duodenal levodopa-carbidopa infusion (Duodopa): report of six patients. Neurol Sci. 2009;30:85–6.
50. Ceravolo R, Cossu G, Bandettini di Poggio M, Santoro L, Barone P, Zibetti M, Frosini D, Nicoletti V, Manganelli F, Iodice R, Picillo M, Merola A, Lopiano L, Paribello A, Manca D, Melis M, Marchese R, Borelli P, Mereu A, Contu P, Abbruzzese G, Bonuccelli U. Neuropathy and levodopa in Parkinson's disease: evidence from a multicenter study. Mov Disord. 2013;28:1391–7.
51. Müller T, van Laar T, Cornblath DR, Odin P, Klostermann F, Grandas FJ, Ebersbach G, Urban PP, Valldeoriola F, Antonini A. Peripheral neuropathy in Parkinson's disease: levodopa exposure and implications for duodenal delivery. Parkinsonism Relat Disord. 2013;19:501–7.
52. Nyholm D, Klangemo K, Johansson A. Levodopa/carbidopa intestinal gel infusion long-term therapy in advanced Parkinson's disease. Eur J Neurol. 2012;19:1079–85.
53. Dewey Jr RB, Hutton JT, LeWitt PA, Factor SA. A randomized, double-blind, placebo-controlled trial of subcutaneously injected apomorphine for parkinsonian off-state events. Arch Neurol. 2001;58:1385–92.
54. García Ruiz PJ, Sesar Ignacio A, Ares Pensado B, et al. Efficacy of long-term continuous subcutaneous apomorphine infusion in advanced Parkinson's disease with motor fluctuations: a multicenter study. Mov Disord. 2008;23:1130–6.

55. Katzenschlager R, Hughes A, Evans A, Manson AJ, Hoffman M, Swinn L, Watt H, Bhatia K, Quinn N, Lees AJ. Continuous subcutaneous apomorphine therapy improves dyskinesias in Parkinson's disease: a prospective study using single-dose challenges. Mov Disord. 2005;20:151–7.
56. Manson AJ, Turner K, Lees AJ. Apomorphine monotherapy in the treatment of refractory motor complications of Parkinson's disease – a long-term follow-up study of 64 patients. Mov Disord. 2002;17:1235–41.
57. Colzi A, Turner K, Lees AJ. Continuous subcutaneous waking day apomorphine in the long term treatment of levodopa induced interdose dyskinesias in Parkinson's disease. J Neurol Neurosurg Psychiatry. 1998;64:573–6.
58. Stocchi F, Vacca L, De Pandis MF, Barbato L, Valente M, Ruggieri S. Subcutaneous continuous apomorphine infusion in fluctuating patients with Parkinson's disease: long-term results. Neurol Sci. 2001;22:93–4.
59. Jenner P. Wearing off, dyskinesia, and the use of continuous drug delivery in Parkinson's disease. Neurol Clin. 2013;31(3 Suppl):S17–35.
60. Martinez-Martin P, Reddy P, Antonini A, Henriksen T, Katzenschlager R, Odin P, Todorova A, Naidu Y, Tluk S, Chandiramani C, Martin A, Chaudhuri KR. Chronic subcutaneous infusion therapy with apomorphine in advanced Parkinson's disease compared to conventional therapy: a real life study of non motor effect. J Parkinsons Dis. 2011;1:197–203.
61. van Laar T, Postma AC, Drent M. Continuous subcutaneous infusion of apomorphine can be used safely in patients with Parkinson's disease and pre-existing visual hallucinations. Parkinsonism Relat Disord. 2010;16:71–2.
62. De Gaspari D, Siri C, Landi A, et al. Clinical and neuropsychological follow up at 12 months in patients with complicated Parkinson's disease treated with subcutaneous apomorphine infusion or deep brain stimulation of the subthalamic nucleus. J Neurol Neurosurg Psychiatry. 2006;77:450–3.
63. Ellis C, Lemmens G, Parkes J, Abbott RJ, Pye IF, Leigh PN, Chaudhuri KR. Use of apomorphine in Parkinsonian patients with neuropsychiatric complications to oral treatment. Parkinsonism Relat Disord. 1997;3:103–7.
64. Lökk J. Lack of information and access to advanced treatment for Parkinson's disease patients. J Multidiscip Healthc. 2011;4:433–9.

Chapter 6
Surgical Options for Levodopa-Induced Dyskinesia in Parkinson's Disease

Renato P. Munhoz and Michael S. Okun

Abstract From the perspective of stereotactic surgery, control of dyskinesia in Parkinson's disease (PD) can be accomplished by reducing the dosage of levodopa through subthalamotomy and subthalamic nucleus (STN) deep brain stimulation (DBS) or by pallidotomy and globus pallidus internus (GPi) DBS. Both STN and GPi interventions seem to be equally effective in controlling the appendicular motor signs of PD; however, only GPi surgery is considered to have a direct effect on dyskinesia. The antidyskinetic effect of STN procedures is in most part related to a reduction of dopaminergic drug dosages. Therefore, the si ne qua non condition for reduction of dyskinesia when STN interventions are considered is their capacity to enable a reduction of levodopa dosage. Pallidal surgery is indicated when dyskinesia is a dose-limiting factor for either maintaining or introducing a more aggressive dopaminergic therapy. Also medications used for the treatment of PD are useful for psychiatric, cognitive, sleep, and other non-motor aspects of the disease; therefore, withdrawal or dose reduction may not be a desired goal. DBS has taking over a central role in surgical treatment of movement disorders, and, in fact, ablative procedures are now considered alternatives, particularly when bilateral procedures are required. Other advantages are the facts that DBS does not produce a brain lesion and that the stimulator can be programmed to induce better therapeutic effects while minimizing adverse effects.

Keywords Parkinson's disease • Dyskinesia • Deep brain stimulation • DBS • Pallidotomy

---

R.P. Munhoz, MD, MSc (✉)
Division of Neurology, Department of Medicine, Toronto Western Hospital, University of Toronto, McLaughlin Pavilion 7th Floor Rm 7-417, 399 Bathurst St., Toronto, ON M5T 2S8, Canada
e-mail: renato.munhoz@uhn.ca

M.S. Okun, MD
Department of Neurology, University of Florida College of Medicine, Room L3-100, McKnight Brain Institute, 1149 Newell Drive, Gainesville, FL 32611, USA

The Center for Movement Disorders and Neurorestoration College of Medicine, University of Florida, Gainesville, FL USA

S.H. Fox, J.M. Brotchie (eds.), *Levodopa-Induced Dyskinesia in Parkinson's Disease*, DOI 10.1007/978-1-4471-6503-3_6

Introduction

Reduction of levodopa-induced dyskinesia (LID) in Parkinson's disease (PD) can be accomplished by providing a significant relief on the motor symptoms of PD through medication optimization and in some cases reducing the dosage of levodopa [classically through subthalamic nucleus (STN) deep brain stimulation (DBS)] or by pallidotomy or globus pallidus internus (GPi) DBS [1]. Currently, DBS has become the preferred stereotactic procedure in PD; however, unilateral ablative surgery continues to be performed and can be quite effective especially when dyskinesia affects only one body side. In this review we will address ablative surgery and DBS.

Ablative Surgery

The indication for any form of stereotactic ablative surgery (SAS) has always been the symptomatic treatment of select motor features of PD as identified during a detailed multidisciplinary workup. As early as the 1950s, the inner segment of the internal globus pallidus (GPi) and the ansa lenticularis have been the common choice for functional neurosurgeons [2]. This approach was advocated and further reinforced following the observation that ligation of the anterior choroidal artery, performed for the treatment of accidental bleeding in a PD patient, resulted in relief of tremor [3, 4]. As this technique (ansa–pallidotomy) became more widely utilized, the results for tremor control were mixed, despite the good outcome for rigidity. As a result, pallidotomy was gradually replaced by thalamotomy, which exerted a more optimal tremor control [5]. The failure to reduce tremor in some cases was likely due to a failure to target the appropriate posteroventral pallidum, which was a limitation of early technology. However, with the advent of levodopa in the 1960s, SAS became gradually less popular and did not reemerge again until the early 1990s [6].

The revival of SAS in PD was driven by the shortcomings of levodopa therapy that were slowly emerging [7]. The first setback was the fact that levodopa, as it is now well understood after decades of use, did not live up to the hopes of preventing disease progression. Also, because of the high doses of levodopa required to control tremor, LID and motor fluctuations started to appear. Medical management was challenging with the balance of relief of parkinsonian motor signs against motor fluctuations and induction of dyskinesia, often with neither being managed optimally [8]. The disappointment with levodopa was coupled with advances from physiological and surgery, and also by the emerging and more precise understanding of the organization of the basal ganglia (BG), better SAS techniques, and use of neuroimaging for accurate target localization. At that time, thalamotomy was reintroduced, and, in addition to improvements in the motor aspects of PD, several authors reported impressive suppression of LID [9].

Svennilson et al. reported that when the posteroventral GPi was lesioned additional benefit to general motor function (interpreted as corresponding to relief of akinesia) could be obtained [10]. Later, Laitinen et al. returned to the initial target and the posteroventral GPi became the preferred surgical treatment for PD patients. In the classical report from 1992, Laitinen's group showed favorable outcomes that included not only robust improvements in the cardinal signs of PD but also significant amelioration of LID [11].

At approximately the same time, anatomic and physiological studies confirmed that the GPi and STN were both overactive in PD, and experimental studies demonstrated that lesions in these structures could improve parkinsonism in animal models [12–14]. The classical basal ganglia model predicted that pallidotomy would worsen dyskinesia; however, this symptom became its most striking benefit, probably as a result of interference with abnormal firing patterns (rather than rates) in circuit neurons [15].

Both the clinical report by Laitinen et al. and the pathophysiological laboratory-based developments in PD reinforced the role of pallidotomy for treatment of LID in PD.

Unilateral Pallidotomy

Although human and animal model studies have revealed convincing evidence that lesions of the pallidal outflow receiving areas of the thalamus (nucleus ventralis oralis and ventralis posterior) can ameliorate LID, lesions of the cerebellar outflow receiving areas (ventralis intermedius nucleus) were in general found to only relieve tremor [28]. Despite these observations, thalamotomies were never consistently used to treat LID, possibly because of the limited results that were observed on other PD symptoms and the fact that it would take a very large anteriorly extended lesion to accomplish suppression of LID.

In the mid-to-late 1990s, several studies reported that unilateral pallidotomy in patients with PD had its most dramatic effect on contralateral dyskinesia. At that time, Lozano et al. [22] published the effect of pallidotomy in 14 patients with rigid akinetic parkinsonism, disabling dyskinesia, motor fluctuations, and intact cognition. Motor improvements in the OFF medication condition were mainly contralateral. The most dramatic improvement was for ON period LID, which was shown to be reduced by 92 % in the contralateral side after 6 months. The typical complications of this procedure, homonymous hemianopia, facial paresis, and hemiparesis, were not observed, except for mild and transient facial droop in three cases (noted by clinicians but not the patients). Two years later, the same group published a series of 40 cases, some with a longer follow-up: 27 for 1 year and 11 for 2 years. Short-term results were similar to their previous study; however, trend analysis revealed a slight worsening of contralateral dyskinesia after the first year and a loss of benefit for ipsilateral dyskinesia by the second year. Age had an impact on OFF-period motor signs, with those older than 65 retaining less improvement after 6 months.

LID responded similarly in the two age groups studied. There were no significant reductions in dopaminergic therapy after surgery. Persistent adverse events included facial weakness (two cases), bulbar deficits (three cases), mild dementia (three cases), and worsening of handwriting in four cases [29]. The findings in the same cohort were published after a much longer follow-up (mean 52 months), showing a sustained improvement in OFF-period contralateral motor signs and in LID. Other than dyskinesia and levodopa-responsive motor signs, no additional characteristics had a significant impact on long-term surgical outcome [17].

In 1998, another group published a preliminary study with a series of 26 PD patients, confirming that the most significant effect following unilateral ventral medial pallidotomy was the reduction of contralateral LID by 67 %, while ipsilateral and axial dyskinesia also improved (both around 50 %) significantly. The improvement in underlying parkinsonism as measured by comparing the Unified Parkinson's Disease Rating Scale (UPDRS) scores in the OFF state before and 3 months after surgery was less robust (27 %). On medication, no significant postoperative improvements in parkinsonism were detected, and antiparkinsonian medication dosages increased by 11 % postoperatively. The presence of disabling LID, therefore, was considered the major indication for this surgical procedure. Two (7.7 %) patients died due to cerebral hemorrhages directly related to surgery, while another 15 % had major complications (significant focal motor and bulbar deficits) [30]. In 2003, the first randomized, prospective controlled trial comparing pallidotomy with the best medical therapy was published. The study included 18 patients in each group, showing, after 6 months, a 32 % improvement of the total UPDRS motor score in the surgical group versus only a 5 % deterioration for those randomized for medical therapy. Mean score improvement in the dyskinesia section of the UPDRS (Part IV) for the surgical group was 45 %, whereas patients kept on medical therapy worsened by 8 %. The study also revealed that LID improved after pallidotomy in all patients, and two-thirds had "complete relief" on the contralateral side. Also, there was a 36 % reduction in ipsilateral dyskinesia severity. Levodopa equivalent doses remained unchanged. There were no fatal outcomes and complications occurred in three cases (16.7 %). This study also showed that the age had a clear relationship with clinical outcome, independent of disease duration, with younger patients showing more improvement. This effect was continuous, with no apparent threshold [31].

The series with the longest follow-up was by Kleiner-Fisman et al. and included ten patients, showing a trend toward significance lasting up to 12 years in contralateral LID [32], and by Hariz and Bergenheim who reviewed 13 of the 38 patients described in Laitinen's original study from 1992. Mean follow-up was 10.5 years (up to 13.5), and the effect of surgery remained consistent for contralateral LID, but varied for the appendicular OFF-period signs. The authors went as far as to consider pallidotomy as a prophylactic measure against LID [33].

Lesion size does not seem to have a significant effect on response [34]; however, the optimal location within the GPi that improved dyskinesia is a matter of controversy. Lesion location and size may be different from that required to ameliorate other PD signs, and some experts still advocate a bigger lesion size for prolonged benefit. While anteromedially placed lesions seem to be better correlated with improvement in contralateral LID, central and posterolateral placed lesions improved OFF parkinsonian signs [35]. However, lesions placed in more ventral locations or anywhere in the posteroventral GPi have been shown to be equally effective [36]. Differences in outcome measures as well as in methods of determining lesion location probably account for many of these discrepancies.

Bilateral Pallidotomy

Bilateral staged and simultaneous pallidotomies produce similar improvements in OFF motor and ON dyskinesia to unilateral procedures, with the possible addition of improvements in truncal dyskinesia, dystonia, and, arguably, selected aspects of gait such as walking speed and freezing [37, 38]. These good results were dampened by reports of unacceptable cognitive and bulbar (mainly speech) adverse effects, described in series that today may be considered not large or sufficiently detailed enough to provide a definitive verdict [39]. Only a few other studies have shown different perspectives [40, 41]. The series by Parkin et al. [42], for instance, showed the results in 115 patients who underwent pallidotomy, 53 of which consisted of bilateral procedures. These authors reported significant effectiveness for bilateral pallidotomy, especially for dyskinesia for up to 12 months, at the expense of worsening of speech in 8 % and salivation in 13 %, figures that were similar to those found for unilateral surgery.

Subthalamotomy

In 1963, Andy et al. described a series of 72 "diencephalic operations" (including bilateral interventions) performed in 58 patients for the treatment of parkinsonian tremor. The authors describe a 75 % reduction of tremor and a low incidence of long-term complications, including apathy in most patients and mild contralateral hemiparesis in 8.6 %. Hemiballism occurred in five patients, all of whom were completely improved 2 months postoperatively. They also concluded that the most efficient lesion was in the posterior subthalamus, including the fields of Forel, the zona incerta, and the prerubral area medial to the STN [43]. Later in the 1990s, with the growing experience and good outcomes of STN DBS, subthalamotomy started to be reappraised. Two small series published in 1997 [44, 45] were encouraging and led to larger and more detailed studies. In 2001, a series of 11 patients were described in an open-label prospective study of unilateral subthalamotomy with a 12-month follow-up period, showing an improvement in mean UPDRS III scores of 39 %, including 50 % for bradykinesia, 70 % for rigidity, and 86.7 % for tremor. Doses of levodopa were not changed during the 12-month period; however, five patients followed up for longer had a mean reduction of 59 % in levodopa equivalent daily dose (LEDD). LID remained unchanged, while lesion-induced hemiballism was significant in only one case, which had to undergo an additional surgical procedure [46]. These patients were later reported after a follow-up of at least 3 years, showing persistent improvements, in some cases for up to 6 years. This study showed that complications were in general related to larger and bilateral lesions [47]. The same group also published a larger series with 89 patients followed up for 36 months, again showing comparable improvements and postoperative hemiballism in 14 patients (15 %), eight of which requiring a pallidotomy. Interestingly, these 14 patients also had significantly higher LID scores before surgery [48]. Another similar observation included 8 patients (bilateral subthalamotomy was performed in four) followed up for 18 months. Although one of these patients died from complications of surgery, the remaining experienced sustained benefits in motor signs and a mean reduction in LID by 75 %. Levodopa requirements were lower in all cases, ranging from a 38 to 66 % daily dose reduction. Hemiballism occurred in 25 % of cases, possibly correlated to the STN lesion size. The authors did not find other complications, including motor, sensory, speech, or cognitive functions [49]. Patel et al. followed 21 patients who underwent unilateral subthalamotomy for at least 12 months. The overall improvement in UPDRS III after 12 months was 54 % (27 % for rigidity, 36.8 % for bradykinesia, and 61 % for tremor). Mean L-dopa dose reduction ranged from 34 to 47 %, with an improvement of dyskinesia score by 51 %. In most cases lesions extended beyond the STN to involve pallidofugal fibers (H2 field of Forel) and the zona incerta. The one patient who presented with refractory and severe lesion-induced hemiballism had the lesion placed more centrally in the STN. Other than that, the only significant complication was cognitive impairment, detected in 9 % of the group [50].

A small comparative study analyzed 16 PD patients randomized to receive bilateral STN DBS, bilateral subthalamotomy, or unilateral subthalamotomy plus contralateral STN DBS, followed for 12 months postoperatively, showing that motor improvement and adverse events, including motor, cognitive, and psychiatric outcomes, were similar [51].

In general, because only a few centers have considerable experience with this procedure, which may have a greater risk and relatively high incidence of persistent dyskinesia, this technique is currently not routinely performed if STN DBS is an available option, despite its proven efficacy. However, subthalamotomy remains an alternative surgical option for patients whose conditions are refractory to pharmacological treatment or those who are unable to receive DBS implants due to medical reasons or access limitations [19].

Studies Comparing Pallidotomy and Deep Brain Stimulation of the GPi or STN

Few studies have compared the efficacy and safety of pallidotomy and DBS of the GPi or STN. An early, non-randomized trial comparing results of pallidotomy, STN, and GPi DBS concluded that GPi DBS had similar effects to pallidotomy, but is safer when bilateral procedures are required. Also, bilateral STN DBS may improve OFF-period motor symptoms to a greater proportion than the other procedures and might also improve ON-period motor function [52]. In 2004, Esselink et al. [53] compared, in a randomized, observer-blind trial, the effect of unilateral pallidotomy and bilateral STN DBS in patients with PD followed up for 6 months, confirming that stimulation was more effective in reducing OFF-period motor signs. In addition, this procedure provided better ON-period motor scores and a greater reduction of dopaminergic drug treatment dosages. Both improved LID and functional scales equally, and the number of adverse events was similar in both groups. The same group also published the results after 4 years with similar findings, except for dopaminergic treatment dosage, which did not significantly differ between groups after the first 12 months [16].

Deep Brain Stimulation

Since the introduction of DBS for the treatment of movement disorders, this procedure has gradually been taking over the central role in SAS. In fact, ablative procedures are now considered alternatives and only used when DBS is not feasible due to technical, travel, patient preference, and economic reasons [54, 55].

Two of the reasons that favor DBS, particularly if bilateral procedures are required, are the facts that it is not intended to produce a brain lesion and that the stimulator can be programmed with respect to several variables, including electrode location, amplitude, frequency, and pulse width, to induce better therapeutic effects while minimizing adverse effects. In the case of PD, DBS electrodes have been placed in two main basal ganglia targets, the GPi and the STN, though other targets are also possible (Tables 6.1 and 6.2) [56].

GPi DBS

The first study to report results of this procedure described three patients in 1994, with the postoperative results described as "excellent," reflecting improvements in all motor signs of the disease, as well as for motor fluctuations and LID [57]. During

Table 6.1 Effects of unilateral pallidotomy, subthalamotomy, bilateral GPi, and STN deep brain stimulation (DBS) on general motor improvement (UPDRS III), dyskinesia (UPDRS IV), and levodopa equivalent daily dose (LEDD)

	Motor improvement (%)	Improvement for dyskinesia (%)	Reduction in LEDD
Unilateral pallidotomy	25–45	45–86	n.s. (0–10 %)
Subthalamotomy	18–50	40–85	23–49 %
GPi DBS	26–43	47–88	n.s. (15–17 %)
STN DBS	25–54	20–83	31–47 %

Refs. [16–27]
Mean improvement after a minimum of 6 months compared to preoperative baseline. Scores reflect the medication off condition, for DBS, stimulation on.
n.s. nonsignificant

Table 6.2 Patient selection and points to be considered when choosing stereotactic surgery for dyskinesia in Parkinson's disease

	Advantages	Disadvantages	Patient profile	Postoperative details
Unilateral pallidotomy	Efficacious	Permanent lesion	Unable to travel	Ipsilateral dyskinesia may not improve significantly, requiring continuing antidyskinetic medical treatment or contralateral GPi DBS
	Less costly than DBS	Not reversible	Live where DBS is too expensive or not available	
	Does not require postoperative programming	Bilateral surgery has higher risk of side effects	Prefer to have chronic hardware	
	No complications related to hardware (infections, malfunction)	Does not allow adjustments to control side effects	High infection risk	
GPi DBS	Direct improvement in dyskinesia	No significant change in drug regimen in many but not all cases	Needs prompt improvement of severe dyskinesia	Ensure that the beneficial effect of L-dopa is not antagonized by stimulation
	Allows *adjustments* in drug regimen	Ventral and dorsal stimulation may induce opposite effects on cardinal motor signs of PD; however, this has not been replicable on all cases	Responds to low-dose L-dopa, but has low threshold for dyskinesia	
	May allow for more aggressive dopaminergic therapy		Has L-dopa responsive non-motor signs	
	Synergistic effect with L-dopa on axial and other symptoms		May be better for patients with preexisting cognitive or psychiatric symptoms	
STN DBS	Allows significant *reduction* in dopaminergic drug dosages	Improvements in dyskinesia depend on reduction of levodopa	Has severe motor fluctuations	Stimulation-induced dyskinesia may appear after a latency of several hours if L-dopa not adjusted
	Effective for OFF-period dystonia	May have negative impact on cognition	Uses higher doses of L-dopa	The electrode that induces dyskinesia is usually the most effective
		More laborious postoperative management	Experiences disabling side effect of dopaminergic treatment	
		May worsen or not improve dyskinesia in brittle dyskinetics	Intact cognition	

the next decade, several descriptions of larger series confirmed these findings. A study with a follow-up of at least 24 months showed that the mean improvements in the UPDRS motor and activities of daily living scores after 12 months were more than 50 %, motor fluctuations were reduced from 40 to 10 %, and the score for LID was reduced to one third. Doses of levodopa tended to remain unchanged. Half of the patients experienced a slight worsening of levodopa and stimulation-resistant gait and bulbar symptoms following 12 months [58]. In 2000, a study by Kumar et al. [21] showed the results seen on a cohort of 22 consecutive cases of PD treated with GPi DBS, 17 of whom had bilateral surgeries. Postoperatively, at 6 months, the motor improvement in the OFF condition reached 31 and 66 % reduction in LID.

The first double-blind, crossover study evaluating the results of GPi and STN DBS in PD was performed in 2001, showing that both procedures induced significant improvements in motor function and dyskinesia (by 58 % for STN and 66 % for GPi DBS); however, the average medication used, measured in levodopa equivalents, decreased significantly more for the STN DBS patients [26]. A study with longer follow-up, mean 48.5 months, showed a 64 % mean improvement in dyskinesia after this period [25]. Finally, another study followed up 11 patients with PD who underwent GPi DBS for up to 5 years, showing that, despite a decline on the motor benefit for the OFF-period scores after 3 years, the improvement in LID was sustained for up to 5 years [59].

STN DBS

STN DBS for advanced PD was first introduced in the 1990s and is currently the most common form of surgical treatment applied for this disorder worldwide. The initial series reported significant improvements in OFF-period tremor, rigidity, and bradykinesia, as well as attenuation of motor fluctuations and LID, associated with a 50 % reduction in dopaminergic treatment dosages [60]. Subsequent studies confirmed these findings. In 2001, a prospective study of 91 patients showed, after 6 months, a robust improvement in all motor signs in the OFF condition, in the percentage of time with good mobility, and no dyskinesia, mean dyskinesia score, as well as a mean reduction in daily levodopa dose equivalents (approximately 60 %) [26]. At this point it became clear that the reduction in dyskinesia could be attributed at least in part to the reduction of levodopa dosage. However, a few studies showed that this may not be the only element in this beneficial effect. A study designed to assess the effect of STN DBS on OFF-period dystonia, and on diphasic and peak dose dyskinesia after a levodopa challenge using the same suprathreshold dose as before surgery with the stimulation on, showed a reduction of OFF-period dystonia by 90 %, and of diphasic dyskinesia by 50 %, and of peak dose dyskinesia by 30 % [61]. The same authors had already reported that chronic STN DBS per se tends to reduce dyskinesia, as opposed to chronic activation of the dopaminergic system with levodopa. The authors speculated that this difference may have been due to the pulsatile nature of levodopa stimulation versus the more continuous activation provided by chronic STN DBS [62]. There was also an important study by Oyama et al. that elegantly showed that dyskinesia could possibly be reduced in both the STN and GPi targets. The authors accounted for medication reduction and showed that in both targets there was a possibility of dyskinesia suppression without medication reduction [63].

Long-term studies of bilateral STN DBS in patients with advanced PD have demonstrated the stability of this therapy over time. A 5-year prospective study of 49 consecutive patients treated with STN DBS noted that OFF-medication motor scores at 5 years were still 54 % better than baseline [20]. Worsening of ON-medication akinesia, speech, postural stability, and freezing of gait was interpreted to be consistent with the natural progression of PD. However, LID benefits persisted, with dyskinesia disability and duration at 5 years being improved by 58 and 71 %, respectively, in comparison with baseline. Similar benefits with respect to dyskinesia were observed in 37 patients followed for 5 years after DBS surgery [64]. Finally, a comprehensive meta-analysis of 921 patients who underwent STN DBS between 1993 and 2004 noted an average reduction in LID of 69.1 % [65].

Mechanisms of Action in Reducing Levodopa-Induced Dyskinesia

Pallidal Stimulation

Restoration of the thalamocortical activity by suppression of the inhibitory output from the pallidum to the ventrolateral thalamus is the suspected mechanism for motor improvement underpinning GPi DBS; however, the cellular mechanisms of high-frequency stimulation are still unknown. The mechanism of GPi DBS in reducing dyskinesia is also not completely understood. The current views of the basal ganglia physiology suggest that inhibition of ventral GPi activity should induce dyskinesia; however, lesioning of the ventral pallidum provides relief of dyskinesia [66]. One of the possible justifications for this apparent paradoxical response is that LID may be more correlated with an abnormal pattern than with the direction and intensity of the neuronal activity within the GPi [66, 67]. Surgical modification of this patterned activity might be accomplished by lesioning (direct neuronal inhibition) or with DBS (indirect inhibition through activation of inhibitory axons close to the electrode). Dyskinesia might also arise from an abnormal balance of activity within different functional zones of the nucleus (ventral versus dorsal GPi), and stimulation may suppress this abnormal activity [66, 68, 69]. Finally, the antidyskinetic effect of GPi DBS may be mediated through effects on the subthalamopallidal tract, which projects to the dorsal GP externus and GPi. Dorsal GPi stimulation might inhibit this projection and would be expected to improve PD symptoms and induce dyskinesia [70].

STN Stimulation

STN DBS mimics the effects of levodopa on parkinsonian motor symptoms and allows reduction of dopaminergic medication, secondarily relieving dyskinesia as medications are reduced or withdrawn postoperatively [63]. However, improvement of dyskinesia is also sometimes observed in the early postoperative period after implantation of electrodes in the STN, even in the absence of a reduction of medications [1]. This indicates a direct antidyskinetic effect of manipulation of the STN (or the vicinity of its dorsal border and perhaps the zona incerta), but long-term relief of dyskinesia generally requires reduction of medications. The specific site of action in stimulation of the STN is unknown. Some data indicate that the best effect can be achieved at the lowest intensity not through stimulation of neurons within the STN, but by stimulation of tissue dorsal to it, which might affect the pallidothalamic bundle, the pallidosubthalamic tract, and/or the zona incerta [71]. Other data indicate that the most effective contact location appears to be within the anterodorsal portion of the STN, although current could spread from this location into the directly superior fields of Forel and zona incerta [72]. The observation that an active DBS contact dorsal to the STN may provide better control of dyskinesia (indicative of a direct antidyskinetic effect) supports the notion that activation of structures dorsal to the STN is important in providing relief of parkinsonian symptoms by DBS of the STN [18]. Overall, the specific mechanisms of action of GPi and STN DBS in suppressing dyskinesia are unknown.

Studies Comparing the Effect of GPi and STN DBS

A few studies have compared the effect of GPi and STN DBS on PD. The first, published in 2001, had a relatively short follow-up period after surgery (3 months) and revealed similar improvements in OFF-period motor parameters, as well as for ON dyskinesia, with the caveat that only the STN group was able to significantly reduce the levodopa equivalent dose [26]. In 2005, a non-randomized extension of this study with 105 patients followed up for at least 3 years showed that, in addition to improvements in all motor signs of parkinsonism in the OFF condition, STN DBS significantly improved OFF dystonia and ON dyskinesia, while GPi had a similar effect on ON dyskinesia with no significant improvement on OFF dystonia. In this study, reduction in postoperative levodopa equivalent doses were significant only for the STN group, in which more than 10 % of patients stopped taking levodopa. These changes were sustained after up to 4 years of follow-up [25]. Moro et al., in an double-blind, non-randomized study with 35 patients who underwent STN DBS and 16 who underwent GPi DBS, found that both procedures induced significant improvements in OFF-period motor signs, ADLs, and ON dyskinesia scores, although only the STN group had a significant reduction in the doses of levodopa. These results were sustained after 6 years of follow-up [73]. A direct comparison of both procedures was published in 2012 [27]. This was a randomized, evaluator-blind study with 198 PD patients followed up for at least 36 months, which concluded that the primary outcome, OFF-period motor improvement (including subscales for each motor sign), was significantly improved, but the improvements were similar, stable over time, and with parallel trends for both targets. The scores for complications of levodopa therapy (UPDRS IV), including dyskinesia, as well as the amount of ON-period time without troublesome LID, were significantly improved for both groups over 36 months, with nonsignificant, but greater decreases in levodopa dosages in the STN group. Finally, one recent double-blind study of 128 PD patients randomized for either form of treatment showed that patients who underwent STN DBS had larger improvements in OFF-period mean UPDRS motor score, mean change in ADLs scores, and mean reduction in medication after surgery. OFF dystonia scores were similarly improved as well as the time in ON phase without dyskinesia. The scores of the dyskinesia rating scales were significantly better 12 months after surgery for those who underwent GPi DBS. This difference probably occurred because the authors assessed patients after 12 months with the same dose of levodopa used at baseline; however, in daily life, they may use lower doses, leading to less LID [24].

Practical Issues: Selection of the Surgical Target, Technique, and Programming

Numerous studies have demonstrated the effectiveness of STN DBS in controlling the appendicular motor signs of PD; however, this procedure is not considered to have as much of a direct effect on the intensity of LID. The antidyskinetic effect of STN DBS has been hypothesized to be related to allowing reduction of dopaminergic drug dosages, with consequent improvement in side effects, including LID. The persistence or worsening of LID after STN DBS is common and is, in fact, indicative of the necessity to reduce the dose of levodopa [74]. Therefore, the si ne qua non condition for reduction of LID when STN DBS is considered is its capacity to enable a reduction of levodopa dosage. If, however, an adequate response of motor symptoms does not occur postoperatively, dyskinesia will remain unchanged. Of importance, STN stimulation not uncommonly induces contralateral dyskinesia, which may be persistent, and in some cases leads to the implantation of rescue GPi leads.

On the other hand, GPi DBS seems to have a direct effect on the reduction of dyskinesia. Patients undergoing this procedure typically cannot tolerate significantly lower doses of levodopa after surgery and still appreciate a marked reduction of dyskinesia. Simplistically, patients who experience a good response of their PD symptoms with levodopa, but whose primary and most significant source of disability is dyskinesia, may benefit from GPi DBS [63]. In other words, GPi DBS can be especially valuable for cases in which LID is a dose-limiting factor for either maintaining or introducing a more aggressive dopaminergic therapy. In addition, levodopa may have a synergistic effect on GPi DBS which is not seen after STN stimulation. Burchiel et al., for instance, in a randomized, double-blind study, comparing the effects of STN and GPi DBS, showed that, in combination with levodopa, UPDRS motor scores were significantly more improved for patients who underwent the pallidal procedure. This combination was also more clinically significant for axial symptoms, which are traditionally considered refractory to either form of treatment alone [75]. Another more recent meta-regression of long-term studies of cases who underwent these procedures confirmed that GPi DBS in combination with levodopa was correlated with better scores for postural instability and gait disorder than STN stimulation plus levodopa [76].

Selection of either target may also be influenced by the fact that medications used for the treatment of PD are useful not only for motor but also for psychiatric, cognitive, sleep, and other non-motor aspects of the disease; therefore, withdrawal or dose reduction may not be a desired goal [77]. Selection of the target should be based on the patient's most disabling symptoms, response, and side effects related to levodopa and the ultimate goals of therapy [78]. If LID is a patient's most disabling symptom, especially if they require more immediate improvement due to its severity and potential morbidity, then GPi DBS should be considered with the knowledge that regardless of changes in medication therapy after surgery there is a high likelihood that dyskinesia will improve. On the other hand, patients undergoing STN DBS must hope for a sufficiently good response after surgery that will allow medications to be sufficiently reduced. If change in parkinsonian motor symptoms after STN DBS is insufficient to guarantee reduction of levodopa dosage, or if its reduction worsens or induces non-motor symptoms, the intervention for dyskinesia may be "unfruitful" [1].

In the case of a patient in whom, in addition to motor signs of parkinsonism, medication side effects other than dyskinesia are a primary source of disability, STN DBS may be more desirable.

In general, when the presence of LID is the main problem and indication for surgery, there are no formal differences in the procedures when compared to situations when the chief complaint is another motor feature of the disease. However, a few minor variables exist. Implantation of leads is typically performed while patients are in the OFF condition to avoid disabling dyskinesia, leading to motion artifacts during preoperative imaging and to better microelectrode recording during the intraoperative procedure [79]. Other variations are used because of possible differential antidyskinetic effects of stimulation at different sites within the GPi as stimulation of two different sites within the nucleus induces different effects on dyskinesia and response to dopaminergic treatment. Studies have shown that stimulation of the most dorsal aspects of the GPi in the OFF period usually leads to improvement of the cardinal signs, especially bradykinesia, while inducing dyskinesia, mimicking the action of levodopa. When deeper (ventral) sites within the nucleus were stimulated, signs worsened. In the ON period, stimulation of the ventral GPi reduced dyskinesia but may have worsened bradykinesia. Stimulation of the intermediate area seems to provide a balance between these two extremes. It is unclear whether these findings have a practical significance, but their existence should be kept in mind during surgical planning, positioning of the lead within the GPi, and during programming sessions [66, 68, 69].

Postoperative Programming: GPi DBS

As a rule, the evaluation of stimulation-related beneficial effects is typically less reliable during the first weeks after electrode implantation, due to the lesion effect of the procedure. Therefore, the initial programming should be performed after at least 2 weeks of surgery. At this time, the patient should be in the OFF-medication condition, after 12 h of dopaminergic drug withdrawal. The first step should focus on achieving the best improvement of the cardinal signs of parkinsonism. The second phase should address the patient during the ON period, under the effect of levodopa, with particular awareness for LID. Therefore, the goal of programming should be attempting to achieve a good relief of PD symptoms in the OFF condition, not associated with the occurrence of dyskinesia in the ON period, and with the highest threshold for side effects of stimulation. This procedure should be performed for all four contacts separately, defining a hierarchy for therapeutic window [80].

In patients whose primary complaint is LID, an additional programming session can be performed in a full ON condition to confirm the adequate beneficial effect of stimulation, but is usually only indicated if there are difficulties suppressing dyskinesia. Special attention must be directed to ensure that beneficial medication effects are not antagonized by stimulation, as well as the OFF-medication symptoms are not exacerbated, since different regions within the GPi may have opposite effects on dyskinesia and on the cardinal signs of parkinsonism, when stimulated. Fortunately, the detrimental effects of stimulation on parkinsonism and the response to levodopa have higher thresholds than the beneficial effects on dyskinesia. As ventral GPi areas may provide good relief of dyskinesia at the expense of loss of beneficial effect of levodopa, a better stimulation response can be detected by using more central contacts, which usually provide good relief of dyskinesia as well as tremor, rigidity, and bradykinesia [68, 81]. It is important to point out that many experts have been unable to replicate the differential effects of programming different contacts in the GPi, and that in general the GPi has been found to be a much easier target to optimize. The GPi target also allows for more flexibility that the STN target, as was recently shown by Weaver et al. in VA study 36-month outcomes [27, 82].

Postoperative Programming: STN DBS

As STN DBS ideally mimics the motor effects of levodopa in many aspects, the main objective of initial programming in cases of dyskinesia relies on providing a significant improvement of the motor signs of parkinsonism and a concomitant decrease in levodopa dosage, which, on average, reaches a 50 % reduction [23]. Therefore, as in the case of GPi DBS, the first programming session should preferentially be performed in patients during the OFF period, holding all medications for PD for 12 h. In fact, most experts that program STN and GPi DBS have patients report to the clinic in an OFF-medication condition, which provides a nearly optimal programming scenario (no bias of medications). This is generally enough for most patients; however, some patients may require longer OFF periods. In difficult cases, after programming for reduction of bradykinesia, tremor, and, especially, rigidity, patients should take their regular doses of levodopa and, in the ON state, be assessed for adverse effects with the combination of stimulation and medications, particularly dyskinesia. The patient should be seen during this first session at the peak effect of levodopa and ideally should have access to expert programming for the next few days, as dyskinesia may appear after a latency period of up to several hours [83, 84]. During the first few weeks and months after surgery, as stimulation is adjusted to provide the best relief of parkinsonian symptoms, medication doses can be slowly titrated downward, and LID tends to improve or resolve. Moreover, dyskinesia has been hypothesized to improve with chronic continuous stimulation due to plastic changes as a direct effect of stimulation, leading to desensitization of the neuronal circuitry underlying to LID. Persistent dyskinesia is generally treated by further reduction of medication [83].

In some instances, especially during the first few weeks after DBS implantation, dyskinesia may be exacerbated, and, in fact, the induction of these involuntary movements in the short term predicts a favorable long-term outcome [63]. Thus, the particular electrode that induces dyskinesia is usually the most effective contact for long-term therapy. In these cases, if reducing levodopa leads to worsening of PD symptoms, medication doses should be kept at the lowest adequate therapeutic level, and stimulation amplitudes or other parameters should be reduced. Over time, the threshold for induction of dyskinesia typically increases, and amplitude can be gradually increased [85]. Finally, if stimulation using the most effective contact precipitates dyskinesia that cannot be controlled except by unacceptable reduction of stimulation intensity, programming the system to use a more proximal contact in a monopolar configuration, or reprogramming to a bipolar configuration, may be necessary. Addition of a contact dorsal to the STN (perhaps in the zona incerta) may also provide better control of dyskinesia [83].

References

1. Follett KA. Comparison of pallidal and subthalamic deep brain stimulation for the treatment of levodopa-induced dyskinesia. Neurosurg Focus. 2004;17(1):E3.
2. Guridi J, Lozano AM. A brief history of pallidotomy. Neurosurgery. 1997;41(5):1169–80.
3. Cooper IS. Surgical alleviation of Parkinsonism; effects of occlusion of the anterior choroidal artery. J Am Geriatr Soc. 1954;2(11):691–718.
4. Rand RW, Stern WE, Orr JK. Parkinsonism; early results of occlusion of the anterior choroidal artery. Calif Med. 1954;81(4):276–8.
5. Gillingham J. Forty-five years of stereotactic surgery for Parkinson's disease: a review. Stereotact Funct Neurosurg. 2000;74(3–4):95–8.
6. Okun MS, Vitek JL. Lesion therapy for Parkinson's disease and other movement disorders: update and controversies. Mov Disord. 2004;19(4):375–89.
7. Speelman JD, Bosch DA. Resurgence of functional neurosurgery for Parkinson's disease: a historical perspective. Mov Disord. 1998;13(3):582–8.
8. Lang AE, Lozano AM. Parkinson's disease. Second of two parts. N Engl J Med. 1998;339:1130–43.
9. Tasker RR, Siqueira J, Hawrylyshyn P, Organ LW. What happened to VIM thalamotomy for Parkinson's disease? Appl Neurophysiol. 1983;46(1–4):68–83.

10. Svennilson E, Torvik A, Lowe R, Leksell L. Treatment of parkinsonism by stereotatic thermolesions in the pallidal region. A clinical evaluation of 81 cases. Acta Psychiatr Scand. 1960;35:358–77.
11. Laitinen LV, Bergenheim AT, Hariz MI. Leksell's posteroventral pallidotomy in the treatment of Parkinson's disease. J Neurosurg. 1992;76(1):53–61.
12. Bergman H, Wichmann T, DeLong MR. Reversal of experimental parkinsonism by lesions of the subthalamic nucleus. Science. 1990;249(4975):1436–8.
13. Aziz TZ, Peggs D, Agarwal E, Sambrook MA, Crossman AR. Subthalamic nucleotomy alleviates parkinsonism in the 1-methyl-4-phenyl-1,2,3,6-tetrahydropyridine (MPTP)-exposed primate. Br J Neurosurg. 1992;6(6):575–82.
14. Hamada I, DeLong MR. Excitotoxic acid lesions of the primate subthalamic nucleus result in reduced pallidal neuronal activity during active holding. J Neurophysiol. 1992;68(5):1859–66.
15. Lozano AM, Lang AE. Pallidotomy for Parkinson's disease. Adv Neurol. 2001;86:413–20.
16. Esselink RA, de Bie RM, de Haan RJ, Lenders MW, Nijssen PC, van Laar T, Schuurman PR, Bosch DA, Speelman JD. Long-term superiority of subthalamic nucleus stimulation over pallidotomy in Parkinson disease. Neurology. 2009;73(2):151–3.
17. Fine J, Duff J, Chen R, Chir B, Hutchison W, Lozano AM, Lang AE. Long-term follow-up of unilateral pallidotomy in advanced Parkinson's disease. N Engl J Med. 2000;342(23): 1708–14.
18. Follett KA, Weaver FM, Stern M, Hur K, Harris CL, Luo P, Marks Jr WJ, Rothlind J, Sagher O, Moy C, Pahwa R, Burchiel K, Hogarth P, Lai EC, Duda JE, Holloway K, Samii A, Horn S, Bronstein JM, Stoner G, Starr PA, Simpson R, Baltuch G, De Salles A, Huang GD, Reda DJ, CSP 468 Study Group. Pallidal versus subthalamic deep-brain stimulation for Parkinson's disease. N Engl J Med. 2010;362(22):2077–91.
19. Jourdain VA, Schechtmann G, Di Paolo T. Subthalamotomy in the treatment of Parkinson's disease: clinical aspects and mechanisms of action. J Neurosurg. 2014;120:140–51.
20. Krack P, Batir A, Van Blercom N, et al. Five-year follow-up of bilateral stimulation of the subthalamic nucleus in advanced Parkinson's disease. N Engl J Med. 2003;349:1925–34.
21. Kumar R, Lang AE, Rodriguez-Oroz MC, Lozano AM, Limousin P, Pollak P, Benabid AL, Guridi J, Ramos E, van der Linden C, Vandewalle A, Caemaert J, Lannoo E, van den Abbeele D, Vingerhoets G, Wolters M, Obeso JA. Deep brain stimulation of the globus pallidus pars interna in advanced Parkinson's disease. Neurology. 2000;55(12 Suppl 6):S34–9.
22. Lozano AM, Lang AE, Galvez-Jimenez N, Miyasaki J, Duff J, Hutchinson WD, Dostrovsky JO. Effect of GPi pallidotomy on motor function in Parkinson's disease. Lancet. 1995; 346(8987):1383–7.
23. Moro E, Scerrati M, Romito LM, Roselli R, Tonali P, Albanese A. Chronic subthalamic nucleus stimulation reduces medication requirements in Parkinson's disease. Neurology. 1999;53(1):85–90.
24. Odekerken VJ, van Laar T, Staal MJ, Mosch A, Hoffmann CF, Nijssen PC, Beute GN, van Vugt JP, Lenders MW, Contarino MF, Mink MS, Bour LJ, van den Munckhof P, Schmand BA, de Haan RJ, Schuurman PR, de Bie RM. Subthalamic nucleus versus globus pallidus bilateral deep brain stimulation for advanced Parkinson's disease (NSTAPS study): a randomised controlled trial. Lancet Neurol. 2013;12(1):37–44.
25. Rodriguez-Oroz MC, Obeso JA, Lang AE, Houeto JL, Pollak P, Rehncrona S, Kulisevsky J, Albanese A, Volkmann J, Hariz MI, Quinn NP, Speelman JD, Guridi J, Zamarbide I, Gironell A, Molet J, Pascual-Sedano B, Pidoux B, Bonnet AM, Agid Y, Xie J, Benabid AL, Lozano AM, Saint-Cyr J, Romito L, Contarino MF, Scerrati M, Fraix V, Van Blercom N. Bilateral deep brain stimulation in Parkinson's disease: a multicentre study with 4 years follow-up. Brain. 2005;128(Pt 10):2240–9.
26. The Deep-Brain Stimulation for Parkinson's Disease Study Group. Deep-brain stimulation of the subthalamic nucleus or the pars interna of the globus pallidus in Parkinson's disease. N Engl J Med. 2001;345:956–63.
27. Weaver FM, Follett KA, Stern M, Luo P, Harris CL, Hur K, Marks Jr WJ, Rothlind J, Sagher O, Moy C, Pahwa R, Burchiel K, Hogarth P, Lai EC, Duda JE, Holloway K, Samii A, Horn S,

Bronstein JM, Stoner G, Starr PA, Simpson R, Baltuch G, De Salles A, Huang GD, Reda DJ, CSP 468 Study Group. Randomized trial of deep brain stimulation for Parkinson disease: thirty-six-month outcomes. Neurology. 2012;79(1):55–65.
28. Narabayashi H, Yokochi F, Nakajima Y. Levodopa-induced dyskinesia and thalamotomy. J Neurol Neurosurg Psychiatry. 1984;47(8):831–9.
29. Lang AE, Lozano AM, Montgomery E, Duff J, Tasker R, Hutchinson W. Posteroventral medial pallidotomy in advanced Parkinson's disease. N Engl J Med. 1997;337(15):1036–42.
30. Samuel M, Caputo E, Brooks DJ, Schrag A, Scaravilli T, Branston NM, Rothwell JC, Marsden CD, Thomas DG, Lees AJ, Quinn NP. A study of medial pallidotomy for Parkinson's disease: clinical outcome, MRI location and complications. Brain. 1998;121(Pt 1):59–75.
31. Vitek JL, Bakay RA, Freeman A, Evatt M, Green J, McDonald W, Haber M, Barnhart H, Wahlay N, Triche S, Mewes K, Chockkan V, Zhang JY, DeLong MR. Randomized trial of pallidotomy versus medical therapy for Parkinson's disease. Ann Neurol. 2003; 53(5):558–69.
32. Kleiner-Fisman G, Lozano A, Moro E, Poon YY, Lang AE. Long-term effect of unilateral pallidotomy on levodopa-induced dyskinesia. Mov Disord. 2010;25(10):1496–8.
33. Hariz MI, Bergenheim AT. A 10-year follow-up review of patients who underwent Leksell's posteroventral pallidotomy for Parkinson disease. J Neurosurg. 2001;94(4):552–8.
34. Goodman SH, Wilkinson S, Overman J, Koller WC, Tröster A, Pahwa R, Lyons K, Kieltyka J, Burns J, Gordon M. Lesion volume and clinical outcome in stereotactic pallidotomy and thalamotomy. Stereotact Funct Neurosurg. 1998;71(4):164–72.
35. Gross RE, Lombardi WJ, Lang AE, Duff J, Hutchison WD, Saint-Cyr JA, Tasker RR, Lozano AM. Relationship of lesion location to clinical outcome following microelectrode-guided pallidotomy for Parkinson's disease. Brain. 1999;122(Pt 3):405–16.
36. Lang AE, Lozano A, Montgomery EB, Tasker RR, Hutchison WD. Posteroventral medial pallidotomy in advanced Parkinson's disease. Adv Neurol. 1999;80:575–83.
37. Pincus MM. Beneficial effect of bilateral pallidotomy on gait is unproven. Arch Neurol. 2000;57(8):1231–2.
38. Siegel KL, Metman LV. Effects of bilateral posteroventral pallidotomy on gait of subjects with Parkinson disease. Arch Neurol. 2000;57(2):198–204.
39. Intemann PM, Masterman D, Subramanian I, DeSalles A, Behnke E, Frysinger R, Bronstein JM. Staged bilateral pallidotomy for treatment of Parkinson disease. J Neurosurg. 2001;94(3):437–44.
40. Favre J, Burchiel KJ, Taha JM, Hammerstad J. Outcome of unilateral and bilateral pallidotomy for Parkinson's disease: patient assessment. Neurosurgery. 2000;46(2):344–53.
41. Counihan TJ, Shinobu LA, Eskandar EN, Cosgrove GR, Penney Jr JB. Outcomes following staged bilateral pallidotomy in advanced Parkinson's disease. Neurology. 2001;56(6): 799–802.
42. Parkin SG, Gregory RP, Scott R, Bain P, Silburn P, Hall B, Boyle R, Joint C, Aziz TZ. Unilateral and bilateral pallidotomy for idiopathic Parkinson's disease: a case series of 115 patients. Mov Disord. 2002;17(4):682–92.
43. Andy OJ, Jurko MF, Sias Jr FR. Subthalamotomy in treatment of parkinsonian tremor. J Neurosurg. 1963;20:860–70.
44. Obeso JA, Alvarez LM, Macias RJ. Lesion of the subthalamic nucleus in Parkinson's disease. Neurology. 1997;48:A138.
45. Gill SS, Heywood P. Bilateral dorsolateral subthalamotomy for advanced Parkinson's disease. Lancet. 1997;350(9086):1224.
46. Alvarez L, Macias R, Guridi J, Lopez G, Alvarez E, Maragoto C, Teijeiro J, Torres A, Pavon N, Rodriguez-Oroz MC, Ochoa L, Hetherington H, Juncos J, DeLong MR, Obeso JA. Dorsal subthalamotomy for Parkinson's disease. Mov Disord. 2001;16(1):72–8.
47. Alvarez L, Macias R, Lopez G, Alvarez E, Pavon N, Rodriguez-Oroz MC, Juncos JL, Maragoto C, Guridi J, Litvan I, Tolosa ES, Koller W, Vitek J, DeLong MR, Obeso JA. Bilateral subthalamotomy in Parkinson's disease: initial and long-term response. Brain. 2005;128(Pt 3):570–83.

48. Alvarez L, Macias R, Pavón N, López G, Rodríguez-Oroz MC, Rodríguez R, Alvarez M, Pedroso I, Teijeiro J, Fernández R, Casabona E, Salazar S, Maragoto C, Carballo M, García I, Guridi J, Juncos JL, DeLong MR, Obeso JA. Therapeutic efficacy of unilateral subthalamotomy in Parkinson's disease: results in 89 patients followed for up to 36 months. J Neurol Neurosurg Psychiatry. 2009;80(9):979–85.
49. Su PC, Tseng HM, Liu HM, Yen RF, Liou HH. Subthalamotomy for advanced Parkinson disease. J Neurosurg. 2002;97(3):598–606.
50. Patel NK, Heywood P, O'Sullivan K, McCarter R, Love S, Gill SS. Unilateral subthalamotomy in the treatment of Parkinson's disease. Brain. 2003;126(Pt 5):1136–45.
51. Merello M, Tenca E, Pérez Lloret S, Martín ME, Bruno V, Cavanagh S, Antico J, Cerquetti D, Leiguarda R. Prospective randomized 1-year follow-up comparison of bilateral subthalamotomy versus bilateral subthalamic stimulation and the combination of both in Parkinson's disease patients: a pilot study. Br J Neurosurg. 2008;22(3):415–22.
52. Kumar R, Lozano AM, Montgomery E, Lang AE. Pallidotomy and deep brain stimulation of the pallidum and subthalamic nucleus in advanced Parkinson's disease. Mov Disord. 1998;13 Suppl 1:73–82.
53. Esselink RA, de Bie RM, de Haan RJ, Lenders MW, Nijssen PC, Staal MJ, Smeding HM, Schuurman PR, Bosch DA, Speelman JD. Unilateral pallidotomy versus bilateral subthalamic nucleus stimulation in PD: a randomized trial. Neurology. 2004;62(2):201–7.
54. Hooper AK, Okun MS, Foote KD, Fernandez HH, Jacobson C, Zeilman P, Romrell J, Rodriguez RL. Clinical cases where lesion therapy was chosen over deep brain stimulation. Stereotact Funct Neurosurg. 2008;86(3):147–52.
55. Munhoz RP, Teive HA, Francisco AN, Raskin S, Rogaeva E. Unilateral pallidotomy in a patient with parkinsonism and G2019S LRRK2 mutation. Mov Disord. 2009;24(5):791–2.
56. Terzic D, Abosch A. Update on deep brain stimulation for Parkinson's disease. J Neurosurg Sci. 2012;56(4):267–77.
57. Siegfried J, Lippitz B. Bilateral chronic electrostimulation of ventroposterolateral pallidum: a new therapeutic approach for alleviating all parkinsonian symptoms. Neurosurgery. 1994;35(6):1126–9.
58. Ghika J, Villemure JG, Fankhauser H, Favre J, Assal G, Ghika-Schmid F. Efficiency and safety of bilateral contemporaneous pallidal stimulation (deep brain stimulation) in levodopa-responsive patients with Parkinson's disease with severe motor fluctuations: a 2-year follow-up review. J Neurosurg. 1998;89(5):713–8.
59. Volkmann J, Allert N, Voges J, Sturm V, Schnitzler A, Freund HJ. Long-term results of bilateral pallidal stimulation in Parkinson's disease. Ann Neurol. 2004;55(6):871–5.
60. Limousin P, Krack P, Pollak P, Benazzouz A, Ardouin C, Hoffmann D, Benabid AL. Electrical stimulation of the subthalamic nucleus in advanced Parkinson's disease. N Engl J Med. 1998;339(16):1105–11.
61. Krack P, Pollak P, Limousin P, Benazzouz A, Deuschl G, Benabid AL. From off-period dystonia to peak-dose chorea. The clinical spectrum of varying subthalamic nucleus activity. Brain. 1999;122(Pt 6):1133–46.
62. Krack P, Limousin P, Benabid AL, Pollak P. Chronic stimulation of subthalamic nucleus improves levodopa-induced dyskinesia in Parkinson's disease. Lancet. 1997;350(9092):1676.
63. Oyama G, Foote KD, Jacobson 4th CE, Velez-Lago F, Go C, Limotai N, Zeilman PR, Romrell J, Wu SS, Neal D, Okun MS. GPi and STN deep brain stimulation can suppress dyskinesia in Parkinson's disease. Parkinsonism Relat Disord. 2012;18(7):814–8.
64. Schupbach M, Gargiulo M, Welter ML, et al. Stimulation of the subthalamic nucleus in Parkinson's disease: a 5 year follow up. J Neurol Neurosurg Psychiatry. 2005;76:1640–4.
65. Kleiner-Fisman G, Herzog J, Fisman D, et al. Subthalamic nucleus deep brain stimulation: summary and meta-analysis of outcomes. Mov Disord. 2006;21 Suppl 14:S290–304.
66. Krack P, Pollak P, Limousin P, et al. Opposite motor effects of pallidal stimulation in Parkinson's disease. Ann Neurol. 1998;43:180–92.
67. Wu YR, Levy R, Ashby P, et al. Does stimulation of the GPi control dyskinesia by activating inhibitory axons? Mov Disord. 2001;16:208–16.

68. Bejjani B, Damier P, Arnulf I, Bonnet AM, Vidailhet M, Dormont D, Pidoux B, Cornu P, Marsault C, Agid Y. Pallidal stimulation for Parkinson's disease. Two targets? Neurology. 1997;49(6):1564–9.
69. Krack P, Pollak P, Limousin P, Hoffmann D, Xie J, Benazzouz A, Benabid AL. Subthalamic nucleus or internal pallidal stimulation in young onset Parkinson's disease. Brain. 1998;121(Pt 3):451–7.
70. Katayama Y, Oshima H, Kano T, Kobayashi K, Fukaya C, Yamamoto T. Direct effect of subthalamic nucleus stimulation on levodopa-induced peak-dose dyskinesia in patients with Parkinson's disease. Stereotact Funct Neurosurg. 2006;84(4):176–9.
71. Plaha P, Ben-Shlomo Y, Patel NK, Gill SS. Stimulation of the caudal zona incerta is superior to stimulation of the subthalamic nucleus in improving contralateral parkinsonism. Brain. 2006;129(Pt 7):1732–47.
72. Saint-Cyr JA, Hoque T, Pereira LC, Dostrovsky JO, Hutchison WD, Mikulis DJ, Abosch A, Sime E, Lang AE, Lozano AM. Localization of clinically effective stimulating electrodes in the human subthalamic nucleus on magnetic resonance imaging. J Neurosurg. 2002;97(5): 1152–66.
73. Moro E, Lozano AM, Pollak P, Agid Y, Rehncrona S, Volkmann J, Kulisevsky J, Obeso JA, Albanese A, Hariz MI, Quinn NP, Speelman JD, Benabid AL, Fraix V, Mendes A, Welter ML, Houeto JL, Cornu P, Dormont D, Tornqvist AL, Ekberg R, Schnitzler A, Timmermann L, Wojtecki L, Gironell A, Rodriguez-Oroz MC, Guridi J, Bentivoglio AR, Contarino MF, Romito L, Scerrati M, Janssens M, Lang AE. Long-term results of a multicenter study on subthalamic and pallidal stimulation in Parkinson's disease. Mov Disord. 2010;25(5): 578–86.
74. Deuschl G, Paschen S, Witt K. Clinical outcome of deep brain stimulation for Parkinson's disease. Handb Clin Neurol. 2013;116C:107–28.
75. Burchiel KJ, Anderson VC, Favre J, Hammerstad JP. Comparison of pallidal and subthalamic nucleus deep brain stimulation for advanced Parkinson's disease: results of a randomized, blinded pilot study. Neurosurgery. 1999;45(6):1375–82.
76. St George RJ, Nutt JG, Burchiel KJ, Horak FB. A meta-regression of the long-term effects of deep brain stimulation on balance and gait in PD. Neurology. 2010;75(14):1292–9.
77. Stacy MA, Murck H, Kroenke K. Responsiveness of motor and nonmotor symptoms of Parkinson disease to dopaminergic therapy. Prog Neuropsychopharmacol Biol Psychiatry. 2010;34(1):57–61.
78. Okun MS, Foote KD. Parkinson's disease DBS: what, when, who and why? The time has come to tailor DBS targets. Expert Rev Neurother. 2010;10(12):1847–57.
79. Bronstein JM, Tagliati M, Alterman RL, Lozano AM, Volkmann J, Stefani A, Horak FB, Okun MS, Foote KD, Krack P, Pahwa R, Henderson JM, Hariz MI, Bakay RA, Rezai A, Marks Jr WJ, Moro E, Vitek JL, Weaver FM, Gross RE, DeLong MR. Deep brain stimulation for Parkinson disease: an expert consensus and review of key issues. Arch Neurol. 2011; 68(2):165.
80. Kumar R. Methods for programming and patient management with deep brain stimulation of the globus pallidus for the treatment of advanced Parkinson's disease and dystonia. Mov Disord. 2002;17 Suppl 3:S198–207.
81. Volkmann J, Herzog J, Kopper F, Deuschl G. Introduction to the programming of deep brain stimulators. Mov Disord. 2002;17 Suppl 3:S181–7.
82. Tagliati M. Turning tables: should GPi become the preferred DBS target for Parkinson disease? Neurology. 2012;79(1):19–20.
83. Krack P, Fraix V, Mendes A, et al. Postoperative management of subthalamic nucleus stimulation for Parkinson's disease. Mov Disord. 2002;17 Suppl 3:S188–97.
84. Mazzone P. Deep brain stimulation in Parkinson's disease: bilateral implantation of globus pallidus and subthalamic nucleus. J Neurosurg Sci. 2003;47(1):47–51.
85. Deuschl G, Fogel W, Hahne M, Kupsch A, Müller D, Oechsner M, Sommer U, Ulm G, Vogt T, Volkmann J. Deep-brain stimulation for Parkinson's disease. J Neurol. 2002;249(Suppl 3):III/36–9.

Chapter 7
Basal Ganglia Circuitry Models of Levodopa-Induced Dyskinesia

Wai Kin D. Ko, Matthieu Bastide, and Erwan Bezard

Abstract L-3,4-dihydroxyphenylalanine (L-DOPA) treatment in Parkinson's disease (PD) patients commonly leads to dyskinesia, a hyperkinetic movement disorder that remains an unsolved clinical problem. The unravelling of key pathophysiological mechanisms in PD and dyskinesia has led to updated models of the basal ganglia motor circuit, capturing nonlinear neuronal information processing in a dynamical network architecture. Our understanding into the functional organization of the basal ganglia motor system is further supported by recent computational models that focus on neuronal activations within distinct closed feedback loops. Together, these models of the basal ganglia circuitry compose a more comprehensive and detailed insight into the diverse neuronal dysfunctions in the pathophysiology of PD and LID.

Keywords L-DOPA-induced dyskinesia • Parkinson's disease • Basal ganglia circuitry models

---

W.K.D. Ko
Department of Neurodegenerative Diseases, Université de Bordeaux, Institut des Maladies Neurodégénératives, UMR 5293, F-33000 146 rue Leo Saignat, Bordeaux, 33076, France

CNRS, Institut des Maladies Neurodégénératives, UMR 5293, Bordeaux, F-33000, France

M. Bastide
Department of Neurodegenerative Diseases, Université de Bordeaux, Institut des Maladies Neurodégénératives, UMR 5293, Bordeaux, France

CNRS, Institut des Maladies Neurodégénératives, UMR 5293, Bordeaux, F-33000, France

E. Bezard (✉)
Motac Neuroscience Ltd, Manchester, UK

Department of Neurodegenerative Diseases, Université de Bordeaux, Institut des Maladies Neurodégénératives, UMR 5293,
Bordeaux, France

CNRS, Institut des Maladies Neurodégénératives, UMR 5293, Bordeaux, F-33000, France
e-mail: erwan.bezard@u-bordeaux2.fr

S.H. Fox, J.M. Brotchie (eds.), *Levodopa-Induced Dyskinesia in Parkinson's Disease*, DOI 10.1007/978-1-4471-6503-3_7

Parkinson's Disease and Levodopa-Induced Dyskinesia

Parkinson's disease (PD) is a progressive neurodegenerative disorder that affects approximately 1 % of the population over the age of 55 years, with highest prevalence in ages of 85 years and over [1]. PD is commonly characterized by a clinical syndrome of motor symptoms (bradykinesia, postural abnormalities, and resting tremor) [2] that occur due to an extensive loss of nigrostriatal neurons which release dopamine [3], a modulatory neurotransmitter of the basal ganglia motor circuit [4].

In the early 1960s, studies showed that dopamine replacement with its immediate metabolic precursor, L-3,4-dihydroxyphenylalanine (L-DOPA), dramatically alleviated PD motor symptoms [5, 6]. Following this, L-DOPA was introduced to PD patients [7] and has since been widely used for the treatment of PD. However, following long-term use of L-DOPA, the initial beneficial effects of treatment are compromised by unpredictable "on-off" fluctuations of therapeutic effects [8, 9], gradual "wearing off" of therapeutic efficacy [10, 11], and L-DOPA-induced dyskinesia (LID) [12]. The latter is a severe hyperkinetic motor complication that is commonly expressed as an idiosyncratic mixture of chorea (irregular flow of muscular movements in rapid and slow phases) and dystonia (slow twisted movements from abnormal muscular contractions) [13]. LID occurs in approximately 90 % of PD patients after 9 years of L-DOPA treatment [14], and once established, dyskinesia is elicited upon each administration of L-DOPA, or dopamine agonist [15]. Moreover, LID increases in severity with further L-DOPA treatment [16] and can become as debilitating as PD itself, causing a negative impact on quality of life [17].

Understanding the pathophysiology of LID is an important step in developing a suitable treatment that can resolve the clinical need of treating dyskinesia. In this review, we discuss the pathophysiology of PD and LID using the basal ganglia circuitry model of the motor circuit. We also describe a recent computational model that demonstrates subtle dysfunctions in neural processing within the basal ganglia following the loss of dopamine. In addition, we highlight recent experimental findings of molecular adaptations that occur in the nuclei outside of the basal ganglia, which may have important roles in the expression of LID.

Basal Ganglia

The basal ganglia are a group of subcortical nuclei that include the striatum (caudate nucleus and putamen), subthalamic nucleus (STN), substantia nigra (pars reticulata, SNr, and pars compacta, SNc), ventral tegmental area, and globus pallidus (internal, GPi, and external, GPe, segments) [18]. These interconnected nuclei are modulated by dopamine and together form a neural network that relays information from the cortex to the thalamus. These so-called corticobasal ganglia-thalamocortical loops functionally convey information for both motor and non-motor processes [4]. Several of these loops exist for motor, oculomotor, associative, limbic, and orbitofrontal functions. Moreover, each loop projects from largely segregated regions of the basal ganglia and thalamus to different cortical target areas of the cerebral hemisphere [19].

Classic "Input" and "Output" Stations of the Basal Ganglia

The striatum is a major input station of the basal ganglia receiving different afferent projections, which include dopaminergic fibers from the midbrain [20], serotonergic fibers from dorsal and medial raphe nucleus [21], noradrenergic fibers from the locus coeruleus [22], acetylcholinergic fibers from the pedunculopontine nucleus, and glutamatergic fibers from the thalamus, STN, and cortex [23, 24]. The glutamatergic fibers of the cortex project massively to the striatum in a somatotopically organized manner [25–27]. In the motor cortico-basal ganglia-thalamo-cortical loop of the primate brain, sensorimotor afferents of the primary motor and somatosensory cortices project to the posterolateral putamen [26, 28]. Here, the dorsal region is occupied by somatotopic representation of the leg, which is followed by the arm, while the facial representation lays most ventral [28]. The putamen projects via γ-aminobutyric acid (GABA)-ergic medium spiny neurons (MSNs) to the GPi/SNr [29, 30], which are the output nuclei of the basal ganglia. These nuclei send GABAergic efferent neurons to the motor nuclei of the thalamus (ventralis anterior and lateralis) and brain stem [19, 31–33]. In turn, the motor nuclei convey excitatory glutamatergic projections to motor-related cortical areas, completing the motor corticobasal ganglia-thalamocortical loop [34–37].

Striatal "Direct" and "Indirect" Pathways

In the 1980s, a model of the basal ganglia circuitry was proposed based on the available anatomical, neurochemical, and electrophysiological data (see Fig. 7.1a) [19, 32]. This now "classic" model describes two main efferent projections from the striatum to the output of the basal ganglia, the so-called direct and indirect pathways. The direct pathway refers to the monosynaptic neuronal connection between the striatum and GPi/SNr. The neurons of this pathway primarily express dopamine D_1 receptors and preproenkephalin-B (PPE-B), an opioid peptide that is subsequently cleaved to produce co-transmitters substance P, dynorphins, leucine-enkephalins, and α-neoendorphin [38]. The "indirect" pathway describes the polysynaptic neuronal connection of the striatum to GPi/SNr. Striatofugal neurons of this pathway project to the GPe and, in turn, the GPe sends GABAergic efferent fibers to the STN. From here, glutamatergic efferent fibers of the STN project to the GPi/SNr. The striatopallidal neurons of the indirect pathway primarily express dopamine D_2 receptors and preproenkephalin-A (PPE-A), an opioid peptide that is subsequently cleaved to enkephalin [38]. Both the direct and indirect pathways are modulated by dopamine, which activates striatonigral neurons of the direct pathway and inhibits striatopallidal neurons of the indirect pathway (see Fig. 7.1a).

Based on segregated pathways, the classic functional model of the basal ganglia describes the processing of neural information in a feed-forward manner for achieving a behavioral outcome. Using the motor circuit as an example, it was suggested that the direct pathway facilitates the execution of desired motor sequences, while the indirect pathway mediates blocking of unwanted motor programs to "smooth" cortical-initiated motor sequences [4, 39–41]. Both the direct and indirect pathways lead to inhibition of the basal ganglia output nuclei for normal motor function. Accordingly, electrophysiological studies of saccadic eye and limb movements in awake monkeys have shown GPi/SNr neurons are tonically active (50–100 Hz) during rest and exhibit reduced activity during movement [42–45].

Basal Ganglia Circuitry in Parkinson's Disease

The classic model of the basal ganglia circuitry has been used to describe the pathophysiology of PD (see Fig. 7.1b) [32, 33, 46]. From this model, PD motor symptoms occur as a result of an imbalance between direct and indirect pathways caused by extensive degeneration of nigrostriatal dopaminergic neurons. While striatonigral neurons of the direct pathway become underactive, striatopallidal neurons of the indirect pathway become overactive leading to inhibition of GPe and subsequent disinhibition of the glutamatergic efferent fibers of the STN [32]. Thus, with loss of dopamine, both pathways lead to increased activation of the GPi/SNr, thereby inhibiting the motor thalamic nuclei. The resulting effect is reduced activation of motor cortical areas, which is seen to occur in the primary sensory motor cortex [47] and supplementary motor area [48] in the parkinsonian state.

In the late 1980s, several groundbreaking studies were conducted that helped uncover key mechanisms in the pathophysiology of PD. In these experiments conducted by Mitchell et al., neuronal metabolic marker 2-deoxyglucose (2-DG) was used to reveal the activity states of the basal ganglia subnuclei in the 1-methyl-4-phenyl-1,2,3,6-tetrahydropyridine (MPTP)-lesioned nonhuman primate (nhp) model of PD. It was found that the STN was hyperactivated, while the GPe, thalamic ventralis anterior, and lateralis nuclei were hyper-inhibited [49–51]. Accordingly, these major discoveries suggested that there was hyperactivation of the basal ganglia output structures in PD [51], which was later confirmed through measurements of electrophysiological activity [52–54] and mRNA expression of neuronal activity marker, cytochrome oxidase subunit I [55]. Additionally, the activation states of the striatofugal pathways in PD have been demonstrated through the expression of striatal PPE precursors, where reports have consistently shown reduced PPE-B and increased PPE-A mRNA expression in the striatum [56–58]. These molecular data demonstrate underactivation of striatonigral neurons and

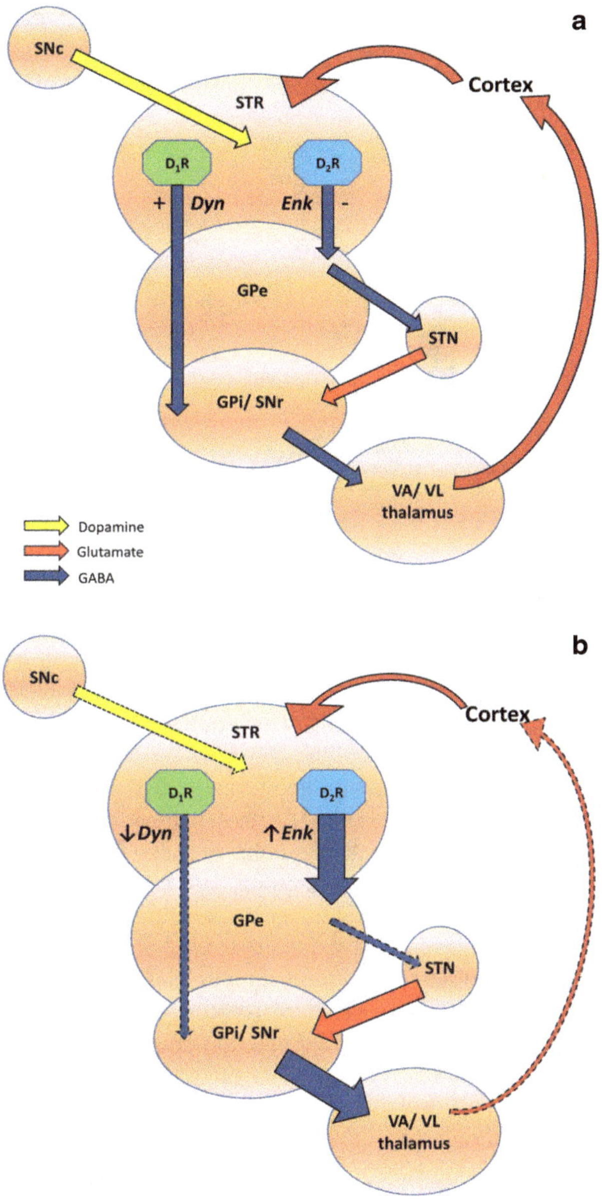

Fig. 7.1 Schematic diagrams of the classic basal ganglia circuitry model illustrating linear, feed-forward information processing. Dopamine mediates opposing functional effects on the two major projection pathways of the striatum for (**a**) normal motor function. Loss of endogenous dopamine in (**b**) Parkinson's disease (PD) causes abnormal neuronal activity leading to reduced excitatory feedback to the cortex. Repeated treatment with L-DOPA in PD induces (**c**) dyskinesia, causing increased activity in the cortex. Arrow size corresponds to activity of neuronal projections. l-DOPA L-3,4-dihydroxyphenylalanine, *D_1R* dopamine D_1 receptor, *D_2R* dopamine D_2 receptor, *Enk* enkephalin, *Dyn* prodynorphin, *STR* striatum, *GPi* internal segment of the globus pallidus, *GPe* external segment of the globus pallidus, *STN* subthalamic nucleus, *SNr* substantia nigra pars reticulata, *SNc* substantia nigra pars compacta, *VA/VL* ventralis anterior and lateralis nuclei

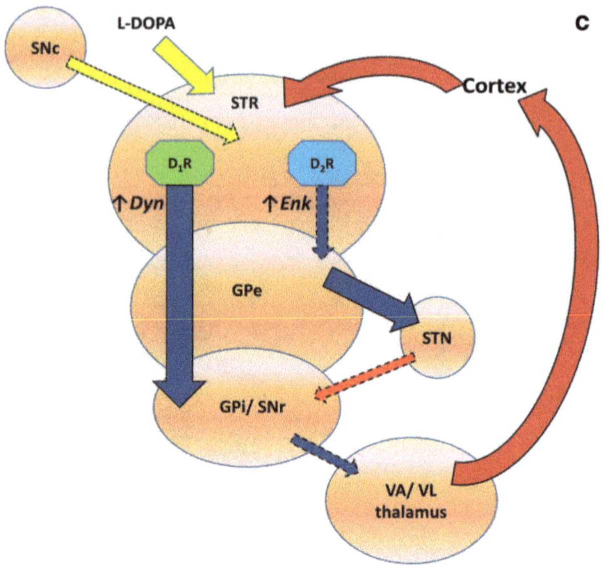

Fig. 7.1 (continued)

hyperactivation of striatopallidal neurons, which both favor the overactivation of the basal ganglia output nuclei in PD.

Following the original neuronal metabolic activity studies, behavioral experiments carried out in MPTP-lesioned nhps further characterized the pathophysiological changes in PD. Fundamentally, these studies revealed the causal role of the STN in production of parkinsonian motor symptoms [59, 60], which were dramatically abolished following surgical or neurochemical (muscimol or kainic acid) lesion of this structure. At the cellular level, subthalamotomy was also shown to reduce the overactivation of the basal ganglia output nuclei in PD [61, 62]. Collectively, these revolutionary findings led to a resurgence of neurosurgical procedures for the treatment of parkinsonism, which included the ablation of the GPi [63, 64] or STN [65], and deep brain stimulation (DBS) of the STN [66–68].

Basal Ganglia Circuitry in Levodopa-Induced Dyskinesia

Early suggestions put forward on the pathophysiology of LID essentially described the opposite functional state to that of PD (see Fig. 7.1c) [69]. Particular emphasis was originally placed on the indirect pathway in the pathogenesis of the dyskinetic state, where it was proposed that the underactivation of the striatopallidal neurons caused disinhibition of the GPe, leading to subsequent over-inhibition of the STN. In turn, disinhibition of thalamic motor nuclei resulted in excessive excitatory input to motor cortical areas, which is found to occur in PD patients expressing LID [70, 71]. Pioneering experiments, again conducted by Mitchell et al. [72] in MPTP-lesioned nhps, showed that at peak dose of dopamine agonist-induced dyskinesia, there was an increased uptake of 2-DG in the STN and GPi, demonstrating that these structures were hyper-inhibited. This study also showed that 2-DG was reduced in the motor thalamic nuclei, reflecting its hyperactivated state in dyskinesia [72].

The role of the direct pathway in the pathogenesis of LID was later emphasized by Bezard et al. [86]. In this key review, it was suggested that underactive/abnormal firing of the basal ganglia output nuclei in dyskinesia [52, 73–78] was primarily caused by overactivated striatonigral neurons of the direct pathway. Indeed, functional hyperactivation of the direct pathway in LID has been demonstrated at the cellular level from (1) dramatic elevations of striatal mRNA expression of PPE-B and prodynorphin [58, 79–82] and (2) supersensitization of striatal dopamine D_1 receptors [83]. In addition, it has been reported that treatment with selective dopamine D_1 receptor agonist, ABT-431, in PD patients elicits dyskinesia to a similar extent to that of L-DOPA [84], supporting the hypothesis of a hyperactivated direct pathway in the pathogenesis of dyskinesia. It should be noted that these data are inline with the mechanism suggested in the classic functional model, whereby an overactivated direct pathway mediates over-inhibition of the basal ganglia output, causing the underactivation of these nuclei (see Fig. 7.1c).

On the contrary, the proposed underactivation of the indirect pathway in the pathophysiology in LID has been, somewhat, inconsistent with several experimental findings, which has presented some limitations of the classic functional model (discussed in more detail in the section below). For example, the underactivation of the indirect pathway due an overactivated GPe is not consistently seen in dyskinetic MPTP-lesioned nhps [55]. In addition, striatopallidal neurons of the indirect pathway are not underactive, as demonstrated by the levels of striatal PPE-A mRNA, which are actually further upregulated, rather than downregulated, in dyskinesia compared to PD [79, 85]. It has since been suggested that increased striatal PPE-A mRNA in LID may occur due to reduced parkinsonism following L-DOPA treatment, rather than LID itself [86]. This is consistent with clinical findings that have shown dopamine D_2 receptor agonists are effective antiparkinsonian agents with a reduced risk of inducing dyskinesia [87].

At this point, it is worth mentioning that the classic functional basal ganglia model has provided an excellent basis for describing the functional mechanisms involved in normal and disease states (see Fig. 7.1a–c). However, the classic model remains too simplistic, and its use is limited when describing the pathophysiological mechanisms in PD and LID. In the next section, we outline some of the main inconsistencies that have arisen between experimental data and the classic functional basal ganglia model.

Developments of the Basal Ganglia Circuitry Model

Progressing from the original descriptions of the classic functional model by Alexander and Crutcher [4], experimental reports have revealed a greater complexity in the neural organization and information processing within the basal ganglia. These data have led to the development of the functional model, which has engaged the highly dynamic nature of neural networking in the basal ganglia circuitry.

Organization and Structure

In the classic model, the separate direct and indirect pathway organization has been widely accepted, but the actual degree of segregation and opposing functional activity of the striatal neuronal projections remains unclear [88]. Firstly, striatofugal axons show consistent collateralization to both the GPe and GPi [89], suggesting an interconnected, rather than segregated, organization. Secondly, the response of the striatofugal pathways to dopamine cannot be simply viewed as an activating or deactivating effect caused exclusively via actions on dopamine D_1 or D_2 receptors, respectively, as (1) a high percentage of striatal MSNs expresses both subtypes of dopamine receptors [90–92] and (2) because dopamine D_1 and D_2 receptor responses are not consistently opposite [93, 94]. It is also worth mentioning that the modulatory effects of dopamine in the basal ganglia are not only restricted to the striatum. In fact, extensive dopaminergic SNc projections are found to innervate most, if not all, of the other basal ganglia subnuclei [95–100]. Thus, taking this into consideration, another level of complexity is added, as these dopaminergic innervations can bypass the feed-forward processing mediated by striatofugal neuronal activity [101–103].

From the early 2000s, the mechanism of neural processing in the basal ganglia has been reevaluated [104, 105]. The concept of a linear feed-forward mechanism, solely based on altered firing rate of each basal ganglia subnucleus, has been found to be inconsistent with preclinical and clinical data, sparking the reorganization of the basal ganglia circuitry. In particular, the model now incorporates the numerous internal feedback loops [104, 106], which have been found to exist through reciprocal connections between the many of the basal ganglia subnuclei [107–110]. This reformed organization of basal ganglia circuitry has brought drastic changes to the arrangement of the classic indirect pathway, which include (1) the GPe as now occupying a central position and being viewed as a key structure for inhibitory modulation of the striatum, GPi and STN (see Fig. 7.2) [111–113], and (2) the STN being considered as another major input station of the basal ganglia, receiving afferent glutamatergic projections from the cortex [114, 115] and thalamus [116], while sending glutamatergic efferent projections to the GPe, GPi/SNr, ventral thalamic nuclei, and pedunculopontine nucleus [117, 118]. Importantly, the overall structural reorganization of the motor circuit now sees the functional dual disynaptic control

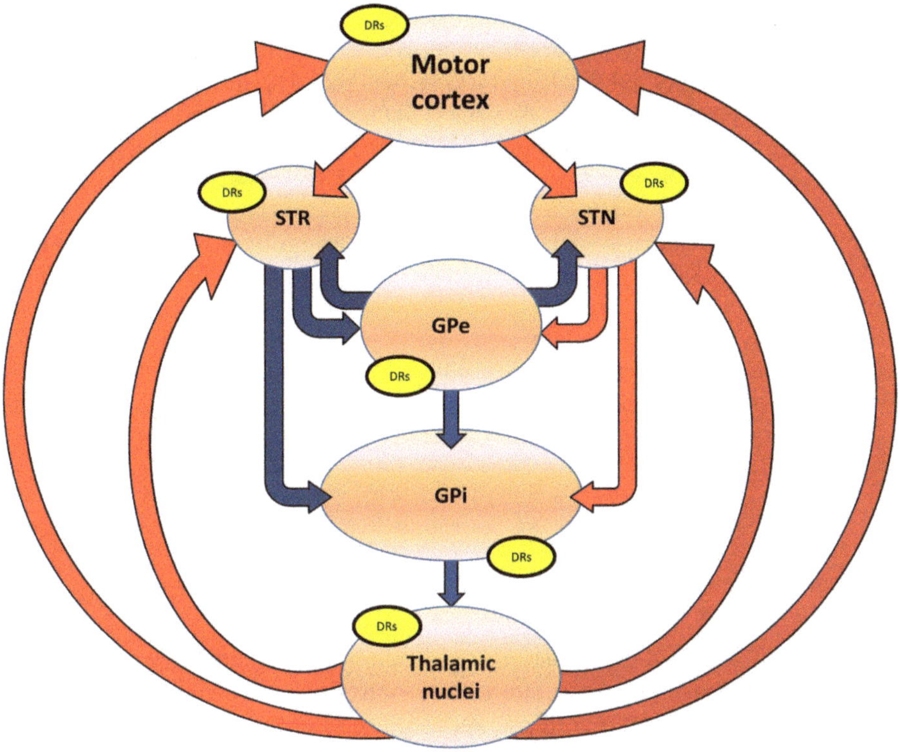

Fig. 7.2 A schematic diagram showing the functional organization of the basal ganglia. This updated model proposed by Obeso et al. [113] illustrates the dual disynaptic control of the internal and external segments of the globus pallidus, originating from corticostriatal and cortico-subthalamic projections. The position of the external segment of the globus pallidus in this model has been emphasized to mediate important inhibitory control of the basal ganglia output nuclei, while modulating the activity of the striatum and subthalamic nucleus via reciprocal connections. *DRs* dopamine receptors, *STR* striatum, *GPi* internal segment of the globus pallidus, *GPe* external segment of the globus pallidus, *STN* subthalamic nucleus

of the GPe and GPi (see Fig. 7.2) [104, 113] via parallel cortical projections to the striatum and STN. The corticostriatal projection uses the striatum to mediate inhibitory control of the pallidal segments, while the cortico-STN projection, also known as the hyper-direct pathway [115], uses the STN to mediate fast excitatory input to these structures [119].

Mechanism of Neural Information Processing

The mechanism of neural processing in the classic functional model, which is based only on the firing rate of each individual subnucleus in the basal ganglia, is unable to explain several major experimental findings in PD and LID. Most notably, simple underactivation of the basal ganglia output nuclei in dyskinesia [73, 76, 77] cannot fully explain the pathogenesis of LID [75]. This is because lesion of the GPi does not result in dyskinesia [120, 121]. In fact, pallidotomy of the internal segment effectively alleviates LID in MPTP-lesioned nhps [122] and PD patients [123, 124], which is opposite to the proposed outcome of the classic functional model. Other major inconsistencies of the classic model have been identified in PD, which include: (1) lesions of the motor thalamic nuclei do not exacerbate parkinsonism in patients [120] and (2) lesions of the GPe do not induce PD motor symptoms [54]. Collectively, these data have indicated that the processing of neural information for motor behavior in the basal ganglia is much more complex than originally thought [19], which cannot rely exclusively on firing rates. Instead, the firing rate model has been developed to incorporate the functional roles of neuronal firing patterns, such as synchronicity and oscillatory activations, in motor function, which are characteristically different in established PD and LID [75, 125–127].

Neuronal Firing Patterns in the Basal Ganglia

The study of neuronal firing patterns is commonly conducted through electrophysiological measurements of single/multiunit activity or local field potentials (LFPs). Single/multiunit recordings show the action potentials of one or multiple neurons, while LFPs typically reflect subthreshold synchronized afferent activations of a larger group of neurons [128]. These sets of neuronal data are analyzed for either synchronous or oscillatory patterns of activity, which are determined using a range of statistical tools on the time and/or frequency domains [129]. While the oscillatory activity can modulate neuronal synchronization in cortical and subcortical regions [130, 131], these patterns of neuronal activity are not mutually exclusive, i.e., synchronized firing can occur in the absence of periodic firing or vice versa [129]. The presence of synchronized activity between networks in distinct neuroanatomical regions is hypothesized to "bind" or "couple" neural ensembles, as part of a wider functional integration process [130–132]. Such synchronized firing is suggested to be a mechanism of neural processing in cortical and thalamic regions. On the contrary, the function of the basal ganglia, as an intermediate in corticobasal ganglia-thalamocortical loop, has been suggested to mediate so-called dimensional reduction, which describes the funnelling of redundant cortical neuronal inputs for efficient action planning [133]. This has been indicated from electrophysiological studies conducted in normal animals that have revealed the activity of neurons in the GPe [134], GPi [135], and STN [136] are generally asynchronous, with approximately 90 % or more of recorded neurons displaying uncorrelated firing patterns [53, 137, 138]. The oscillations in spike activity of MSNs in the striatum are typically weak [139, 140].

Abnormal Neuronal Firing Patterns in the Pathophysiology of PD and LID

Following original reports of abnormal neuronal firing patterns in experimental parkinsonism [53, 141], patterns in burst firing, synchronization, and oscillatory activity have been well studied in PD and LID. Initial experiments conducted single-unit measurements of neuronal activity in the subnuclei of the basal ganglia in MPTP-lesioned nhps and showed that the incidence of burst firing was increased in the GPe, GPi, and STN [53, 141, 142]. Additionally, neurons within these structures were found to demonstrate hyper-synchronized oscillatory activity that was characteristically <30 Hz (within the β (beta)-band) [53, 135, 143]. Similar findings have been reported in clinical studies, following measurements of LFPs via macroelectrodes during neurosurgery [125]. Although these data are not directly comparable to single-unit measurements, LFPs in the STN and GPi of PD patients in the off-state showed dominant low-frequency (<30 Hz) oscillations, with increased coherence in activity (at 6 and 20 Hz) between these structures. The origin of these low-frequency oscillations in PD has been suggested to arise from an abnormal network effect following extensive loss of dopamine in the basal ganglia, which may occur from imbalanced activity between the direct and hyper-direct pathways [105] or due to rebound firing of STN [144–146] caused by abnormal inhibitory input from the GPe [147–149]. The STN is likely to impose enhanced β (beta)-band oscillations on the GPi through its direct synaptic connection, which then reverberates through the motor corticobasal ganglia-thalamocortical loop, as indicated from coherent β (beta)-band oscillations between the basal ganglia subnuclei and motor cortical regions [150–153].

The functional relevance of enhanced β (beta)-band oscillations in the motor corticobasal ganglia-thalamocortical loop has been postulated to disrupt information processing, contributing to the expression of PD motor symptoms [154]. In-line with this suggestion, low-frequency (5–20 Hz) stimulations of the STN typically worsen akinesia in PD patients [155–157], while neuronal firing of GPi cells at 4–6 Hz in MPTP-lesioned nhps [53] and PD patients [158] has been correlated to the frequency of resting tremor. Moreover, studies have shown that treatment with dopaminergic agents in PD suppresses the low-frequency β (beta)-band oscillations in the basal ganglia [125, 159, 160] and motor cortical regions [151, 152], causing several marked changes in neuronal activity, such as the uncoupling of high-frequency oscillations (HFO) (>300 Hz) to low-frequency β (beta)-band oscillations [160] and a shift to a new prominent peak in activity at ~70 Hz (γ (gamma)-band) [125], as parkinsonism is alleviated. The presence of γ (gamma)-band oscillations (70–85 Hz) in the basal ganglia, particularly the STN, may be reflective of an improved motor state in PD as clinical studies have shown (1) HFS of the STN >70 Hz alleviates PD motor symptoms in patients [155] which also suppresses low-frequency β (beta)-band oscillations in the GPi [161] and (2) increased coherence between the STN and GPi in the γ (gamma)-band frequency following L-DOPA treatment in PD patients, which also augments with movement [151]. However, in

PD patients that exhibit dyskinesia following treatment with dopaminergic medication, different patterns of neuronal activities are induced [127, 162]. While LFPs in the STN of these patients do show increased (17.8 %) logarithmic power of activity in the γ (gamma)-frequency range, there is a more striking increment (77.6 %) in activity at 4–10 Hz (θ (theta)/α (alpha)-band) that is specifically associated with dyskinesia [127, 163]. Clinical data have also revealed that the coherence between the STN and GPi at <10 Hz is increased in the dyskinetic state [162].

As discussed above, neuronal patterns of burst firing, synchronization, and oscillatory activity within the motor corticobasal ganglia-thalamocortical loop are associated with different motor states. It has been postulated that tonic levels of endogenous dopamine in the normal basal ganglia may mediate desynchronized neuronal firing for the processing of motor commands [126, 164]. However, in PD, when there is extensive loss of dopamine, these motor commands are not, or inefficiently, processed within the basal ganglia. Prominent changes, such as enhanced hyper-synchronization and oscillatory patterns of firing at β (beta)-band frequencies, may represent increased threshold levels of activity, acting as a "barrier" that ultimately impedes information processing for movement [126, 164]. On the contrary, hyper-synchronization of neuronal activity at θ (theta)/α (alpha)-band frequencies, as recorded in LID [127, 162, 163], may allow for release of involuntary motor sequences that become expressed as dyskinesia. Thus, neurosurgical procedures for symptomatic treatments of PD and LID can be viewed as a method of alleviating, or resetting, abnormal subcortical activations [165], allowing the resumption of neuronal processing for motor programming in the absence of an endogenous dopamine tone. In the next section, we discuss a recently developed computational model of the basal ganglia circuit for action selection, which describes the loss of motor function and emergence of low-frequency oscillations following striatal dopamine denervation.

Computational Models of the Basal Ganglia Motor Circuit

Although the precise functional consequences of abnormal neuronal firing patterns in the basal ganglia in parkinsonian and dyskinetic states remain unclear, recent advances have been made in our understanding of neural processing in the expression of the motor symptoms. Studies conducted by Boraud's group identified key functional changes in the basal ganglia output nuclei that related to the onset of parkinsonism [166]. Their work demonstrated that hyper-synchronized β (beta)-band oscillations in the GPi occurred following the establishment of PD motor symptoms in MPTP-lesioned nhps [166]. In the same study, the authors identified that onset of experimental parkinsonism was closely related to a shift in the firing profile of GPi neurons, where there was an increased (~1.5-fold) proportion of excitable neurons and a decreased (~0.5-fold) number of functional inhibitory neurons [166]. Moreover, in an earlier study conducted by the same research group, GPi neurons in MPTP-lesioned nhps were found to have altered firing activities that were related to the spatiotemporal aspects of motor processing [167]. In these MPTP-treated nhps, it was shown that the number of GPi neurons responsive to manipulated limb movements was increased (~5-fold), demonstrating a loss of somatosensory selectivity [167]. Neurons within the GPi also displayed premature firing in relation to onset of muscular activity, suggesting dysfunctional neural processing for movement [167]. These key experimental findings indicated that changes in neural activity at the level of the GPi could be instrumental in disrupting motor processing in action selection, leading to the motor disabilities seen in PD.

In 2006, Boraud's team put forward a dynamic computational model of the basal ganglia network that described the neural processing for action selection from closed feedback loops (see Fig. 7.3) [105]. Interestingly, in-line with experimental findings in MPTP-treated nhps [166], this computational model described how reduced dopamine levels caused loss of action selection that correlated with a shift in the proportion of activated neurons in the GPi, prior to the development of synchronized low-frequency oscillations within the basal ganglia [105]. This functional model of the basal ganglia uses two main feedback loops that are each arranged in a somatotopic manner [168–171]: (i) the hyperdirect pathway (cortex-STN-GPi-thalamus-cortex) [114] and (ii) the direct

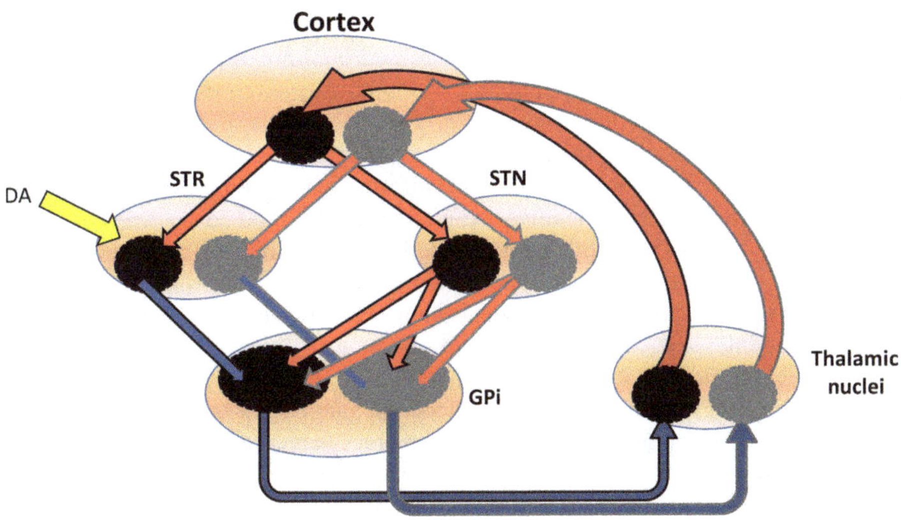

Fig. 7.3 A schematic diagram of basal ganglia neuronal connections in a model for action selection proposed by Leblois et al. [105]. In this circuit, two neuronal populations (*black and gray*), each composed of (i) cortex-STR-GPi-thalamus-cortex and (ii) cortex-STN-GPi-thalamus-cortex pathways compete to execute action selection. This is achieved when activity in of one cortical population overcomes threshold activity in the other cortical population. Importantly, projections from the STN cross over, modulating GPi neuronal activity in the competing loop. Dopamine in the striatum mediates potentiation of corticostriatal synapses, strengthening the activity of a specific striatal projection. This can lead to increased feedback to the cortex in the same neuronal population to cause action selection. *DA* dopamine, *STR* striatum, *GPi* internal segment of the globus pallidus, *STN* subthalamic nucleus

pathway (cortex-striatum-GPi-thalamus-cortex). These two pathways have opposing effects on cortical activity; the hyper-direct inhibits the thalamic nuclei which causes reduced cortical firing (global negative), while the direct pathway disinhibits the thalamic nuclei leading to increased cortical firing (global positive). Leblois et al. [105] described two somatotopic channels from one corticobasal ganglia-thalamocortical loop (each composed of two feedback loops) that act in parallel and compete to execute action selection (see Fig. 7.3) [170–172]. This is achieved when the activity of one cortical population overcomes the threshold activity of that in the other. Importantly, the projections from the STN to the GPi have pivotal roles in the execution of a motor program, as the STN mediates "cross-path" activity, modulating GPi neuronal activity in its own loop and also in the competing circuit [105]. This functional model also describes the effects of dopamine in the striatum, where it mediates potentiation of corticostriatal synapses in sensorimotor regions [173], strengthening direct pathway activity for biasing selection of the motor program in the corresponding circuit. When dopamine levels are normal, i.e., 100 %, the computational model demonstrates how action selection can occur as "symmetry breaking," the transition of activity when one neuronal population becomes greater than the other, takes place. Such a situation arises when motor planning information, sent from sensorimotor cortical areas, produces strong positive feedback activity in the direct pathway and inhibitory feedback activity in the hyper-direct pathway, causing asymmetric activations of neuronal populations in the GPi. In turn, cortical activity in one population is enhanced, while the other is attenuated leading to the selection of an action [105].

An extension to the computational model proposed by Leblois et al. has recently been described for the processing of neural information in the basal ganglia for two level decision-making (i.e., cognitive and motor) [174]. Based on electrophysiological data in nhps [175], the updated model demonstrated how task-related decisions made at the cognitive level can influence the motor level for action selection. The model architecture of the updated model is more sophisticated, describing two action selection modules, i.e., one cognition and one motor, which act in parallel. Each action selection module arises from distinct regions of the cortex, consisting of the direct and hyper-direct pathways, in a corticobasal ganglia-thalamocortical loop. For each loop, channels composed of separate ensembles in cortical areas are representative of decision choices that compete for action selection [174]. This computational model incorporates the idea of multiple corticobasal ganglia-thalamocortical loops for different aspects of neural processing [4], which interact in the striatum as afferent fibers converge in specific overlapping regions [174]. In this model, symmetry breaking for action selection can be initiated by internal noise prior to learning, which is followed by dopamine-mediated effects at corticostriatal synapses. Subsequently, synaptic gain in the direct pathway at the striatal level mediates positive feedback of that channel, while negative feedback of the hyper-direct pathway suppresses competing channels [174], in a center-surround inhibitory fashion [176]. Thus, activities of both circuits promote action selection of a specific channel.

These novel dynamic computational models of neural networks may prove to be important tools in the study of basal ganglia disorders. In the original computational model, Leblois et al. [105] demonstrated that striatal dopamine denervation leads to complete loss of action selection ability. Initial changes included a marked reduction in the ratio of inhibited GPi neurons by the direct pathway, which occurred following ~30 % dopamine loss [105]. As a result, feedback from the direct pathway was reduced, preventing the mechanism of symmetry breaking. Interestingly, after approximately 70–80 % striatal dopamine denervation, the inability of the direct pathway to counteract negative feedback of the hyper-direct pathway resulted in synchronized oscillatory neuronal activity (frequency of 10–12 Hz) [105]. As these predictions are in-line with experimental findings in PD [166], the model provides an excellent tool for studying the pathogenesis of disease states, with the advantage of incorporating more parallel loops and additional anatomical subnuclei [105, 174]. Although the pathophysiological changes in LID have yet to be modelled in these computational models, it would be particularly interesting to investigate whether synchronized θ (theta)/α (alpha)-band oscillations are produced in the basal ganglia in the dyskinetic state, as reported in patients [127, 162]. Speculatively, the dyskinesia could be modelled by incorporating pathophysiological hallmarks of LID, such as dysfunctional LTP at corticostriatal synapses [177, 178]. If this is possible, the current computational model could help elucidate the precise consequences of abnormal neuronal oscillatory activations in the basal ganglia subnuclei on action selection or identify the subtle changes in firing activities that lead to the expression of dyskinesia. In addition, we suggest that future basal ganglia models should be extended for describing the pathophysiology of LID. This is because recent studies have demonstrated molecular and functional adaptations associated with the expression of dyskinesia also occur in anatomical regions beyond the subnuclei of the basal ganglia. In a recent study conducted by Halje et al. [179], it was shown that dyskinetic motor symptoms in the unilateral 6-hydroxydopamine (6-OHDA)-lesioned rat model of LID were alleviated as abnormal motor cortical oscillations (80 Hz) were attenuated, following application of a dopamine D_1 receptor antagonist to specific cortical regions. These data, as well as our recent findings [180], highlight the need to look beyond the basal ganglia subnuclei for functional changes that can directly impact motor function.

Additional Nuclei in the Pathophysiology of LID

Molecular changes in the pathophysiology of LID have been well studied in the basal ganglia subnuclei [181], but little remains known of the adaptations that occur in other structures. A previous report identified the bed nucleus of the stria terminalis (BST) was hyperactivated in dyskinetic MPTP-lesioned nhps [182], suggesting a potential role of this structure in the pathophysiology of LID. Using the unilateral 6-OHDA-lesioned rat model of LID, we recently investigated the molecular adaptations in the whole brain by quantifying the expression of four immediate early genes (IEGs) (ΔFosB, ARC, FRA2, Zif268/EGR1) [180]. We found that dyskinesia severity in L-DOPA-treated unilateral 6-OHDA-lesioned rats correlated to the overexpression of these specific IEGs in the following structures: oval (oBST), juxta (jBST), and medial (mBST) bed nucleus of the stria terminalis, lateral habenula (lHb), pontine nuclei (Pn), and cuneiform nucleus (CnF). Such molecular adaptations in these nuclei could stem from irregular activities of afferent fibers. For example, serotonergic afferents [183] to the oBST and jBST may facilitate unregulated fluctuations of dopamine release in LID [184–187], which is likely to cause an abnormal functional state of these nuclei [188]. The molecular adaptations in the lHb nuclei, a structure that projects to different monoaminergic regions including the serotonergic dorsal and medial raphe, could be involved in the aberrant release of dopamine from serotonergic (5-HT) terminals in LID, contributing to the pathogenesis of dyskinesia [189–192]. Further studies are currently being conducted to fully elucidate the functional roles of these additional nuclei in the pathogenesis of dyskinesia. It is important to note that recent findings from our group and others [179] highlight the need to evaluate regions outside of the basal ganglia for fully uncovering the pathophysiological mechanisms in LID.

Conclusions

The functional basal ganglia circuitry model for describing the pathophysiology of PD and LID has developed quite considerably over the past few decades. Critical evaluation of functional mechanisms has proved an important step in progressing from the original descriptions of basic box-arrow circuitry and feed-forward information processing [4] to more updated basal ganglia models, which have captured complex neural network connections [104] and the dynamic nonlinear neuronal processing in disease states [105]. While our understanding of the pathophysiological mechanisms of PD and LID motor symptoms remains incomplete, the road to uncovering subtle dysfunctional neuronal processes will undoubtedly be guided by accurately modelling the latest experimental findings. Recent technological advancements that allow for the simultaneous measurements of single-unit neuronal activity, whole body kinematics, and muscular activities in freely moving nhps [193] are likely to be at the forefront of relating specific motor abnormalities that occur in PD and LID to abnormal neural processing in the basal ganglia and other anatomical regions. By striving to understand the complex mechanisms involved, we hope to make solid progress in the development of novel clinical treatments for PD and LID, to ultimately improve the quality of life of patients suffering from these movement disorders.

References

1. de Rijk MC, Tzourio C, Breteler MM, Dartigues JF, Amaducci L, Lopez-Pousa S, et al. Prevalence of parkinsonism and Parkinson's disease in Europe: the EUROPARKINSON Collaborative Study. European Community Concerted Action on the Epidemiology of Parkinson's disease. J Neurol Neurosurg Psychiatry. 1997;62(1):10–5.

2. Marsden CD. Parkinson's disease. J Neurol Neurosurg Psychiatry. 1994;57(6):672–81.
3. Ehringer H, Hornykiewicz O. Distribution of noradrenaline and dopamine (3-hydroxytyramine) in the human brain and their behavior in diseases of the extrapyramidal system. Klin Wochenschr. 1960;38:1236–9.
4. Alexander GE, Crutcher MD. Functional architecture of basal ganglia circuits: neural substrates of parallel processing. Trends Neurosci. 1990;13(7):266–71.
5. Carlsson A, Lindqvist M, Magnusson T. 3,4-Dihydroxyphenylalanine and 5-hydroxytryptophan as reserpine antagonists. Nature. 1957;180(4596):1200.
6. Friedhoff AJ, Hekimian L, Alpert M, Tobach E. Dihydroxyphenylalanine in extrapyramidal disease. JAMA. 1963;184:285–6.
7. Birkmayer W, Hornykiewicz O. The L-3,4-dioxyphenylalanine (DOPA)-effect in Parkinson-akinesia. Wien Klin Wochenschr. 1961;73:787–8.
8. Fahn S. "On-off" phenomenon with levodopa therapy in Parkinsonism. Clinical and pharmacologic correlations and the effect of intramuscular pyridoxine. Neurology. 1974;24(5):431–41.
9. Duvoisin RC. Variations in the "on-off" phenomenon. Adv Neurol. 1974;5:339–40.
10. Shoulson I, Glaubiger GA, Chase TN. On-off response. Clinical and biochemical correlations during oral and intravenous levodopa administration in parkinsonian patients. Neurology. 1975;25(12):1144–8.
11. Marsden CD, Parkes JD. "On-off" effects in patients with Parkinson's disease on chronic levodopa therapy. Lancet. 1976;1(7954):292–6.
12. Cotzias GC, Papavasiliou PS, Gellene R. Modification of Parkinsonism–chronic treatment with L-dopa. N Engl J Med. 1969;280(7):337–45.
13. Fahn S. The spectrum of levodopa-induced dyskinesias. Ann Neurol. 2000;47(4 Suppl 1):S2–9; discussion S-11.
14. Ahlskog JE, Muenter MD. Frequency of levodopa-related dyskinesias and motor fluctuations as estimated from the cumulative literature. Mov Disord. 2001;16(3):448–58.
15. Rascol O. Medical treatment of levodopa-induced dyskinesias. Ann Neurol. 2000;47(4 Suppl 1):S179–88.
16. Fahn S, Oakes D, Shoulson I, Kieburtz K, Rudolph A, Lang A, et al. Levodopa and the progression of Parkinson's disease. N Engl J Med. 2004;351(24):2498–508.
17. Chapuis S, Ouchchane L, Metz O, Gerbaud L, Durif F. Impact of the motor complications of Parkinson's disease on the quality of life. Mov Disord. 2005;20(2):224–30.
18. Crossman AR, Neary D. Neuroanatomy: an illustrated colour text. Edinburgh: Churchill Livingstone; 2000.
19. Alexander GE, DeLong MR, Strick PL. Parallel organization of functionally segregated circuits linking basal ganglia and cortex. Annu Rev Neurosci. 1986;9:357–81.
20. Lavoie B, Smith Y, Parent A. Dopaminergic innervation of the basal ganglia in the squirrel monkey as revealed by tyrosine hydroxylase immunohistochemistry. J Comp Neurol. 1989;289(1):36–52.
21. Lavoie B, Parent A. Immunohistochemical study of the serotoninergic innervation of the basal ganglia in the squirrel monkey. J Comp Neurol. 1990;299(1):1–16.
22. Delfs JM, Zhu Y, Druhan JP, Aston-Jones GS. Origin of noradrenergic afferents to the shell subregion of the nucleus accumbens: anterograde and retrograde tract-tracing studies in the rat. Brain Res. 1998;806(2):127–40.
23. Groenewegen HJ, Galis-de Graaf Y, Smeets WJ. Integration and segregation of limbic cortico-striatal loops at the thalamic level: an experimental tracing study in rats. J Chem Neuroanat. 1999;16(3):167–85.
24. Castle M, Aymerich MS, Sanchez-Escobar C, Gonzalo N, Obeso JA, Lanciego JL. Thalamic innervation of the direct and indirect basal ganglia pathways in the rat: Ipsi- and contralateral projections. J Comp Neurol. 2005;483(2):143–53.
25. Kunzle H. Bilateral projections from precentral motor cortex to the putamen and other parts of the basal ganglia. An autoradiographic study in Macaca fascicularis. Brain Res. 1975;88(2):195–209.

26. Kunzle H. Projections from the primary somatosensory cortex to basal ganglia and thalamus in the monkey. Exp Brain Res. 1977;30(4):481–92.
27. McGeorge AJ, Faull RL. The organization of the projection from the cerebral cortex to the striatum in the rat. Neuroscience. 1989;29(3):503–37.
28. Romanelli P, Esposito V, Schaal DW, Heit G. Somatotopy in the basal ganglia: experimental and clinical evidence for segregated sensorimotor channels. Brain Res Brain Res Rev. 2005;48(1):112–28.
29. Somogyi P, Smith AD. Projection of neostriatal spiny neurons to the substantia nigra. Application of a combined Golgi-staining and horseradish peroxidase transport procedure at both light and electron microscopic levels. Brain Res. 1979;178(1):3–15.
30. Somogyi P, Bolam JP, Totterdell S, Smith AD. Monosynaptic input from the nucleus accumbens–ventral striatum region to retrogradely labelled nigrostriatal neurones. Brain Res. 1981;217(2):245–63.
31. Carpenter MB, Nakano K, Kim R. Nigrothalamic projections in the monkey demonstrated by autoradiographic technics. J Comp Neurol. 1976;165(4):401–15.
32. Albin RL, Young AB, Penney JB. The functional anatomy of basal ganglia disorders. Trends Neurosci. 1989;12(10):366–75.
33. DeLong MR. Primate models of movement disorders of basal ganglia origin. Trends Neurosci. 1990;13(7):281–5.
34. Clavier RM, Atmadja S, Fibiger HC. Nigrothalamic projections in the rat as demonstrated by orthograde and retrograde tracing echniques. Brain Res Bull. 1976;1(4):379–84.
35. Herkenham M. The afferent and efferent connections of the ventromedial thalamic nucleus in the rat. J Comp Neurol. 1979;183(3):487–517.
36. Deniau JM, Chevalier G. The lamellar organization of the rat substantia nigra pars reticulata: distribution of projection neurons. Neuroscience. 1992;46(2):361–77.
37. Deniau JM, Mailly P, Maurice N, Charpier S. The pars reticulata of the substantia nigra: a window to basal ganglia output. Prog Brain Res. 2007;160:151–72.
38. Gerfen CR. The neostriatal mosaic: multiple levels of compartmental organization in the basal ganglia. Annu Rev Neurosci. 1992;15:285–320.
39. Chevalier G, Deniau JM. Disinhibition as a basic process in the expression of striatal functions. Trends Neurosci. 1990;13(7):277–80.
40. Graybiel AM. Neurotransmitters and neuromodulators in the basal ganglia. Trends Neurosci. 1990;13(7):244–54.
41. Agid Y. Parkinson's disease: pathophysiology. Lancet. 1991;337(8753):1321–4.
42. Hikosaka O, Wurtz RH. Visual and oculomotor functions of monkey substantia nigra pars reticulata. I. Relation of visual and auditory responses to saccades. J Neurophysiol. 1983;49(5):1230–53.
43. Hikosaka O, Wurtz RH. Visual and oculomotor functions of monkey substantia nigra pars reticulata. II. Visual responses related to fixation of gaze. J Neurophysiol. 1983;49(5):1254–67.
44. Hikosaka O, Wurtz RH. Visual and oculomotor functions of monkey substantia nigra pars reticulata. III. Memory-contingent visual and saccade responses. J Neurophysiol. 1983;49(5):1268–84.
45. Turner RS, Anderson ME. Context-dependent modulation of movement-related discharge in the primate globus pallidus. J Neurosci. 2005;25(11):2965–76.
46. Penney Jr JB, Young AB. Speculations on the functional anatomy of basal ganglia disorders. Annu Rev Neurosci. 1983;6:73–94.
47. Rascol O, Sabatini U, Chollet F, Celsis P, Montastruc JL, Marc-Vergnes JP, et al. Supplementary and primary sensory motor area activity in Parkinson's disease. Regional cerebral blood flow changes during finger movements and effects of apomorphine. Arch Neurol. 1992;49(2):144–8.
48. Bezard E, Crossman AR, Gross CE, Brotchie JM. Structures outside the basal ganglia may compensate for dopamine loss in the presymptomatic stages of Parkinson's disease. FASEB J. 2001;15(6):1092–4.

49. Crossman AR, Mitchell IJ, Sambrook MA. Regional brain uptake of 2-deoxyglucose in N-methyl-4-phenyl-1,2,3,6-tetrahydropyridine (MPTP)-induced parkinsonism in the macaque monkey. Neuropharmacology. 1985;24(6):587–91.
50. Mitchell IJ, Cross AJ, Sambrook MA, Crossman AR. N-methyl-4-phenyl-1,2,3,6-tetrahydropyridine-induced parkinsonism in the monkey: neurochemical pathology and regional brain metabolism. J Neural Transm Suppl. 1986;20:41–6.
51. Mitchell IJ, Clarke CE, Boyce S, Robertson RG, Peggs D, Sambrook MA, et al. Neural mechanisms underlying parkinsonian symptoms based upon regional uptake of 2-deoxyglucose in monkeys exposed to 1-methyl-4-phenyl-1,2,3,6-tetrahydropyridine. Neuroscience. 1989;32(1):213–26.
52. Filion M, Tremblay L, Bedard PJ. Effects of dopamine agonists on the spontaneous activity of globus pallidus neurons in monkeys with MPTP-induced parkinsonism. Brain Res. 1991;547(1):152–61.
53. Bergman H, Wichmann T, Karmon B, DeLong MR. The primate subthalamic nucleus. II. Neuronal activity in the MPTP model of parkinsonism. J Neurophysiol. 1994;72(2):507–20.
54. Soares J, Kliem MA, Betarbet R, Greenamyre JT, Yamamoto B, Wichmann T. Role of external pallidal segment in primate parkinsonism: comparison of the effects of 1-methyl-4-phenyl-1,2,3,6-tetrahydropyridine-induced parkinsonism and lesions of the external pallidal segment. J Neurosci. 2004;24(29):6417–26.
55. Vila M, Levy R, Herrero MT, Ruberg M, Faucheux B, Obeso JA, et al. Consequences of nigrostriatal denervation on the functioning of the basal ganglia in human and nonhuman primates: an in situ hybridization study of cytochrome oxidase subunit I mRNA. J Neurosci. 1997;17(2):765–73.
56. Gerfen CR, McGinty JF, Young 3rd WS. Dopamine differentially regulates dynorphin, substance P, and enkephalin expression in striatal neurons: in situ hybridization histochemical analysis. J Neurosci. 1991;11(4):1016–31.
57. Henry B, Crossman AR, Brotchie JM. Characterization of enhanced behavioral responses to L-DOPA following repeated administration in the 6-hydroxydopamine-lesioned rat model of Parkinson's disease. Exp Neurol. 1998;151(2):334–42.
58. Ravenscroft P, Chalon S, Brotchie JM, Crossman AR. Ropinirole versus L-DOPA effects on striatal opioid peptide precursors in a rodent model of Parkinson's disease: implications for dyskinesia. Exp Neurol. 2004;185(1):36–46.
59. Bergman H, Wichmann T, DeLong MR. Reversal of experimental parkinsonism by lesions of the subthalamic nucleus. Science. 1990;249(4975):1436–8.
60. Aziz TZ, Peggs D, Sambrook MA, Crossman AR. Lesion of the subthalamic nucleus for the alleviation of 1-methyl-4-phenyl-1,2,3,6-tetrahydropyridine (MPTP)-induced parkinsonism in the primate. Mov Disord. 1991;6(4):288–92.
61. Wichmann T, Bergman H, DeLong MR. The primate subthalamic nucleus. III. Changes in motor behavior and neuronal activity in the internal pallidum induced by subthalamic inactivation in the MPTP model of parkinsonism. J Neurophysiol. 1994;72(2):521–30.
62. Guridi J, Herrero MT, Luquin MR, Guillen J, Ruberg M, Laguna J, et al. Subthalamotomy in parkinsonian monkeys. Behavioural and biochemical analysis. Brain. 1996;119(Pt 5):1717–27.
63. Laitinen LV, Bergenheim AT, Hariz MI. Leksell's posteroventral pallidotomy in the treatment of Parkinson's disease. J Neurosurg. 1992;76(1):53–61.
64. Baron MS, Vitek JL, Bakay RA, Green J, Kaneoke Y, Hashimoto T, et al. Treatment of advanced Parkinson's disease by posterior GPi pallidotomy: 1-year results of a pilot study. Ann Neurol. 1996;40(3):355–66.
65. Gill SS, Heywood P. Bilateral dorsolateral subthalamotomy for advanced Parkinson's disease. Lancet. 1997;350(9086):1224.
66. Benabid AL, Pollak P, Gross C, Hoffmann D, Benazzouz A, Gao DM, et al. Acute and long-term effects of subthalamic nucleus stimulation in Parkinson's disease. Stereotact Funct Neurosurg. 1994;62(1–4):76–84.

67. Limousin P, Pollak P, Benazzouz A, Hoffmann D, Broussolle E, Perret JE, et al. Bilateral subthalamic nucleus stimulation for severe Parkinson's disease. Mov Disord. 1995;10(5):672–4.
68. Limousin P, Krack P, Pollak P, Benazzouz A, Ardouin C, Hoffmann D, et al. Electrical stimulation of the subthalamic nucleus in advanced Parkinson's disease. N Engl J Med. 1998;339(16):1105–11.
69. Crossman AR. A hypothesis on the pathophysiological mechanisms that underlie levodopa- or dopamine agonist-induced dyskinesia in Parkinson's disease: implications for future strategies in treatment. Mov Disord. 1990;5(2):100–8.
70. Brooks DJ, Piccini P, Turjanski N, Samuel M. Neuroimaging of dyskinesia. Ann Neurol. 2000;47(4 Suppl 1):S154–8; discussion S8–9.
71. Rascol O, Sabatini U, Brefel C, Fabre N, Rai S, Senard JM, et al. Cortical motor overactivation in parkinsonian patients with L-dopa-induced peak-dose dyskinesia. Brain. 1998;121(Pt 3):527–33.
72. Mitchell IJ, Boyce S, Sambrook MA, Crossman AR. A 2-deoxyglucose study of the effects of dopamine agonists on the parkinsonian primate brain. Implications for the neural mechanisms that mediate dopamine agonist-induced dyskinesia. Brain. 1992;115(Pt 3):809–24.
73. Papa SM, Desimone R, Fiorani M, Oldfield EH. Internal globus pallidus discharge is nearly suppressed during levodopa-induced dyskinesias. Ann Neurol. 1999;46(5):732–8.
74. Boraud T, Bezard E, Guehl D, Bioulac B, Gross C. Effects of L-DOPA on neuronal activity of the globus pallidus externalis (GPe) and globus pallidus internalis (GPi) in the MPTP-treated monkey. Brain Res. 1998;787(1):157–60.
75. Boraud T, Bezard E, Bioulac B, Gross CE. Dopamine agonist-induced dyskinesias are correlated to both firing pattern and frequency alterations of pallidal neurones in the MPTP-treated monkey. Brain. 2001;124(Pt 3):546–57.
76. Merello M, Balej J, Delfino M, Cammarota A, Betti O, Leiguarda R. Apomorphine induces changes in GPi spontaneous outflow in patients with Parkinson's disease. Mov Disord. 1999;14(1):45–9.
77. Lozano AM, Lang AE, Levy R, Hutchison W, Dostrovsky J. Neuronal recordings in Parkinson's disease patients with dyskinesias induced by apomorphine. Ann Neurol. 2000;47(4 Suppl 1):S141–6.
78. Stefani A, Stanzione P, Bassi A, Mazzone P, Vangelista T, Bernardi G. Effects of increasing doses of apomorphine during stereotaxic neurosurgery in Parkinson's disease: clinical score and internal globus pallidus activity. Short communication. J Neural Transm. 1997;104(8–9):895–904.
79. Cenci MA, Lee CS, Bjorklund A. L-DOPA-induced dyskinesia in the rat is associated with striatal overexpression of prodynorphin- and glutamic acid decarboxylase mRNA. Eur J Neurosci. 1998;10(8):2694–706.
80. Henry B, Crossman AR, Brotchie JM. Effect of repeated L-DOPA, bromocriptine, or lisuride administration on preproenkephalin-A and preproenkephalin-B mRNA levels in the striatum of the 6-hydroxydopamine-lesioned rat. Exp Neurol. 1999;155(2):204–20.
81. Lundblad M, Picconi B, Lindgren H, Cenci MA. A model of L-DOPA-induced dyskinesia in 6-hydroxydopamine lesioned mice: relation to motor and cellular parameters of nigrostriatal function. Neurobiol Dis. 2004;16(1):110–23.
82. Aubert I, Guigoni C, Li Q, Dovero S, Bioulac BH, Gross CE, et al. Enhanced preproenkephalin-B-derived opioid transmission in striatum and subthalamic nucleus converges upon globus pallidus internalis in L-3,4-dihydroxyphenylalanine-induced dyskinesia. Biol Psychiatry. 2007;61(7):836–44.
83. Aubert I, Guigoni C, Hakansson K, Li Q, Dovero S, Barthe N, et al. Increased D1 dopamine receptor signaling in levodopa-induced dyskinesia. Ann Neurol. 2005;57(1):17–26.
84. Rascol O, Nutt JG, Blin O, Goetz CG, Trugman JM, Soubrouillard C, et al. Induction by dopamine D1 receptor agonist ABT-431 of dyskinesia similar to levodopa in patients with Parkinson disease. Arch Neurol. 2001;58(2):249–54.

85. Calon F, Birdi S, Rajput AH, Hornykiewicz O, Bedard PJ, Di Paolo T. Increase of preproenkephalin mRNA levels in the putamen of Parkinson disease patients with levodopa-induced dyskinesias. J Neuropathol Exp Neurol. 2002;61(2):186–96.
86. Bezard E, Brotchie JM, Gross CE. Pathophysiology of levodopa-induced dyskinesia: potential for new therapies. Nat Rev Neurosci. 2001;2(8):577–88.
87. Rascol O, Brooks DJ, Korczyn AD, De Deyn PP, Clarke CE, Lang AE, et al. Development of dyskinesias in a 5-year trial of ropinirole and L-dopa. Mov Disord. 2006;21(11):1844–50.
88. Nadjar A, Brotchie JM, Guigoni C, Li Q, Zhou SB, Wang GJ, et al. Phenotype of striatofugal medium spiny neurons in parkinsonian and dyskinetic nonhuman primates: a call for a reappraisal of the functional organization of the basal ganglia. J Neurosci. 2006;26(34):8653–61.
89. Levesque M, Parent A. The striatofugal fiber system in primates: a reevaluation of its organization based on single-axon tracing studies. Proc Natl Acad Sci U S A. 2005;102(33):11888–93.
90. Surmeier DJ, Song WJ, Yan Z. Coordinated expression of dopamine receptors in neostriatal medium spiny neurons. J Neurosci. 1996;16(20):6579–91.
91. Yung KK, Smith AD, Levey AI, Bolam JP. Synaptic connections between spiny neurons of the direct and indirect pathways in the neostriatum of the rat: evidence from dopamine receptor and neuropeptide immunostaining. Eur J Neurosci. 1996;8(5):861–9.
92. Aizman O, Brismar H, Uhlen P, Zettergren E, Levey AI, Forssberg H, et al. Anatomical and physiological evidence for D1 and D2 dopamine receptor colocalization in neostriatal neurons. Nat Neurosci. 2000;3(3):226–30.
93. Kerr JN, Wickens JR. Dopamine D-1/D-5 receptor activation is required for long-term potentiation in the rat neostriatum in vitro. J Neurophysiol. 2001;85(1):117–24.
94. Nicola SM, Hopf FW, Hjelmstad GO. Contrast enhancement: a physiological effect of striatal dopamine? Cell Tissue Res. 2004;318(1):93–106.
95. Joel D, Weiner I. The connections of the dopaminergic system with the striatum in rats and primates: an analysis with respect to the functional and compartmental organization of the striatum. Neuroscience. 2000;96(3):451–74.
96. Prensa L, Cossette M, Parent A. Dopaminergic innervation of human basal ganglia. J Chem Neuroanat. 2000;20(3–4):207–13.
97. Smith Y, Kieval JZ. Anatomy of the dopamine system in the basal ganglia. Trends Neurosci. 2000;23(10 Suppl):S28–33.
98. Sanchez-Gonzalez MA, Garcia-Cabezas MA, Rico B, Cavada C. The primate thalamus is a key target for brain dopamine. J Neurosci. 2005;25(26):6076–83.
99. Smith Y, Villalba R. Striatal and extrastriatal dopamine in the basal ganglia: an overview of its anatomical organization in normal and Parkinsonian brains. Mov Disord. 2008;23 Suppl 3:S534–47.
100. Rommelfanger KS, Wichmann T. Extrastriatal dopaminergic circuits of the Basal Ganglia. Front Neuroanat. 2010;4:139.
101. Kreiss DS, Anderson LA, Walters JR. Apomorphine and dopamine D(1) receptor agonists increase the firing rates of subthalamic nucleus neurons. Neuroscience. 1996;72(3):863–76.
102. Francois C, Savy C, Jan C, Tande D, Hirsch EC, Yelnik J. Dopaminergic innervation of the subthalamic nucleus in the normal state, in MPTP-treated monkeys, and in Parkinson's disease patients. J Comp Neurol. 2000;425(1):121–9.
103. Jan C, Francois C, Tande D, Yelnik J, Tremblay L, Agid Y, et al. Dopaminergic innervation of the pallidum in the normal state, in MPTP-treated monkeys and in parkinsonian patients. Eur J Neurosci. 2000;12(12):4525–35.
104. Obeso JA, Rodriguez-Oroz MC, Rodriguez M, Lanciego JL, Artieda J, Gonzalo N, et al. Pathophysiology of the basal ganglia in Parkinson's disease. Trends Neurosci. 2000;23(10 Suppl):S8–19.
105. Leblois A, Boraud T, Meissner W, Bergman H, Hansel D. Competition between feedback loops underlies normal and pathological dynamics in the basal ganglia. J Neurosci. 2006;26(13):3567–83.

106. McHaffie JG, Stanford TR, Stein BE, Coizet V, Redgrave P. Subcortical loops through the basal ganglia. Trends Neurosci. 2005;28(8):401–7.
107. Haber SN, Fudge JL, McFarland NR. Striatonigrostriatal pathways in primates form an ascending spiral from the shell to the dorsolateral striatum. J Neurosci. 2000;20(6):2369–82.
108. Mallet N, Micklem BR, Henny P, Brown MT, Williams C, Bolam JP, et al. Dichotomous organization of the external globus pallidus. Neuron. 2012;74(6):1075–86.
109. Sato F, Lavallee P, Levesque M, Parent A. Single-axon tracing study of neurons of the external segment of the globus pallidus in primate. J Comp Neurol. 2000;417(1):17–31.
110. Shink E, Bevan MD, Bolam JP, Smith Y. The subthalamic nucleus and the external pallidum: two tightly interconnected structures that control the output of the basal ganglia in the monkey. Neuroscience. 1996;73(2):335–57.
111. Obeso JA, Rodriguez-Oroz MC, Javier Blesa F, Guridi J. The globus pallidus pars externa and Parkinson's disease. Ready for prime time? Exp Neurol. 2006;202(1):1–7.
112. Kita H. Globus pallidus external segment. Prog Brain Res. 2007;160:111–33.
113. Obeso JA, Rodriguez-Oroz MC, Benitez-Temino B, Blesa FJ, Guridi J, Marin C, et al. Functional organization of the basal ganglia: therapeutic implications for Parkinson's disease. Mov Disord. 2008;23 Suppl 3:S548–59.
114. Nambu A, Tokuno H, Hamada I, Kita H, Imanishi M, Akazawa T, et al. Excitatory cortical inputs to pallidal neurons via the subthalamic nucleus in the monkey. J Neurophysiol. 2000;84(1):289–300.
115. Nambu A, Tokuno H, Takada M. Functional significance of the cortico-subthalamo-pallidal 'hyperdirect' pathway. Neurosci Res. 2002;43(2):111–7.
116. Lanciego JL, Gonzalo N, Castle M, Sanchez-Escobar C, Aymerich MS, Obeso JA. Thalamic innervation of striatal and subthalamic neurons projecting to the rat entopeduncular nucleus. Eur J Neurosci. 2004;19(5):1267–77.
117. Mena-Segovia J, Bolam JP, Magill PJ. Pedunculopontine nucleus and basal ganglia: distant relatives or part of the same family? Trends Neurosci. 2004;27(10):585–8.
118. Rico AJ, Barroso-Chinea P, Conte-Perales L, Roda E, Gomez-Bautista V, Gendive M, et al. A direct projection from the subthalamic nucleus to the ventral thalamus in monkeys. Neurobiol Dis. 2010;39(3):381–92.
119. Kita H, Tachibana Y, Nambu A, Chiken S. Balance of monosynaptic excitatory and disynaptic inhibitory responses of the globus pallidus induced after stimulation of the subthalamic nucleus in the monkey. J Neurosci. 2005;25(38):8611–9.
120. Marsden CD, Obeso JA. The functions of the basal ganglia and the paradox of stereotaxic surgery in Parkinson's disease. Brain. 1994;117(Pt 4):877–97.
121. Obeso JA, Rodriguez MC, DeLong MR. Basal ganglia pathophysiology. A critical review. Adv Neurol. 1997;74:3–18.
122. Iravani MM, Costa S, Al-Bargouthy G, Jackson MJ, Zeng BY, Kuoppamaki M, et al. Unilateral pallidotomy in 1-methyl-4-phenyl-1,2,3,6-tetrahydropyridine-treated common marmosets exhibiting levodopa-induced dyskinesia. Eur J Neurosci. 2005;22(6):1305–18.
123. Lang AE. Surgery for levodopa-induced dyskinesias. Ann Neurol. 2000;47(4 Suppl 1):S193–9; discussion S9–202.
124. Parkin SG, Gregory RP, Scott R, Bain P, Silburn P, Hall B, et al. Unilateral and bilateral pallidotomy for idiopathic Parkinson's disease: a case series of 115 patients. Mov Disord. 2002;17(4):682–92.
125. Brown P, Oliviero A, Mazzone P, Insola A, Tonali P, Di Lazzaro V. Dopamine dependency of oscillations between subthalamic nucleus and pallidum in Parkinson's disease. J Neurosci. 2001;21(3):1033–8.
126. Brown P. Oscillatory nature of human basal ganglia activity: relationship to the pathophysiology of Parkinson's disease. Mov Disord. 2003;18(4):357–63.
127. Alonso-Frech F, Zamarbide I, Alegre M, Rodriguez-Oroz MC, Guridi J, Manrique M, et al. Slow oscillatory activity and levodopa-induced dyskinesias in Parkinson's disease. Brain. 2006;129(Pt 7):1748–57.

128. Magill PJ, Sharott A, Bevan MD, Brown P, Bolam JP. Synchronous unit activity and local field potentials evoked in the subthalamic nucleus by cortical stimulation. J Neurophysiol. 2004;92(2):700–14.
129. Boraud T, Bezard E, Bioulac B, Gross CE. From single extracellular unit recording in experimental and human Parkinsonism to the development of a functional concept of the role played by the basal ganglia in motor control. Prog Neurobiol. 2002;66(4):265–83.
130. Engel AK, Singer W. Temporal binding and the neural correlates of sensory awareness. Trends Cogn Sci. 2001;5(1):16–25.
131. Engel AK, Fries P, Singer W. Dynamic predictions: oscillations and synchrony in top-down processing. Nat Rev Neurosci. 2001;2(10):704–16.
132. Mackay WA. Synchronized neuronal oscillations and their role in motor processes. Trends Cogn Sci. 1997;1(5):176–83.
133. Bar-Gad I, Morris G, Bergman H. Information processing, dimensionality reduction and reinforcement learning in the basal ganglia. Prog Neurobiol. 2003;71(6):439–73.
134. Raz A, Vaadia E, Bergman H. Firing patterns and correlations of spontaneous discharge of pallidal neurons in the normal and the tremulous 1-methyl-4-phenyl-1,2,3,6-tetrahydropyridine vervet model of parkinsonism. J Neurosci. 2000;20(22):8559–71.
135. Nini A, Feingold A, Slovin H, Bergman H. Neurons in the globus pallidus do not show correlated activity in the normal monkey, but phase-locked oscillations appear in the MPTP model of parkinsonism. J Neurophysiol. 1995;74(4):1800–5.
136. Wichmann T, Bergman H, DeLong MR. The primate subthalamic nucleus. I. Functional properties in intact animals. J Neurophysiol. 1994;72(2):494–506.
137. Bar-Gad I, Heimer G, Ritov Y, Bergman H. Functional correlations between neighboring neurons in the primate globus pallidus are weak or nonexistent. J Neurosci. 2003;23(10):4012–6.
138. Bergman H, Feingold A, Nini A, Raz A, Slovin H, Abeles M, et al. Physiological aspects of information processing in the basal ganglia of normal and parkinsonian primates. Trends Neurosci. 1998;21(1):32–8.
139. Berke JD, Okatan M, Skurski J, Eichenbaum HB. Oscillatory entrainment of striatal neurons in freely moving rats. Neuron. 2004;43(6):883–96.
140. Dejean C, Gross CE, Bioulac B, Boraud T. Synchronous high-voltage spindles in the cortex-basal ganglia network of awake and unrestrained rats. Eur J Neurosci. 2007;25(3):772–84.
141. Filion M, Tremblay L. Abnormal spontaneous activity of globus pallidus neurons in monkeys with MPTP-induced parkinsonism. Brain Res. 1991;547(1):142–51.
142. Wichmann T, Soares J. Neuronal firing before and after burst discharges in the monkey basal ganglia is predictably patterned in the normal state and altered in parkinsonism. J Neurophysiol. 2006;95(4):2120–33.
143. Heimer G, Rivlin-Etzion M, Bar-Gad I, Goldberg JA, Haber SN, Bergman H. Dopamine replacement therapy does not restore the full spectrum of normal pallidal activity in the 1-methyl-4-phenyl-1,2,3,6-tetra-hydropyridine primate model of Parkinsonism. J Neurosci. 2006;26(31):8101–14.
144. Bevan MD, Wilson CJ, Bolam JP, Magill PJ. Equilibrium potential of GABA(A) current and implications for rebound burst firing in rat subthalamic neurons in vitro. J Neurophysiol. 2000;83(5):3169–72.
145. Hallworth NE, Bevan MD. Globus pallidus neurons dynamically regulate the activity pattern of subthalamic nucleus neurons through the frequency-dependent activation of postsynaptic GABAA and GABAB receptors. J Neurosci. 2005;25(27):6304–15.
146. Bevan MD, Hallworth NE, Baufreton J. GABAergic control of the subthalamic nucleus. Prog Brain Res. 2007;160:173–88.
147. Cruz AV, Mallet N, Magill PJ, Brown P, Averbeck BB. Effects of dopamine depletion on information flow between the subthalamic nucleus and external globus pallidus. J Neurophysiol. 2011;106(4):2012–23.

148. Tachibana Y, Iwamuro H, Kita H, Takada M, Nambu A. Subthalamo-pallidal interactions underlying parkinsonian neuronal oscillations in the primate basal ganglia. Eur J Neurosci. 2011;34(9):1470–84.
149. Schwab BC, Heida T, Zhao Y, Marani E, van Gils SA, van Wezel RJ. Synchrony in Parkinson's disease: importance of intrinsic properties of the external globus pallidus. Front Syst Neurosci. 2013;7:60.
150. Marsden JF, Limousin-Dowsey P, Ashby P, Pollak P, Brown P. Subthalamic nucleus, sensorimotor cortex and muscle interrelationships in Parkinson's disease. Brain. 2001;124(Pt 2):378–88.
151. Cassidy M, Mazzone P, Oliviero A, Insola A, Tonali P, Di Lazzaro V, et al. Movement-related changes in synchronization in the human basal ganglia. Brain. 2002;125(Pt 6):1235–46.
152. Williams D, Tijssen M, Van Bruggen G, Bosch A, Insola A, Di Lazzaro V, et al. Dopamine-dependent changes in the functional connectivity between basal ganglia and cerebral cortex in humans. Brain. 2002;125(Pt 7):1558–69.
153. Shimamoto SA, Ryapolova-Webb ES, Ostrem JL, Galifianakis NB, Miller KJ, Starr PA. Subthalamic nucleus neurons are synchronized to primary motor cortex local field potentials in Parkinson's disease. J Neurosci. 2013;33(17):7220–33.
154. Gatev P, Darbin O, Wichmann T. Oscillations in the basal ganglia under normal conditions and in movement disorders. Mov Disord. 2006;21(10):1566–77.
155. Moro E, Esselink RJ, Xie J, Hommel M, Benabid AL, Pollak P. The impact on Parkinson's disease of electrical parameter settings in STN stimulation. Neurology. 2002;59(5):706–13.
156. Timmermann L, Wojtecki L, Gross J, Lehrke R, Voges J, Maarouf M, et al. Ten-Hertz stimulation of subthalamic nucleus deteriorates motor symptoms in Parkinson's disease. Mov Disord. 2004;19(11):1328–33.
157. Fogelson N, Kuhn AA, Silberstein P, Limousin PD, Hariz M, Trottenberg T, et al. Frequency dependent effects of subthalamic nucleus stimulation in Parkinson's disease. Neurosci Lett. 2005;382(1–2):5–9.
158. Hutchison WD, Lozano AM, Tasker RR, Lang AE, Dostrovsky JO. Identification and characterization of neurons with tremor-frequency activity in human globus pallidus. Exp Brain Res. 1997;113(3):557–63.
159. Kuhn AA, Tsui A, Aziz T, Ray N, Brucke C, Kupsch A, et al. Pathological synchronisation in the subthalamic nucleus of patients with Parkinson's disease relates to both bradykinesia and rigidity. Exp Neurol. 2009;215(2):380–7.
160. Lopez-Azcarate J, Tainta M, Rodriguez-Oroz MC, Valencia M, Gonzalez R, Guridi J, et al. Coupling between beta and high-frequency activity in the human subthalamic nucleus may be a pathophysiological mechanism in Parkinson's disease. J Neurosci. 2010;30(19):6667–77.
161. Brown P, Mazzone P, Oliviero A, Altibrandi MG, Pilato F, Tonali PA, et al. Effects of stimulation of the subthalamic area on oscillatory pallidal activity in Parkinson's disease. Exp Neurol. 2004;188(2):480–90.
162. Foffani G, Ardolino G, Meda B, Egidi M, Rampini P, Caputo E, et al. Altered subthalamo-pallidal synchronisation in parkinsonian dyskinesias. J Neurol Neurosurg Psychiatry. 2005;76(3):426–8.
163. Meissner W, Ravenscroft P, Reese R, Harnack D, Morgenstern R, Kupsch A, et al. Increased slow oscillatory activity in substantia nigra pars reticulata triggers abnormal involuntary movements in the 6-OHDA-lesioned rat in the presence of excessive extracellular striatal dopamine. Neurobiol Dis. 2006;22(3):586–98.
164. Brown P. Abnormal oscillatory synchronisation in the motor system leads to impaired movement. Curr Opin Neurobiol. 2007;17(6):656–64.
165. Brown P, Eusebio A. Paradoxes of functional neurosurgery: clues from basal ganglia recordings. Mov Disord. 2008;23(1):12–20; quiz 158.
166. Leblois A, Meissner W, Bioulac B, Gross CE, Hansel D, Boraud T. Late emergence of synchronized oscillatory activity in the pallidum during progressive Parkinsonism. Eur J Neurosci. 2007;26(6):1701–13.

167. Leblois A, Meissner W, Bezard E, Bioulac B, Gross CE, Boraud T. Temporal and spatial alterations in GPi neuronal encoding might contribute to slow down movement in Parkinsonian monkeys. Eur J Neurosci. 2006;24(4):1201–8.
168. Parent A, Hazrati LN. Functional anatomy of the basal ganglia. I. The cortico-basal ganglia-thalamo-cortical loop. Brain Res Brain Res Rev. 1995;20(1):91–127.
169. Kayahara T, Nakano K. Pallido-thalamo-motor cortical connections: an electron microscopic study in the macaque monkey. Brain Res. 1996;706(2):337–42.
170. Hoover JE, Strick PL. The organization of cerebellar and basal ganglia outputs to primary motor cortex as revealed by retrograde transneuronal transport of herpes simplex virus type 1. J Neurosci. 1999;19(4):1446–63.
171. Kelly RM, Strick PL. Macro-architecture of basal ganglia loops with the cerebral cortex: use of rabies virus to reveal multisynaptic circuits. Prog Brain Res. 2004;143:449–59.
172. Deniau JM, Menetrey A, Charpier S. The lamellar organization of the rat substantia nigra pars reticulata: segregated patterns of striatal afferents and relationship to the topography of corticostriatal projections. Neuroscience. 1996;73(3):761–81.
173. Calabresi P, Centonze D, Bernardi G. Electrophysiology of dopamine in normal and denervated striatal neurons. Trends Neurosci. 2000;23(10 Suppl):S57–63.
174. Guthrie M, Leblois A, Garenne A, Boraud T. Interaction between cognitive and motor cortico-basal ganglia loops during decision making: a computational study. J Neurophysiol. 2013;109(12):3025–40.
175. Pasquereau B, Nadjar A, Arkadir D, Bezard E, Goillandeau M, Bioulac B, et al. Shaping of motor responses by incentive values through the basal ganglia. J Neurosci. 2007;27(5):1176–83.
176. Mink JW. The basal ganglia: focused selection and inhibition of competing motor programs. Prog Neurobiol. 1996;50(4):381–425.
177. Picconi B, Centonze D, Hakansson K, Bernardi G, Greengard P, Fisone G, et al. Loss of bidirectional striatal synaptic plasticity in L-DOPA-induced dyskinesia. Nat Neurosci. 2003;6(5):501–6.
178. Belujon P, Lodge DJ, Grace AA. Aberrant striatal plasticity is specifically associated with dyskinesia following levodopa treatment. Mov Disord. 2010;25(11):1568–76.
179. Halje P, Tamte M, Richter U, Mohammed M, Cenci MA, Petersson P. Levodopa-induced dyskinesia is strongly associated with resonant cortical oscillations. J Neurosci. 2012;32(47):16541–51.
180. Bastide MF, Dovero S, Charron G, Porras G, Gross CE, Fernagut PO, et al. Immediate-early gene expression in structures outside the basal ganglia is associated to l-DOPA-induced dyskinesia. Neurobiol Dis. 2013;62:179–92.
181. Cenci MA, Konradi C. Maladaptive striatal plasticity in L-DOPA-induced dyskinesia. Prog Brain Res. 2010;183:209–33.
182. Guigoni C, Li Q, Aubert I, Dovero S, Bioulac BH, Bloch B, et al. Involvement of sensorimotor, limbic, and associative basal ganglia domains in L-3,4-dihydroxyphenylalanine-induced dyskinesia. J Neurosci. 2005;25(8):2102–7.
183. Phelix CF, Liposits Z, Paull WK. Monoamine innervation of bed nucleus of stria terminalis: an electron microscopic investigation. Brain Res Bull. 1992;28(6):949–65.
184. Cenci MA, Lundblad M. Post- versus presynaptic plasticity in L-DOPA-induced dyskinesia. J Neurochem. 2006;99(2):381–92.
185. Carta M, Carlsson T, Munoz A, Kirik D, Bjorklund A. Involvement of the serotonin system in L-dopa-induced dyskinesias. Parkinsonism Relat Disord. 2008;14 Suppl 2:S154–8.
186. Navailles S, Bioulac B, Gross C, De Deurwaerdere P. Serotonergic neurons mediate ectopic release of dopamine induced by L-DOPA in a rat model of Parkinson's disease. Neurobiol Dis. 2010;38(1):136–43.
187. Carta M, Bezard E. Contribution of pre-synaptic mechanisms to L-DOPA-induced dyskinesia. Neuroscience. 2011;198:245–51.
188. Krawczyk M, Georges F, Sharma R, Mason X, Berthet A, Bezard E, et al. Double-dissociation of the catecholaminergic modulation of synaptic transmission in the oval bed nucleus of the stria terminalis. J Neurophysiol. 2011;105(1):145–53.

189. Carta M, Carlsson T, Kirik D, Bjorklund A. Dopamine released from 5-HT terminals is the cause of L-DOPA-induced dyskinesia in parkinsonian rats. Brain. 2007;130(Pt 7):1819–33.
190. Carta M, Carlsson T, Munoz A, Kirik D, Bjorklund A. Serotonin-dopamine interaction in the induction and maintenance of L-DOPA-induced dyskinesias. Prog Brain Res. 2008;172:465–78.
191. Rylander D, Parent M, O'Sullivan SS, Dovero S, Lees AJ, Bezard E, et al. Maladaptive plasticity of serotonin axon terminals in levodopa-induced dyskinesia. Ann Neurol. 2010;68(5):619–28.
192. Navailles S, Bioulac B, Gross C, De Deurwaerdere P. Chronic L-DOPA therapy alters central serotonergic function and L-DOPA-induced dopamine release in a region-dependent manner in a rat model of Parkinson's disease. Neurobiol Dis. 2011;41(2):585–90.
193. Vollenweider I, Lang Y, Borton D, Ko D, Li Q, Courtine G, et al. Translational analysis platform for neuromotor disease research and therapeutic validation: application to Parkinson's disease. In: Society of Neuroscience conference. San Diego; 2013. Poster 241.18/O1.

Chapter 8
Features and Mechanisms of Diphasic Dyskinesia in Parkinson's Disease

Marcelo Merello, Inés Trigo Damas, and José A. Obeso

Abstract Levodopa-induced dyskinesia is normally assessed based on the course of the appearance of their symptoms. Diphasic dyskinesia (DD) usually appears at the beginning and at the end, but not at the peak, of the levodopa effect in long-term treated Parkinson's disease patients. The most commonly affected subjects with this form of dyskinesia are those who have an early onset of the disease, approaching 20 % of globally treated parkinsonian patients. Typically, they are present in the lower limbs and exhibit rhythmic and sometime stereotypic movement patterns. In the past, DD were a serious management problem and often associated with severe dysautonomic manifestations. Current pharmacological trends to avoid high levodopa use have reduced the incidence of very troublesome DD. When severe, surgical approach may be considered, since pallidotomy typically resolves the movements. In a broader sense, the net predominance of the lower limb in DD is a fascinating mystery of which resolution could lead to important advances in the functional anatomy of the basal ganglia and Parkinson's disease.

Keywords Diphasic dyskinesia • Low dose levodopa • Graft induced dyskinesia • Serotonin • Dystonia improvement dystonia • Stereotyped lower limb dyskinesia

---

M. Merello, MD, PhD
Raul Carrea Institute for Neurological Research (FLENI),
Universidad Católica Argentina, CONICET,
Montaneses 2325, Buenos Aires (1428), Argentina

Raul Carrea Institute for Neurological Research, Buenos Aires, Argentina
e-mail: mmerello@fleni.org.ar

I. Trigo Damas, PhD • J.A. Obeso, MD, PhD (✉)
Movement Disorders Laboratory, Neuroscience Department, Center for Applied Medical Research (CIMA), University of Navarra, Pamplona, Spain
e-mail: jobeso@unav.es

S.H. Fox, J.M. Brotchie (eds.), *Levodopa-Induced Dyskinesia in Parkinson's Disease*, DOI 10.1007/978-1-4471-6503-3_8

Introduction

Dopamine replacement therapy using levodopa usually shows an initial period where the response to the medication is adequate and motor manifestations are well controlled. However, motor complications (fluctuations and dyskinesia mainly) arise in most patients within 5–10 years of levodopa treatment [1, 2].

Levodopa-induced dyskinesia (LID) has different presentation patterns [3], namely, peak-dose dyskinesia, diphasic dyskinesia, and "off"-period dystonia [4–7]. Typically, peak-dose dyskinesia shows choreiform-type movements in the most affected side of the body as fleeting or purposeless fidgety-type movements that can be exaggerated by emotional states or stress [8, 9], whereas OFF dystonia consists of spasms and dystonic postures commonly affecting one foot and with early-morning predominance. Diphasic dyskinesia (DD) also labeled as "beginning and end of dose" or DID (for dystonia improvement dystonia) is less common than the other categories but becomes extremely disabled and a major management problem. In addition, some unique features of DD make this complication extremely interesting in terms of understanding basal ganglia pathophysiology.

History and Major Features of Diphasic Dyskinesia

A diphasic pattern (Fig. 8.1) of some LIDs was first recognized by Tolosa et al. [10, 11] but described and labeled as a distinct subtype of LID by Lhermitte and Agid [10] and by Muenter et al. [12, 13]. The essential and intriguing feature of DD is that the abnormal movements occur when levodopa plasma levels are ascending and descending and, accordingly, can cease during peak plasma levels [2, 12, 14]. This makes DD the only secondary effect of a drug (at least for those acting on the nervous system) which is not increased but ameliorated by increasing the dose, thus truly a paradoxical effect.

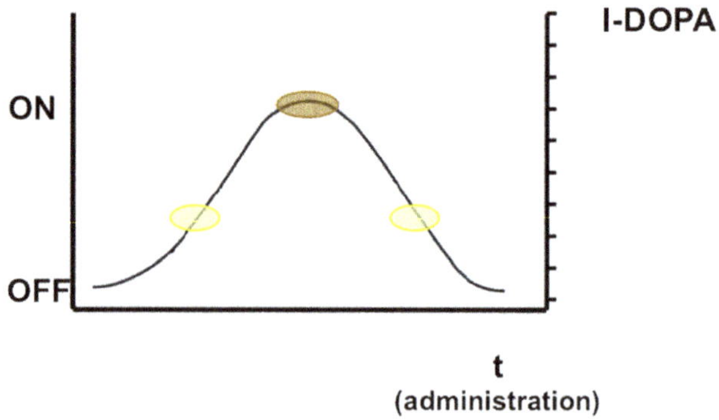

Fig. 8.1 Time period of diphasic dyskinesia. *Yellow ellipses* indicate the appearance of diphasic dyskinesia during the wearing ON and OFF of levodopa levels in plasma. *Brown ellipse* indicates the peak of dose period when the subjects show normal motor behavior

Characteristically, DD affects younger patients, and they occur when patients are in the transitional state from the OFF to the ON period (Fig. 8.1). In other words, DD occurs at a time when dopaminergic activity has to reach a threshold which is below the one needed to obtain an anti-parkinsonian benefit. The latter, of course, coincides with peak levodopa concentrations. The phenomenology of DD has been a matter of discussion and confusion for years [6, 15]. Many different labels (dystonia, choreoathetosis, ballism, stereotyped movements, etc.) have been used to describe DD. However, this is essentially a semantic problem. In the majority of cases, careful observation reveals that DD typically begins by the lower limbs and has a repetitive, slow (2–3 Hz), flexion–extension character [3, 9, 16], whereas the upper half of the body remains parkinsonian. A typical combination is someone kicking the leg while exhibiting a typical parkinsonian tremor in the upper limb of the same body side. The lower limb movements of DD may become quite large, involve more proximal segments, and loss their repetitive stereotypic character, thus leading to the use of ballism or chorea, but, in fact, the reciprocal activation of agonist/antagonist muscles is maintained in most instances [4]. In a minor proportion of patients, the predominant manifestation may be dystonic posturing of the limb. In some patients, there may be more complex manifestations like "silly gait" or "drummer gait" whereby patients do seem to be performing a voluntary action [17].

In addition, in a relatively large proportion of patients, the dyskinesia does not cease completely as the patient goes into the ON stage but becomes different, typically choreic or dystonic movements [18] which have contributed to the confusing terminology.

The true incidence of DD is not well ascertained as the movements often pass unnoticed when mild and short lasting. It is estimated that about 20 % of Parkinson's disease (PD) patients show DD severe enough to be clinically relevant [15]. Luquin et al. [19] studied the pattern of dyskinesia among 168 patients. Ninety-four percent of patients showed ON-period dyskinesia and 18.5 % had a diphasic presentation. OFF-period dystonia was noted in 35.7 % of patients (Table 8.1).

In the 1970s and 1980s, when many current treatments were not available, patients with DD could become extremely disabled entering into a "dyskinetic status" (thus also called "dyskinesia without benefit") accompanied by pain and autonomic changes which considerably decreased the patients' quality of life [6, 15, 20]. Indeed, in some few instances the patient died from systemic complications.

Do Patients with DD Correspond to a Different PD Subtype?

Why is it that only a small group of patients display DD? It is possible that the phenomena may exist in all patients but reasons related to the velocity in turning ON may be affecting the process, such that some patients exert DD while others do not. DD usually affects patients with early-onset disease [21, 22]. The risk of developing dyskinesia or wearing OFF is closely linked to levodopa dose [23]; however no single work has studied demographic or disease-related differences between both types of dyskinesia.

Table 8.1 Characteristics and established observations of diphasic dyskinesia

Occur at the beginning and end of the levodopa effect
Disappear with a higher dose of levodopa or with a "rescue" dose of subcutaneous apomorphine
May be considered as the only drug-induced side effect that disappears when *increasing* the dose of the offensive drug
Characteristically, the leg is the body part predominantly involved
Occurs only in a subgroup of PD patients: factors distinguishing who will or will not develop DD are not recognized
Are abolished by either pallidotomy or thalamotomy
Different movement disorders may be seen as an expression of DD, such as dystonia, ballism, or stereotyped movements. Rhythmic, alternating movements of the legs are the most common manifestation
The nomenclature used in the literature has been variable and confusing over time
Some patients treated with continuous delivery of apomorphine or lisuride subcutaneously and also with graft-induced dyskinesias exhibit similar repetitive movements and failed to reach a full anti-parkinsonian response, imitating the movements of DD but without the beginning–end of dose pattern
Subthalamic nucleus surgery may induce dyskinesia in some patients which is phenomenologically similar to the typical repetitive moments of DD

What Can Graft-Induced Dyskinesia Tell Us About DD Pathophysiology and the Putative Role of Serotonin in Its Genesis?

Recently, a new and occasionally severe form of dyskinesia that persists for prolonged periods of time following withdrawal of dopaminergic medication has been reported in patients with PD, who have undergone fetal nigral transplantation, so-called graft-induced dyskinesia or OFF-levodopa posttransplant dyskinesia. These movements contrast with classical LID, which typically disappears within hours after stopping the medication and its phenomenology resemble typical DD [24]. Thus, like DD, OFF dyskinesias are asymmetric, rhythmic, alternating, stereotypic movements that predominantly affected the legs. The basis of OFF-medication dyskinesia that develops following fetal nigral transplantation is not known. It has been proposed that they may be due to graft overgrowth with excess dopamine production due to prolonged periods of preoperative culturing of cells. However, posttransplant OFF medication more often occurred in association with incomplete (or negligible) improvement in the OFF motor scores, indicating that the graft-derived increases in dopaminergic activity was insufficient to achieve a full anti-parkinsonian effect but perhaps enough to trigger dyskinesia that resembles DD but was rather maintained. A similar situation can be observed in PD patients treated with continuous infusion of levodopa or apomorphine [25, 26]. Similarly, STN-DBS can also elicit a DD pattern of movements, which can be overcome by increasing the current voltage [27].

More recently, experimental evidence indicating a pivotal role of the striatal serotonergic innervation in the genesis of OFF-drug dyskinesia has been generated by using the 6-OHDA lesion rat model [28, 29]. Serotonin neuron-rich fetal ventral mesencephalic (FVM) tissue grafts exacerbated both the severity and duration of abnormal involuntary movements in this model with no motor function improvement, whereas dopamine neuron-rich grafts produced less dyskinesia and substantial functional improvement. Carlsson et al. [29] suggested that with an equal ratio of dopamine to serotonin neurons in the PD brain, remaining dopamine neurons can buffer dopaminergic activity from serotonergic nerve terminals.

Serotonergic axons arborize as densely and as widely as dopaminergic neurons in the striatum, and their proximal location allows for a complex serotonin–dopamine interaction. Serotonin receptors have diffuse striatal morphology and are expressed by many non-dopaminergic neurons, including medium-sized spiny neurons and cholinergic interneurons. However, there is a higher representation of 5HT receptors in the ventral striatum compared with dorsal striatum.

The similarities between OFF-levodopa posttransplant dyskinesias and DD, the role of serotonin in stereotypies and persistent behaviors, as well as the selective preference of the leg consistent to the differential expression of 5HT receptors between dorsal and ventral striatum make the serotoninergic hypothesis of DD trustworthy for further studies.

Does DD Share the Same Pathophysiology with Peak-Dose Dyskinesia?

Despite the established correlation with low levodopa plasma levels, the pathophysiology of DD is uncertain. Whereas it is true that increasing levodopa dose can cause a significant reduction or even disappearance of DD, this is typically short lasting and gives way to generalized peak-dose dyskinesia. Equally, the use of apomorphine injections as a rescue therapy for patients with DD is also accompanied by aggravation of dyskinesias. Thus, Durif et al. [30] reported that apomorphine bolus administration reduced DD in three patients, but the improvement was rapidly followed by peak-dose dyskinesia. They noticed that in two of those patients the apomorphine boosts were efficient just at the first morning dose but not during the rest of the day.

The available pharmacological observations and the distinct clinical features of DD would suggest they represent a different subtype of LID and not just a fragment of peak-dose dyskinesias. Recently, a genetic study [31] showed that DD was associated with the DRD3 p.S9G polymorphism variant suggesting a specific genetic susceptibility and differentiated mechanism for both DD and peak-dose LID.

However, evidence in the opposite direction, i.e., that both DD and peak-dose LID have similar pathophysiological mechanisms, has been generated recently by neurophysiological studies of local field potentials. In PD patients undergoing deep brain stimulation (DBS), the implantation of electrodes allows the recording of field potential activity from various subcortical structures. In summary, in the OFF state, the STN shows a peak of abnormal activity in the low beta range (between 10 and 30 Hz), which disappears in the ON medication state [32, 33]. The disappearance of the low beta peak is accompanied in around 30 % of patients by an increase in gamma activity (60–80 Hz) [34]. Related studies have also shown evidence of changes in higher frequencies (200–400 Hz) between both states, linked to complex interactions with the beta activities [35]. In the ON state, an additional peak in the theta/alpha range (4–10 Hz) has been associated with the presence of levodopa-induced dyskinesia and impulse control disorders [36]. In specific patients, continuous recording throughout a levodopa cycle has shown the same oscillatory activity (around 8 Hz) and STN topography while exhibiting DD and "peak-dose" dyskinesia [34, 35].

Animal models support the findings from DBS in patients mentioned above [37, 38]. In the MPTP (1-methyl-4-phenyl-1,2,3,6-tetrahydropyridine) monkey model, activity in the beta range has been recognized in the STN and GPi. In this model, the peak of such synchronous spiking activity appeared between 10 and 12 Hz coinciding with the frequency of tremor typically observed in African green or vervet monkeys, which are the only ones exhibiting tremor at rest after MPTP. Accordingly, the MPTP monkey model supports, by and large, the findings in PD patients (indicating that severe striatal dopamine depletion leads to increased synchronization in the basal ganglia in the low beta range). Indeed, synchronization between oscillatory local field potentials (LFPs) and single neuronal action potential activity within a given nucleus, i.e., STN, is enhanced in the parkinsonian state, quite unlike the normal state.

Additionally, studies conducted in the 6-OHDA (6-hydroxydopamine) unilaterally lesioned rat have also shown a peak of beta activity (essentially around 30–35 Hz in the awake animal and around 20 Hz in the anesthetized rat) in the STN, SNpr, and GP and in the motor cortex. Thus, accordingly, enhanced beta activity within the basal ganglia in the "indirect" circuit and between the motor cortex and the STN is considered a net characteristic of the unilaterally dopamine-depleted rat. Administration of dopaminergic agents, such as apomorphine or levodopa, attenuates beta activity, further mimicking the observations in PD patients. More in detail, theta/alpha range LFP power (4–10 Hz) is increased in 6-OHDA-lesioned groups following a single levodopa administration [39–41]. Interestingly, in a recent study by Halje et al. [42] with rats as well, they showed a tight association between presence of LID and strong resonant high-frequency oscillations in the primary motor cortex of the lesioned (parkinsonian) hemisphere. The frequency was essentially the same, 80 Hz in all animals and in all experiments during the whole dyskinetic period. These oscillations were also found in striatal recordings indicating the possibility of a downstream pathway from the cortex. Also, these oscillations were not found in the non-parkinsonian hemisphere.

Thus, according to the above findings, the neuronal activity associated with DD and peak-dose dyskinesias might be the same. Furthermore, both types of LID are equally eliminated by pallidotomy and thalamotomy, suggesting marked similarities in the neuronal circuits conveying the signals leading to the abnormal movements. Recently, Alegre et al. [43] described a theta peak in a group of patients showing DD, very similar to the one present during peak-dose dyskinesias, suggesting that they are highly related neurophysiological phenomena. These findings may be considered as the first physiological evidence supporting the notion that diphasic dyskinesia is the beginning of ON or peak-dose dyskinesia restricted to mechanisms controlling the lower limbs, thus representing the initial manifestation of levodopa-induced dyskinesia. However, this similarity does not rule out that some distinct pathophysiological mechanisms may underlie either type and account for some of the clinical features. Recently, Filipovic et al. [44] studied the effect of rTMS in motor cortex (M1) in a single patient who suffered from both DD and peak-dose dyskinesia to evaluate the potential differences in the responses to the stimulation. They found that the effect was more pronounced for the DD. This preliminary observation could suggest differential cortical mechanisms.

Are Proprioceptive Deficits Involved in DD Generation?

It has been hypothesized that joint position sensation becomes phasically diminished after levodopa intake [45]. The brain, receiving weak positional information, would then facilitate exaggerated or adventitious searching movements in an attempt to keep abreast of the deployment of body parts. This pathophysiology would have some relation with the pseudo-chorea seen in response to peripheral deafferentation and may be specially involved in the typically diphasic dyskinetic gait usually seen in many patients and described by Ruzicka et al. [17] as "silly walks" which may mimic psychogenic gaits. Characteristically, those gait patterns occurred at the beginning and end of each levodopa dose effect with some dystonia followed by ballistic kicking and stamping. The gait patterns presented indeed look bizarre but also remarkably similar across patients. Recurring characteristics include "stepping with kicks," "high knee elevations," and ballistic "stamping" of the lower limb, generally on the side first affected by PD. In addition, dystonic postures of the contralateral foot and ipsilateral arm were present. Another important element was that the timing of these abnormal movements in relation to the intake and clinical efficacy of dopaminergic medication clearly had a diphasic pattern: "silly gaits" developed shortly following a levodopa dose, at the beginning of the therapeutic effect. When the levodopa effect increased further to reach a full ON state, the dyskinesias largely disappeared, and gait improved considerably, consistent with previous descriptions of diphasic levodopa-induced dyskinesias [17].

O'Suilleabhain et al. [46] intended to determine if levodopa and dopamine agonists have acute depressant effects on joint position awareness and if such effects differ between dyskinetic and nondyskinetic patients. Unfortunately, the authors did not discriminate between DD and peak dose. They found that when ON, the 17 PD patients tended to score worse than controls for each of the tests of proprioception, with differences reaching significance for the elbow discrimination and matching tests but not the spatial recall.

Does the Predilection of Leg Involvement in DD Represent Differential Somatotopic Sensitivity to Levodopa?

The predilection for the leg and stereotypic movements represents a challenge in terms of basal ganglia pathophysiology. In patients with DD, typically the OFF spreads rostro-caudally (from face to feet), whereas the ON follows the opposite caudo-rostral spreading (feet first, face latest) [16]. This might suggest an initial preponderance for striatal dopaminergic depletion in the dorsal motor putamen, i.e., the lower limb region.

In PD the loss of dopaminergic neurons in the nigrostriatal system results in a dorsoventral gradient of striatal denervation with greater loss of dopaminergic terminals in the dorsolateral striatum compared to the ventromedial part [47, 48]. The striatum is somatotopically organized being the dorsolateral part the zone corresponding to the leg. If early loss of dopaminergic innervation to the striatum affects and denervates the dorsal putamen first, and if LID is associated with hypersensitivity of striatal dopaminergic receptors, one may expect that lower levodopa plasma and DA striatal levels are needed to trigger dyskinesia in the lower limb. However, why DD can stop with higher dopaminergic activity is enigmatic. We can only speculate and provide here a hypothesis to explain the observations. We propose that different levels of striatal dopaminergic activity are needed to reach normal motor control of distinct body parts; the dopaminergic threshold for the lower limb would be lower than for the face and arm (Fig. 8.2). This explains why the foot area switches ON first and OFF later than rostral body parts and also that relatively low levodopa levels become “excessive,” triggering dyskinesias. However, why do the

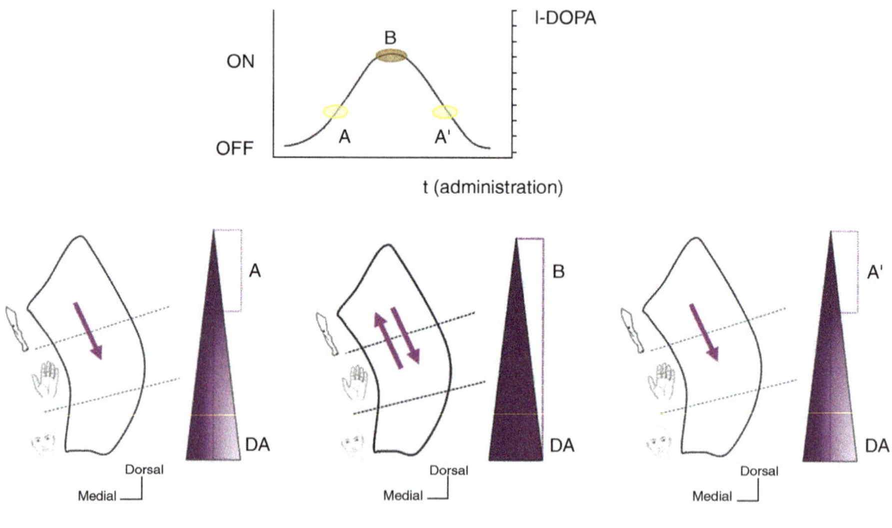

Fig. 8.2 Functional modulation of different topographical regions of the posterior putamen according to the dopaminergic gradient during the time period of diphasic dyskinesias

leg movements stop with peak dopaminergic activity? One putative explanation would be that in a normally operating striatum and motor circuit, activation of the upper limbs leads to reduced facilitation and excitability of neurons engage in lower limb mobility. In PD patients with DD, once the physiological tone of the upper limbs and face is restored, coinciding with higher levodopa levels, the lower limb excitability would be reduced, increasing the threshold for dyskinesias, which would no longer be present in the affected leg. This is suggested by the natural observation that most skilful motor activities (i.e., writing, painting, shooting, etc.) are performed while still and vice versa; walking or running is not typically associated with fine manual performance.

Finally, it is pertinent to question why DD has a net predominance to show repetitive, stereotypic movements and walking behaviors. It would appear as if the nigrostriatal denervation pattern of such PD patients would render the motor system especially sensitive to *release* such patterns with low dopaminergic stimulation. Maybe the very automatic nature of stepping and walking requires the least dopaminergic modulation.

Is There Any Difference in the Therapeutic Response Between DD and Peak Dyskinesia?

So far, functional surgery represents the most effective treatment for DD which may be completely resolved after either pallidotomy [49, 50] or thalamotomy [51]. STN-DBS studies have shown responses ranging from 30 to 40 % when evaluated separately from peak dose [52]. OFF-period dystonia, associated with neuronal hyperactivity in the STN, is directly affected by STN stimulation and disappears immediately. The effect of chronic high-frequency stimulation of the STN on diphasic and peak-dose dyskinesias is more complex and is related directly to the functional inhibition of the STN and indirectly to the replacement of the pulsatile dopaminergic stimulation by continuous functional inhibition of the STN.

Conclusions

DD is common in PD patients treated with levodopa, but over the last two decades the severity of this motor complication has waned, progressively becoming a rare clinical problem. The particular sensitivity of a proportion of PD patients to develop DD is not understood, and a recent study suggested genotypic differences compared with PD patients showing "peak-dose" dyskinesias only. The net predominance of the lower limb in DD is a fascinating mystery of which resolution could lead to important advances in the functional anatomy of the basal ganglia and the understanding of the onset of nigrostriatal degeneration in PD.

References

1. Del Sorbo F, Albanese A. Levodopa-induced dyskinesias and their management. J Neurol. 2008;255 Suppl 4:32–41.
2. Manson A, Stirpe P, et al. Levodopa-induced-dyskinesias clinical features, incidence, risk factors, management and impact on quality of life. J Parkinsons Dis. 2012;2(3):189–98.
3. Luquin MR, Vaamonde J, et al. Levodopa and 3-O-methyldopa plasma levels in parkinsonian patients with stable and fluctuating motor response. Clin Neuropharmacol. 1989;12(1): 46–54.
4. Obeso JA, Grandas F, et al. Motor complications associated with chronic levodopa therapy in Parkinson's disease. Neurology. 1989;39(11 Suppl 2):11–9.
5. Nutt JG. Levodopa-induced dyskinesia: review, observations, and speculations. Neurology. 1990;40(2):340–5.
6. Fahn S. The spectrum of levodopa-induced dyskinesias. Ann Neurol. 2000;47(4 Suppl 1):S2–9; discussion S9–11.
7. Schrag A, Quinn N. Dyskinesias and motor fluctuations in Parkinson's disease. A community-based study. Brain. 2000;123(Pt 11):2297–305.
8. Murphy MJ. L-dopa induced dyskinesias in asymmetrical parkinsonism. Acta Neurol Latinoam. 1979;25(1–2):47–52.
9. Marconi R, Lefebvre-Caparros D, et al. Levodopa-induced dyskinesias in Parkinson's disease phenomenology and pathophysiology. Mov Disord. 1994;9(1):2–12.
10. Lhermitte F, Agid Y, et al. "Beginning and end of dose" dyskinesias caused by L-DOPA. Rev Neurol (Paris). 1977;133(5):297–308.
11. Tolosa ES, Maritn WE, et al. Letter: Dyskinesias during levodopa therapy. Lancet. 1975;1(7921):1381–2.
12. Muenter MD, Sharpless NS, et al. Patterns of dystonia ("I-D-I" and "D-I-D-") in response to l-dopa therapy for Parkinson's disease. Mayo Clin Proc. 1977;52(3):163–74.
13. Muenter MD, Tyce GM. L-dopa therapy of Parkinson's disease: plasma L-dopa concentration, therapeutic response, and side effects. Mayo Clin Proc. 1971;46(4):231–9.
14. Lhermitte F, Agid Y, et al. Onset and end-of-dose levodopa-induced dyskinesias. Possible treatment by increasing the daily doses of levodopa. Arch Neurol. 1978;35(5):261–3.
15. Jankovic J. Motor fluctuations and dyskinesias in Parkinson's disease: clinical manifestations. Mov Disord. 2005;20 Suppl 11:S11–6.
16. Vidailhet M, Bonnet AM, et al. Do parkinsonian symptoms and levodopa-induced dyskinesias start in the foot? Neurology. 1994;44(9):1613–6.
17. Ruzicka E, Zarubova K, et al. "Silly walks" in Parkinson's disease: unusual presentation of dopaminergic-induced dyskinesias. Mov Disord. 2011;26(9):1782–4.
18. Merello M, Lees AJ. Beginning-of-dose motor deterioration following the acute administration of levodopa and apomorphine in Parkinson's disease. J Neurol Neurosurg Psychiatry. 1992;55(11):1024–6.
19. Luquin MR, Scipioni O, et al. Levodopa-induced dyskinesias in Parkinson's disease: clinical and pharmacological classification. Mov Disord. 1992;7(2):117–24.
20. Chapuis S, Ouchchane L, et al. Impact of the motor complications of Parkinson's disease on the quality of life. Mov Disord. 2005;20(2):224–30.
21. Barbeau A. Letter: Diphasic dyskinesia during levodopa therapy. Lancet. 1975;1(7909):756.
22. Quinn N, Critchley P, et al. Young onset Parkinson's disease. Mov Disord. 1987;2(2):73–91.
23. Warren Olanow C, Kieburtz K, et al. Factors predictive of the development of Levodopa-induced dyskinesia and wearing-off in Parkinson's disease. Mov Disord. 2013;28(8): 1064–71.
24. Olanow CW, Gracies JM, et al. Clinical pattern and risk factors for dyskinesias following fetal nigral transplantation in Parkinson's disease: a double blind video-based analysis. Mov Disord. 2009;24(3):336–43.

25. Quinn N, Marsden CD, et al. Complicated response fluctuations in Parkinson's disease: response to intravenous infusion of levodopa. Lancet. 1982;2(8295):412–5.
26. Vaamonde J, Luquin MR, et al. Subcutaneous lisuride infusion in Parkinson's disease. Response to chronic administration in 34 patients. Brain. 1991;114(Pt 1B):601–17.
27. Krack P, Pollak P, et al. From off-period dystonia to peak-dose chorea. The clinical spectrum of varying subthalamic nucleus activity. Brain. 1999;122(Pt 6):1133–46.
28. Carlsson T, Carta M, et al. Serotonin neuron transplants exacerbate L-DOPA-induced dyskinesias in a rat model of Parkinson's disease. J Neurosci. 2007;27(30):8011–22.
29. Carlsson T, Carta M, et al. Impact of grafted serotonin and dopamine neurons on development of L-DOPA-induced dyskinesias in parkinsonian rats is determined by the extent of dopamine neuron degeneration. Brain. 2009;132(Pt 2):319–35.
30. Durif F, Deffond D, et al. Apomorphine and diphasic dyskinesia. Clin Neuropharmacol. 1994;17(1):99–102.
31. Lee JY, Cho J, et al. Differential genetic susceptibility in diphasic and peak-dose dyskinesias in Parkinson's disease. Mov Disord. 2011;26(1):73–9.
32. Bergman H, Wichmann T, et al. The primate subthalamic nucleus. II. Neuronal activity in the MPTP model of parkinsonism. J Neurophysiol. 1994;72(2):507–20.
33. Brown P, Oliviero A, et al. Dopamine dependency of oscillations between subthalamic nucleus and pallidum in Parkinson's disease. J Neurosci. 2001;21(3):1033–8.
34. Alonso-Frech F, Zamarbide I, et al. Slow oscillatory activity and levodopa-induced dyskinesias in Parkinson's disease. Brain. 2006;129(Pt 7):1748–57.
35. Lopez-Azcarate J, Tainta M, et al. Coupling between beta and high-frequency activity in the human subthalamic nucleus may be a pathophysiological mechanism in Parkinson's disease. J Neurosci. 2010;30(19):6667–77.
36. Rodriguez-Oroz MC, Lopez-Azcarate J, et al. Involvement of the subthalamic nucleus in impulse control disorders associated with Parkinson's disease. Brain. 2011;134(Pt 1):36–49.
37. Brown P, Williams D. Basal ganglia local field potential activity: character and functional significance in the human. Clin Neurophysiol. 2005;116(11):2510–9.
38. Williams D, Kuhn A, et al. The relationship between oscillatory activity and motor reaction time in the parkinsonian subthalamic nucleus. Eur J Neurosci. 2005;21(1):249–58.
39. Meissner W, Ravenscroft P, et al. Increased slow oscillatory activity in substantia nigra pars reticulata triggers abnormal involuntary movements in the 6-OHDA-lesioned rat in the presence of excessive extracellular striatal dopamine. Neurobiol Dis. 2006;22(3):586–98.
40. Quiroga-Varela A, Walters JR, et al. What basal ganglia changes underlie the parkinsonian state? The significance of neuronal oscillatory activity. Neurobiol Dis. 2013;58:242–8.
41. Alam M, Capelle HH, et al. Effect of deep brain stimulation on levodopa-induced dyskinesias and striatal oscillatory local field potentials in a rat model of Parkinson's disease. Brain Stimul. 2014;7(1):13–20.
42. Halje P, Tamte M, et al. Levodopa-induced dyskinesia is strongly associated with resonant cortical oscillations. J Neurosci. 2012;32(47):16541–51.
43. Alegre M, Lopez-Azcarate J, et al. Subthalamic activity during diphasic dyskinesias in Parkinson's disease. Mov Disord. 2012;27(9):1178–81.
44. Filipovic SR, Bhatia KP, et al. 1-Hz repetitive transcranial magnetic stimulation and diphasic dyskinesia in Parkinson's disease. Mov Disord. 2013;28(2):245–6.
45. Moore AP. Impaired sensorimotor integration in parkinsonism and dyskinesia: a role for corollary discharges? J Neurol Neurosurg Psychiatry. 1987;50(5):544–52.
46. O'Suilleabhain P, Bullard J, et al. Proprioception in Parkinson's disease is acutely depressed by dopaminergic medications. J Neurol Neurosurg Psychiatry. 2001;71(5):607–10.
47. Blesa J, Pifl C, et al. The nigrostriatal system in the presymptomatic and symptomatic stages in the MPTP monkey model: a PET, histological and biochemical study. Neurobiol Dis. 2012;48(1):79–91.
48. Kordower JH, Olanow CW, et al. Disease duration and the integrity of the nigrostriatal system in Parkinson's disease. Brain. 2013;136(Pt 8):2419–31.

49. Lozano AM, Lang AE, et al. Effect of GPi pallidotomy on motor function in Parkinson's disease. Lancet. 1995;346(8987):1383–7.
50. Baron MS, Vitek JL, et al. Treatment of advanced Parkinson's disease by posterior GPi pallidotomy: 1-year results of a pilot study. Ann Neurol. 1996;40(3):355–66.
51. Narabayashi H, Yokochi F, et al. Levodopa-induced dyskinesia and thalamotomy. J Neurol Neurosurg Psychiatry. 1984;47(8):831–9.
52. Krack P, Hamel W, et al. Surgical treatment of Parkinson's disease. Curr Opin Neurol. 1999;12(4):417–25.

Chapter 9
Pharmacological Properties of Levodopa

Philippe Huot

Abstract Therapy with L-3,4-dihydroxyphenylalanine (levodopa, L-DOPA), the immediate precursor of dopamine, is the most effective treatment for Parkinson's disease. However, despite its undeniable antiparkinsonian efficacy, L-DOPA administration does not recapitulate dopaminergic transmission as it occurs under physiological conditions, notably because of short plasma half-life and variable absorption. Hence, in the non-parkinsonian state, dopamine levels in the striatum are constantly maintained above a threshold and can further increase, in a stimulus-dependent manner. In contrast, in PD, dopamine levels are markedly reduced and transiently increased with each administration of L-DOPA, leading to variable levels of striatal dopamine, alternating between peaks and troughs. Such fluctuations in dopamine levels lead to discontinuous, pulsatile stimulation of dopamine receptors within the striatum, which is thought to be a key determinant underlying the dyskinetic state. Therapeutic approaches that would produce continuous drug delivery (CDD) and ensuing continuous dopaminergic stimulation (CDS) have been actively sought. After a brief historical review, this chapter summarizes the biochemical and pharmacokinetic properties of L-DOPA. Results of studies assessing the effect of CDD/CDS paradigms on dyskinesia development and on the severity of dyskinesia once it has developed are then presented.

Keywords L-DOPA • Pharmacokinetics • Pulsatility • Continuous dopaminergic stimulation • Dyskinesia

---

Disclosures: None

P. Huot
Division of Neurology, Department of Pharmacology, Faculty of Medicine, Université de Montréal and Centre Hospitalier de l'Université de Montréal, 1560 rue Sherbrooke Est, Room GR-1177, Montréal, QC, Canada
e-mail: p.huot@umontreal.ca

S.H. Fox, J.M. Brotchie (eds.), *Levodopa-Induced Dyskinesia in Parkinson's Disease*, DOI 10.1007/978-1-4471-6503-3_9

A Brief History of L-DOPA in Parkinson's Disease

Dopamine was first synthesized in 1910 [1], and its precursor, D,L-3,4-dihydroxyphenylalanine (D,L-DOPA), was synthesized the following year [2]. The enzyme aromatic L-amino acid decarboxylase (AADC or DOPA decarboxylase) was discovered in 1938 [3]. An anti-akinetic effect of D,L-DOPA was demonstrated in 1957, when it was administered to rabbits that had received the vesicle-depleting agent reserpine [4], a finding that would be later reproduced in human [5]. At the same time, it was also demonstrated that reserpine induces dopamine depletion that is reversed by L-3,4-dihydroxyphenylalanine (levodopa, L-DOPA) administration [6]. In their groundbreaking study, Ehringer and Hornykiewicz discovered that dopamine levels were reduced in the striatum of patients suffering from Parkinson's disease (PD) [7]. Hornykiewicz later correlated most of PD motor symptoms with striatal dopamine depletion [8]. Approximately at the same time, a reduction of dopamine excretion in the urine of PD patients was reported [9].

These seminal studies led to the introduction of dopamine replacement therapy, in the early 1960s, with the dopamine precursor L-DOPA. In 1961, Birkmayer and Hornykiewicz administered intravenous L-DOPA to PD patients and found an important reversal of parkinsonian disability [10]. The following year, Barbeau et al. reported an improvement of parkinsonism, mostly rigidity, after the administration of oral L-DOPA to PD patients [11] and would later report further observations regarding L-DOPA efficacy [12, 13]. In the second half of the 1960s, Cotzias et al. reported the usefulness of high-dose L-DOPA to alleviate parkinsonism, mostly bradykinesia and rigidity [14–17]. In 1969, a double-blind study encompassing a placebo treatment established the efficacy of L-DOPA [18].

L-DOPA was first administered without a peripherally acting AADC inhibitor, such as carbidopa or benserazide, and the doses of L-DOPA required to achieve therapeutic efficacy were higher than those used today. AADC inhibitors were added to L-DOPA during the second half of the 1960s and first half of the 1970s [19–21]. AADC inhibitors allowed to reduce the doses of L-DOPA, a faster onset of antiparkinsonian benefit and a decrease in side effects of dopamine on the cardiovascular and gastrointestinal systems, such as hypotension, nausea, vomiting, and anorexia [16, 22–24].

Biochemistry of L-DOPA

Tyrosine hydroxylase (tyrosine 3-monooxygenase, TH) is the rate-limiting enzyme in the formation of L-DOPA, dopamine, and catecholamines. TH requires oxygen and the cofactors tetrahydrobiopterin and iron to metabolize L-tyrosine into L-DOPA [25]. The role of TH is not so critical in PD, where patients are administered L-DOPA directly. L-DOPA is transformed into dopamine by the enzyme AADC, which requires pyridoxal phosphate as a cofactor [26, 27]. AADC is also present in serotonergic neurons, where it converts 5-hydroxytryptophan into 5-hydroxytryptamine (serotonin, 5-HT) [27, 28]. AADC is also encountered outside of the brain [29], which is why L-DOPA is administered with a peripherally acting AADC inhibitor that does not cross the blood-brain barrier, thereby allowing for more L-DOPA to enter the brain. AADC inhibition allows reducing L-DOPA doses required by as much as 60–80 % [30, 31]. At least 50–75 mg of an AADC inhibitor, sometimes more, are necessary on a daily basis [1, 32].

L-DOPA is absorbed in the duodenum and proximal jejunum by active transport via the large neutral amino acid (LNAA) system [33, 34]. L-DOPA is also transported actively in the brain via the LNAA system [35, 36] present in the endothelial cells and astrocytes [37]. Because L-DOPA enters the body and the brain by active transport, it has to compete with dietary proteins and amino acids [38], and high protein intake, even in the absence of fluctuations of L-DOPA plasma levels, can reduce L-DOPA antiparkinsonian action. For instance, in a study where L-DOPA was delivered intraduodenally, motor performance declined, despite stable plasma L-DOPA levels, following oral protein intake [39]. In another study, where L-DOPA was delivered intravenously, the administration of a high-protein meal reduced the antiparkinsonian efficacy of L-DOPA, without altering its plasma concentration [40]. That last study also showed that meals interfere with the absorption of oral L-DOPA and reduce L-DOPA peak plasma levels [40]. Compared with plasma fluctuations of L-DOPA, plasma fluctuations of LNAA throughout the day are small and are believed to contribute to about 10 % of the variability of L-DOPA entry within the brain [41, 42].

Once in the brain, L-DOPA is transformed into dopamine by the enzyme AADC [43], as mentioned above. AADC is found in catecholaminergic neurons in the brain [44], notably dopaminergic neurons from the substantia nigra and their projections, as well as in 5-HT neurons from the raphe complex and their striatal projections [45, 46]. Some interneurons within the striatum also harbor the enzyme AADC [47, 48]. These three neuronal populations represent sites where L-DOPA is metabolized in dopamine in the striatum. In PD, with the degeneration of the nigrostriatal system, L-DOPA is transformed in dopamine mostly by raphe-striatal 5-HT neurons and, to a lesser extent, striatal intrinsic AADC-containing interneurons.

Following synaptic release, dopamine exerts its affinity through interaction with dopamine D_{1-5} receptors [49, 50], to which it binds with moderate/high affinity. Dopamine also binds to some non-dopaminergic receptors, as well as to monoaminergic transporters [51]. The dopamine transporter (DAT) transports dopamine back into the presynaptic neuron [52, 53]. Dopamine is also transported back to the presynaptic neuron, albeit to a lesser extent, by the plasma membrane monoamine transporter (PMAT) [54, 55]. Within the presynaptic neuron, dopamine can either be recycled into the vesicles via the vesicular monoaminergic transporter type 2 ($VMAT_2$) [56, 57] or degraded. When dopamine is not reuptaken by the presynaptic neuron or repackaged in presynaptic vesicles, it undergoes degradation. Several enzymes participate in the catabolism of dopamine. Monoamine oxidase B (MAO-B) and catechol-O-methyltransferase (COMT) are important enzymes in the process, and the inhibition of their activity is an efficacious way to extend the duration of L-DOPA antiparkinsonian action [58–62]. Monoamine oxidase A (MAO-A), aldehyde dehydrogenase, and aldehyde reductase are also involved in dopamine breakdown [63].

MAOs are intracellular enzymes localized on the outer mitochondrial membrane [64, 65], and both MAO-A and MAO-B metabolize dopamine equally within the human brain [66]. MAO-A metabolizes dopamine within neurons, while MAO-B metabolizes dopamine within both neuronal and glial cells [67]. The MAOs transform dopamine into 3,4-dihydroxyphenylacetic acid (DOPAC) [8, 68]; 3,4-dihydroxyphenylacetaldehyde (DOPAL) and 3,4-dihydroxyphenylethanol (DOPET) are intermediate products in the metabolism of dopamine and are transformed into DOPAC by alcohol dehydrogenase and aldehyde dehydrogenase, respectively [69, 70]. A fraction of dopamine is also autoxidated by interaction with molecular oxygen [71].

There are two isoforms of COMT [72]. COMT is a cytoplasmic and intranuclear enzyme [73, 74] that is found within both neurons and glia [75]. COMT is also encountered in the extracellular space [76]. COMT transforms L-DOPA into 3-O-methyldopa (3-OMD) and dopamine into 3-methoxytyramine (3-MT) [73, 77]. 3-MT is then converted to homovanillic acid (HVA) by the MAO and, to a lesser extent, the aldehyde dehydrogenase, whereas DOPAC is converted to HVA by the COMT [78]. Other metabolic pathways also exist where dopamine is metabolized, notably by dopamine beta-hydroxylase in noradrenergic neurons [79], but these will not be reviewed here.

Pharmacokinetics of L-DOPA

In the clinic, L-DOPA half-life (T½) is short and lasts 1.5–2 h [80–85]. After oral administration, L-DOPA plasma levels are maximal (Tmax) approximately 1 h after intake, although the unpredictable absorption discussed above introduces variability [84, 86]. Tmax and the maximal plasma levels (Cmax) are variable and depend on the dose administered, as well as on the time since last meal or the gastric-emptying time [87]. Following the administration of L-DOPA/AADC inhibitor 100/25 mg, L-DOPA plasma levels reach 6–10 nmol/ml [85, 88, 89], whereas plasma levels reach 13–17 nmol/ml following the administration of L-DOPA/AADC inhibitor 200/50 mg [84, 89]. The threshold for minimal clinical effect is estimated to be around 7 nmol/ml [89, 90]. L-DOPA plasma levels are 10–15-fold higher than L-DOPA levels in the ventricular cerebrospinal fluid [91]. About 25–50 % of L-DOPA in the plasma is bound to proteins [92, 93]; part of this variability comes from the techniques employed to determine protein-bound and unbound L-DOPA [94]. L-DOPA clearance is ≈0.4 l/min [95].

COMT inhibitors are commonly used in the clinic to extend the duration of L-DOPA antiparkinsonian action, on-time. Tolcapone increases the area under the curve (AUC) when administered with L-DOPA, but it does not increase plasma Cmax [96]. Entacapone also increases L-DOPA AUC by 20–40 % and prolongs

L-DOPA T½ by ≈40 % [97–99], without changing L-DOPA Cmax or Tmax [99]. Few studies assessing the effects of selegiline on L-DOPA pharmacokinetic parameters have been performed. Selegiline may increase L-DOPA T½ by as much as ≈90 % [95], without increasing Cmax [100]. To the best knowledge of this author, studies assessing the effect of rasagiline on L-DOPA pharmacokinetic parameters have not been performed in clinical settings. However, in a microdialysis study performed in the rat, rasagiline increased L-DOPA-derived dopamine concentrations by almost twofold [101]. Whether such elevations are achieved in the clinic remains to be demonstrated.

Continuous Dopaminergic Stimulation

In the non-parkinsonian, normal state, dopamine release is both tonic and phasic, with dopamine levels always remaining above a certain threshold [102, 103]. In the denervated striatum of PD, especially late in disease course, where few nigrostriatal dopaminergic neurons remain and the "buffering capacity" of the DAT has disappeared [104, 105], tonic dopamine release has disappeared, and dopamine release becomes mostly pulsatile, following each dose of L-DOPA, and parallels L-DOPA pharmacokinetic profile, the "short-duration response" [106]. However, early in disease process, repeated administration of L-DOPA leads to a sustained antiparkinsonian effect that persists hours/days after discontinuation of L-DOPA, the "long-duration response," which, at this point, may account for as much as one third to one half of the antiparkinsonian benefit conferred by L-DOPA [107, 108].

In agreement with reduced buffering and storage capacity of the nigrostriatal system in dyskinesia and possibly indicative of a more severe disease, DAT binding levels in the putamen of PD patients with dyskinesia are lower than those of PD patients without dyskinesia [109]. While a reduction of DAT levels may lead to higher synaptic dopamine concentration and, as a corollary, greater antiparkinsonian benefit, it may also enhance peak dopamine levels associated with each L-DOPA administration [110], thereby exacerbating the fluctuations in dopamine levels. In addition to the DAT, alterations in vesicular dopamine content and vesicular dopamine release might also contribute to the pulsatility of dopamine transmission in the parkinsonian striatum [111]. A third factor underlying the variability in dopamine levels might come from raphe-striatal 5-HT neurons which contain the enzyme AADC and, in advanced PD, constitute the main dopaminergic input to the striatum [28, 112–116]. However, 5-HT fibers do not contain the autoregulatory mechanisms required for physiological dopamine release, and this aberrant, compensatory dopamine release has been identified as a causative element in the pathophysiology of dyskinesia [117–120]. All of these changes lead to higher peaks and lower troughs, increasing dyskinesia severity with a shorter duration of L-DOPA antiparkinsonian action [121, 122].

Pulsatile, nonphysiological dopamine release and dopamine receptor stimulation are regarded as important factors in the induction and maintenance of dyskinesia, possibly by inducing abnormal plasticity within the striatum [123], which paved the way to the concepts of "continuous dopaminergic stimulation" (CDS) and "continuous drug delivery" (CDD), where antiparkinsonian therapy would be delivered constantly, leading to uninterrupted dopamine receptor stimulation, more akin to the physiological state. In CDD and CDS, the stimulation of dopaminergic receptors would not be a pharmacokinetic phenomenon occurring, for instance, with each L-DOPA intake, but would rather be continuous temporally and more constant in terms of plasma and brain levels of L-DOPA/dopamine achieved [124, 125].

Continuous Dopaminergic Stimulation and the Development of Dyskinesia

The concept of CDS has led to the use of longer-acting dopamine agonists as antiparkinsonian therapy in the early stages of the disease [126–128], although most of dopamine agonists available do not provide continuous antiparkinsonian benefit and their use does not prevent dyskinesia induction. However, in preclinical studies, the administration of short-acting dopamine agonists to previously untreated parkinsonian animals elicited more severe dyskinesia than longer-acting dopamine agonists or continuous infusion of dopamine agonists. For instance, in the 6-hydroxydopamine (6-OHDA)-lesioned rat, the short-acting dopamine agonist apomorphine elicited more severe dyskinesia than the longer-acting dopamine agonists pramipexole and pergolide [129]. In contrast, in the 1-methyl-4-phenyl-1,2,3,6-tetrahydropyridine (MPTP)-lesioned marmoset, initiating dopaminergic therapy with either apomorphine or pergolide led to dyskinesia of the same severity [130]. However, in the MPTP-lesioned marmoset, de novo administration of the dopamine agonists apomorphine or rotigotine once or twice daily induced more severe dyskinesia than de novo continuous administration of these two drugs [131, 132]. Similar results were obtained in the MPTP-lesioned macaque, where once daily apomorphine injections led to dyskinesia development, whereas continuous apomorphine infusion did not, even after 6 months [132].

These preclinical studies provide evidence that continuous stimulation of dopamine receptors with dopamine agonists elicits less dyskinesia than pulsatile dopamine agonist administration. However, the interpretation of studies comparing de novo dopamine agonist with de novo L-DOPA administration is more complex, as the dopamine agonists clinically available all display greater affinity for dopamine D_2/D_3 over D_1 receptors compared with L-DOPA and dopamine, which exhibit similar affinity for D_1/D_2 receptors [133–140] (Table 9.1). Because of this preferential D_2 affinity of dopamine agonists, any effect on dyskinesia development when dopamine replacement therapy is commenced with dopamine agonists instead of L-DOPA cannot be solely attributed to a longer duration of action. Notwithstanding this preferential D_2/D_3 affinity of dopamine agonists, there is some evidence, albeit some controversy persists, that starting therapy with a dopamine agonist delays the onset of dyskinesia when compared to L-DOPA.

Table 9.1 Dopamine binding affinity

Target	Affinity (Ki, nM)	References
D_1	2,340–4,900	[137, 197]
D_2	7.5–710	[139, 198]
D_3	5.6–29	[198, 199]
D_4	28	[139]
D_5	228–1,080	[137, 197]
Trace amine-associated receptor 1	422	[200]
5-HT_{1A}	1,400 (EC_{50})	[201]
DAT	890–1,200	[202]
NET	139	[203]
SERT	240,000	[204]
$VMAT_2$	1,400	[205]

In one study evaluating ropinirole monotherapy (056 study), early PD patients were treated initially with ropinirole, which successfully delayed dyskinesia onset when compared to initial therapy with L-DOPA [141, 142]. However, the dyskinesia-sparing effect of ropinirole was lost when L-DOPA was added [142], but 10 years after treatment introduction, the odds of exhibiting severe dyskinesia were greater in patients initially part of the L-DOPA group [143]. Similarly, in the CALM-PD (Comparison of the Agonist pramipexole versus Levodopa on Motor complications of Parkinson's Disease) study, significantly less subjects treated initially with pramipexole exhibited dyskinesia than subjects treated initially with L-DOPA after a 2-year follow-up [144]. After a 6-year open-label follow-up, the majority of patients initially randomized to pramipexole were taking L-DOPA, but significantly less patients initially receiving pramipexole were experiencing dyskinesia compared with patients who started directly on L-DOPA therapy [145]. Similar to ropinirole and pramipexole, in another study, initial treatment with bromocriptine for 3 years delayed the emergence of dyskinesia when compared to initial treatment with L-DOPA [146]. In the Sydney Multicentre Study of Parkinson's disease, patients were initially randomized to receive low-dose bromocriptine or L-DOPA. No dyskinesia developed while patients were treated solely with bromocriptine, but L-DOPA had to be commenced in most of the bromocriptine-treated patients, and after 5 years, there was no difference in the prevalence or severity of dyskinesia, regardless of the initial treatment arm [147]. A 15-year follow-up encompassing one third of the patients from the initial cohort showed that 94 % of subjects were afflicted by dyskinesia [148]. The results of the Sydney Multicentre Study do not, however, argue against the CDS concept, as the half-life of bromocriptine, although longer than L-DOPA, is still relatively short [80].

Another study performed in patients not optimally treated 3 years after disease onset provided complementary results to the above trials. In that study, ropinirole prolonged release was added to L-DOPA in a subset of patients, while L-DOPA was increased in the other subgroup. Patients on ropinirole prolonged release developed significantly less dyskinesia than patients whose L-DOPA dose was increased [149].

In contrast to the studies cited above, this last trial was not a de novo study with a dopaminergic agonist and indicates that even when pulsatile treatment has been established, continuous stimulation of dopamine receptors might attenuate the process leading to the development of dyskinesia.

In clinical settings, however, it is not always possible to initiate dopaminergic therapy with a dopamine agonist or even to add a dopamine agonist to patients, for instance, when they have severe disease or psychiatric manifestations. It was suggested that combining L-DOPA with entacapone would increase L-DOPA half-life and result in less pulsatility and elicit less dyskinesia than L-DOPA alone. Preclinical studies performed in the 6-OHDA-lesioned rat [150] and the MPTP-lesioned marmoset [151, 152] suggested that it might indeed be an effective way to attenuate dyskinesia development, although another study performed in the MPTP-lesioned monkey suggested otherwise [153]. In the clinic, the STRIDE-PD (STalevo Reduction in Dyskinesia Evaluation in Parkinson's Disease) study was a 134-week clinical trial that aimed at determining if initiating L-DOPA therapy in combination with entacapone would attenuate dyskinesia development compared with commencing L-DOPA without entacapone. In contrast to what was expected, time to dyskinesia onset was shorter in patients taking L-DOPA/carbidopa/entacapone compared with patients taking L-DOPA/carbidopa, and at the end of the study, more patients had developed dyskinesia in the L-DOPA/carbidopa/entacapone group [154]. It is noteworthy that the STRIDE-PD study was not a pure de novo study, as several patients enrolled were taking dopamine agonists prior to treatment allocation. In addition, patients who developed dyskinesia were taking higher L-DOPA equivalent doses, and a follow-up of the STRIDE-PD patients for 208 weeks established that higher L-DOPA dose is an important factor to develop dyskinesia [155], in agreement with the previous literature [156]. In contrast, in the FIRST-STEP (Favorability of Immediate-Release carbidopa/levodopa vs STalevo; Short-Term comparison in Early Parkinson's) study, no difference in time to dyskinesia onset could be demonstrated after 39 weeks between L-DOPA/carbidopa/entacapone and L-DOPA/carbidopa treatments [157]. Although the STRIDE-PD and FIRST-STEP studies do not provide evidence to support the CDS concept, they do not disprove it either, because of differences in L-DOPA equivalent doses between dyskinetic and non-dyskinetic patients. Moreover, in these two studies, treatments were administered three or four times a day, which may not be sufficient to provide continuous coverage, and perhaps more frequent, smaller doses would have led to different outcomes.

Transdermal rotigotine is possibly the only clinically available pharmacological tool that enables CDD with constant plasma levels round the clock [158]. Transdermal rotigotine was administered to patients with early PD in the context of controlled trials, and no dyskinesia was reported after 41 weeks [159, 160]. In an open-label extension of one of these studies, a few patients developed dyskinesia while receiving rotigotine without L-DOPA [161]. This is an important finding that needs to be discussed. Indeed, CDD may not necessarily lead to CDS, even with constant plasma levels, as factors such as receptor desensitization and internalization may result in variability in dopamine receptor stimulation [158, 162], which perhaps explains why a subset of patients receiving continuous rotigotine therapy developed dyskinesia. Another explanation is that CDD and ensuing CDS may be effective at preventing dyskinesia in a subset of patients, perhaps with a lesser degree of nigrostriatal denervation. Another possibility though is that CDS may simply be a way to delay the emergence of dyskinesia, as several other factors have also been identified to play a role in dyskinesia [163], and perhaps dyskinesia will remain inevitable until a way to target several, if not all, of the etiological factors simultaneously is discovered.

Continuous Dopaminergic Stimulation and Established Dyskinesia

Whether CDS can attenuate dyskinesia once the dyskinetic phenotype is well established is also unclear and varies according to studies. Preclinical studies have hinted that CDS effectively alleviates established dyskinesia, but there is more variability in the clinic. For instance, in the MPTP-lesioned marmoset, continuous infusion of rotigotine [164] or administration of the long-acting dopamine agonist cabergoline [165] alleviated established dyskinesia.

A micronized suspension of L-DOPA (20 mg/ml) and carbidopa (5 mg/ml) in methylcellulose gel provides chemical and physical stability in addition to high L-DOPA concentrations and is now utilized in clinic to infuse L-DOPA/carbidopa continuously within the duodenum or proximal jejunum [166, 167]. Clinical trials using L-DOPA intestinal gel have provided mixed results with regard to dyskinesia. In two small trials, continuous intraintestinal L-DOPA delivery over a 12-h period daily for 6 months reduced the severity of established dyskinesia [168], whereas it did not diminish dyskinesia severity when L-DOPA was delivered over a 14-h period daily for 18 months [169]. However, in a recent randomized, controlled, double-blind, double-dummy multicentre study where L-DOPA was administered over a 16-h period daily for 12 weeks, the duration of on-time with troublesome dyskinesia was not significantly reduced, and no effect on dyskinesia severity was noted, although patients enrolled all had low baseline dyskinesia and an effect on dyskinesia severity was not the primary end point of the study [170]. Nevertheless, this lack of effect on dyskinesia severity was a finding difficult to reconcile with the CDS concept, and it was suggested that a 3-month period may not be long enough to reverse the molecular changes characterizing the dyskinetic state [171]. In agreement with that possibility, an open-label study where L-DOPA was infused in the jejunum over 16 h daily for 54 weeks found a reduction of on-time with troublesome dyskinesia [172]. Importantly, continuous intraintestinal delivery of L-DOPA may not completely abolish the need for oral antiparkinsonian medication [173], in which case a certain pulsatility of L-DOPA administration and ensuing dopamine plasma and brain concentrations would remain. Under such circumstances, the intraintestinal L-DOPA infusion might be akin to the tonic dopamine release evoked above, while oral L-DOPA might be akin to phasic dopamine release and perhaps such a mode of dopamine replacement therapy would mimic more closely the physiological state, although this remains speculative.

However, intraintestinal infusion of L-DOPA over a 16-h period daily probably does not totally eliminate pulsatile stimulation of dopamine receptors, even if delivered at a constant rate in the absence of concurrent oral dopaminergic therapy. Hence, as seen above, despite stable plasma levels, meals and protein intake interfere with brain entry of L-DOPA, and a certain variability of L-DOPA delivered to the brain appears inevitable. In addition, there will be a time, when the intestinal pump is switched off, where dopamine levels will fall to a relative trough and, when the intestinal pump is switched back on, a relative peak will occur. As such, any drug delivery paradigm that does not provide round-the-clock L-DOPA delivery may always encompass a certain degree of pulsatility, and perhaps this is why the effects of intraintestinal L-DOPA infusion on dyskinesia severity are variable. Accordingly, clinical studies in which dopaminergic agents were administered without interruption for extended periods of time have shown a reduction of previously established dyskinesia. For instance, continuous, uninterrupted intravenous infusion of the dopamine agonist lisuride over a 3-month period led to a significant reduction of dyskinesia severity [174]. Shorter continuous, uninterrupted intravenous administration of dopaminergic agents is also sufficient to modulate the dyskinesia threshold, although the optimal duration has not been established. Continuous intravenous administration of L-DOPA over 48 h did not reduce the dyskinesia threshold in one study [175], while 7–12 days may be sufficient [176]. Of course, further studies, encompassing more patients, are needed to confirm the results of these trials where dopaminergic agents were infused uninterruptedly.

In a placebo-controlled study where transdermal rotigotine was administered to PD patients for 24 weeks, the duration of on-time with troublesome dyskinesia was unchanged, while the duration of on-time without troublesome dyskinesia was increased in the rotigotine arm vs. the placebo arm, possibly reflecting greater duration of antiparkinsonian benefit. In this study, adjusting L-DOPA doses was permitted, notably to reduce dyskinesia severity, if needed [177]; as such, the conclusions that can be drawn on continuous rotigotine delivery on dyskinesia intensity from the trial are limited. In another study, transdermal rotigotine also increased the duration of on-time without troublesome dyskinesia, here again possibly reflecting an effect related to on-time rather than a reduction of dyskinesia per se, although more patients experienced dyskinesia in the rotigotine group and had to have their L-DOPA dose reduced than in the placebo group [178]. In an open-label extension of these studies, continuous rotigotine administration did not prevent the development of dyskinesia in patients previously devoid of dyskinesia, which developed at the incidence of 4–8 % per patient-year over 4–6 years [179]. The results of these studies with transdermal rotigotine coupled with L-DOPA are difficult to interpret when it comes to CDS. As rotigotine is a preferential D_3/D_2 dopamine agonist, perhaps these receptors were stimulated constantly, while D_1 receptors might have been stimulated in a discontinuous way with each L-DOPA administration. Although such an explanation remains speculative, it could explain, at least partly, why continuous rotigotine administration did not alleviate or prevent the development of dyskinesia.

A concern that was raised with technologies leading to CDD such as intraintestinal administration of L-DOPA or dopamine agonists is whether or not tolerance will occur over time, especially if drugs are administered uninterruptedly over 24 h [180, 181].

The Quest for Continuous L-DOPA Delivery

Currently, two strategies are available clinically to achieve CDD, transdermal rotigotine, which delivers drug for 24 h, and intraintestinal L-DOPA infusion, which delivers L-DOPA for 12–16 h daily [162]. Although transdermal drug delivery may be a good way to achieve CDS, the technology is available only with rotigotine, and intraintestinal delivery of L-DOPA, necessitating a surgical procedure, is too invasive to be offered as a treatment in the early stages of PD, where oral therapy is available. Research is ongoing to develop an oral formulation of L-DOPA that would provide constant delivery and stable plasma levels over a prolonged period of time.

IPX066 is a dual immediate- and extended-release formulation of L-DOPA/carbidopa that maintains L-DOPA plasma levels above 50 % of Cmax for ≈4.0 h, in contrast to ≈1.4 h for immediate-release L-DOPA/carbidopa [82]. In a study performed in L-DOPA-naïve PD patients, IPX066 induced dyskinesia in about 5 % of patients over a 30-week time period [182], which seems comparable to figures obtained when immediate-release L-DOPA/carbidopa is administered [142]. In PD patients with motor complications, IPX066 significantly increased the duration of both on-time without dyskinesia and on-time without troublesome dyskinesia in a 13-week phase III study [183].

L-DOPA prodrugs that would be absorbed anywhere in the intestine could theoretically provide constant L-DOPA delivery and maintain plasma and brain levels constant [184]. XP21279 is a prodrug of L-DOPA that is being actively absorbed throughout the gastrointestinal tract. A sustained-release XP21279–carbidopa bilayer tablet was developed [185]. When administered to PD patients in a randomized, double-blind phase II crossover study with regular L-DOPA/carbidopa, XP21279 increased the duration of on-time without troublesome dyskinesia but also worsened existing, or led to the development of, dyskinesia in 21 % of patients taking the drug; it is noteworthy that patients were already taking L-DOPA prior to enter the study [186]. Based on these results, switching to XP21279 may not be a way to prevent the emergence of dyskinesia in patients previously treated with immediate-release L-DOPA.

Studies presented so far only as abstracts have reported data on a sustained-release "accordion" L-DOPA/carbidopa pill [187, 188]. This accordion pill has better absorption and pharmacokinetic profiles than regular L-DOPA/carbidopa and leads to more constant plasma levels. In a phase II multicentre study, troublesome dyskinesia was decreased in some patients [188].

Polymer-coated sustained-release L-DOPA/carbidopa tablets have been developed [189–191] that could theoretically lead to prolonged L-DOPA release, but they still need to be tested in PD patients.

L-DOPA patches and rods that would enable continuous and stable delivery of L-DOPA have been suggested as potential approaches to achieve CDD and CDS, but difficulties with L-DOPA solubility, stability, and propensity to oxidize limit the use of these therapeutic paradigms at the moment [192, 193]. Such a patch (ND0611) was developed with L-DOPA ethyl ester as the active ingredient [166, 194]. A phase I/IIa study assessing the safety and tolerability of continuous subcutaneous delivery of L-DOPA/carbidopa is currently underway (clinicaltrials.gov identifier: NCT01725802) [195].

Concluding Remarks

L-DOPA is the most efficacious antiparkinsonian agent, and with disease progression, virtually all patients ultimately end up taking it [196]. However, its limited absorption and short half-life do not recapitulate all of the features of dopaminergic transmission under physiological conditions, but rather lead to pulsatile, discontinuous stimulation of dopamine receptors, which is thought to be a core factor in the development of dyskinesia. CDS is regarded as a way to achieve sustained stimulation of dopamine receptors; however, the pharmacological tools to achieve CDS remain limited and may not reproduce the physiological state either. For instance, although transdermal rotigotine allows for round-the-clock stimulation of dopaminergic receptors, unlike dopamine, it exhibits a strong preference for D_3, then D_2, and D_1 receptors; as such, it does not completely mimic the action of the endogenous transmitter. Intestinal delivery of L-DOPA/carbidopa is certainly attractive, but may not fully recapitulate physiological conditions either. For instance, in the normal state, movement itself can influence dopamine release and ensuing dopamine levels [158]; such a tight regulation of dopamine is currently not possible with the available technologies. Moreover, intraintestinal delivery of L-DOPA requires an abdominal surgery and is an invasive approach that may be difficult to justify in early PD where a good control of symptoms can be achieved with oral dopamine agonists or L-DOPA, although these treatments lead to pulsatile dopamine receptor stimulation and, eventually, motor complications. Research is ongoing to develop oral therapies that will lead to constant L-DOPA delivery and more steady plasma levels, hopefully allowing for more physiological dopamine receptor stimulation and preventing the emergence of dyskinesia.

Acknowledgments This work was supported by the Department of Pharmacology and Faculty of Medicine of Université de Montréal.

References

1. Fahn S. The history of dopamine and levodopa in the treatment of Parkinson's disease. Mov Disord. 2008;23 Suppl 3:S497–508.
2. Funk C. Synthesis of dl-3:4-dihydroxyphenylalanine. J Chem Soc. 1911;99:554–7.
3. Holtz P, Heise R, Ludtke K. Fermentativer abbau von Ldioxyphenylalanin (Dopa) durch niere. Naunyn-Schmiedeberg's Arch Exp Path Pharmak. 1938;191:87–118.
4. Carlsson A, Lindqvist M, Magnusson T. 3,4-Dihydroxyphenylalanine and 5-hydroxytryptophan as reserpine antagonists. Nature. 1957;180:1200.
5. Degkwitz R, Frowein R, Kulenkampff C, Mohs U. Uber die wirkungen des L-dopa beim menschen und deren beeinflussung durch Reserpin, Chlorpromazin, Iproniazid und Vitamin B6. Klin Wochenschr. 1960;38:120–3.
6. Carlsson A, Lindqvist M, Magnusson T, Waldeck B. On the presence of 3-hydroxytyramine in brain. Science. 1958;127:471.
7. Ehringer H, Hornykiewicz O. [Distribution of noradrenaline and dopamine (3-hydroxytyramine) in the human brain and their behavior in diseases of the extrapyramidal system.]. Klin Wochenschr. 1960;38:1236–9.
8. Hornykiewicz O. Dopamine (3-hydroxytyramine) and brain function. Pharmacol Rev. 1966;18:925–64.
9. Barbeau A, Murphy GF, Sourkes TL. Excretion of dopamine in diseases of basal ganglia. Science. 1961;133:1706–7.
10. Birkmayer W, Hornykiewicz O. [The L-3,4-dioxyphenylalanine (DOPA)-effect in Parkinson-akinesia]. Wien Klin Wochenschr. 1961;73:787–8.
11. Barbeau A, Sourkes TL, Murphy GF. Les catécholamines dans la maladie de Parkinson. In: de Ajuriaguerra J, editor. Monoamines et système nerveux central. Geneva: Georg & Cie SA; 1962. p. 247–69.
12. Barbeau A. L-dopa therapy in Parkinson's disease: a critical review of nine years' experience. Can Med Assoc J. 1969;101:59–68.
13. Barbeau A. [Treatment of Parkinson's disease by L-DOPA]. Union Med Can. 1969;98:183–6.
14. Cotzias GC. Parkinsonism and dopa. J Chronic Dis. 1969;22:297–301.
15. Cotzias GC, Papavasiliou PS, Gellene R. L-dopa in parkinson's syndrome. N Engl J Med. 1969;281:272.
16. Cotzias GC, Papavasiliou PS, Gellene R. Modification of Parkinsonism–chronic treatment with L-dopa. N Engl J Med. 1969;280:337–45.
17. Cotzias GC, Van Woert MH, Schiffer LM. Aromatic amino acids and modification of parkinsonism. N Engl J Med. 1967;276:374–9.
18. Yahr MD, Duvoisin RC, Schear MJ, Barrett RE, Hoehn MM. Treatment of parkinsonism with levodopa. Arch Neurol. 1969;21:343–54.
19. Marsden CD, Parkes JD, Rees JE. A year's comparison of treatment of patients with parkinson's disease with levodopa combined with carbidopa versus treatment with levodopa alone. Lancet. 1973;2:1459–62.
20. Marsden CD, Barry PE, Parkes JD, Zilkha KJ. Treatment of Parkinson's disease with levodopa combined with L-alpha-methyldopahydrazine, an inhibitor of extracerebral DOPA decarboxylase. J Neurol Neurosurg Psychiatry. 1973;36:10–4.
21. Barbeau A, Gillo-Joffroy L, Mars H. Treatment of Parkinson's disease with levodopa and Ro 4-4602. Clin Pharmacol Ther. 1971;12:353–9.
22. Editorial: Dopa decarboxylase inhibitors. Br Med J. 1974;4:250–1.
23. Birkmayer W, Linauer W, Mentasti M, Riederer P. 2-year experiences with a combination treatment of Parkinsonism with L-dopa and a decarboxylase inhibitor (Benserazid, Ro 4–4602). Wien Med Wochenschr. 1974;124:340–4.
24. Sweet RD, McDowell FH, Wasterlain CG, Stern PH. Treatment of "on-off effect" with a dopa decarboxylase inhibitor. Arch Neurol. 1975;32:560–3.

25. Nagatsu T. Tyrosine hydroxylase: human isoforms, structure and regulation in physiology and pathology. Essays Biochem. 1995;30:15–35.
26. Opacka-Juffry J, Brooks DJ. L-dihydroxyphenylalanine and its decarboxylase: new ideas on their neuroregulatory roles. Mov Disord. 1995;10:241–9.
27. Sumi-Ichinose C, Ichinose H, Takahashi E, Hori T, Nagatsu T. Molecular cloning of genomic DNA and chromosomal assignment of the gene for human aromatic L-amino acid decarboxylase, the enzyme for catecholamine and serotonin biosynthesis. Biochemistry. 1992;31:2229–38.
28. Tison F, Normand E, Jaber M, Aubert I, Bloch B. Aromatic L-amino-acid decarboxylase (DOPA decarboxylase) gene expression in dopaminergic and serotoninergic cells of the rat brainstem. Neurosci Lett. 1991;127:203–6.
29. Christenson JG, Dairman W, Udenfriend S. On the identity of DOPA decarboxylase and 5-hydroxytryptophan decarboxylase (immunological titration-aromatic L-amino acid decarboxylase-serotonin-dopamine-norepinephrine). Proc Natl Acad Sci USA. 1972;69:343–7.
30. Hauser RA. Levodopa: past, present, and future. Eur Neurol. 2009;62:1–8.
31. Nutt JG, Woodward WR, Anderson JL. The effect of carbidopa on the pharmacokinetics of intravenously administered levodopa: the mechanism of action in the treatment of parkinsonism. Ann Neurol. 1985;18:537–43.
32. LeWitt PA. Levodopa for the treatment of Parkinson's disease. N Engl J Med. 2008;359:2468–76.
33. Wade DN, Mearrick PT, Morris JL. Active transport of L-dopa in the intestine. Nature. 1973;242:463–5.
34. LeWitt PA. Levodopa therapeutics: new treatment strategies. Neurology. 1993;43:S31–7.
35. Kageyama T, Nakamura M, Matsuo A, Yamasaki Y, Takakura Y, Hashida M, et al. The 4F2hc/LAT1 complex transports L-DOPA across the blood-brain barrier. Brain Res. 2000;879:115–21.
36. Pardridge WM. Transport of small molecules through the blood-brain barrier: biology and methodology. Adv Drug Deliv Rev. 1995;15:5–36.
37. Inyushin MY, Huertas A, Kucheryavykh YV, Kucheryavykh LY, Tsydzik V, Sanabria P, et al. L-DOPA uptake in astrocytic endfeet enwrapping blood vessels in rat brain. Parkinsons Dis. 2012;2012:321406.
38. Mena I, Cotzias GC. Protein intake and treatment of Parkinson's disease with levodopa. N Engl J Med. 1975;292:181–4.
39. Frankel JP, Kempster PA, Bovingdon M, Webster R, Lees AJ, Stern GM. The effects of oral protein on the absorption of intraduodenal levodopa and motor performance. J Neurol Neurosurg Psychiatry. 1989;52:1063–7.
40. Nutt JG, Woodward WR, Hammerstad JP, Carter JH, Anderson JL. The "on-off" phenomenon in Parkinson's disease. Relation to levodopa absorption and transport. N Engl J Med. 1984;310:483–8.
41. Nutt JG, Woodward WR, Carter JH, Trotman TL. Influence of fluctuations of plasma large neutral amino acids with normal diets on the clinical response to levodopa. J Neurol Neurosurg Psychiatry. 1989;52:481–7.
42. Eriksson T, Magnusson T, Carlsson A, Linde A, Granerus AK. "On-off" phenomenon in Parkinson's disease: correlation to the concentration of dopa in plasma. J Neural Transm. 1984;59:229–40.
43. Berry MD, Juorio AV, Li XM, Boulton AA. Aromatic L-amino acid decarboxylase: a neglected and misunderstood enzyme. Neurochem Res. 1996;21:1075–87.
44. Beltramo M, Krieger M, Calas A, Franzoni MF, Thibault J. Aromatic amino acid decarboxylase (AADC) immunohistochemistry in vertebrate brainstem with an antiserum raised against AADC made in E. coli. Brain Res Bull. 1993;32:123–32.
45. Karasawa N, Hayashi M, Yamada K, Nagatsu I, Iwasa M, Takeuchi T, et al. Tyrosine hydroxylase (TH)- and aromatic-L-amino acid decarboxylase (AADC)-immunoreactive neurons of

the common marmoset (Callithrix jacchus) brain: an immunohistochemical analysis. Acta Histochem Cytochem. 2007;40:83–92.
46. Arai R, Karasawa N, Nagatsu I. Aromatic L-amino acid decarboxylase is present in serotonergic fibers of the striatum of the rat. A double-labeling immunofluorescence study. Brain Res. 1996;706:177–9.
47. Huot P, Parent A. Dopaminergic neurons intrinsic to the striatum. J Neurochem. 2007;101:1441–7.
48. Meredith GE, Farrell T, Kellaghan P, Tan Y, Zahm DS, Totterdell S. Immunocytochemical characterization of catecholaminergic neurons in the rat striatum following dopamine-depleting lesions. Eur J Neurosci. 1999;11:3585–96.
49. Missale C, Nash SR, Robinson SW, Jaber M, Caron MG. Dopamine receptors: from structure to function. Physiol Rev. 1998;78:189–225.
50. Beaulieu JM, Gainetdinov RR. The physiology, signaling, and pharmacology of dopamine receptors. Pharmacol Rev. 2011;63:182–217.
51. Huot P, Fox SH, Brotchie JM. The serotonergic system in Parkinson's disease. Prog Neurobiol. 2011;95:163–212.
52. Torres GE, Gainetdinov RR, Caron MG. Plasma membrane monoamine transporters: structure, regulation and function. Nat Rev Neurosci. 2003;4:13–25.
53. Torres GE. The dopamine transporter proteome. J Neurochem. 2006;97 Suppl 1:3–10.
54. Engel K, Zhou M, Wang J. Identification and characterization of a novel monoamine transporter in the human brain. J Biol Chem. 2004;279:50042–9.
55. Dahlin A, Xia L, Kong W, Hevner R, Wang J. Expression and immunolocalization of the plasma membrane monoamine transporter in the brain. Neuroscience. 2007;146: 1193–211.
56. Zheng G, Dwoskin LP, Crooks PA. Vesicular monoamine transporter 2: role as a novel target for drug development. AAPS J. 2006;8:E682–92.
57. Stoessl AJ, Martin WW, McKeown MJ, Sossi V. Advances in imaging in Parkinson's disease. Lancet Neurol. 2011;10:987–1001.
58. Parkinson Study Group. A randomized placebo-controlled trial of rasagiline in levodopa-treated patients with Parkinson disease and motor fluctuations: the PRESTO study. Arch Neurol. 2005;62:241–8.
59. Rascol O, Brooks DJ, Melamed E, Oertel W, Poewe W, Stocchi F, et al. Rasagiline as an adjunct to levodopa in patients with Parkinson's disease and motor fluctuations (LARGO, Lasting effect in Adjunct therapy with Rasagiline Given Once daily, study): a randomised, double-blind, parallel-group trial. Lancet. 2005;365:947–54.
60. Fenelon G, Gimenez-Roldan S, Montastruc JL, Bermejo F, Durif F, Bourdeix I, et al. Efficacy and tolerability of entacapone in patients with Parkinson's disease treated with levodopa plus a dopamine agonist and experiencing wearing-off motor fluctuations. A randomized, double-blind, multicentre study. J Neural Transm. 2003;110:239–51.
61. Ferreira JJ, Rascol O, Poewe W, Sampaio C, Rocha JF, Nunes T, et al. A double-blind, randomized, placebo and active-controlled study of nebicapone for the treatment of motor fluctuations in Parkinson's disease. CNS Neurosci Ther. 2010;16:337–47.
62. Adler CH, Singer C, O'Brien C, Hauser RA, Lew MF, Marek KL, et al. Randomized, placebo-controlled study of tolcapone in patients with fluctuating Parkinson disease treated with levodopa-carbidopa. Tolcapone Fluctuator Study Group III. Arch Neurol. 1998;55: 1089–95.
63. Eisenhofer G, Kopin IJ, Goldstein DS. Catecholamine metabolism: a contemporary view with implications for physiology and medicine. Pharmacol Rev. 2004;56:331–49.
64. Gottowik J, Malherbe P, Lang G, Da Prada M, Cesura AM. Structure/function relationships of mitochondrial monoamine oxidase A and B chimeric forms. Eur J Biochem. 1995;230:934–42.
65. Shih JC, Chen K, Ridd MJ. Monoamine oxidase: from genes to behavior. Annu Rev Neurosci. 1999;22:197–217.

66. O'Carroll AM, Fowler CJ, Phillips JP, Tobbia I, Tipton KF. The deamination of dopamine by human brain monoamine oxidase. Specificity for the two enzyme forms in seven brain regions. Naunyn Schmiedebergs Arch Pharmacol. 1983;322:198–202.
67. Youdim MB, Edmondson D, Tipton KF. The therapeutic potential of monoamine oxidase inhibitors. Nat Rev Neurosci. 2006;7:295–309.
68. Anden NE, Roos BE, Werdinius B. On the occurrence of homovanillic acid in brain and cerebrospinal fluid and its determination by a fluorometric method. Life Sci. 1963;7: 448–58.
69. Meiser J, Weindl D, Hiller K. Complexity of dopamine metabolism. Cell Commun Signal. 2013;11:34.
70. Jinsmaa Y, Florang VR, Rees JN, Anderson DG, Strack S, Doorn JA. Products of oxidative stress inhibit aldehyde oxidation and reduction pathways in dopamine catabolism yielding elevated levels of a reactive intermediate. Chem Res Toxicol. 2009;22:835–41.
71. Hermida-Ameijeiras A, Mendez-Alvarez E, Sanchez-Iglesias S, Sanmartin-Suarez C, Soto-Otero R. Autoxidation and MAO-mediated metabolism of dopamine as a potential cause of oxidative stress: role of ferrous and ferric ions. Neurochem Int. 2004;45:103–16.
72. Tenhunen J, Salminen M, Lundstrom K, Kiviluoto T, Savolainen R, Ulmanen I. Genomic organization of the human catechol O-methyltransferase gene and its expression from two distinct promoters. Eur J Biochem. 1994;223:1049–59.
73. Axelrod J. Metabolism of epinephrine and other sympathomimetic amines. Physiol Rev. 1959;39:751–76.
74. Ulmanen I, Peranen J, Tenhunen J, Tilgmann C, Karhunen T, Panula P, et al. Expression and intracellular localization of catechol O-methyltransferase in transfected mammalian cells. Eur J Biochem. 1997;243:452–9.
75. Matsumoto M, Weickert CS, Akil M, Lipska BK, Hyde TM, Herman MM, et al. Catechol O-methyltransferase mRNA expression in human and rat brain: evidence for a role in cortical neuronal function. Neuroscience. 2003;116:127–37.
76. Mannisto PT, Kaakkola S. Catechol-O-methyltransferase (COMT): biochemistry, molecular biology, pharmacology, and clinical efficacy of the new selective COMT inhibitors. Pharmacol Rev. 1999;51:593–628.
77. van Laar T. Levodopa-induced response fluctuations in patients with Parkinson's disease: strategies for management. CNS Drugs. 2003;17:475–89.
78. Kopin IJ. Catecholamine metabolism: basic aspects and clinical significance. Pharmacol Rev. 1985;37:333–64.
79. Rush RA, Geffen LB. Dopamine beta-hydroxylase in health and disease. Crit Rev Clin Lab Sci. 1980;12:241–77.
80. Cedarbaum JM. Clinical pharmacokinetics of anti-parkinsonian drugs. Clin Pharmacokinet. 1987;13:141–78.
81. Gancher ST, Nutt JG, Woodward WR. Peripheral pharmacokinetics of levodopa in untreated, stable, and fluctuating parkinsonian patients. Neurology. 1987;37:940–4.
82. Hauser RA, Ellenbogen AL, Metman LV, Hsu A, O'Connell MJ, Modi NB, et al. Crossover comparison of IPX066 and a standard levodopa formulation in advanced Parkinson's disease. Mov Disord. 2011;26:2246–52.
83. Nagai M, Kubo M, Nishikawa N, Nomoto M. Fluctuation in plasma entacapone concentrations in accordance with variable plasma levodopa concentrations. Parkinsonism Relat Disord. 2010;16:697–9.
84. Nutt JG, Woodward WR, Beckner RM, Stone CK, Berggren K, Carter JH, et al. Effect of peripheral catechol-O-methyltransferase inhibition on the pharmacokinetics and pharmacodynamics of levodopa in parkinsonian patients. Neurology. 1994;44:913–9.
85. Tohgi H, Abe T, Yamazaki K, Saheki M, Takahashi S, Tsukamoto Y. Effects of the catechol-O-methyltransferase inhibitor tolcapone in Parkinson's disease: correlations between concentrations of dopaminergic substances in the plasma and cerebrospinal fluid and clinical improvement. Neurosci Lett. 1995;192:165–8.

86. Olanow CW, Gauger LL, Cedarbaum JM. Temporal relationships between plasma and cerebrospinal fluid pharmacokinetics of levodopa and clinical effect in Parkinson's disease. Ann Neurol. 1991;29:556–9.
87. Nomoto M, Nishikawa N, Nagai M, Yabe H, Nakatsuka A, Moritoyo H, et al. Inter- and intra-individual variation in L-dopa pharmacokinetics in the treatment of Parkinson's disease. Parkinsonism Relat Disord. 2009;15 Suppl 1:S21–4.
88. Davis TL, Roznoski M, Burns RS. Acute effects of COMT inhibition on L-DOPA pharmacokinetics in patients treated with carbidopa and selegiline. Clin Neuropharmacol. 1995;18: 333–7.
89. Dethy S, Laute MA, Van Blercom N, Damhaut P, Goldman S, Hildebrand J. Microdialysis-HPLC for plasma levodopa and metabolites monitoring in parkinsonian patients. Clin Chem. 1997;43:740–4.
90. Nutt JG, Woodward WR. Levodopa pharmacokinetics and pharmacodynamics in fluctuating parkinsonian patients. Neurology. 1986;36:739–44.
91. Woodward WR, Olanow CW, Beckner RM, Hauser RA, Gauger LL, Cedarbaum JM, et al. The effect of L-dopa infusions with and without phenylalanine challenges in parkinsonian patients: plasma and ventricular CSF L-dopa levels and clinical responses. Neurology. 1993;43:1704–8.
92. Dizdar N, Kullman A, Norlander B, Olsson JE, Kagedal B. Human pharmacokinetics of L-3,4-dihydroxyphenylalanine studied with microdialysis. Clin Chem. 1999;45:1813–20.
93. Rizzo V, Memmi M, Moratti R, Melzi d'Eril G, Perucca E. Concentrations of L-dopa in plasma and plasma ultrafiltrates. J Pharm Biomed Anal. 1996;14:1043–6.
94. Melzi d'Eril G, Moratti R, Perucca E. Total and non-protein-bound fractions of 3,4-dihydroxyphenylalanine. Clin Chem. 1998;44:895.
95. Roberts J, Waller DG, O'Shea N, Macklin BS, Renwick AG. The effect of selegiline on the peripheral pharmacokinetics of levodopa in young volunteers. Br J Clin Pharmacol. 1995;40:404–6.
96. Napolitano A, Del Dotto P, Petrozzi L, Dell'Agnello G, Bellini G, Gambaccini G, et al. Pharmacokinetics and pharmacodynamics of L-Dopa after acute and 6-week tolcapone administration in patients with Parkinson's disease. Clin Neuropharmacol. 1999;22:24–9.
97. Ruottinen HM, Rinne UK. A double-blind pharmacokinetic and clinical dose-response study of entacapone as an adjuvant to levodopa therapy in advanced Parkinson's disease. Clin Neuropharmacol. 1996;19:283–96.
98. Ruottinen HM, Rinne UK. Effect of one month's treatment with peripherally acting catechol-O-methyltransferase inhibitor, entacapone, on pharmacokinetics and motor response to levodopa in advanced parkinsonian patients. Clin Neuropharmacol. 1996;19:222–33.
99. Kaakkola S, Teravainen H, Ahtila S, Rita H, Gordin A. Effect of entacapone, a COMT inhibitor, on clinical disability and levodopa metabolism in parkinsonian patients. Neurology. 1994;44:77–80.
100. Cedarbaum JM, Silvestri M, Clark M, Harts A, Kutt H. L-deprenyl, levodopa pharmacokinetics, and response fluctuations in Parkinson's disease. Clin Neuropharmacol. 1990;13: 29–35.
101. Sader-Mazbar O, Loboda Y, Rabey MJ, Finberg JP. Increased L-DOPA-derived dopamine following selective MAO-A or -B inhibition in rat striatum depleted of dopaminergic and serotonergic innervation. Br J Pharmacol. 2013;170:999–1013.
102. Goto Y, Otani S, Grace AA. The Yin and Yang of dopamine release: a new perspective. Neuropharmacology. 2007;53:583–7.
103. Grace AA. The tonic/phasic model of dopamine system regulation: its relevance for understanding how stimulant abuse can alter basal ganglia function. Drug Alcohol Depend. 1995;37:111–29.
104. Sohn YH, Metman LV, Bravi D, Linfante I, Aotsuka A, Mouradian MM, et al. Levodopa peak response time reflects severity of dopamine neuron loss in Parkinson's disease. Neurology. 1994;44:755–7.

105. LeWitt PA, Jennings D, Lyons KE, Pahwa R, Rabinowicz AL, Wang J, et al. Pharmacokinetic-pharmacodynamic crossover comparison of two levodopa extension strategies. Mov Disord. 2009;24:1319–24.
106. Nutt JG. Pharmacokinetics and pharmacodynamics of levodopa. Mov Disord. 2008;23 Suppl 3:S580–4.
107. Zappia M, Oliveri RL, Bosco D, Nicoletti G, Branca D, Caracciolo M, et al. The long-duration response to L-dopa in the treatment of early PD. Neurology. 2000;54:1910–5.
108. Nutt JG, Holford NH. The response to levodopa in Parkinson's disease: imposing pharmacological law and order. Ann Neurol. 1996;39:561–73.
109. Troiano AR, de la Fuente-Fernandez R, Sossi V, Schulzer M, Mak E, Ruth TJ, et al. PET demonstrates reduced dopamine transporter expression in PD with dyskinesias. Neurology. 2009;72:1211–6.
110. de la Fuente-Fernandez R, Sossi V, Huang Z, Furtado S, Lu JQ, Calne DB, et al. Levodopa-induced changes in synaptic dopamine levels increase with progression of Parkinson's disease: implications for dyskinesias. Brain. 2004;127:2747–54.
111. de la Fuente-Fernandez R, Schulzer M, Mak E, Calne DB, Stoessl AJ. Presynaptic mechanisms of motor fluctuations in Parkinson's disease: a probabilistic model. Brain. 2004;127: 888–99.
112. Arai R, Karasawa N, Geffard M, Nagatsu I. L-DOPA is converted to dopamine in serotonergic fibers of the striatum of the rat: a double-labeling immunofluorescence study. Neurosci Lett. 1995;195:195–8.
113. Arai R, Karasawa N, Geffard M, Nagatsu T, Nagatsu I. Immunohistochemical evidence that central serotonin neurons produce dopamine from exogenous L-DOPA in the rat, with reference to the involvement of aromatic L-amino acid decarboxylase. Brain Res. 1994;667: 295–9.
114. Gershanik OS, Heikkila RE, Duvoisin RC. The role of serotonin in the action of L-dopa in an animal model of parkinsonism. Neurology. 1978;29(Suppl):S553.
115. Maeda T, Nagata K, Yoshida Y, Kannari K. Serotonergic hyperinnervation into the dopaminergic denervated striatum compensates for dopamine conversion from exogenously administered l-DOPA. Brain Res. 2005;1046:230–3.
116. Tanaka H, Kannari K, Maeda T, Tomiyama M, Suda T, Matsunaga M. Role of serotonergic neurons in L-DOPA-derived extracellular dopamine in the striatum of 6-OHDA-lesioned rats. Neuroreport. 1999;10:631–4.
117. Carta M, Carlsson T, Kirik D, Bjorklund A. Dopamine released from 5-HT terminals is the cause of L-DOPA-induced dyskinesia in parkinsonian rats. Brain. 2007;130:1819–33.
118. Carta M, Carlsson T, Munoz A, Kirik D, Bjorklund A. Involvement of the serotonin system in L-dopa-induced dyskinesias. Parkinsonism Relat Disord. 2008;14 Suppl 2:S154–8.
119. Carta M, Carlsson T, Munoz A, Kirik D, Bjorklund A. Serotonin-dopamine interaction in the induction and maintenance of L-DOPA-induced dyskinesias. Prog Brain Res. 2008;172: 465–78.
120. Nevalainen N, Af Bjerken S, Lundblad M, Gerhardt GA, Stromberg I. Dopamine release from serotonergic nerve fibers is reduced in L-DOPA-induced dyskinesia. J Neurochem. 2011;118:12–23.
121. Schapira AH, Emre M, Jenner P, Poewe W. Levodopa in the treatment of Parkinson's disease. Eur J Neurol. 2009;16:982–9.
122. Obeso JA, Rodriguez-Oroz MC, Chana P, Lera G, Rodriguez M, Olanow CW. The evolution and origin of motor complications in Parkinson's disease. Neurology. 2000;55:S13–20; discussion S1-3.
123. Grace AA. Physiology of the normal and dopamine-depleted basal ganglia: insights into levodopa pharmacotherapy. Mov Disord. 2008;23 Suppl 3:S560–9.
124. Stocchi F. The therapeutic concept of continuous dopaminergic stimulation (CDS) in the treatment of Parkinson's disease. Parkinsonism Relat Disord. 2009;15 Suppl 3:S68–71.
125. Stocchi F. The hypothesis of the genesis of motor complications and continuous dopaminergic stimulation in the treatment of Parkinson's disease. Parkinsonism Relat Disord. 2009;15 Suppl 1:S9–15.

126. Schapira AH, Agid Y, Barone P, Jenner P, Lemke MR, Poewe W, et al. Perspectives on recent advances in the understanding and treatment of Parkinson's disease. Eur J Neurol. 2009;16:1090–9.
127. Olanow CW, Obeso JA, Stocchi F. Continuous dopamine-receptor treatment of Parkinson's disease: scientific rationale and clinical implications. Lancet Neurol. 2006;5:677–87.
128. Stocchi F, Olanow CW. Continuous dopaminergic stimulation in early and advanced Parkinson's disease. Neurology. 2004;62:S56–63.
129. Papathanou M, Rose S, McCreary A, Jenner P. Induction and expression of abnormal involuntary movements is related to the duration of dopaminergic stimulation in 6-OHDA-lesioned rats. Eur J Neurosci. 2011;33:2247–54.
130. Maratos EC, Jackson MJ, Pearce RK, Cannizzaro C, Jenner P. Both short- and long-acting D-1/D-2 dopamine agonists induce less dyskinesia than L-DOPA in the MPTP-lesioned common marmoset (Callithrix jacchus). Exp Neurol. 2003;179:90–102.
131. Stockwell KA, Scheller D, Rose S, Jackson MJ, Tayarani-Binazir K, Iravani MM, et al. Continuous administration of rotigotine to MPTP-treated common marmosets enhances antiparkinsonian activity and reduces dyskinesia induction. Exp Neurol. 2009;219:533–42.
132. Bibbiani F, Costantini LC, Patel R, Chase TN. Continuous dopaminergic stimulation reduces risk of motor complications in parkinsonian primates. Exp Neurol. 2005;192:73–8.
133. Newman-Tancredi A, Cussac D, Audinot V, Nicolas JP, De Ceuninck F, Boutin JA, et al. Differential actions of antiparkinson agents at multiple classes of monoaminergic receptor. II. Agonist and antagonist properties at subtypes of dopamine D(2)-like receptor and alpha(1)/alpha(2)-adrenoceptor. J Pharmacol Exp Ther. 2002;303:805–14.
134. Millan MJ, Maiofiss L, Cussac D, Audinot V, Boutin JA, Newman-Tancredi A. Differential actions of antiparkinson agents at multiple classes of monoaminergic receptor. I. A multivariate analysis of the binding profiles of 14 drugs at 21 native and cloned human receptor subtypes. J Pharmacol Exp Ther. 2002;303:791–804.
135. Scheller D, Ullmer C, Berkels R, Gwarek M, Lubbert H. The in vitro receptor profile of rotigotine: a new agent for the treatment of Parkinson's disease. Naunyn Schmiedebergs Arch Pharmacol. 2009;379:73–86.
136. Alachkar A, Brotchie JM, Jones OT. Binding of dopamine and 3-methoxytyramine as l-DOPA metabolites to human alpha(2)-adrenergic and dopaminergic receptors. Neurosci Res. 2010;67:245–9.
137. Sunahara RK, Guan HC, O'Dowd BF, Seeman P, Laurier LG, Ng G, et al. Cloning of the gene for a human dopamine D5 receptor with higher affinity for dopamine than D1. Nature. 1991;350:614–9.
138. Sokoloff P, Giros B, Martres MP, Bouthenet ML, Schwartz JC. Molecular cloning and characterization of a novel dopamine receptor (D3) as a target for neuroleptics. Nature. 1990;347:146–51.
139. Van Tol HH, Bunzow JR, Guan HC, Sunahara RK, Seeman P, Niznik HB, et al. Cloning of the gene for a human dopamine D4 receptor with high affinity for the antipsychotic clozapine. Nature. 1991;350:610–4.
140. Newman-Tancredi A, Audinot V, Chaput C, Verriele L, Millan MJ. [35S]Guanosine-5′-O-(3-thio)triphosphate binding as a measure of efficacy at human recombinant dopamine D4.4 receptors: actions of antiparkinsonian and antipsychotic agents. J Pharmacol Exp Ther. 1997;282:181–91.
141. Rascol O, Brooks DJ, Korczyn AD, De Deyn PP, Clarke CE, Lang AE. A five-year study of the incidence of dyskinesia in patients with early Parkinson's disease who were treated with ropinirole or levodopa. 056 Study Group. N Engl J Med. 2000;342:1484–91.
142. Rascol O, Brooks DJ, Korczyn AD, De Deyn PP, Clarke CE, Lang AE, et al. Development of dyskinesias in a 5-year trial of ropinirole and L-dopa. Mov Disord. 2006;21:1844–50.
143. Hauser RA, Rascol O, Korczyn AD, Jon Stoessl A, Watts RL, Poewe W, et al. Ten-year follow-up of Parkinson's disease patients randomized to initial therapy with ropinirole or levodopa. Mov Disord. 2007;22:2409–17.
144. Parkinson Study Group. Pramipexole vs levodopa as initial treatment for Parkinson disease: A randomized controlled trial. Parkinson Study Group. JAMA. 2000;284:1931–8.

145. Parkinson Study Group. Long-term effect of initiating pramipexole vs levodopa in early Parkinson disease. Arch Neurol. 2009;66:563–70.
146. Montastruc JL, Rascol O, Senard JM, Rascol A. A randomised controlled study comparing bromocriptine to which levodopa was later added, with levodopa alone in previously untreated patients with Parkinson's disease: a five year follow up. J Neurol Neurosurg Psychiatry. 1994;57:1034–8.
147. Hely MA, Morris JG, Reid WG, O'Sullivan DJ, Williamson PM, Rail D, et al. The Sydney Multicentre Study of Parkinson's disease: a randomised, prospective five year study comparing low dose bromocriptine with low dose levodopa-carbidopa. J Neurol Neurosurg Psychiatry. 1994;57:903–10.
148. Hely MA, Morris JG, Reid WG, Trafficante R. Sydney Multicenter Study of Parkinson's disease: non-L-dopa-responsive problems dominate at 15 years. Mov Disord. 2005;20:190–9.
149. Watts RL, Lyons KE, Pahwa R, Sethi K, Stern M, Hauser RA, et al. Onset of dyskinesia with adjunct ropinirole prolonged-release or additional levodopa in early Parkinson's disease. Mov Disord. 2010;25:858–66.
150. Marin C, Aguilar E, Obeso JA. Coadministration of entacapone with levodopa attenuates the severity of dyskinesias in hemiparkinsonian rats. Mov Disord. 2006;21:646–53.
151. Smith LA, Jackson MJ, Al-Barghouthy G, Rose S, Kuoppamaki M, Olanow W, et al. Multiple small doses of levodopa plus entacapone produce continuous dopaminergic stimulation and reduce dyskinesia induction in MPTP-treated drug-naive primates. Mov Disord. 2005;20:306–14.
152. Zubair M, Jackson MJ, Tayarani-Binazir K, Stockwell KA, Smith LA, Rose S, et al. The administration of entacapone prevents L-dopa-induced dyskinesia when added to dopamine agonist therapy in MPTP-treated primates. Exp Neurol. 2007;208:177–84.
153. Smith LA, Jackson MJ, Hansard MJ, Maratos E, Jenner P. Effect of pulsatile administration of levodopa on dyskinesia induction in drug-naive MPTP-treated common marmosets: effect of dose, frequency of administration, and brain exposure. Mov Disord. 2003;18:487–95.
154. Stocchi F, Rascol O, Kieburtz K, Poewe W, Jankovic J, Tolosa E, et al. Initiating levodopa/carbidopa therapy with and without entacapone in early Parkinson disease: the STRIDE-PD study. Ann Neurol. 2010;68:18–27.
155. Warren Olanow C, Kieburtz K, Rascol O, Poewe W, Schapira AH, Emre M, et al. Factors predictive of the development of Levodopa-induced dyskinesia and wearing-off in Parkinson's disease. Mov Disord. 2013;28:1064–71.
156. Hauser RA, McDermott MP, Messing S. Factors associated with the development of motor fluctuations and dyskinesias in Parkinson disease. Arch Neurol. 2006;63:1756–60.
157. Hauser RA, Panisset M, Abbruzzese G, Mancione L, Dronamraju N, Kakarieka A. Double-blind trial of levodopa/carbidopa/entacapone versus levodopa/carbidopa in early Parkinson's disease. Mov Disord. 2009;24:541–50.
158. Jenner P. Preventing and controlling dyskinesia in Parkinson's disease–a view of current knowledge and future opportunities. Mov Disord. 2008;23 Suppl 3:S585–98.
159. Watts RL, Jankovic J, Waters C, Rajput A, Boroojerdi B, Rao J. Randomized, blind, controlled trial of transdermal rotigotine in early Parkinson disease. Neurology. 2007;68:272–6.
160. Giladi N, Boroojerdi B, Korczyn AD, Burn DJ, Clarke CE, Schapira AH. Rotigotine transdermal patch in early Parkinson's disease: a randomized, double-blind, controlled study versus placebo and ropinirole. Mov Disord. 2007;22:2398–404.
161. Giladi N, Boroojerdi B, Surmann E. The safety and tolerability of rotigotine transdermal system over a 6-year period in patients with early-stage Parkinson's disease. J Neural Transm. 2013;120:1321–9.
162. Jenner P, McCreary AC, Scheller DK. Continuous drug delivery in early- and late-stage Parkinson's disease as a strategy for avoiding dyskinesia induction and expression. J Neural Transm. 2011;118:1691–702.
163. Huot P, Johnston TH, Koprich JB, Fox SH, Brotchie JM. The pharmacology of L-DOPA-induced dyskinesia in Parkinson's disease. Pharmacol Rev. 2013;65:171–222.

164. Stockwell KA, Scheller DK, Smith LA, Rose S, Iravani MM, Jackson MJ, et al. Continuous rotigotine administration reduces dyskinesia resulting from pulsatile treatment with rotigotine or L-DOPA in MPTP-treated common marmosets. Exp Neurol. 2010;221:79–85.
165. Hadj Tahar A, Gregoire L, Bangassoro E, Bedard PJ. Sustained cabergoline treatment reverses levodopa-induced dyskinesias in parkinsonian monkeys. Clin Neuropharmacol. 2000;23:195–202.
166. Hauser RA. Future treatments for Parkinson's disease: surfing the PD pipeline. Int J Neurosci. 2011;121 Suppl 2:53–62.
167. Samanta J, Hauser RA. Duodenal levodopa infusion for the treatment of Parkinson's disease. Expert Opin Pharmacother. 2007;8:657–64.
168. Stocchi F, Vacca L, Ruggieri S, Olanow CW. Intermittent vs continuous levodopa administration in patients with advanced Parkinson disease: a clinical and pharmacokinetic study. Arch Neurol. 2005;62:905–10.
169. Puente V, De Fabregues O, Oliveras C, Ribera G, Pont-Sunyer C, Vivanco R, et al. Eighteen month study of continuous intraduodenal levodopa infusion in patients with advanced Parkinson's disease: impact on control of fluctuations and quality of life. Parkinsonism Relat Disord. 2010;16:218–21.
170. Olanow CW, Kieburtz K, Odin P, Espay AJ, Standaert DG, Fernandez HH, et al. Continuous intrajejunal infusion of levodopa-carbidopa intestinal gel for patients with advanced Parkinson's disease: a randomised, controlled, double-blind, double-dummy study. Lancet Neurol. 2014;13:141–9.
171. Rascol O. Jejunal levodopa infusion in Parkinson's disease. Lancet Neurol. 2014;13:128–9.
172. Fernandez HH, Vanagunas A, Odin P, Espay AJ, Hauser RA, Standaert DG, et al. Levodopa-carbidopa intestinal gel in advanced Parkinson's disease open-label study: interim results. Parkinsonism Relat Disord. 2013;19:339–45.
173. Nilsson D, Nyholm D, Aquilonius SM. Duodenal levodopa infusion in Parkinson's disease–long-term experience. Acta Neurol Scand. 2001;104:343–8.
174. Baronti F, Mouradian MM, Davis TL, Giuffra M, Brughitta G, Conant KE, et al. Continuous lisuride effects on central dopaminergic mechanisms in Parkinson's disease. Ann Neurol. 1992;32:776–81.
175. Juncos JL, Mouradian MM, Fabbrini G, Serrati C, Chase TN. Levodopa methyl ester treatment of Parkinson's disease. Neurology. 1987;37:1242–5.
176. Mouradian MM, Heuser IJ, Baronti F, Chase TN. Modification of central dopaminergic mechanisms by continuous levodopa therapy for advanced Parkinson's disease. Ann Neurol. 1990;27:18–23.
177. LeWitt PA, Lyons KE, Pahwa R. Advanced Parkinson disease treated with rotigotine transdermal system: PREFER Study. Neurology. 2007;68:1262–7.
178. Poewe WH, Rascol O, Quinn N, Tolosa E, Oertel WH, Martignoni E, et al. Efficacy of pramipexole and transdermal rotigotine in advanced Parkinson's disease: a double-blind, double-dummy, randomised controlled trial. Lancet Neurol. 2007;6:513–20.
179. LeWitt PA, Boroojerdi B, Surmann E, Poewe W. Rotigotine transdermal system for long-term treatment of patients with advanced Parkinson's disease: results of two open-label extension studies, CLEOPATRA-PD and PREFER. J Neural Transm. 2013;120:1069–81.
180. Gershanik O, Jenner P. Moving from continuous dopaminergic stimulation to continuous drug delivery in the treatment of Parkinson's disease. Eur J Neurol. 2012;19:1502–8.
181. Obeso JA, Grandas F, Herrero MT, Horowski R. The role of pulsatile versus continuous dopamine receptor stimulation for functional recovery in Parkinson's disease. Eur J Neurosci. 1994;6:889–97.
182. Pahwa R, Lyons KE, Hauser RA, Fahn S, Jankovic J, Pourcher E, et al. Randomized trial of IPX066, carbidopa/levodopa extended release, in early Parkinson's disease. Parkinsonism Relat Disord. 2014;20:142–8.
183. Hauser RA, Hsu A, Kell S, Espay AJ, Sethi K, Stacy M, et al. Extended-release carbidopa-levodopa (IPX066) compared with immediate-release carbidopa-levodopa in patients with

Parkinson's disease and motor fluctuations: a phase 3 randomised, double-blind trial. Lancet Neurol. 2013;12:346–56.
184. Di Stefano A, Sozio P, Cerasa LS. Antiparkinson prodrugs. Molecules. 2008;13:46–68.
185. Lewitt PA, Ellenbogen A, Chen D, Lal R, McGuire K, Zomorodi K, et al. Actively transported levodopa prodrug XP21279: a study in patients with Parkinson disease who experience motor fluctuations. Clin Neuropharmacol. 2012;35:103–10.
186. Lewitt PA, Huff FJ, Hauser RA, Chen D, Lissin D, Zomorodi K, et al. Double-blind study of the actively transported levodopa prodrug XP21279 in Parkinson's disease. Mov Disord. 2014;29:75–82.
187. LeWitt PA, Friedman H, Giladi N, Gurevich T, Shabtai H, Djaldetti R, et al. Accordion pill carbidopa/levodopa for improved treatment of advanced Parkinson's disease symptoms [abstract]. Mov Disord. 2012;27 Suppl 1:S130–1.
188. LeWitt PA, Friedman H, Giladi N, Gurevich T, Shabtai H, Djaldetti R, et al. Sustained-release carbidopa-levodopa (accordian pill) in patients with advanced Parkinson's disease: pharmacokinetic and clinical experience [abstract]. Mov Disord. 2013;28 Suppl 1:S177.
189. Goole J, Van Gansbeke B, Pilcer G, Deleuze P, Blocklet D, Goldman S, et al. Pharmacoscintigraphic and pharmacokinetic evaluation on healthy human volunteers of sustained-release floating minitablets containing levodopa and carbidopa. Int J Pharm. 2008;364:54–63.
190. Goole J, Amighi K, Vanderbist F. Evaluation and floating enhancement of levodopa sustained release floating minitablets coated with insoluble acrylic polymer. Drug Dev Ind Pharm. 2008;34:827–33.
191. Goole J, Deleuze P, Vanderbist F, Amighi K. New levodopa sustained-release floating minitablets coated with insoluble acrylic polymer. Eur J Pharm Biopharm. 2008;68:310–8.
192. Olanow CW. Levodopa/dopamine replacement strategies in Parkinson's disease–future directions. Mov Disord. 2008;23 Suppl 3:S613–22.
193. Kankkunen T, Huupponen I, Lahtinen K, Sundell M, Ekman K, Kontturi K, et al. Improved stability and release control of levodopa and metaraminol using ion-exchange fibers and transdermal iontophoresis. Eur J Pharm Biopharm. 2002;16:273–80.
194. Stocchi F. Therapy for Parkinson's disease: what is in the pipeline? Neurotherapeutics. 2014;11:24–33.
195. Olanow CW, Schapira AH. Therapeutic prospects for Parkinson disease. Ann Neurol. 2013;74:337–47.
196. Nutt JG, Obeso JA, Stocchi F. Continuous dopamine-receptor stimulation in advanced Parkinson's disease. Trends Neurosci. 2000;23:S109–15.
197. Tiberi M, Caron MG. High agonist-independent activity is a distinguishing feature of the dopamine D1B receptor subtype. J Biol Chem. 1994;269:27925–31.
198. Freedman SB, Patel S, Marwood R, Emms F, Seabrook GR, Knowles MR, et al. Expression and pharmacological characterization of the human D3 dopamine receptor. J Pharmacol Exp Ther. 1994;268:417–26.
199. Burris KD, Pacheco MA, Filtz TM, Kung MP, Kung HF, Molinoff PB. Lack of discrimination by agonists for D2 and D3 dopamine receptors. Neuropsychopharmacology. 1995;12:335–45.
200. Borowsky B, Adham N, Jones KA, Raddatz R, Artymyshyn R, Ogozalek KL, et al. Trace amines: identification of a family of mammalian G protein-coupled receptors. Proc Natl Acad Sci USA. 2001;98:8966–71.
201. Oz M, Zhang L, Rotondo A, Sun H, Morales M. Direct activation by dopamine of recombinant human 5-HT1A receptors: comparison with human 5-HT2C and 5-HT3 receptors. Synapse. 2003;50:303–13.
202. Giros B, el Mestikawy S, Godinot N, Zheng K, Han H, Yang-Feng T, et al. Cloning, pharmacological characterization, and chromosome assignment of the human dopamine transporter. Mol Pharmacol. 1992;42:383–90.

203. Pacholczyk T, Blakely RD, Amara SG. Expression cloning of a cocaine- and antidepressant-sensitive human noradrenaline transporter. Nature. 1991;350:350–4.
204. Barker EL, Moore KR, Rakhshan F, Blakely RD. Transmembrane domain I contributes to the permeation pathway for serotonin and ions in the serotonin transporter. J Neurosci. 1999;19:4705–17.
205. Erickson JD, Schafer MK, Bonner TI, Eiden LE, Weihe E. Distinct pharmacological properties and distribution in neurons and endocrine cells of two isoforms of the human vesicular monoamine transporter. Proc Natl Acad Sci USA. 1996;93:5166–71.

Chapter 10
Dopamine Receptors and Levodopa-Induced Dyskinesia

Vincent A. Jourdain, Nicolas Morin, and Thérèse Di Paolo

Abstract This chapter reviews preclinical and relevant clinical studies investigating the role and contribution of dopamine (DA) receptor subtypes in the pathophysiology of L-3,4-dihydroxyphenylalanine (L-DOPA)-induced dyskinesias (LID) in parkinsonian patients and animal models. Altered dopaminergic neurotransmission in the basal ganglia are observed in LID. Two conditions are necessary for their appearance: (1) loss of DA in nigrostriatal pathway and (2) treatment with L-DOPA or DA agonists, the basis of replacement therapy. LID are clearly more complex than a hypersensitivity due to a simple increase in the density of striatal DA receptors. The development and expression of LID are related to increases in the activity of D_1, D_2, and D_3 receptors, while the contribution of the activity of D_4 and D_5 receptors remains unexplored. In clinical trials with PD patients, some factors have been identified to increase the risk of developing LID such as high doses of L-DOPA or DA agonist treatment, abnormal and pulsatile stimulation of DA receptors, activation of a specific DA receptor subtype (D_1 vs. D_2/D_3), and polymorphisms of the DA receptor subtypes (D_1, D_2). DA receptors interact with receptors of several other neurotransmitters. The implications of these interactions in the development and expression of LID in PD patients and animal models need further investigation to find novel drug targets.

Keywords Parkinson's disease • L-DOPA-induced dyskinesia • Animal models • Parkinsonian patients • Motor complications • Dopamine receptor • Sensitization • Receptor supersensitivity • Dopamine signaling pathways

---

Vincent A. Jourdain and Nicolas Morin are co-first authors and have contributed equally in the writing of this chapter

V.A. Jourdain, PhD • N. Morin, B.Pharm, MSc • T. Di Paolo, PhD (✉)
Neuroscience Research Unit, Centre de recherche du CHU de Québec,
2705 Laurier Boulevard, Quebec, QC G1V 4G2, Canada

Faculty of Pharmacy, Laval University, Quebec, QC, Canada
e-mail: therese.dipaolo@crchul.ulaval.ca

S.H. Fox, J.M. Brotchie (eds.), *Levodopa-Induced Dyskinesia in Parkinson's Disease*, DOI 10.1007/978-1-4471-6503-3_10

Introduction

Parkinson's disease (PD) is the most common neurodegenerative movement disorder and is likely to increase due to the aging of populations [1, 2]. PD involves principally the death of dopamine (DA) neurons in the substantia nigra *pars compacta* (SNc), but other neurotransmitters and neuromodulators are also affected [3].

The treatment of motor symptoms of PD with the DA precursor, L-3,4-dihydroxyphenylalanine (L-DOPA), introduced 50 years ago still remains a very effective medication. However, various complications including motor fluctuations and abnormal involuntary movements, such as L-DOPA-induced dyskinesias (LID), limit the quality of life in PD patients and can be very difficult to manage [4–6]. LID are irreversible or at least persistent, and this suggests that dopaminergic drugs can permanently or persistently modify the brain response to DA. Indeed, once LID have appeared and L-DOPA treatment is withdrawn, the first dose after several weeks of drug holiday will trigger them again [7]. No drug is yet approved for dyskinesias, aside from a modest benefit with amantadine in some PD patients [8]. Selective D_2/D_3 DA agonists have less potential to induce motor complications compared to L-DOPA [9, 10]. However, even if such DA agonists exert an antiparkinsonian effect, they are less potent than L-DOPA to control motor symptoms of PD [11]. Hence, as disease progresses, parkinsonian patients initiated with DA agonist monotherapy will eventually require L-DOPA, and after 10–15 years, their motor complications appear similar as they would have been if started initially on L-DOPA therapy [12, 13]. These observations suggest that disease progression plays a major role in the onset and the development of LID rather than the type of dopaminergic drug treatment used.

The mechanisms involved in the occurrence of LID are complex and have been investigated in numerous studies using animal models and parkinsonian patients. The loss of DA in the nigrostriatal pathway and the chronic administration of L-DOPA, or DA agonists, are two necessary conditions for their appearance. However, LID are clearly more complex than hypersensitivity due to a simple increase in the density of striatal DA receptors. Moreover, L-DOPA can induce a sensitization to dopaminergic response. Hence, multiple changes in DA receptors located in the basal ganglia and in their respective signaling pathways have been observed, including, but not restricted to, the modulation of the expression and the activity of subtypes of DA receptors, G proteins, effectors, protein kinases, transcription factors, etc. The development of LID seems to be related to increases in the activity of D_1, D_2, and D_3 receptor subtypes, while the contribution of the activity of D_4 and D_5 receptors remains unexplored.

Much information has been gained from animal models, especially from the 6-hydroxydopamine (6-OHDA)-lesioned rat and the 1-methyl-4-phenyl-1,2,3,6-tetrahydropyridine (MPTP) nonhuman primate models [14]. However, differences in the basal ganglia dopaminergic system between animal models and human brain are observed. Hence, the rodent basal ganglia show anatomical differences compared to the human and nonhuman primates. For instance, the caudate nucleus and putamen are components of the striatum which are fused in rodents, whereas they are separated by the internal capsule in primates [15, 16]. Moreover, the segregation of the so-called direct (D_1 receptor-related) and indirect (D_2 receptor-related) pathways of the basal ganglia is well documented in rodents, but their separation is not as clear in primates [17]. Hence, in primates both D_1 and D_2 receptor agonists can induce dyskinesias [18], whereas in rodents the contribution of the direct pathway with an increased activity of D_1 receptors has been more associated with LID [19]. Moreover, most striatal output neurons in rodent and primate brains can extensively collateralize and send collateral projections to every striatal output target [20–22]. Thus, DA transmission in the basal ganglia is more complex than the simplistic model of a complete segregation between D_1 and D_2 DA receptors.

Much remains to be learned from dopaminergic systems and biochemical processes that underlie the development of LID and how dopaminergic and non-dopaminergic drugs can be used to avoid the initiation of LID in early PD, to prevent or inhibit their expression in later stages of the disease, and to reverse the priming process through a normalization of the basal ganglia function. This chapter presents preclinical and relevant clinical studies reviewing the role and contribution of DA receptor subtypes and their signaling in the pathophysiology of LID. The translational values of the animal models will be discussed with salient examples of clinical results.

Classification of Dopamine Receptors and Their Distribution in the Basal Ganglia

DA binds to one of the five different subtypes of G protein-coupled DA receptors, D_1–D_5 [23]. DA receptors regulate cAMP-protein kinase A through G protein-mediated signaling [24]. The D_1 class of receptors (D_1 and D_5) couples mostly to $G\alpha_s/G\alpha_{olf}$ and stimulates production of cAMP and activity of protein kinase A [25, 26]. In contrast, the D_2 class (D_2, D_3, and D_4) couples $G\alpha_i/G\alpha_o$, regulates production of cAMP negatively, and modulates intracellular Ca^{2+} levels [27, 28].

Both D_1 and D_2 receptors are highly expressed by striatal medium spiny neurons (MSN) and are the most studied in PD and LID as compared to the other subtypes [29, 30]. D_1 and D_2 receptors are present at lower levels in the cortex as compared to the striatum [31]. D_1 receptors are expressed in striatofugal neurons containing substance P and dynorphin that project to the substantia nigra *pars reticulata* and to the internal globus pallidus, which constitute the direct striatal output pathway [32]. By contrast, D_2 receptors are predominantly localized in striatofugal neurons expressing enkephalin, which project to the external globus pallidus, constituting the indirect pathway [33, 34]. As compared to the postsynaptic D_1 receptors, D_2 receptors are also localized on presynaptic nigrostriatal dopaminergic terminals, on the SNc neurons, and on presynaptic corticostriatal terminals where they can inhibit striatal glutamate release [29, 35, 36]. In humans and nonhuman primates, D_3 receptors are mostly found in the nucleus accumbens and in the striatum and are also localized in the internal globus pallidus, anterior thalamus, amygdala, hippocampus, and cortex [37–39]. In the human striatum, there is approximately one D_3 receptor for two D_2 receptors, and D_3 receptor can colocalize with both D_1 and D_2 receptors [38, 40]. The distribution of D_3 receptors in rodents is not similar to the pattern observed in human and nonhuman primate brains [40, 41]. This receptor is undetectable in the internal globus pallidus of rodents [40]. Moreover, a lower D_3/D_2 ratio (approximately 1/20) is found in the rodent's striatum as compared to human and nonhuman primate brains [38, 40]. The D_4 receptor is found in extrastriatal areas, namely, the septum, cortex, hippocampus, and thalamus [31, 42]. The D_5 receptor is localized in the neck of dendritic spines in striatal medium spiny neurons and cholinergic interneurons [43]. The D_5 receptor is also found in the cortex, hippocampus, substantia nigra, cerebellum, and thalamus [42, 44].

Contribution of Dopamine Receptors in L-DOPA-Induced Dyskinesias

Dopamine Receptor Supersensivity and Dopamine Sensitization

Denervation-induced supersensitivity of DA receptors was initially recognized as a plausible mechanism of LID. Numerous studies evaluated the density of D_1, D_2, and D_3 receptors by autoradiography in the brain of human and animal models (summarized in Tables 10.1 and 10.2). LID were found to be clearly more complex than a hypersensitivity due to a simple increase in the density of striatal DA receptors. If hypersensitivity of DA receptors were present in untreated PD patients and were the cause of LID, LID would appear with the first dose of L-DOPA. However, LID usually do not emerge at first exposure to L-DOPA, but rather develop gradually over years of treatment.

D_1-Like Family of Dopamine Receptors

Early studies using dopaminergic drugs showed that D_1 receptor agonists were as effective as D_2 receptor agonists in improving parkinsonian symptoms in primates while inducing less dyskinesia [113–115]. More recently, the use of D_1 receptor agonists in 6-OHDA rats demonstrated their dyskinesiogenic effects and pharmacological blockade of D_1 receptor is more effective than D_2 receptor antagonists in reducing LID [116–118]. However, this concept was already known for some time. In fact, it has been shown several years before that dyskinesia could develop in drug-naive MPTP primates by a chronic administration of a full D_1 receptor agonist [119]. Furthermore, genetic knockout of D_1 receptors completely blocks LID in parkinsonian mice, whereas D_2 receptor knockout mice develop LID similar to wild-type mice [120].

Table 10.1 Changes in striatal dopamine (DA) receptors and intracellular pathways associated with MPTP lesion and L-DOPA treatment inducing dyskinesias in nonhuman primates

DA markers	Effect of lesion and treatment by technique			References
D_1 receptor	*Autoradiography and homogenate binding*			
	MPTP	No change		[45–53]
			Posterior	[54]
		Increase		[55–59]
			Anterior	[54]
		Decrease[a]		[54]
	MPTP + L-DOPA			
		No change		[46, 54]
		Increase		[47, 52, 59]
		Decrease		[55]
	In situ hybridization			
	MPTP	No change	Anterior	[58]
			Posterior	[60, 61]
		Decrease	Anterior	[47, 60–62]
			Posterior	[58, 62]
	MPTP + L-DOPA	No change	Anterior	[47][b]
			Posterior	[61]
		Increase	Anterior	[47][b]
		Decrease	Anterior	[61]
D_2 receptor	*Autoradiography and homogenate binding*			
	MPTP	No change		[48, 49, 56, 63]
			Posterior	[54]
		Increase		[45, 46, 50–53, 55, 57, 64–67]
			Anterior	[54]
	MPTP + L-DOPA			
		No change		[48, 52]
		Decrease		[46, 51, 55, 65]
	In situ hybridization			
	MPTP	No change		[58, 67]
		Increase		[47, 60, 61, 68]
	MPTP + L-DOPA	No change		[47, 68]
		Decrease		[61]
D_3 receptor	*Autoradiography*			
	MPTP	No change		[69]
			Posterior	[70]
		Decrease		[37, 52, 71]
			Anterior	[70]
	MPTP + L-DOPA			
		No change		[69]
		Increase		[37, 52, 71]
	In situ hybridization			
	MPTP	Increase (early)		[67]

(continued)

Table 10.1 (continued)

DA markers	Effect of lesion and treatment by technique		References
Akt/GSK3	*Western blot*		
	MPTP	Decrease phosphorylation	[72]
	MPTP+L-DOPA	Increase phosphorylation	[72]
ERK1/2	*Western blot*		
	MPTP	No change	[73]
		Increase phosphorylation	[74]
	MPTP+L-DOPA	Increase phosphorylation	[73]
		Decrease phosphorylation	[74]
DARPP-32	*Immunohistochemistry*		
	MPTP	Decrease	[75]
	Western blot		
	MPTP	No change	[47, 73]
	MPTP+L-DOPA	Increase phosphorylation	[47, 73]
ΔFosB	*Western blot*		
	MPTP	Increase	[76, 77]
	MPTP+L-DOPA	Increase	[77]

[a]An agonist radioligand was used for this particular autoradiography, whereas antagonists were used for the others listed

[b]Non-dyskinetic animals showed no change, whereas dyskinetic animals showed increases

Table 10.2 Changes in striatal dopamine (DA) receptors and intracellular pathways associated with Parkinson's disease and L-DOPA treatment inducing dyskinesias

DA markers	Technique	Effect of disease compared to controls	Effect of L-DOPA compared to controls	References
D_1 receptor	*Homogenate binding and autoradiography*			
	[^{3}H]SCH-23390	Increase	No change	[78]
			No change	[79–84]
			Caudate: decrease Putamen: no change	[85]
			Increase	[86]
	PET			
	[^{11}C]SCH-23390	No change		[87]
		No change	No change	[88, 89]
			No change	[90]

(continued)

Table 10.2 (continued)

DA markers	Technique	Effect of disease compared to controls	Effect of L-DOPA compared to controls	References
D_2 receptor	*Homogenate binding and autoradiography*			
	[^{3}H]haloperidol	Increase	No change	[91]
			Caudate: No change Putamen: Increase	[80, 92]
	[^{3}H]spiperone	Increase	No change	[93]
			No change	[79, 81–83, 94]
		No change	No change	[95, 96]
			Caudate: decrease Putamen: no change	[85]
	[^{3}H]raclopride		Increase	[86]
	[^{125}I]epidepride		Increase	[97]
	[^{3}H]CV205-502		No change	[84]
	PET			
	[^{11}C]raclopride	Increase		[98, 99]
		Increase	No change	[100, 101]
		Increase	Caudate: decrease Putamen: no change	[90, 102]
		No change		[103]
	[^{11}C]methylspiperone		No change	[104]
		No change		[103]
	[^{123}I]IBZM	No change		[105]
		No change	Decrease	[106]
			Increase	[107]
		Increase		[108]
D_3 receptors	*Homogenate binding and autoradiography*			
	[^{3}H]7-OH-DPAT		Decrease	[86]
			No change	[109]
	[^{125}I]trans-7-OH-PIPAT		Decrease	[97]
	mRNA		No change	[109]
	PET [^{11}C]-(+)-PHNO	Decrease		[110]
Akt and pAkt(Ser473)	*Western blot*		Decrease	[111]
DARPP-32	*Western blot*		Decrease	[79, 112]

PET positron emission tomography

Several autoradiographic investigations of D_1 receptors were performed in vivo and on postmortem tissues of animal models (Table 10.1) and in PD humans (Table 10.2), but no general consensus emerges [121]. These differences may be due to various experimental assays and the subregion of the striatum measured. In primates, striatal D_1 receptor specific binding remains unchanged [45–54] or increases [54–59] after MPTP lesion. L-DOPA treatment in MPTP monkeys induces increases [47, 52, 59] or no change in D_1 receptors [46, 48, 54]. The administration of D_1 receptor agonists in monkeys increases D_1 receptors binding [47–49, 62], whereas D_1 receptor specific binding returns to control values with D_2 receptor agonists or the combination of D_1 and D_2 receptor agonists [54, 122]. In humans, D_1 receptors are unaffected whether PD patients are treated or not with L-DOPA [88, 89] or decreased by 20 % with long-term L-DOPA treatment as measured by PET [123]. D_1 receptor mRNA is reported to remain unchanged [58, 60, 61] or decrease after exposure to MPTP [47, 58, 60–62]. This decrease is reversed with the administration of L-DOPA [47, 61] or with pulsatile administration of D_1 receptor agonists in the caudal striatum [62]. On the other hand, D_2 receptor agonists exert no change on MPTP-induced decrease of D_1 receptor mRNA [58, 60].

Although the association between the expression of D_1 receptors and LID is unclear, the sensitivity of D_1 receptors as measured by GTPγS binding is reported to be linearly related to the severity of LID [47]. L-DOPA induces a decrease of D_1 receptor sensitivity in non-dyskinetic MPTP monkeys, whereas its sensitivity is strongly increased in dyskinetic animals [47]. Such change in sensitivity of D_1 receptors in LID may depend on its subcellular distribution. Indeed, in non-dyskinetic animals killed at the peak of L-DOPA plasma levels (1 h), D_1 receptors are located at the synaptic membrane, while, by contrast, they are present at both synaptic and cytoplasmic membranes of striatofugal neurons in dyskinetic monkeys [124]. The activation of D_1 receptors by exogenous L-DOPA produces a proteasome chymotrypsin-like catalytic hypoactivity, which, in turn, leads to a D_1 receptor abnormal trafficking in striatal neurons by an impairment of receptor degradation [125]. Lentiviral overexpression of striatal G protein-coupled receptor kinase 6 leads to an internalization of D_1 receptors and decreases in primates with established LID [126]. On the other hand, in 6-OHDA rats killed 45 min after a single dose of L-DOPA, D_1 receptors were internalized in the cytoplasm compared to normal rats treated with L-DOPA [127]. Similar findings were found in PD patients, with a preferential cytoplasmatic localization of D_1 receptors after chronic L-DOPA when compared to healthy controls [128]. Though the subcellular distribution of D_1 receptors is altered with L-DOPA, it is not clear what is the consequence of this distribution in the development, priming process, and expression of LID.

The stimulation of D_1 receptors activates the phosphorylation of the DA- and cAMP-regulated phosphoprotein 32 kDa (DARPP-32) [129]. DARPP-32 is unaffected by L-DOPA treatment in normal mice, but L-DOPA produces increases in its phosphorylation in a DA-depleted state [130]. Long-term potentiation (LTP) and long-term depression (LTD) are two types of synaptic plasticity that modify neurotransmission efficacy in striatofugal neurons [131]. LTP is lost in parkinsonian animals [132], and L-DOPA treatment restores corticostriatal LTP, but not in those displaying LID [133]. The activation of the D_1 receptor/DARPP-32 pathway is important in the induction of LTP and LTD since both are lost in mice lacking DARPP-32 [134].

D_1 receptors interact with the ionotropic glutamate receptors NMDA at the postsynaptic striatal level and may form hetero-oligomeric complexes [135]. This interaction influences trafficking, signaling, and desensitization of both receptors [136, 137], and such complexes are lost in 6-OHDA rats displaying abnormal involuntary movements (AIMs) [138]. The extracellular signal-regulated kinase (ERK) is an important intracellular protein involved in LTP [139] and thought to be associated with the LID priming process [73]. Interestingly, ERK signaling is part of the intracellular pathways of both NMDA and D_1 receptors [140], and its dephosphorylation is regulated by the protein phosphatase 1 [141]. The activation of D_1 receptors induces phosphorylation of ERK in normal [142] and in DA-depleted rodents' brains [143], and there is a correlation between phosphorylated ERK and LID [118, 120, 130, 144]. Phosphorylation of ERK is, at least, dependent of abnormal activation of DARPP-32 and its phosphorylated state being greatly reduced in mice lacking DARPP-32 receiving L-DOPA [130]. In addition, LID are reduced with pharmacological inhibition of ERK intracellular signaling [145, 146]. Lastly, Ras-GRF1 is a neuronal specific activator of ERK signaling [140, 147]. Its genetic deletion abolishes D_1 receptor-induced phosphorylation of ERK [147] and reverts LID in mice and monkeys [148] but at the cost of reduced locomotor activity [148]. It remains however important to consider that ERK is not restricted to LID, but is also involved in many other functions [149].

The family of transcription factors *fos/jun* has been extensively studied [150]. Dopaminomimetic agents induce the expression of c-jun, c-fos, ΔFosB, and FosB in striatal neurons in normal animals [151, 152] and hemiparkinsonian animals [153] and require, at least, the activation of D_1 but not D_2 receptors [152]. In monkeys, ΔFosB is increased with MPTP and remains elevated for several months after exposure to MPTP [76]. Pulsatile administration of the short-acting D_1 receptor agonist SKF-82958 upregulated further ΔFosB in MPTP monkeys and was associated with the development of LID, thus indicating an involvement of D_1 receptors in the priming process leading to their expression [76]. On the other hand, SKF-82958 administered continuously through a minipump or chronic treatment with a D_2 receptor agonist did not induce LID and no change in ΔFosB [154]. Such dopaminergic-induced elevation in ΔFosB remains upregulated several weeks after treatment discontinuation [155]. In addition, there is a positive correlation between the number of cells immunoreactive to FosB/ΔFosB and the severity of AIMs in 6-OHDA rats [156]. Genetic overexpression of ΔFosB in normal rats leads to involuntary movements similar to L-DOPA-induced AIMs observed in 6-OHDA rats [157]. Genetic overexpression of JunD, a dominant negative inhibitor of ΔFosB, reduces LID in MPTP primates [77]. Moreover, the D_1 antagonist SCH-23390, as well as antidyskinetic agents, is associated with a decreased FosB/ΔFosB expression in dyskinetic animals [118, 158]. The long-term increase in ΔFosB [155], and possibly the activation of ERK pathway, may contribute to the "memory for LID" [159] and could explain partially why L-DOPA drug holiday has few or no effect on LID in patients [7]. More information on D_1 receptor-mediated abnormal transmission in LID can be found in [160].

D_1 and adenosine A_1 receptors are known to form functionally interacting complexes in cortical neurons and basal ganglia [161, 162]. Simultaneous pretreatment with A_1 and D_1 receptors agonists in D_1/A_1 cells was shown to decrease D_1 receptor adenylyl cyclase signaling [163]. The adenosinergic involvement in LID is discussed in another chapter of the present textbook. Concerning the other DA receptor in the D_1-like subfamily, namely, the D_5 receptor, there is, to our knowledge, no study on its involvement in PD and LID.

D_2-Like Family of Dopamine Receptors

Though D_1 receptors have been classically thought as being responsible for priming and expression of LID and have thus received more attention [160, 164, 165], D_2 receptors also contribute to LID. In fact, once primed to express LID, D_2 agonists will trigger AIMs in 6-OHDA rats [166, 167] as well as LID in MPTP monkeys [148, 168] and PD patients [169]. Moreover, a comparison of several D_1 and D_2 agonists in MPTP monkeys with established LID demonstrated that D_2 agonists produce higher dyskinesia than their D_1 counterparts [113]. Furthermore, the administration of the D_2/D_3 agonist (+)-PHNO may also lead rapidly to the development of LID in MPTP monkeys [170]. Once primed, these monkeys displayed LID that were unaffected by the addition of the D_1 antagonist SCH-23390 to (+)-PHNO [170, 171]. Primed with (+)-PHNO, MPTP monkeys will display LID, as severe as those elicited with (+)-PHNO, when administered with the D_1 agonist CY-208243 [171].

Compared to D_1 receptors, autoradiography studies on D_2 receptors show more consistent results (Table 10.1). Nevertheless, here also differences among studies are reported and maybe due to the experimental assay and the subregion of the striatum investigated. Levels of D_2 receptors in the striatum remain unchanged [48, 49, 54, 56, 63, 95, 96, 103, 105, 106] or were increased with MPTP lesion in monkeys [45, 46, 50–55, 57, 64–67] and in untreated PD patients [79–83, 90–94, 98–102, 108]. Such DA denervation-induced increases in the expression of D_2 receptors are reversed with the administration of L-DOPA [46, 51, 55, 65, 84, 104] or were not affected [48, 52]. The administration of D_2 receptor agonists reduces the MPTP-induced upregulation of D_2 receptors [54, 57], but not as efficiently as L-DOPA, whereas D_1 receptor agonists have no effects on D_2 receptors or can produce an increase [49]. Similar observations were made in de novo PD patients as measured by PET scan [98, 172]. In PD patients for whom dopaminergic treatment was discontinued after subthalamic deep brain stimulation, the L-DOPA-induced downregulation of D_2 receptors was reversed [101]. D_2 receptor mRNA increases in the posterior striatum after exposure to MPTP [47, 60, 61, 68] or remains unchanged [58, 67]. The MPTP-induced upregulation of striatal D_2 receptor mRNA is completely reversed by L-DOPA treatment in monkeys [61] or unaffected [47, 68], whereas the administration of D_2 receptor agonists will either decrease (pulsatile) or reverse (continuous) the upregulation [60]. To our knowledge, D_2 receptor trafficking alterations in PD and LID has not yet been established.

The regulator of G-protein signaling (RSG) 9–2 is known to inhibit D_2 adenylyl cyclase-dependent intracellular signaling in the basal ganglia [173]. Its genetic deletion in mice causes greater locomotor responses to the D_1/D_2 agonist apomorphine [173]. The overexpression of RSG9-2 in MPTP monkeys diminishes LID, as well as D_2 agonist-induced dyskinesia, but at the cost of decreased antiparkinsonian activity [174]. On the other hand, the striatal D_2 receptor-regulated Akt/GSK3 signaling cascade, which is independent from adenylyl cyclase, contributes to neurodegenerative disease including PD [175]. It was recently shown in MPTP monkeys that L-DOPA treatment with or without antidyskinetic drugs induced a prolonged phosphorylation of both Akt and GSK3 [72]. Furthermore, the severity of LID was correlated with their respective levels of phosphorylation in the posterior putamen [72]. As demonstrated for D_1 receptors, intracellular signaling pathways in D_2 receptor expressing neurons are also impaired in LID and may represent other targets for pharmacological treatments.

D_2 and adenosine A_{2A} receptors are known to form functional hetero-oligomers [162, 176]. Long-term administration of A_{2A} or D_2 agonists induces an internalization and desensitization of the D_2/A_{2A} complex [176], whereas D_2 antagonists trigger an increase in D_2/A_{2A} complex immunoreactivity [177]. Moreover, A_{2A} and the glutamate metabotropic mGlu5 receptors are also known to interact [178].

Compared to D_1 and D_2 receptors, striatal D_3 receptors are much less abundant [179] and unaffected by DA denervation in rats [180]. D_3 receptor specific binding decreases in MPTP monkeys [37, 52, 70, 71] and PD patients [86, 97, 110]. However, its expression highly increases with L-DOPA treatment or D_1-like agonists and suggests an involvement of D_3 receptors in sensitization to L-DOPA [37, 52, 70, 71, 181]. Furthermore, the administration of a D_3 receptor agonist potentiates the behavioral response to D_1 receptor stimulation in 6-OHDA rats [182]. On the other hand, no change in D_3 specific binding and mRNA was observed after MPTP and L-DOPA in common marmosets [69] and in humans [109], indicating species differences. Nevertheless, the administration of a selective D_3 receptor agonist in L-DOPA-PRIMED MPTP monkeys elicited LID comparable to those induced by apomorphine [183]. Antagonizing D_3 receptors may help in reducing the development of LID but not those already established, as demonstrated in MPTP marmosets [184]. In MPTP macaques, the D_3 antagonist nafadotride reduced established LID but at the cost of decreased antiparkinsonian benefits of L-DOPA [37]. D_3 receptors seem to be involved in LID, but further studies are needed to establish its role in the development, priming, and expression of LID.

To our knowledge, no investigation of D_4 receptors has yet been done on postmortem tissues of PD patients or MPTP monkeys. However, the addition of the selective and potent D_4 receptor antagonist L-745,870 to L-DOPA was recently associated with a reduced severity of LID and an increased on-time duration without disabling LID in MPTP monkeys [185]. Though its antidyskinetic effects remain to be elucidated, the relative abundance of D_4 receptors on GABAergic neurons of the pallidum compared to the striatum [186] suggests an involvement of the nigropallidal pathway, the latter being relatively spared in MPTP monkeys [187]. It also remains unknown if D_4 receptors are involved in the priming process.

Dopamine–Dopamine Receptor Hetero-oligomers

The so-called direct D_1 receptor-related and the indirect D_2 receptor-related pathways (subject covered in the previous chapter) have opposite effects [188] and may be an important factor to consider in LID. In rodents, it has long been thought that both pathways were well segregated [189]. It was however shown that most [190], if not virtually all, DA neurons express both D_1 and D_2 receptors in rodents [191]. It is noteworthy that a significant proportion of MSN coexpress both D_1 and D_2 receptor proteins in monkeys [21]. On the other hand, roughly 25 % of neurons in rats [192] and up to 5 % in primates [33] coexpress both receptor mRNAs. Such coexpression of D_1 and D_2 receptors suggests a functional cross talk between these two receptors. Indeed, the administration of D_2 agonists potentiates the effects of immediate early genes upon D_1 receptors stimulation [193]. Moreover, D_1 and D_2 receptors can form functional hetero-oligomers and may account for the D_1–D_2 receptor synergetic effects [194]. In fact, the D_1/D_2 hetero-oligomer produces a functional unit for calcium generation, which is not observed with the stimulation of D_1 or D_2 receptor homo-oligomer [195]. A rapid desensitization of such D_1/D_2 hetero-oligomer may occur with the stimulation of D_1 receptors with the specific agonist SKF-83822 or a pretreatment with the D_1/D_2 hetero-oligomer agonist SKF-83959 [196]. D_1 and D_3 receptors are coexpressed in several striatal neurons in rats [197] and may also physically interact and form functional hetero-oligomers [198]. Hetero-oligomerization between the D_1 and D_3 receptors abolishes the D_1 agonist-induced cytoplasmatic sequestration [199]. Behavioral and locomotor implications of D_1/D_3 hetero-oligomer have been reviewed [200] and deserve more attention. The D_1/D_2 and D_1/D_3 hetero-oligomers offer a new exploratory path of research for LID, and more studies are needed to fully understand their possible involvement in LID.

Although most of the binding, in situ hybridization, and PET studies described in the present section did not show a clear correlation between DA receptor supersensitivity and LID (Tables 10.1 and 10.2), it should be considered that receptor supersensitivity may not be strictly related to changes in receptor protein or mRNA levels [201, 202]. Changes in DA-related intracellular pathways are to be taken into account in the development, priming process, and expression of LID. Furthermore, DA receptor subtypes interact with one another and with receptors from other neurotransmitters. Hence, LID are by far more complex than one would expect from classical biochemical concepts.

Relevant Clinical Studies in Parkinsonian Patients on Dopamine Receptors

The implications of the DA receptor subtypes in the pathophysiology of PD and LID have been investigated in major clinical studies in PD patients. Studies verified whether the use of higher doses of L-DOPA or DA agonist, where DA receptors are probably more stimulated, provides a better antiparkinsonian response but may increase the risk of developing LID. In the Deprenyl and Tocopherol Antioxidant Therapy of Parkinsonism (DATATOP) controlled and multicenter trial, a mean daily L-DOPA dose of 338 mg was not associated with LID in parkinsonian patients, while at higher dose (mean of 387 mg), LID were observed at the same follow-up time [203]. Moreover, the Earlier vs. Later L-DOPA Therapy in PD (ELLDOPA) study, a prospective, double-blind, and placebo-controlled trial, was conducted to address the efficacy and the safety of different doses of L-DOPA in 317 previously untreated PD patients (<14 days of dopaminergic drugs) for 40 weeks [204]. While providing a better antiparkinsonian effect, higher doses of L-DOPA were associated with a dose-dependent effect on LID [204]. Moreover, a community-based study, where 87 PD patients were treated with L-DOPA with a >10-year follow-up, showed that the development of LID was increased with disease duration and severity, but LID were also related to the duration and the dose of L-DOPA treatment [205]. In normal brains, striatal DA concentrations are constant, there is a continuous stimulation of MSN, and the activity of the basal ganglia is normalized. However, L-DOPA is not able to maintain constant synaptic and extrasynaptic DA concentrations in PD. Hence, DA receptors are not constantly exposed to DA. L-DOPA, which already promotes fluctuations in brain DA concentration levels, is not able to completely normalize basal ganglia activity. It has been showed in a [^{11}C]raclopride PET study that patients with peak-dose LID had larger 1 h increases in synaptic DA levels than non-dyskinetic PD patients [206]. Inversely, diphasic dyskinesias, which are different from the peak-dose LID (which are associated with high L-DOPA doses), can occur with low doses of L-DOPA and can be attenuated by increasing L-DOPA doses [207]. These studies suggest a relation between L-DOPA doses (level of stimulation of DA receptors) and the risk of developing LID.

The development of LID is also probably due to an abnormal and pulsatile stimulation of DA receptors. This has been verified with dopaminergic drugs that are able to stimulate more continuously DA receptors, consequently decreasing the risk of developing LID [8, 208–210]. Hence, long-acting DA agonists and continuous infusion of L-DOPA can reduce already established LID or reduce the development of LID as compared to L-DOPA-treated PD patients. The ergot derivatives (bromocriptine, pergolide, and cabergoline), long-acting D_2 receptor agonists, were the first DA agonists approved, and they reduce dyskinesia when combined to L-DOPA or used as monotherapy [211–217]. Similar results were also obtained with nonergot DA agonists such as pramipexole and ropinirole where these agonists improved parkinsonian disability and reduce established LID in advanced PD state [218, 219]. Interestingly, many prospective double-blind randomized controlled trials evaluated the risk of developing LID in previously untreated early PD patients when treated chronically with L-DOPA or DA agonists. In a 5-year study including 268 PD patients, ropinirole reduced the development of LID as compared to the L-DOPA group (45 % vs. 20 %) [10]. Similarly, in the prospective randomized multicenter and double-blind CALM-PD (Comparison of the Agonist Pramipexole with L-DOPA on Motor Complications of Parkinson's Disease) study, pramipexole reduced the risk of developing LID as compared to L-DOPA (31 % vs. 10 %) after 2 years of treatment [9]. The authors of this same study obtained similar results at 6-year follow-up (54 % with L-DOPA vs. 24 %

with pramipexole) [220]. However, a later subanalysis of the ropinirole study and CALM-PD showed that delaying the introduction of L-DOPA did not prevent the development of LID [221, 222]. Thus, the use of a long-acting DA agonist may mask LID but does not prevent the development of LID once L-DOPA treatment is started.

Another strategy to avoid fluctuating DA concentrations in the basal ganglia and reducing already established LID is to add catechol-O-methyltransferase (COMT) inhibitors to L-DOPA treatment [223–226]. The COMT inhibitor entacapone was shown to inhibit the expression of LID in monkeys [227], while in the multicenter and double-blind clinical trial Stalevo Reduction in Dyskinesia Evaluation in Parkinson's disease (STRIDE-PD), an increase of LID was observed with entacapone [228]. The increase of LID observed might be related to the choice of a non-appropriate 3.5 h interval between Stalevo administrations [228]. Concerning the monoamine oxidase B (MAO-B) inhibitors, in Lasting effect in Adjunct therapy with Rasagiline Given Once daily (PRESTO) and A Randomized Placebo-Controlled Trial of Rasagiline in Levodopa-Treated Patients With Parkinson Disease and Motor Fluctuations (LARGO) clinical studies, the use of rasagiline 1 mg/day and rasagiline 1 mg/day + entacapone 200 mg, respectively, in adjunct to L-DOPA was associated with increases in LID [229, 230]. Thus, DA brain levels were probably too high, enhancing global dopaminergic activity and increasing the risk of developing LID. It has also been hypothesized that the development of LID might be related to the activation of a specific DA receptor subtype. In rodent models, the contribution of the direct pathway with an increased activity of D_1 receptors has been associated with AIMs [19]. It was also initially thought that the stimulation of D_2 rather than D_1 receptors with a specific agonist could reduce or prevent the development of LID. However, in humans and nonhuman primates, the use of short-acting and specific agonists of D_1, D_2, and D_3 receptor subtypes seems to be implicated in the development of LID [37, 113, 119, 231, 232]. For instance, the selective D_1 agonist prodrug, ABT-431, was administered as monotherapy in advanced PD patients with fluctuating response to L-DOPA and induced similar antiparkinsonian benefits and dyskinesia as L-DOPA [232]. Hence, this study shows that D_1 receptor agonists are as likely to produce dyskinesias as L-DOPA.

Moreover, genetic variations in DA receptors have been identified to play a role in the occurrence of peak-dose LID. A case–control study comparing sporadic PD patients and control subjects was conducted to evaluate three polymorphisms involving the D_1 receptor gene and one intronic short tandem repeat polymorphism of the D_2 receptor gene [233]. Polymorphisms of the D_1 receptor gene were not associated with the risk of developing PD or peak-dose LID, while the 15 allele of the polymorphism of the D_2 receptor gene was more frequent in parkinsonian than in control subjects [233]. The frequency of both the 13 allele and the 14 allele of the D_2 receptor gene polymorphism was higher in non-dyskinetic than in the dyskinetic PD subjects, while the risk reduction of developing peak-dose LID for PD subjects carrying at least 1 of the 13 or 14 alleles was 72 % with respect to the PD subjects who did not carry these alleles [233]. In another cohort study, genetic factors related to the D_2 receptor polymorphic status were found to have a protective effect in the development of LID in men, but not in women [234]. Moreover, more recent studies also suggest that gene polymorphisms in DA receptor subtypes might be implicated in the development of LID [235, 236].

Conclusion

Studies of LID in parkinsonian patients and animal models show multiple changes in the basal ganglia dopaminergic systems in the activity and the modulation of DA receptors and their signaling pathways. The mechanisms involved in the occurrence of PD, DA depletion, and LID are complex and involve numerous neurotransmitters. LID are by far more complex than one would expect from classical biochemical concepts. DA receptor supersensitivity may not be strictly related to changes in receptor protein or mRNA levels. Moreover, LID may not be associated with a specific DA receptor subtype but are generated less by dopaminergic drugs with long half-life. It is important to continue to identify which proteins are up- or down-regulated in direct and indirect output pathways of the basal ganglia implicated in LID. More studies with PD patients and animal models are needed to better understand the implication of DA receptors subtypes and their interactions with other neurotransmitters and their receptors in the development of LID to improve the effectiveness of present treatments or to develop new therapies against the LID.

Acknowledgments This work was supported by a grant from the Canadian Institutes of Health Research to T.D.P. V.A.J. received a studentship from the Fonds d'Enseignement et de Recherche of the Faculté de Pharmacie of Université Laval and a studentship from the Centre de Recherche en Endocrinologie Moléculaire et Oncologique et en Génomique Humaine. N.M. received a professional health-care studentship from the Fonds de la Recherche en Santé du Québec.

References

1. de Lau LM, Breteler MM. Epidemiology of Parkinson's disease. Lancet Neurol. 2006;5:525–35.
2. Siderowf A, Stern M. Update on Parkinson disease. Ann Intern Med. 2003;138:651–8.
3. Jellinger KA. Pathology of Parkinson's disease. Changes other than the nigrostriatal pathway. Mol Chem Neuropathol. 1991;14:153–97.
4. Fabbrini G, Brotchie JM, Grandas F, Nomoto M, Goetz CG. Levodopa-induced dyskinesias. Mov Disord. 2007;22:1379–89.
5. Gottwald MD, Aminoff MJ. Therapies for dopaminergic-induced dyskinesias in Parkinson disease. Ann Neurol. 2011;69:919–27.
6. Meissner WG, Frasier M, Gasser T, Goetz CG, Lozano A, Piccini P, et al. Priorities in Parkinson's disease research. Nat Rev Drug Discov. 2011;10:377–93.
7. Mayeux R, Stern Y, Mulvey K, Cote L. Reappraisal of temporary levodopa withdrawal ("drug holiday") in Parkinson's disease. N Engl J Med. 1985;313:724–8.
8. Olanow CW, Stern MB, Sethi K. The scientific and clinical basis for the treatment of Parkinson disease (2009). Neurology. 2009;72:S1–136.
9. Parkinson Study Group. Pramipexole vs levodopa as initial treatment for Parkinson disease: a randomized controlled trial. JAMA. 2000;284:1931–8.
10. Rascol O, Brooks DJ, Korczyn AD, De Deyn PP, Clarke CE, Lang AE. A five-year study of the incidence of dyskinesia in patients with early Parkinson's disease who were treated with ropinirole or levodopa. 056 Study Group. N Engl J Med. 2000;342:1484–91.
11. Olanow CW. The scientific basis for the current treatment of Parkinson's disease. Annu Rev Med. 2004;55:41–60.

12. Parkinson Study Group. Long-term effect of initiating pramipexole vs levodopa in early Parkinson disease. Arch Neurol. 2009;66:563–70.
13. Katzenschlager R, Head J, Schrag A, Ben-Shlomo Y, Evans A, Lees AJ. Fourteen-year final report of the randomized PDRG-UK trial comparing three initial treatments in PD. Neurology. 2008;71:474–80.
14. Morin N, Jourdain VA, Di Paolo T. Modeling dyskinesia in animal models of Parkinson disease. Exp Neurol. 2014;256:105–16.
15. Martin RF, Bowden DM. Primate brain maps: structure of the macaque brain. Amsterdam: Elsevier Science B.V; 2000.
16. Paxinos G, Watson C. The rat brain in stereotaxic coordinates. 4th ed. London: Academic; 2007.
17. Parent A, Lévesque M, Parent M. A re-evaluation of the current model of the basal ganglia. Parkinsonism Relat Disord. 2001;7:193–8.
18. Blanchet PJ, Calon F, Morissette M, Hadj Tahar A, Belanger N, Samadi P, et al. Relevance of the MPTP primate model in the study of dyskinesia priming mechanisms. Parkinsonism Relat Disord. 2004;10:297–304.
19. Cenci MA, Ohlin KE, Rylander D. Plastic effects of L-DOPA treatment in the basal ganglia and their relevance to the development of dyskinesia. Parkinsonism Relat Disord. 2009;15 Suppl 3:S59–63.
20. Levesque M, Parent A. The striatofugal fiber system in primates: a reevaluation of its organization based on single-axon tracing studies. Proc Natl Acad Sci. 2005;102:11888–93.
21. Nadjar A, Brotchie JM, Guigoni C, Li Q, Zhou S-B, Wang G-J, et al. Phenotype of striatofugal medium spiny neurons in parkinsonian and dyskinetic nonhuman primates: a call for a reappraisal of the functional organization of the basal ganglia. J Neurosci. 2006;26:8653–61.
22. Wu Y, Richard S, Parent A. The organization of the striatal output system: a single-cell juxtacellular labeling study in the rat. Neurosci Res. 2000;38:49–62.
23. Beaulieu JM, Gainetdinov RR. The physiology, signaling, and pharmacology of dopamine receptors. Pharmacol Rev. 2011;63:182–217.
24. Beaulieu JM, Gainetdinov RR, Caron MG. The Akt-GSK-3 signaling cascade in the actions of dopamine. Trends Pharmacol Sci. 2007;28:166–72.
25. Civelli O, Bunzow JR, Grandy DK. Molecular diversity of the dopamine receptors. Annu Rev Pharmacol Toxicol. 1993;33:281–307.
26. Gingrich JA, Caron MG. Recent advances in the molecular biology of dopamine receptors. Annu Rev Neurosci. 1993;16:299–321.
27. Greengard P. The neurobiology of slow synaptic transmission. Science (New York, NY). 2001;294:1024–30.
28. Missale C, Nash SR, Robinson SW, Jaber M, Caron MG. Dopamine receptors: from structure to function. Physiol Rev. 1998;78:189–225.
29. Hurd YL, Suzuki M, Sedvall GC. D1 and D2 dopamine receptor mRNA expression in whole hemisphere sections of the human brain. J Chem Neuroanat. 2001;22:127–37.
30. Sokoloff P, Schwartz JC. Novel dopamine receptors half a decade later. Trends Pharmacol Sci. 1995;16:270–5.
31. Meador-Woodruff JH, Damask SP, Wang J, Haroutunian V, Davis KL, Watson SJ. Dopamine receptor mRNA expression in human striatum and neocortex. Neuropsychopharmacology. 1996;15:17–29.
32. DeLong MR, Wichmann T. Circuits and circuit disorders of the basal ganglia. Arch Neurol. 2007;64:20–4.
33. Aubert I, Ghorayeb I, Normand E, Bloch B. Phenotypical characterization of the neurons expressing the D1 and D2 dopamine receptors in the monkey striatum. J Comp Neurol. 2000;418:22–32.
34. Koprich JB, Johnston TH, Huot P, Fox SH, Brotchie JM. New insights into the organization of the basal ganglia. Curr Neurol Neurosci Rep. 2009;9:298–304.

35. Cepeda C, Hurst RS, Altemus KL, Flores-Hernandez J, Calvert CR, Jokel ES, et al. Facilitated glutamatergic transmission in the striatum of D2 dopamine receptor-deficient mice. J Neurophysiol. 2001;85:659–70.
36. Wang H, Pickel VM. Dopamine D2 receptors are present in prefrontal cortical afferents and their targets in patches of the rat caudate-putamen nucleus. J Comp Neurol. 2002;442:392–404.
37. Bezard E, Ferry S, Mach U, Stark H, Leriche L, Boraud T, et al. Attenuation of levodopa-induced dyskinesia by normalizing dopamine D3 receptor function. Nat Med. 2003;9:762–7.
38. Murray AM, Ryoo HL, Gurevich E, Joyce JN. Localization of dopamine D3 receptors to mesolimbic and D2 receptors to mesostriatal regions of human forebrain. Proc Natl Acad Sci. 1994;91:11271–5.
39. Suzuki M, Hurd YL, Sokoloff P, Schwartz JC, Sedvall G. D3 dopamine receptor mRNA is widely expressed in the human brain. Brain Res. 1998;779:58–74.
40. Gurevich EV, Himes JW, Joyce JN. Developmental regulation of expression of the D3 dopamine receptor in rat nucleus accumbens and islands of Calleja. J Pharmacol Exp Ther. 1999;289:587–98.
41. Bouthenet ML, Souil E, Martres MP, Sokoloff P, Giros B, Schwartz JC. Localization of dopamine D3 receptor mRNA in the rat brain using in situ hybridization histochemistry: comparison with dopamine D2 receptor mRNA. Brain Res. 1991;564:203–19.
42. Meador-Woodruff JH, Grandy DK, Van Tol HH, Damask SP, Little KY, Civelli O, et al. Dopamine receptor gene expression in the human medial temporal lobe. Neuropsychopharmacology. 1994;10:239–48.
43. Kruusmägi M, Kumar S, Zelenin S, Brismar H, Aperia A, Scott L. Functional differences between D1 and D5 revealed by high resolution imaging on live neurons. Neuroscience. 2009;164:463–9.
44. Khan ZU, Gutierrez A, Martin R, Penafiel A, Rivera A, de la Calle A. Dopamine D5 receptors of rat and human brain. Neuroscience. 2000;100:689–99.
45. Alexander GM, Brainard DL, Gordon SW, Hichens M, Grothusen JR, Schwartzman RJ. Dopamine receptor changes in untreated and (+)-PHNO-treated MPTP parkinsonian primates. Brain Res. 1991;547:181–9.
46. Alexander GM, Schwartzman RJ, Grothusen JR, Brainard L, Gordon SW. Changes in brain dopamine receptors in MPTP parkinsonian monkeys following L-dopa treatment. Brain Res. 1993;625:276–82.
47. Aubert I, Guigoni C, Håkansson K, Li Q, Dovero S, Barthe N, et al. Increased D1 dopamine receptor signaling in levodopa-induced dyskinesia. Ann Neurol. 2005;57:17–26.
48. Calon F, Goulet M, Blanchet PJ, Martel JC, Piercey MF, Bédard PJ, et al. Levodopa or D2 agonist induced dyskinesia in MPTP monkeys: correlation with changes in dopamine and GABAA receptors in the striatopallidal complex. Brain Res. 1995;680:43–52.
49. Goulet M, Grondin R, Blanchet PJ, Bédard PJ, Di Paolo T. Dyskinesias and tolerance induced by chronic treatment with a D1 agonist administered in pulsatile or continuous mode do not correlate with changes of putaminal D1 receptors in drug-naive MPTP monkeys. Brain Res. 1996;719:129–37.
50. Graham WC, Clarke CE, Boyce S, Sambrook MA, Crossman AR, Woodruff GN. Autoradiographic studies in animal models of hemi-parkinsonism reveal dopamine D2 but not D1 receptor supersensitivity. II. Unilateral intra-carotid infusion of MPTP in the monkey (Macaca fascicularis). Brain Res. 1990;514:103–10.
51. Graham WC, Sambrook MA, Crossman AR. Differential effect of chronic dopaminergic treatment on dopamine D1 and D2 receptors in the monkey brain in MPTP-induced parkinsonism. Brain Res. 1993;602:290–303.
52. Guigoni C, Aubert I, Li Q, Gurevich VV, Benovic JL, Ferry S, et al. Pathogenesis of levodopa-induced dyskinesia: focus on D1 and D3 dopamine receptors. Parkinsonism Relat Disord. 2005;11 Suppl 1:S25–9.

53. Przedborski S, Jackson-Lewis V, Popilskis S, Kostic V, Levivier M, Fahn S, et al. Unilateral MPTP-induced parkinsonism in monkeys. A quantitative autoradiographic study of dopamine D1 and D2 receptors and re-uptake sites. Neurochirurgie. 1991;37:377–82.
54. Gagnon C, Gomez-Mancilla B, Markstein R, Bédard PJ, Di Paolo T. Effect of adding the D-1 agonist CY 208–243 to chronic bromocriptine treatment of MPTP-monkeys: regional changes of brain dopamine receptors. Prog Neuropsychopharmacol Biol Psychiatry. 1995;19:667–76.
55. Gagnon C, Bedard PJ, Di Paolo T. Effect of chronic treatment of MPTP monkeys with dopamine D-1 and/or D- 2 receptor agonists. Eur J Pharmacol. 1990;178:115–20.
56. Gnanalingham KK, Smith LA, Hunter AJ, Jenner P, Marsden CD. Alterations in striatal and extrastriatal D-1 and D-2 dopamine receptors in the MPTP-treated common marmoset: an autoradiographic study. Synapse. 1993;14:184–94.
57. Gomez-Mancilla B, Boucher R, Gagnon C, Di Paolo T, Markstein R, Bédard PJ. Effect of adding the D1 agonist CY 208–243 to chronic bromocriptine treatment. I: Evaluation of motor parameters in relation to striatal catecholamine content and dopamine receptors. Mov Disord. 1993;8:144–50.
58. Goulet M, Grondin R, Morissette M, Maltais S, Falardeau P, Bédard PJ, et al. Regulation by chronic treatment with cabergoline of dopamine D1 and D2 receptor levels and their expression in the striatum of Parkinsonian-monkeys. Prog Neuropsychopharmacol Biol Psychiatry. 2000;24:607–17.
59. Rioux L, Frohna PA, Joyce JN, Schneider JS. The effects of chronic levodopa treatment on pre- and postsynaptic markers of dopaminergic function in striatum of parkinsonian monkeys. Mov Disord. 1997;12:148–58.
60. Goulet M, Morissette M, Calon F, Blanchet PJ, Falardeau P, Bédard PJ, et al. Continuous or pulsatile chronic D2 dopamine receptor agonist (U91356A) treatment of drug-naive 4-phenyl-1,2,3,6-tetrahydropyridine monkeys differentially regulates brain D1 and D2 receptor expression: in situ hybridization histochemical analysis. Neuroscience. 1997;79:497–507.
61. Morissette M, Goulet M, Calon F, Falardeau P, Blanchet PJ, Bédard PJ, et al. Changes of D1 and D2 dopamine receptor mRNA in the brains of monkeys lesioned with 1-methy-4-phenyl-1,2,3,6-tetrahydropyridine: correction with chronic administration of L-3,4-dihydroxyphenylalanine. Mol Pharmacol. 1996;50:1073–9.
62. Grondin R, Goulet M, Morissette M, Bédard PJ, Di Paolo T. Dopamine D1 receptor mRNA and receptor levels in the striatum of MPTP monkeys chronically treated with SKF-82958. Eur J Pharmacol. 1999;378:259–63.
63. Pifl C, Nanoff C, Schingnitz G, Schütz W, Hornykiewicz O. Sensitization of dopamine-stimulated adenylyl cyclase in the striatum of 1-methyl-4-phenyl-1,2,3,6-tetrahydropyridine-treated rhesus monkeys and patients with idiopathic Parkinson's disease. J Neurochem. 1992;58:1997–2004.
64. Elsworth JD, Brittan MS, Taylor JR, Sladek JR, Redmond DE, Innis RB, et al. Upregulation of striatal D2 receptors in the MPTP-treated vervet monkey is reversed by grafts of fetal ventral mesencephalon: an autoradiographic study. Brain Res. 1998;795:55–62.
65. Falardeau P, Bouchard S, Bedard PJ, Boucher R, Di Paolo T. Behavioral and biochemical effect of chronic treatment with D-1 and/or D-2 dopamine agonists in MPTP monkeys. Eur J Pharmacol. 1988;150:59–66.
66. Joyce JN, Marshall JF, Bankiewicz KS, Kopin IJ, Jacobowitz DM. Hemiparkinsonism in a monkey after unilateral internal carotid artery infusion of 1-methyl-4-phenyl-1,2,3,6-tetrahydropyridine (MPTP) is associated with regional ipsilateral changes in striatal dopamine D-2 receptor density. Brain Res. 1986;382:360–4.
67. Todd RD, Carl J, Harmon S, O'Malley KL, Perlmutter JS. Dynamic changes in striatal dopamine D2 and D3 receptor protein and mRNA in response to 1-methyl-4-phenyl-1,2,3,6-tetrahydropyridine (MPTP) denervation in baboons. J Neurosci. 1996;16:7776–82.
68. Herrero MT, Augood SJ, Asensi H, Hirsch EC, Agid Y, Obeso JA, et al. Effects of L-DOPA-therapy on dopamine D2 receptor mRNA expression in the striatum of MPTP-intoxicated parkinsonian monkeys. Brain Res Mol Brain Res. 1996;42:149–55.

69. Hurley MJ, Jolkkonen J, Stubbs CM, Jenner P, Marsden CD. Dopamine D3 receptors in the basal ganglia of the common marmoset and following MPTP and L-DOPA treatment. Brain Res. 1996;709:259–64.
70. Morissette M, Goulet M, Grondin R, Blanchet P, Bedard P, Di Paolo T, et al. Associative and limbic regions of monkey striatum express high levels of dopamine D3 receptors: effects of MPTP and dopamine agonist replacement therapies. Eur J Neurosci. 1998;10:2565–73.
71. Quik M, Police S, He L, Di Monte DA, Langston JW. Expression of D(3) receptor messenger RNA and binding sites in monkey striatum and substantia nigra after nigrostriatal degeneration: effect of levodopa treatment. Neuroscience. 2000;98:263–73.
72. Morissette M, Samadi P, Tahar AH, Bélanger N, Di Paolo T. Striatal Akt/GSK3 signaling pathway in the development of L-Dopa-induced dyskinesias in MPTP monkeys. Prog Neuropsychopharmacol Biol Psychiatry. 2010;34:446–54.
73. Santini E, Sgambato-Faure V, Li Q, Savasta M, Dovero S, Fisone G, et al. Distinct changes in cAMP and extracellular signal-regulated protein kinase signalling in L-DOPA-induced dyskinesia. PLoS One. 2010;5:e12322.
74. Bezard E, Gross CE, Qin L, Gurevich VV, Benovic JL, Gurevich EV. L-DOPA reverses the MPTP-induced elevation of the arrestin2 and GRK6 expression and enhanced ERK activation in monkey brain. Neurobiol Dis. 2005;18:323–35.
75. Petzinger GM, Fisher B, Hogg E, Abernathy A, Arevalo P, Nixon K, et al. Behavioral motor recovery in the 1-methyl-4-phenyl-1,2,3,6-tetrahydropyridine-lesioned squirrel monkey (Saimiri sciureus): changes in striatal dopamine and expression of tyrosine hydroxylase and dopamine transporter proteins. J Neurosci Res. 2006;83:332–47.
76. Doucet JP, Nakabeppu Y, Bédard PJ, Hope BT, Nestler EJ, Jasmin BJ, et al. Chronic alterations in dopaminergic neurotransmission produce a persistent elevation of deltaFosB-like protein(s) in both the rodent and primate striatum. Eur J Neurosci. 1996;8:365–81.
77. Berton O, Guigoni C, Li Q, Bioulac BH, Aubert I, Gross CE, et al. Striatal overexpression of DeltaJunD resets L-DOPA-induced dyskinesia in a primate model of Parkinson disease. Biol Psychiatry. 2009;66:554–61.
78. Raisman R, Cash R, Ruberg M, Javoy-Agid F, Agid Y. Binding of [3H]SCH 23390 to D-1 receptors in the putamen of control and parkinsonian subjects. Eur J Pharmacol. 1985;113:467–8.
79. Cash R, Raisman R, Ploska A, Agid Y. Dopamine D-1 receptor and cyclic AMP-dependent phosphorylation in Parkinson's disease. J Neurochem. 1987;49:1075–83.
80. Joyce JN. Differential response of striatal dopamine and muscarinic cholinergic receptor subtypes to the loss of dopamine. III. Results in Parkinson's disease cases. Brain Res. 1993;600:156–60.
81. Palacios JM, Camps M, Cortes R, Probst A. Mapping dopamine receptors in the human brain. J Neural Transm Suppl. 1988;27:227–35.
82. Pierot L, Desnos C, Blin J, Raisman R, Scherman D, Javoy-Agid F, et al. D1 and D2-type dopamine receptors in patients with Parkinson's disease and progressive supranuclear palsy. J Neurol Sci. 1988;86:291–306.
83. Pimoule C, Schoemaker H, Reynolds GP, Langer SZ. [3H]SCH 23390 labeled D1 dopamine receptors are unchanged in schizophrenia and Parkinson's disease. Eur J Pharmacol. 1985;114:235–7.
84. Cortes R, Camps M, Gueye B, Probst A, Palacios JM. Dopamine receptors in human brain: autoradiographic distribution of D1 and D2 sites in Parkinson syndrome of different etiology. Brain Res. 1989;483:30–8.
85. Rinne JO, Laihinen A, Lonnberg P, Marjamaki P, Rinne UK. A post-mortem study on striatal dopamine receptors in Parkinson's disease. Brain Res. 1991;556:117–22.
86. Piggott MA, Marshall EF, Thomas N, Lloyd S, Court JA, Jaros E, et al. Striatal dopaminergic markers in dementia with Lewy bodies, Alzheimer's and Parkinson's diseases: rostrocaudal distribution. Brain. 1999;122(Pt 8):1449–68.
87. Rinne JO, Laihinen A, Nagren K, Bergman J, Solin O, Haaparanta M, et al. PET demonstrates different behaviour of striatal dopamine D-1 and D-2 receptors in early Parkinson's disease. J Neurosci Res. 1990;27:494–9.

88. Shinotoh H, Hirayama K, Tateno Y. Dopamine D1 and D2 receptors in Parkinson's disease and striatonigral degeneration determined by PET. Adv Neurol. 1993;60:488–93.
89. Shinotoh H, Inoue O, Hirayama K, Aotsuka A, Asahina M, Suhara T, et al. Dopamine D1 receptors in Parkinson's disease and striatonigral degeneration: a positron emission tomography study. J Neurol Neurosurg Psychiatry. 1993;56:467–72.
90. Turjanski N, Lees AJ, Brooks DJ. In vivo studies on striatal dopamine D1 and D2 site binding in L-dopa-treated Parkinson's disease patients with and without dyskinesias. Neurology. 1997;49:717–23.
91. Lee T, Seeman P, Rajput A, Farley IJ, Hornykiewicz O. Receptor basis for dopaminergic supersensitivity in Parkinson's disease. Nature. 1978;273:59–61.
92. Bokobza B, Ruberg M, Scatton B, Javoy-Agid F, Agid Y. [3H]spiperone binding, dopamine and HVA concentrations in Parkinson's disease and supranuclear palsy. Eur J Pharmacol. 1984;99:167–75.
93. Guttman M, Seeman P. L-dopa reverses the elevated density of D2 dopamine receptors in Parkinson's diseased striatum. J Neural Transm. 1985;64:93–103.
94. Guttman M, Seeman P, Reynolds GP, Riederer P, Jellinger K, Tourtellotte WW. Dopamine D2 receptor density remains constant in treated Parkinson's disease. Ann Neurol. 1986;19:487–92.
95. Quik M, Spokes EG, Mackay AV, Bannister R. Alterations in [3H]spiperone binding in human caudate nucleus, substantia nigra and frontal cortex in the Shy-Drager syndrome and Parkinson's disease. J Neurol Sci. 1979;43:429–37.
96. Rinne UK, Lonnberg P, Koskinen V. Dopamine receptors in the Parkinsonian brain. J Neural Transm. 1981;51:97–106.
97. Ryoo HL, Pierrotti D, Joyce JN. Dopamine D3 receptor is decreased and D2 receptor is elevated in the striatum of Parkinson's disease. Mov Disord. 1998;13:788–97.
98. Rinne JO, Laihinen A, Rinne UK, Nagren K, Bergman J, Ruotsalainen U. PET study on striatal dopamine D2 receptor changes during the progression of early Parkinson's disease. Mov Disord. 1993;8:134–8.
99. Rinne JO, Laihinen A, Ruottinen H, Ruotsalainen U, Nagren K, Lehikoinen P, et al. Increased density of dopamine D2 receptors in the putamen, but not in the caudate nucleus in early Parkinson's disease: a PET study with [11C]raclopride. J Neurol Sci. 1995;132:156–61.
100. Antonini A, Schwarz J, Oertel WH, Pogarell O, Leenders KL. Long-term changes of striatal dopamine D2 receptors in patients with Parkinson's disease: a study with positron emission tomography and [11C]raclopride. Mov Disord. 1997;12:33–8.
101. Thobois S, Vingerhoets F, Fraix V, Xie-Brustolin J, Mollion H, Costes N, et al. Role of dopaminergic treatment in dopamine receptor down-regulation in advanced Parkinson disease: a positron emission tomographic study. Arch Neurol. 2004;61:1705–9.
102. Brooks DJ, Ibanez V, Sawle GV, Playford ED, Quinn N, Mathias CJ, et al. Striatal D2 receptor status in patients with Parkinson's disease, striatonigral degeneration, and progressive supranuclear palsy, measured with 11C-raclopride and positron emission tomography. Ann Neurol. 1992;31:184–92.
103. Kaasinen V, Ruottinen HM, Nagren K, Lehikoinen P, Oikonen V, Rinne JO. Upregulation of putaminal dopamine D2 receptors in early Parkinson's disease: a comparative PET study with [11C] raclopride and [11C]N-methylspiperone. J Nucl Med. 2000;41:65–70.
104. Hagglund J, Aquilonius SM, Eckernas SA, Hartvig P, Lundquist H, Gullberg P, et al. Dopamine receptor properties in Parkinson's disease and Huntington's chorea evaluated by positron emission tomography using 11C-N-methyl-spiperone. Acta Neurol Scand. 1987;75:87–94.
105. Schwarz J, Tatsch K, Arnold G, Gasser T, Trenkwalder C, Kirsch CM, et al. 123I-iodobenzamide-SPECT predicts dopaminergic responsiveness in patients with de novo parkinsonism. Neurology. 1992;42:556–61.
106. Brucke T, Podreka I, Angelberger P, Wenger S, Topitz A, Kufferle B, et al. Dopamine D2 receptor imaging with SPECT: studies in different neuropsychiatric disorders. J Cereb Blood Flow Metab. 1991;11:220–8.

107. Laulumaa V, Kuikka JT, Soininen H, Bergstrom K, Lansimies E, Riekkinen P. Imaging of D2 dopamine receptors of patients with Parkinson's disease using single photon emission computed tomography and iodobenzamide I 123. Arch Neurol. 1993;50:509–12.
108. Verstappen CC, Bloem BR, Haaxma CA, Oyen WJ, Horstink MW. Diagnostic value of asymmetric striatal D2 receptor upregulation in Parkinson's disease: an [123I]IBZM and [123I] FP-CIT SPECT study. Eur J Nucl Med Mol Imaging. 2007;34:502–7.
109. Hurley MJ, Stubbs CM, Jenner P, Marsden CD. D3 receptor expression within the basal ganglia is not affected by Parkinson's disease. Neurosci Lett. 1996;214:75–8.
110. Boileau I, Guttman M, Rusjan P, Adams JR, Houle S, Tong J, et al. Decreased binding of the D3 dopamine receptor-preferring ligand [11C]-(+)-PHNO in drug-naive Parkinson's disease. Brain. 2009;132:1366–75.
111. Timmons S, Coakley MF, Moloney AM, O'Neill C. Akt signal transduction dysfunction in Parkinson's disease. Neurosci Lett. 2009;467:30–5.
112. Cash R, Raisman R, Cervera P, Javoy-Agid F, Agid Y. Pitfalls in membrane binding sites studies in post-mortem human brain. J Recept Res. 1987;7:527–54.
113. Blanchet P, Bedard PJ, Britton DR, Kebabian JW. Differential effect of selective D-1 and D-2 dopamine receptor agonists on levodopa-induced dyskinesia in 1-methyl-4-phenyl-1,2,3,6-tetrahydropyridine- exposed monkeys. J Pharmacol Exp Ther. 1993;267:275–9.
114. Goulet M, Madras BK. D(1) dopamine receptor agonists are more effective in alleviating advanced than mild parkinsonism in 1-methyl-4-phenyl-1,2,3, 6-tetrahydropyridine-treated monkeys. J Pharmacol Exp Ther. 2000;292:714–24.
115. Grondin R, Bédard PJ, Britton DR, Shiosaki K. Potential therapeutic use of the selective dopamine D1 receptor agonist, A-86929: an acute study in parkinsonian levodopa-primed monkeys. Neurology. 1997;49:421–6.
116. Delfino M, Kalisch R, Czisch M, Larramendy C, Ricatti J, Taravini IRE, et al. Mapping the effects of three dopamine agonists with different dyskinetogenic potential and receptor selectivity using pharmacological functional magnetic resonance imaging. Neuropsychopharmacology. 2007;32:1911–21.
117. Taylor JL, Bishop C, Walker PD. Dopamine D1 and D2 receptor contributions to L-DOPA-induced dyskinesia in the dopamine-depleted rat. Pharmacol Biochem Behav. 2005;81:887–93.
118. Westin JE, Vercammen L, Strome EM, Konradi C, Cenci MA. Spatiotemporal pattern of striatal ERK1/2 phosphorylation in a rat model of L-DOPA–induced dyskinesia and the role of dopamine D1 receptors. Biol Psychiatry. 2007;62:800–10.
119. Blanchet PJ, Grondin R, Bedard PJ. Dyskinesia and wearing-off following dopamine D1 agonist treatment in drug-naive 1-methyl-4-phenyl-1,2,3,6-tetrahydropyridine-lesioned primates. Mov Disord. 1996;11:91–4.
120. Darmopil S, Martín AB, De Diego IR, Ares S, Moratalla R. Genetic inactivation of dopamine D1 but not D2 receptors inhibits L-DOPA-induced dyskinesia and histone activation. Biol Psychiatry. 2009;66:603–13.
121. Playford ED, Brooks DJ. In vivo and in vitro studies of the dopaminergic system in movement disorders. Cerebrovasc Brain Metab Rev. 1992;4:144–71.
122. Gagnon C, Gomez-Mancilla B, Bédard PJ, Di Paolo T, Chronic CY. 208–243 treatment of MPTP-monkeys causes regional changes of dopamine and GABAA receptors. Neurosci Lett. 1993;163:31–5.
123. Brooks DJ. PET studies and motor complications in Parkinson's disease. Trends Neurosci. 2000;23:S101–8.
124. Guigoni C, Doudnikoff E, Li Q, Bloch B, Bezard E. Altered D(1) dopamine receptor trafficking in parkinsonian and dyskinetic non-human primates. Neurobiol Dis. 2007;26:452–63.
125. Berthet A, Bezard E, Porras G, Fasano S, Barroso-Chinea P, Dehay B, et al. L-DOPA impairs proteasome activity in parkinsonism through D1 dopamine receptor. J Neurosci. 2012;32:681–91.
126. Ahmed MR, Berthet A, Bychkov E, Porras G, Li Q, Bioulac BH, et al. Lentiviral overexpression of GRK6 alleviates L-dopa-induced dyskinesia in experimental Parkinson's disease. Sci Transl Med. 2010;2:28ra.

127. Muriel M-P, Orieux G, Hirsch EC. Levodopa but not ropinirole induces an internalization of D1 dopamine receptors in parkinsonian rats. Mov Disord. 2002;17:1174–9.
128. Muriel MP, Bernard V, Levey AI, Laribi O, Abrous DN, Agid Y, et al. Levodopa induces a cytoplasmic localization of D1 dopamine receptors in striatal neurons in Parkinson's disease. Ann Neurol. 1999;46:103–11.
129. Fienberg AA, Hiroi N, Mermelstein PG, Song W, Snyder GL, Nishi A, et al. DARPP-32: regulator of the efficacy of dopaminergic neurotransmission. Science (New York, NY). 1998;281:838–42.
130. Santini E, Valjent E, Usiello A, Carta M, Borgkvist A, Girault J-A, et al. Critical involvement of cAMP/DARPP-32 and extracellular signal-regulated protein kinase signaling in L-DOPA-induced dyskinesia. J Neurosci. 2007;27:6995–7005.
131. Calabresi P, Di Filippo M, Ghiglieri V, Picconi B. Molecular mechanisms underlying levodopa-induced dyskinesia. Mov Disord. 2008;23 Suppl 3:S570–9.
132. Centonze D, Gubellini P, Picconi B, Calabresi P, Giacomini P, Bernardi G. Unilateral dopamine denervation blocks corticostriatal LTP. J Neurophysiol. 1999;82:3575–9.
133. Picconi B, Centonze D, Håkansson K, Bernardi G, Greengard P, Fisone G, et al. Loss of bidirectional striatal synaptic plasticity in L-DOPA-induced dyskinesia. Nat Neurosci. 2003;6:501–6.
134. Calabresi P, Gubellini P, Centonze D, Picconi B, Bernardi G, Chergui K, et al. Dopamine and cAMP-regulated phosphoprotein 32 kDa controls both striatal long-term depression and long-term potentiation, opposing forms of synaptic plasticity. J Neurosci. 2000;20:8443–51.
135. Fiorentini C, Missale C. Oligomeric assembly of dopamine D1 and glutamate NMDA receptors: molecular mechanisms and functional implications. Biochem Soc Trans. 2004;32:1025–8.
136. Fiorentini C, Gardoni F, Spano P, Di Luca M, Missale C. Regulation of dopamine D1 receptor trafficking and desensitization by oligomerization with glutamate N-methyl-D-aspartate receptors. J Biol Chem. 2003;278:20196–202.
137. Lee FJS, Xue S, Pei L, Vukusic B, Chéry N, Wang Y, et al. Dual regulation of NMDA receptor functions by direct protein-protein interactions with the dopamine D1 receptor. Cell. 2002;111:219–30.
138. Fiorentini C, Rizzetti MC, Busi C, Bontempi S, Collo G, Spano P, et al. Loss of synaptic D1 dopamine/N-methyl-D-aspartate glutamate receptor complexes in L-DOPA-induced dyskinesia in the rat. Mol Pharmacol. 2006;69:805–12.
139. Sweatt JD. The neuronal MAP kinase cascade: a biochemical signal integration system subserving synaptic plasticity and memory. J Neurochem. 2001;76:1–10.
140. Krapivinsky G, Krapivinsky L, Manasian Y, Ivanov A, Tyzio R, Pellegrino C, et al. The NMDA receptor is coupled to the ERK pathway by a direct interaction between NR2B and RasGRF1. Neuron. 2003;40:775–84.
141. Valjent E, Pascoli V, Svenningsson P, Paul S, Enslen H, Corvol J-C, et al. Regulation of a protein phosphatase cascade allows convergent dopamine and glutamate signals to activate ERK in the striatum. Proc Natl Acad Sci. 2005;102:491–6.
142. Valjent E, Corvol JC, Pages C, Besson MJ, Maldonado R, Caboche J. Involvement of the extracellular signal-regulated kinase cascade for cocaine-rewarding properties. J Neurosci. 2000;20:8701–9.
143. Gerfen CR, Miyachi S, Paletzki R, Brown P. D1 dopamine receptor supersensitivity in the dopamine-depleted striatum results from a switch in the regulation of ERK1/2/MAP kinase. J Neurosci. 2002;22:5042–54.
144. Pavon N, Martín AB, Mendialdua A, Moratalla R. ERK phosphorylation and FosB expression are associated with L-DOPA-induced dyskinesia in hemiparkinsonian mice. Biol Psychiatry. 2006;59:64–74.
145. Lindgren HS, Ohlin KE, Cenci MA. Differential involvement of D1 and D2 dopamine receptors in L-DOPA-induced angiogenic activity in a rat model of Parkinson's disease. Neuropsychopharmacology. 2009;34:2477–88.

146. Schuster S, Nadjar A, Guo JT, Li Q, Ittrich C, Hengerer B, et al. The 3-hydroxy-3-methylglutaryl-CoA reductase inhibitor lovastatin reduces severity of L-DOPA-induced abnormal involuntary movements in experimental Parkinson's disease. J Neurosci. 2008;28:4311–6.
147. Fasano S, D'Antoni A, Orban PC, Valjent E, Putignano E, Vara H, et al. Ras-guanine nucleotide-releasing factor 1 (Ras-GRF1) controls activation of extracellular signal-regulated kinase (ERK) signaling in the striatum and long-term behavioral responses to cocaine. Biol Psychiatry. 2009;66:758–68.
148. Fasano S, Bezard E, D'Antoni A, Francardo V, Indrigo M, Qin L, et al. Inhibition of Ras-guanine nucleotide-releasing factor 1 (Ras-GRF1) signaling in the striatum reverts motor symptoms associated with L-dopa-induced dyskinesia. Proc Natl Acad Sci. 2010;107: 21824–9.
149. Murphy LO, Smith S, Chen R-H, Fingar DC, Blenis J. Molecular interpretation of ERK signal duration by immediate early gene products. Nat Cell Biol. 2002;4:556–64.
150. Durchdewald M, Angel P, Hess J. The transcription factor Fos: a Janus-type regulator in health and disease. Histol Histopathol. 2009;24:1451–61.
151. Asin KE, Wirtshafter D, Nikkel A. Amphetamine induces Fos-like immunoreactivity in the striatum of primates. Brain Res. 1996;719:138–42.
152. Graybiel AM, Moratalla R, Robertson HA. Amphetamine and cocaine induce drug-specific activation of the c-fos gene in striosome-matrix compartments and limbic subdivisions of the striatum. Proc Natl Acad Sci. 1990;87:6912–6.
153. Svenningsson P, Arts J, Gunne L, Andrén PE. Acute and repeated treatment with L-DOPA increase c-jun expression in the 6-hydroxydopamine-lesioned forebrain of rats and common marmosets. Brain Res. 2002;955:8–15.
154. Calon F, Grondin R, Morissette M, Goulet M, Blanchet PJ, Di Paolo T, et al. Molecular basis of levodopa-induced dyskinesias. Ann Neurol. 2000;47:S70–8.
155. Andersson M, Westin JE, Cenci MA. Time course of striatal DeltaFosB-like immunoreactivity and prodynorphin mRNA levels after discontinuation of chronic dopaminomimetic treatment. Eur J Neurosci. 2003;17:661–6.
156. Levandis G, Bazzini E, Armentero M-T, Nappi G, Blandini F. Systemic administration of an mGluR5 antagonist, but not unilateral subthalamic lesion, counteracts l-DOPA-induced dyskinesias in a rodent model of Parkinson's disease. Neurobiol Dis. 2008;29:161–8.
157. Cao X, Yasuda T, Uthayathas S, Watts RL, Mouradian MM, Mochizuki H, et al. Striatal overexpression of DeltaFosB reproduces chronic levodopa-induced involuntary movements. J Neurosci. 2010;30:7335–43.
158. Padovan-Neto FE, Ferreira NR, de Oliveira-Tavares D, de Aguiar D, da Silva CA, Raisman-Vozari R, et al. Anti-dyskinetic effect of the neuronal nitric oxide synthase inhibitor is linked to decrease of FosB/DeltaFosB expression. Neurosci Lett. 2013;541:126–31.
159. Cenci MA, Lundblad M. Post- versus presynaptic plasticity in L-DOPA-induced dyskinesia. J Neurochem. 2006;99:381–92.
160. Feyder M, Bonito-Oliva A, Fisone G. L-DOPA-induced dyskinesia and abnormal signaling in striatal medium spiny neurons: focus on dopamine D1 receptor-mediated transmission. Front Behav Neurosci. 2011;5:71.
161. Fuxe K, Ferré S, Zoli M, Agnati LF. Integrated events in central dopamine transmission as analyzed at multiple levels. Evidence for intramembrane adenosine A2A/dopamine D2 and adenosine A1/dopamine D1 receptor interactions in the basal ganglia. Brain Res Brain Res Rev. 1998;26:258–73.
162. Fuxe K, Ferré S, Genedani S, Franco R, Agnati LF. Adenosine receptor-dopamine receptor interactions in the basal ganglia and their relevance for brain function. Physiol Behav. 2007;92:210–7.
163. Ginés S, Hillion J, Torvinen M, Le Crom S, Casado V, Canela EI, et al. Dopamine D1 and adenosine A1 receptors form functionally interacting heteromeric complexes. Proc Natl Acad Sci. 2000;97:8606–11.

164. Berthet A, Bezard E. Dopamine receptors and L-dopa-induced dyskinesia. Parkinsonism Relat Disord. 2009;15 Suppl 4:S8–12.
165. Murer MG, Moratalla R. Striatal signaling in L-DOPA-induced dyskinesia: common mechanisms with drug abuse and long term memory involving D1 dopamine receptor stimulation. Front Neuroanat. 2011;5:51.
166. Drake JD, Kibuuka LN, Dimitrov KD, Pollack AE. Abnormal involuntary movement (AIM) expression following D2 dopamine agonist challenge is determined by the nature of prior dopamine receptor stimulation (priming) in 6-hydroxydopamine lesioned rats. Pharmacol Biochem Behav. 2013;105:26–33.
167. Dupre KB, Eskow KL, Negron G, Bishop C. The differential effects of 5-HT(1A) receptor stimulation on dopamine receptor-mediated abnormal involuntary movements and rotations in the primed hemiparkinsonian rat. Brain Res. 2007;1158:135–43.
168. Samadi P, Grégoire L, Bédard PJ. Opioid antagonists increase the dyskinetic response to dopaminergic agents in parkinsonian monkeys: interaction between dopamine and opioid systems. Neuropharmacology. 2003;45:954–63.
169. Schapira AHV, Barone P, Hauser RA, Mizuno Y, Rascol O, Busse M, et al. Extended-release pramipexole in advanced Parkinson disease: a randomized controlled trial. Neurology. 2011;77:767–74.
170. Luquin MR, Laguna J, Obeso JA. Selective D2 receptor stimulation induces dyskinesia in parkinsonian monkeys. Ann Neurol. 1992;31:551–4.
171. Gomez-Mancilla B, Bédard PJ. Effect of chronic treatment with (+)-PHNO, a D2 agonist in MPTP-treated monkeys. Exp Neurol. 1992;117:185–8.
172. Antonini A, Schwarz J, Oertel WH, Beer HF, Madeja UD, Leenders KL. [11C]raclopride and positron emission tomography in previously untreated patients with Parkinson's disease: influence of L-dopa and lisuride therapy on striatal dopamine D2-receptors. Neurology. 1994;44:1325–9.
173. Rahman Z, Schwarz J, Gold SJ, Zachariou V, Wein MN, Choi KH, et al. RGS9 modulates dopamine signaling in the basal ganglia. Neuron. 2003;38:941–52.
174. Gold SJ, Hoang CV, Potts BW, Porras G, Pioli E, Kim KW, et al. RGS9-2 negatively modulates L-3,4-dihydroxyphenylalanine-induced dyskinesia in experimental Parkinson's disease. J Neurosci. 2007;27:14338–48.
175. Aleyasin H, Rousseaux MWC, Marcogliese PC, Hewitt SJ, Irrcher I, Joselin AP, et al. DJ-1 protects the nigrostriatal axis from the neurotoxin MPTP by modulation of the AKT pathway. Proc Natl Acad Sci. 2010;107:3186–91.
176. Hillion J, Canals M, Torvinen M, Casadó V, Scott R, Terasmaa A, et al. Coaggregation, cointernalization, and codesensitization of adenosine A2A receptors and dopamine D2 receptors. J Biol Chem. 2002;277:18091–7.
177. Torvinen M, Torri C, Tombesi A, Marcellino D, Watson S, Lluís C, et al. Trafficking of adenosine A2A and dopamine D2 receptors. J Mol Neurosci. 2005;25:191–200.
178. Bogenpohl JW, Ritter SL, Hall RA, Smith Y. Adenosine A2A receptor in the monkey basal ganglia: ultrastructural localization and colocalization with the metabotropic glutamate receptor 5 in the striatum. J Comp Neurol. 2012;520:570–89.
179. Lévesque D, Diaz J, Pilon C, Martres MP, Giros B, Souil E, et al. Identification, characterization, and localization of the dopamine D3 receptor in rat brain using 7-[3H]hydroxy-N, N-di-n-propyl-2-aminotetralin. Proc Natl Acad Sci. 1992;89:8155–9.
180. Lévesque D, Martres MP, Diaz J, Griffon N, Lammers CH, Sokoloff P, et al. A paradoxical regulation of the dopamine D3 receptor expression suggests the involvement of an anterograde factor from dopamine neurons. Proc Natl Acad Sci. 1995;92:1719–23.
181. Bordet R, Ridray S, Carboni S, Diaz J, Sokoloff P, Schwartz JC. Induction of dopamine D3 receptor expression as a mechanism of behavioral sensitization to levodopa. Proc Natl Acad Sci. 1997;94:3363–7.
182. Pilla M, Perachon S, Sautel F, Garrido F, Mann A, Wermuth CG, et al. Selective inhibition of cocaine-seeking behaviour by a partial dopamine D3 receptor agonist. Nature. 1999;400: 371–5.

183. Blanchet PJ, Konitsiotis S, Chase TN. Motor response to a dopamine D3 receptor preferring agonist compared to apomorphine in levodopa-primed 1-methyl-4-phenyl-1,2,3,6-tetrahydropyridine monkeys. J Pharmacol Exp Ther. 1997;283:794–9.
184. Visanji NP, Fox SH, Johnston T, Reyes G, Millan MJ, Brotchie JM. Dopamine D3 receptor stimulation underlies the development of L-DOPA-induced dyskinesia in animal models of Parkinson's disease. Neurobiol Dis. 2009;35:184–92.
185. Huot P, Johnston TH, Koprich JB, Aman A, Fox SH, Brotchie JM. L-745,870 reduces L-DOPA-induced dyskinesia in the 1-methyl-4-phenyl-1,2,3,6-tetrahydropyridine-lesioned macaque model of Parkinson's disease. J Pharmacol Exp Ther. 2012;342:576–85.
186. Mrzljak L, Bergson C, Pappy M, Huff R, Levenson R, Goldman-Rakic PS. Localization of dopamine D4 receptors in GABAergic neurons of the primate brain. Nature. 1996;381:245–8.
187. Parent A, Lavoie B, Smith Y, Bédard P. The dopaminergic nigro-pallidal projection in primates: distinct cellular origin and relative sparing in MPTP-treated monkeys. Adv Neurol. 1990;53:111–6.
188. Albin RL, Young AB, Penney JB. The functional anatomy of basal ganglia disorders. Trends Neurosci. 1989;12:366–75.
189. Gerfen CR, Engber TM, Mahan LC, Susel Z, Chase TN, Monsma FJ, et al. D1 and D2 dopamine receptor-regulated gene expression of striatonigral and striatopallidal neurons. Science (New York, NY). 1990;250:1429–32.
190. Surmeier DJ, Reiner A, Levine MS, Ariano MA. Are neostriatal dopamine receptors co-localized? Trends Neurosci. 1993;16:299–305.
191. Aizman O, Brismar H, Uhlén P, Zettergren E, Levey AI, Forssberg H, et al. Anatomical and physiological evidence for D1 and D2 dopamine receptor colocalization in neostriatal neurons. Nat Neurosci. 2000;3:226–30.
192. Lester J, Fink S, Aronin N, DiFiglia M. Colocalization of D1 and D2 dopamine receptor mRNAs in striatal neurons. Brain Res. 1993;621:106–10.
193. Gerfen CR, Keefe KA, Gauda EB. D1 and D2 dopamine receptor function in the striatum: coactivation of D1- and D2-dopamine receptors on separate populations of neurons results in potentiated immediate early gene response in D1-containing neurons. J Neurosci. 1995;15:8167–76.
194. Rashid AJ, So CH, Kong MM, Furtak T, El-Ghundi M, Cheng R, et al. D1-D2 dopamine receptor heterooligomers with unique pharmacology are coupled to rapid activation of Gq/11 in the striatum. Proc Natl Acad Sci. 2007;104:654–9.
195. So CH, Verma V, Alijaniaram M, Cheng R, Rashid AJ, O'Dowd BF, et al. Calcium signaling by dopamine D5 receptor and D5-D2 receptor hetero-oligomers occurs by a mechanism distinct from that for dopamine D1-D2 receptor hetero-oligomers. Mol Pharmacol. 2009;75:843–54.
196. Verma V, Hasbi A, O'Dowd BF, George SR. Dopamine D1-D2 receptor Heteromer-mediated calcium release is desensitized by D1 receptor occupancy with or without signal activation: dual functional regulation by G protein-coupled receptor kinase 2. J Biol Chem. 2010;285:35092–103.
197. Schwartz JC, Diaz J, Bordet R, Griffon N, Perachon S, Pilon C, et al. Functional implications of multiple dopamine receptor subtypes: the D1/D3 receptor coexistence. Brain Res Brain Res Rev. 1998;26:236–42.
198. Marcellino D, Ferré S, Casadó V, Cortés A, Le Foll B, Mazzola C, et al. Identification of dopamine D1-D3 receptor heteromers. Indications for a role of synergistic D1-D3 receptor interactions in the striatum. J Biol Chem. 2008;283:26016–25.
199. Fiorentini C, Busi C, Gorruso E, Gotti C, Spano P, Missale C. Reciprocal regulation of dopamine D1 and D3 receptor function and trafficking by heterodimerization. Mol Pharmacol. 2008;74:59–69.
200. Fiorentini C, Busi C, Spano P, Missale C. Dimerization of dopamine D1 and D3 receptors in the regulation of striatal function. Curr Opin Pharmacol. 2010;10:87–92.

201. Kostrzewa RM. Dopamine receptor supersensitivity. Neurosci Biobehav Rev. 1995;19:1–17.
202. Mileson BE, Lewis MH, Mailman RB. Dopamine receptor "supersensitivity" occurring without receptor up-regulation. Brain Res. 1991;561:1–10.
203. Impact of deprenyl and tocopherol treatment on Parkinson's disease in DATATOP patients requiring levodopa. Parkinson Study Group. Ann Neurol. 1996;39:37–45.
204. Fahn S, Oakes D, Shoulson I, Kieburtz K, Rudolph A, Lang A, et al. Levodopa and the progression of Parkinson's disease. N Engl J Med. 2004;351:2498–508.
205. Schrag A, Quinn N. Dyskinesias and motor fluctuations in Parkinson's disease. A community-based study. Brain. 2000;123(Pt 11):2297–305.
206. de la Fuente-Fernandez R, Sossi V, Huang Z, Furtado S, Lu JQ, Calne DB, et al. Levodopa-induced changes in synaptic dopamine levels increase with progression of Parkinson's disease: implications for dyskinesias. Brain. 2004;127:2747–54.
207. Muenter MD, Sharpless NS, Tyce GM, Darley FL. Patterns of dystonia ("I-D-I" and "D-I-D-") in response to l-dopa therapy for Parkinson's disease. Mayo Clin Proc. 1977;52:163–74.
208. Chase TN, Baronti F, Fabbrini G, Heuser IJ, Juncos JL, Mouradian MM. Rationale for continuous dopaminomimetic therapy of Parkinson's disease. Neurology. 1989;39:7–10; discussion 9.
209. Olanow CW, Obeso JA, Stocchi F. Continuous dopamine-receptor treatment of Parkinson's disease: scientific rationale and clinical implications. Lancet Neurol. 2006;5:677–87.
210. Olanow W, Schapira AH, Rascol O. Continuous dopamine-receptor stimulation in early Parkinson's disease. Trends Neurosci. 2000;23:S117–26.
211. Bracco F, Battaglia A, Chouza C, Dupont E, Gershanik O, Marti Masso JF, et al. The long-acting dopamine receptor agonist cabergoline in early Parkinson's disease: final results of a 5-year, double-blind, levodopa-controlled study. CNS Drugs. 2004;18:733–46.
212. Hoehn MM, Elton RL. Low dosages of bromocriptine added to levodopa in Parkinson's disease. Neurology. 1985;35:199–206.
213. Kartzinel R, Teychenne P, Gillespie MM, Perlow M, Gielen AC, Sadowsky DA, et al. Bromocriptine and levodopa (with or without carbidopa) in parkinsonism. Lancet. 1976;2:272–5.
214. Lieberman A, Kupersmith M, Estey E, Goldstein M. Treatment of parkinson's disease with bromocriptine. N Engl J Med. 1976;295:1400–4.
215. Montastruc JL, Rascol O, Senard JM, Rascol A. A randomised controlled study comparing bromocriptine to which levodopa was later added, with levodopa alone in previously untreated patients with Parkinson's disease: a five year follow up. J Neurol Neurosurg Psychiatry. 1994;57:1034–8.
216. Oertel WH, Wolters E, Sampaio C, Gimenez-Roldan S, Bergamasco B, Dujardin M, et al. Pergolide versus levodopa monotherapy in early Parkinson's disease patients: The PELMOPET study. Mov Disord. 2006;21:343–53.
217. Olanow CW, Fahn S, Muenter M, Klawans H, Hurtig H, Stern M, et al. A multicenter double-blind placebo-controlled trial of pergolide as an adjunct to Sinemet in Parkinson's disease. Mov Disord. 1994;9:40–7.
218. Lieberman A, Olanow CW, Sethi K, Swanson P, Waters CH, Fahn S, et al. A multicenter trial of ropinirole as adjunct treatment for Parkinson's disease. Ropinirole Study Group. Neurology. 1998;51:1057–62.
219. Lieberman A, Ranhosky A, Korts D. Clinical evaluation of pramipexole in advanced Parkinson's disease: results of a double-blind, placebo-controlled, parallel-group study. Neurology. 1997;49:162–8.
220. Holloway RG, Shoulson I, Fahn S, Kieburtz K, Lang A, Marek K, et al. Pramipexole vs levodopa as initial treatment for Parkinson disease: a 4-year randomized controlled trial. Arch Neurol. 2004;61:1044–53.
221. Constantinescu R, Romer M, McDermott MP, Kamp C, Kieburtz K. Impact of pramipexole on the onset of levodopa-related dyskinesias. Mov Disord. 2007;22:1317–9.
222. Rascol O, Brooks DJ, Korczyn AD, De Deyn PP, Clarke CE, Lang AE, et al. Development of dyskinesias in a 5-year trial of ropinirole and L-dopa. Mov Disord. 2006;21:1844–50.

223. Olanow CW, Obeso JA. Pulsatile stimulation of dopamine receptors and levodopa-induced motor complications in Parkinson's disease: implications for the early use of COMT inhibitors. Neurology. 2000;55:S72–7; discussion S8–81.
224. Olanow CW, Stocchi F. COMT inhibitors in Parkinson's disease: can they prevent and/or reverse levodopa-induced motor complications? Neurology. 2004;62:S72–81.
225. Sawle GV, Burn DJ, Morrish PK, Lammertsma AA, Snow BJ, Luthra S, et al. The effect of entacapone (OR-611) on brain [18 F]-6-L-fluorodopa metabolism: implications for levodopa therapy of Parkinson's disease. Neurology. 1994;44:1292–7.
226. Schapira AH, Obeso JA, Olanow CW. The place of COMT inhibitors in the armamentarium of drugs for the treatment of Parkinson's disease. Neurology. 2000;55:S65–8; discussion S9–71.
227. Smith LA, Jackson MJ, Al-Barghouthy G, Rose S, Kuoppamaki M, Olanow W, et al. Multiple small doses of levodopa plus entacapone produce continuous dopaminergic stimulation and reduce dyskinesia induction in MPTP-treated drug-naive primates. Mov Disord. 2005;20:306–14.
228. Stocchi F, Rascol O, Kieburtz K, Poewe W, Jankovic J, Tolosa E, et al. Initiating levodopa/carbidopa therapy with and without entacapone in early Parkinson disease: the STRIDE-PD study. Ann Neurol. 2010;68:18–27.
229. Parkinson Study Group. A randomized placebo-controlled trial of rasagiline in levodopa-treated patients with Parkinson disease and motor fluctuations: the PRESTO study. Arch Neurol. 2005;62:241–8.
230. Rascol O, Brooks DJ, Melamed E, Oertel W, Poewe W, Stocchi F, et al. Rasagiline as an adjunct to levodopa in patients with Parkinson's disease and motor fluctuations (LARGO, Lasting effect in Adjunct therapy with Rasagiline Given Once daily, study): a randomised, double-blind, parallel-group trial. Lancet. 2005;365:947–54.
231. Maratos EC, Jackson MJ, Pearce RK, Cannizzaro C, Jenner P. Both short- and long-acting D-1/D-2 dopamine agonists induce less dyskinesia than L-DOPA in the MPTP-lesioned common marmoset (Callithrix jacchus). Exp Neurol. 2003;179:90–102.
232. Rascol O, Nutt JG, Blin O, Goetz CG, Trugman JM, Soubrouillard C, et al. Induction by dopamine D1 receptor agonist ABT-431 of dyskinesia similar to levodopa in patients with Parkinson disease. Arch Neurol. 2001;58:249–54.
233. Oliveri RL, Annesi G, Zappia M, Civitelli D, Montesanti R, Branca D, et al. Dopamine D2 receptor gene polymorphism and the risk of levodopa-induced dyskinesias in PD. Neurology. 1999;53:1425–30.
234. Zappia M, Annesi G, Nicoletti G, Arabia G, Annesi F, Messina D, et al. Sex differences in clinical and genetic determinants of levodopa peak-dose dyskinesias in Parkinson disease: an exploratory study. Arch Neurol. 2005;62:601–5.
235. Lee JY, Cho J, Lee EK, Park SS, Jeon BS. Differential genetic susceptibility in diphasic and peak-dose dyskinesias in Parkinson's disease. Mov Disord. 2011;26:73–9.
236. Wang J, Liu ZL, Chen B. Association study of dopamine D2, D3 receptor gene polymorphisms with motor fluctuations in PD. Neurology. 2001;56:1757–9.

Chapter 11
The Serotonergic System in Levodopa-Induced Dyskinesia

Elisabetta Tronci, Camino Fidalgo, and Manolo Carta

Abstract An increasing body of experimental evidence suggests that serotonergic neurons play a major role in the production of levodopa-derived dopamine when dopaminergic neurons have degenerated, and that unregulated release of dopamine from serotonergic neurons is responsible for the appearance of levodopa-induced dyskinesia (LID) in animal models of Parkinson's disease (PD).

Promising preclinical findings show that the activation of 5-HT1 receptors, induced by the administration of 5-HT1A and/or 5-HT1B receptor agonists, suppressed LID in 6-OHDA-lesioned rat, as well as in MPTP-treated nonhuman primate models of PD, suggesting a possible clinical application. This chapter will provide an overview of these preclinical findings concerning the role of serotonergic neurons and serotonergic receptors in the appearance of LID, with a brief review of relevant clinical studies.

Keywords Levodopa • LID • Dopamine • Serotonin • 5-HT receptors

The Serotonergic System in Parkinson's Disease

Serotonin (5-hydroxytryptamine, 5-HT) is an important modulator of the central nervous system (CNS), and its action is mediated by a large variety of receptor subtypes. Serotonin is synthesized from L-tryptophan by a two-step reaction: the tryptophan hydroxylase enzyme generates 5-hydroxytryptophan (5-HTP), which is then converted to serotonin by the L-amino acid decarboxylase enzyme (AADC). The serotonergic system originates from the raphe nuclei and is one of the most widely distributed, innervating virtually all regions of the CNS and

E. Tronci • C. Fidalgo • M. Carta (✉)
Section of Physiology, Department of Biomedical Sciences, University of Cagliari, Cittadella Universitaria, SS554 km 4.5, Monserrato 09042, Italy
e-mail: manolocarta@unica.it

S.H. Fox, J.M. Brotchie (eds.), *Levodopa-Induced Dyskinesia in Parkinson's Disease*, DOI 10.1007/978-1-4471-6503-3_11

participating in mechanisms of cognition, feeding and satiety, mood and emotion, circadian and sleep-wake cycle regulation, pain, and motor functions [1, 2]. An increasing body of experimental evidence demonstrates the involvement of the serotonergic system in modulating the function of basal ganglia circuits and its interaction with the dopaminergic system. Both serotonergic and dopaminergic systems are affected in neurodegenerative disorders such as Parkinson's disease (PD) and have been implicated in the pathophysiology of depression and schizophrenia [3]. A role of the serotonergic system in the regulation of motor function is suggested by the dense serotonergic input received by areas such as the striatum, substantia nigra pars reticulata, and globus pallidus [4, 5]. Interestingly, postmortem studies of patients with PD have shown loss of serotonergic markers in the caudate, as well as hypothalamus and frontal cortex [6, 7]. However, as also seen in PET imaging studies, the degree of serotonergic terminal loss appears to be less severe than that affecting the dopaminergic system, and there is no correlation with motor disability, or dyskinesia [8]. On the other hand, it has been suggested that the partial loss of serotonergic innervation may contribute to the development of depression in PD patients [9], although consistent evidence is lacking [5].

Serotonin exerts its actions via specific receptors, which are located in most of the brain, especially in the hippocampus, basal ganglia, and striatum. To date, 14 distinct subtypes of the serotonergic receptors have been identified [10, 11]; among all serotonergic receptors, some have been shown to participate in the regulation of motor function and/or induction of dyskinesia in PD, such as the 5-HT1A, 5-HT1B, 5-HT2A, 5-HT2C, and the 5-HT3 receptors.

5-HT1A receptors are located somato-dendritically in the dorsal raphe nuclei, where they regulate cell firing [12]. 5-HT1A receptors have been also identified postsynaptically in other brain regions, such as the cerebral cortex, striatum, and subthalamic nucleus [13], where they serve to control release of other neurotransmitters, such as glutamate [14]. Interestingly, the activation of postsynaptic 5-HT1A receptors located at prefrontal cortex has also been shown to affect serotonin neuron activity through an indirect loop to the raphe nuclei [15–17]. 5-HT1B receptors are also expressed as autoreceptors at serotonergic terminals in target areas, where they contribute to control serotonin release. Moreover, these receptors are also localized in non-serotonergic neurons, such as the striatal medium spiny neurons, and regulate GABA release [18–20].

Among the other 5-HT receptors, the 5-HT2A receptors are involved in a variety of behaviors and diseases including anxiety, hyperactivity disorders, aggression, and social interaction. Moreover, they play a role in drug addiction and in the mechanism of action of anti-psychotic drugs. It has been shown that 5-HT2A receptors are abundant in the neocortex, striatum, and nucleus accumbens, where they appear to have a role in the emergence of motor complications [21, 22].

Another interesting serotonergic receptor that has been shown to play a role in the regulation of motor functions is the 5-HT2C receptor. This receptor subtype shows high and moderate expression in limbic areas and basal ganglia, respectively. The activation or blockade of 5-HT2C receptors results in an opposite effect on dopamine release; indeed, their activation induces a reduction of dopamine release along the dopaminergic pathway, while their blockade increases dopamine release along the meso-cortico-limbic as well as nigrostriatal pathways [23, 24]. Despite the influence of 5-HT2C receptors in modulating dopaminergic system, very little is known on their role in PD and dyskinesia.

5-HT3 receptors are mostly localized in limbic areas and brainstem nuclei, while they show low expression in the basal ganglia areas. Unlike other serotonergic receptors, 5-HT3 receptors are ion channels. They have been found to be involved in anxiety, schizophrenia, learning, and attention, as well as in craving and pain. Although 5-HT3 receptors have been reported to play a role in controlling striatal dopamine release [25], their involvement in PD and dyskinesia is poorly studied.

Despite the important role of serotonergic receptors in regulating motor functions, selective agonists for these receptors were generally unsuccessful for the treatment of parkinsonian motor symptoms [26]; on the other hand, the serotonergic receptors have been implicated in the development of motor complications induced by drug treatment in parkinsonian patients, such as levodopa-induced dyskinesia (LID).

Involvement of the Serotonergic System in Levodopa-Induced Dyskinesia

The therapeutic efficacy of levodopa during the first few years of treatment is conceivably due to the presence of a sufficient number of spared dopamine neurons that can provide conversion of levodopa and mediate physiological release of dopamine; however, with the progression of the disease, most of dopaminergic neurons are lost, and levodopa-derived dopamine is produced in non-dopaminergic elements, including the serotonergic terminals. In fact, serotonergic neurons are able to convert exogenous levodopa to dopamine and store and release dopamine in an activity-dependent manner in a number of experimental conditions, both in vitro and in vivo [27–29]. Indeed, serotonergic neurons share the same enzymatic machinery as dopaminergic terminals (aromatic amino acid decarboxylase and vesicular monoamine transporter 2 enzymes) and are able to convert levodopa to dopamine and to mediate storage of dopamine into synaptic vesicles.

It is conceivable to think that conversion of levodopa takes place into serotonergic neurons also in earlier stages of disease; however, as long as there are sufficient spared dopaminergic terminals, the contribution of serotonergic neurons may be beneficial, as dopaminergic terminals provide a buffering system for the levodopa-derived dopamine (through the dopamine transporter). By contrast, as the disease progresses and spared dopaminergic terminals are lost, the contribution of serotonergic neurons becomes detrimental (see Fig. 11.1).

In fact, unlike dopaminergic neurons, serotonergic neurons cannot regulate the extracellular levels of dopamine due to the lack of an autoregulatory feedback mechanism for dopamine release. As a consequence, levodopa-derived dopamine is released in uncontrolled way following levodopa administration. This will act in

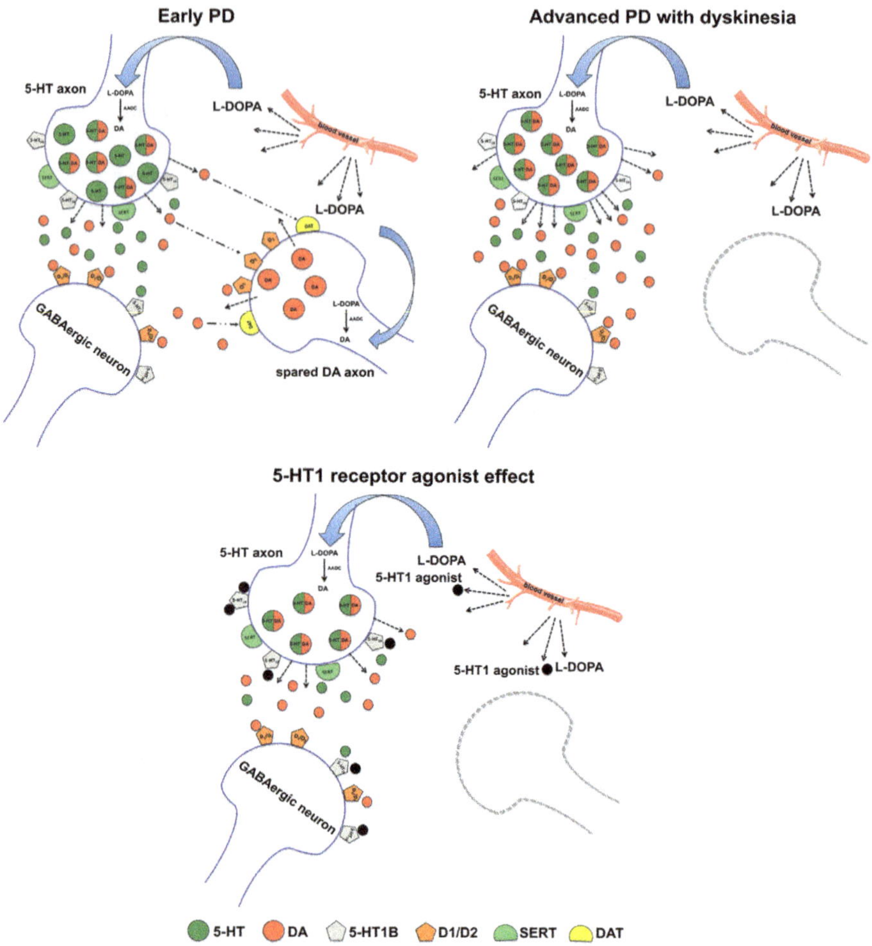

Fig. 11.1 In the early stage of PD, the ability to release dopamine is partially preserved due to the presence of a sufficient number of spared dopaminergic neurons that have not yet degenerated. At this stage, serotonergic terminals may contribute to levodopa-derived dopamine release, due to an ability to convert exogenous levodopa to dopamine and store and release dopamine in an activity-dependent manner. However, their contribution may be beneficial as synaptic dopamine levels are kept within a physiological range due to the presence of an efficient buffering system provided by the spared dopamine terminals (through the dopamine transporter). With the progression of the disease, when most of dopaminergic terminals are lost, the contribution of serotonergic neurons becomes detrimental due to the lack of autoregulatory feedback mechanisms able to regulate synaptic dopamine levels. The uncontrolled release of dopamine from serotonergic neurons will act in concert with the intermittent nature of the orally administered levodopa to produce swings in synaptic dopamine levels and pulsatile stimulation of striatal dopaminergic receptors, which is responsible for the onset of dyskinesia. 5-HT1 receptor agonists will act to reduce levodopa-derived dopamine release from serotonin neurons, which will dampen swings in synaptic dopamine levels

concert with the intermittent nature of the orally administered levodopa, to cause swing in synaptic dopamine levels, which is responsible for pulsatile stimulation of striatal dopaminergic receptors and aberrant downstream signaling cascade (see Fig. 11.1). In line with this view, it has been shown that removal of the forebrain serotonergic innervation by the selective toxin 5,7-dihydroxy-tryptamine (5,7-DHT) produces an almost complete suppression of LID [30, 31]. Dramatic reduction of levodopa-derived striatal dopamine levels appears to account for the anti-dyskinetic effect of 5,7-DHT lesions [29].

Support for the important role of serotonergic neurons in dyskinesia also came from a rat PET imaging study, where the administration of the 5-HT1A receptor agonist 8-OH-DPAT (which reduced LID) reversed levodopa-induced decrease of [(11) C]raclopride binding and increase of extracellular dopamine [32].

Interestingly, Rylander and colleagues have demonstrated a correlation between the density of serotonergic terminals in the striatum and the severity of LID across different species. In fact, dyskinetic rats and primates, as well as postmortem tissue of PD patients, have shown a growth-promoting effect on the striatal serotonergic fibers, accompanied by an increase of serotonin transporter (SERT) expression, induced by levodopa treatment. Regenerative sprouting of the serotonergic neurons was also accompanied by increased BDNF (Brain Derived Neurotrophic Factor) levels and increased BDNF-induced levodopa-derived dopamine release in rat striatal slices [33]. This is interesting as intracerebral delivery of BDNF has previously been shown to promote serotonergic terminal sprouting [34].

Further evidence for the involvement of serotonergic neurons in the appearance of LID came from recent experimental studies, where increased serotonergic tone by administration of either selective serotonin reuptake blockers (SSRIs) or the serotonergic precursor 5-hydroxytryptophan significantly reduced LID expression in hemiparkinsonian rats, without compromising the levodopa therapeutic efficacy [35–37].

Swings in synaptic dopamine levels, rather than high dopamine levels, are suggested to be a key factor for the development of LID. Indeed, postsynaptic dopaminergic receptors are highly plastic and can adapt to either low or high synaptic dopamine levels by altering the efficiency of intracellular signal transduction; by contrast, rapid changes in synaptic neurotransmitter levels, with high dopaminergic receptor occupancy (following each levodopa dose), followed by dramatic reduction (few hours after each levodopa dose), would impair this ability. In agreement with this view, it has been shown that dyskinetic patients present higher synaptic dopamine levels 1 h after levodopa administration compared to stable responders, while this difference disappeared few hours later [38]. Furthermore, dyskinesia is significantly dampened in advanced dyskinetic patients receiving continuous intraduodenal infusion of levodopa, which is effective in reducing swing in synaptic dopamine levels [39].

All together, these experimental findings provide strong evidences supporting the important role of serotonergic neurons in the induction and expression of LID, at least in animal models of PD.

5-HT1A and 5-HT1B Receptor Agonists in the Treatment of Dyskinesia

The activation of serotonergic autoreceptors is a physiological mechanism meant to avoid excessive synaptic neurotransmitter release. Thus, while in normal conditions this mechanism is able to keep synaptic serotonin levels within a physiological range, under levodopa treatment in parkinsonian states, the activation of serotonergic autoreceptors is also able to inhibit serotonergic neuron-derived dopamine release. In fact, accumulating evidence support the efficacy of 5-HT1 receptor agonists for the treatment of LID, with the 5-HT1A receptors having received most of the attention, as they are able to control serotonin neuron firing and release (See Fig. 11.1).

Several 5-HT1A receptor agonists have displayed acute and chronic anti-dyskinetic effects in both animal models and in clinical studies [30, 31, 40–44]. Unfortunately, some of these studies have also shown that the administration of 5-HT1 agonists is associated with the induction of side effects, including worsening of the anti-parkinsonian efficacy of levodopa (See Table 11.1). For instance, the partial 5-HT1A receptor agonist sarizotan demonstrated potent efficacy in reducing dyskinesias in primate and rodent models of PD and in idiopathic PD patients in early open-label studies [41]; however, the anti-dyskinetic efficacy of sarizotan could not be shown to be significant compared to placebo in a following study, where the drug increased also off-time duration [45]. Moreover, in rats, high doses of sarizotan resulted in the induction of "serotonin syndrome" components, i.e., body posture associated with motor depression. Furthermore, the selective 5-HT1A receptor agonist 8-OH-DPAT inhibited LID in dyskinetic MPTP primates but only with increased motor disability [46, 47]. The partial, non-selective 5-HT1A receptor agonist buspirone was found to reduce LID in patients [48], but two other studies have found that this effect was again associated with a reduction of the anti-parkinsonian efficacy of levodopa [49, 50].

It has recently been demonstrated that 5-HT1B receptor agonists can also elicit anti-dyskinetic effects in animal models of PD [30, 51, 52]. However, to date, no clinical trials have been performed with these agonists.

It should be stressed that the anti-dyskinetic effect of 5-HT1 receptor agonists, as seen in the studies reported above, is likely not exclusively due to the activation of 5-HT1 autoreceptors. In fact, the activation of postsynaptic 5-HT1A and 5-HT1B receptors has also been shown to produce anti-dyskinetic effect by reducing striatal release of glutamate and GABA, respectively [30, 53–55]. Accordingly, relatively high doses of 5-HT1 receptor agonists have been shown to reduce dyskinesia induced by direct dopamine receptor agonists, such as apomorphine, which is not dependent on serotonergic neuron release [30, 52, 56].

Recent results have also shown that simultaneous activation of 5-HT1A and 5-HT1B receptors, using low doses of 8-OH-DPAT and CP-94253, respectively, produced a synergistic effect on suppression of LID in 6-OHDA-lesioned rats and in MPTP-treated macaques, with near to full inhibition at doses of the two drugs

Table 11.1 5-HT1 receptor agonists for the treatment of LID

Reference	Drug name	Acute/chronic treatment	Animal model/ patients	Efficacy against LID	Effect on therapeutic action of levodopa
[40]	Sarizotan	Acute	6-OHDA-lesioned rats	no	
		Acute	*Macaca fascicularis*	yes	
[41]	Sarizotan	Acute	Patients in moderate-advance state of PD	yes	
[30]	8-OH-DPAT +CP-94253	Acute	6-OHDA-lesioned rats	yes	Not affected
[31]	8-OH-DPAT	Acute	6-OHDA-lesioned rats	yes	Increased
[42]	Buspirone	Acute	6-OHDA-lesioned rats	yes	
		Chronic	6-OHDA-lesioned rats	yes	
[44]	Eltoprazine	Chronic	6-OHDA-lesioned rats	yes	Partial reduction
		Acute	*Macaca fascicularis*	yes	Partial reduction
[43]	Anpirtoline	Acute	6-OHDA-lesioned rats	yes	Not affected
		Acute	*Macaca fascicularis*	yes	Partial reduction

The effects of 5-HT1 receptor agonists on LID and levodopa-induced therapeutic effects are summarized in this table

that were ineffective when given individually [30, 56]. In contrast, the same combined doses did not affect dyskinesia induced by apomorphine, suggesting that the observed effect is mainly due to the activation of presynaptic receptors [30]. Accordingly, combination of 8-OH-DPAT and CP-94253 was found to reduce extracellular dopamine levels following levodopa administration [57].

Recently, the mixed 5-HT1A/5-HT1B receptor agonist eltoprazine has been characterized for its anti-dyskinetic properties, both in rat and monkey models of PD. Eltoprazine, developed for the treatment of aggression, exhibited a safe toxicological profile and lack of serious side effects [58, 59], and is currently investigated in a clinical trial for the treatment of attention-deficit hyperactivity disorder (ADHD) (ClinicalTrials.gov Identifier: NCT01266174). In the rat 6-OHDA as well as macaque MPTP lesion models of PD, eltoprazine displayed high effectiveness in blocking LID at doses that were ineffective to reduce dyskinesia induced by apomorphine, supporting a presynaptic effect of this drug; however, the anti-dyskinetic effect of eltoprazine was accompanied by a partial reduction of the levodopa therapeutic efficacy [44, 60]. Similarly, anpirtoline, a mixed 5-HT1A/5-HT1B receptor agonist, with higher affinity for the 5-HT1B receptor, was very effective in reducing dyskinesia in rats and monkeys but at the expense of PD score at significantly effective doses [43].

Thus, as seen with other selective 5-HT1 receptor agonists, the maintenance of the levodopa therapeutic effect may represent a concern in this approach, not only for eltoprazine but, possibly, for any drug with similar profile. In spite of this, eltoprazine is under investigation in a phase 2 double-blind clinical trial employing a limited number of patients, with encouraging preliminary results (see http://www.psychogenics.com/press2012.html).

Based on the preclinical observation that combination of 5-HT1A and 5-HT1B receptor agonists showed anti-dyskinetic effects, without worsening levodopa therapeutic properties [56], it is also possible that the ideal compound has to possess the right affinity for the two serotonergic autoreceptors to be clinically useful. Nevertheless, the narrow therapeutic window is a concern and may require a careful titration of the selected compound in each patient.

A plausible explanation for the worsening of the levodopa therapeutic effect observed after the administration of 5-HT1 receptor agonists is the advanced stage of dopaminergic neuron degeneration characterizing the animal models employed in these investigations, as well as the patients recruited in the sarizotan trial. In fact, under this condition, the serotonergic neurons may mediate not only the pro-dyskinetic effect of levodopa but also its residual therapeutic efficacy. If so, the anti-dyskinetic effect of 5-HT1 receptor agonists should unavoidably lead to parallel reduction of the therapeutic effect of levodopa. Therefore, 5-HT1 receptor agonists may find better clinical efficacy in a situation of less severe dopaminergic neuron degeneration, where spared dopamine neurons can mediate the effect of levodopa, and silencing of serotonergic neurons should be less detrimental. PET imaging studies could be useful to identify patients retaining some residual dopaminergic innervation that may be more likely to benefit from 5-HT1 receptor agonists. On the other hand, for the same reason, these patients are also less likely to suffer from severe dyskinesia.

It should also be mentioned that sarizotan, like other 5-HT1 receptor agonists, such as buspirone, has antagonistic activity for the dopaminergic D2 receptors that may have contributed to the negative outcome of the clinical trial [61, 62]. Moreover, these compounds are able to target only one autoreceptor subtype, and sufficient anti-dyskinetic efficacy may be obtained at relatively high doses of the drugs, which are likely to also affect postsynaptic 5-HT1 receptors.

Therefore, new large clinical investigations employing more selective 5-HT1A/5-HT1B receptor agonists are warranted to clarify the role of serotonin neurons in the appearance of LID in PD patients and to verify whether pharmacological silencing of serotonin neurons is a feasible therapeutic approach for the treatment of LID.

5-HT2A Receptors Antagonists in the Treatment of Dyskinesia

Serotonergic 5-HT2A receptors are localized at postsynaptic level, and in general they exert an excitatory effect. The role of 5-HT2A receptors in the expression of dyskinesia has been recently confirmed in a study by Riahi et al. [63] where they showed an increase at striatal level of these receptors in levodopa dyskinetic compared to non-dyskinetic monkeys, suggesting a possible use of drugs targeting these receptors for the treatment of dyskinesia. Preclinical and clinical studies have demonstrated the effect of drugs acting on 5-HT2A receptors in controlling levodopa-induced motor complications [64, 65]; interestingly, one clinical study reported the ability of aripiprazole, a 5-HT2A receptor antagonist and partial 5-HT1A and dopamine D2 receptor agonist, to reduce LID without worsening motor performance in PD patients [66]; however, it is difficult to establish the contribution of each receptor subtype, given that the effect may also be due to the activation of 5-HT1A receptors. In MPTP-treated primates, the selective 5-HT2A inverse agonist pimavanserin was recently demonstrated to reduce LID by 36 % without worsening motor scores [67]. However, it has also been reported that the selective 5-HT2A antagonist ritanserin alleviated LID but at the expense of levodopa-induced therapeutic action [68, 69]. By contrast, the selective 5-HT2A antagonist volinanserin was not effective in reducing LID in 6-OHDA-lesioned rats [70].

Further work is required to establish whether 5-HT2A antagonists could be useful in dyskinetic patients.

SERT as a Possible Target for Dyskinesia

Recent evidence has demonstrated the implication of the SERT in the pathophysiology of LID; Rylander et al. [33] observed a significant upregulation of SERT expression in the striatum of dyskinetic animals, both in rats and nonhuman primates, as well as in dyskinetic patients, which provided support to the idea that the serotonergic system is involved in the appearance of LID also in PD patients. Accordingly, few preclinical studies have shown a significant reduction of dyskinesia after blockade of the serotonin transporter by SSRIs. Thus, acute [35] and chronic [37, 71] administration of the serotonin reuptake inhibitors citalopram and paroxetine in levodopa-dyskinetic rats resulted in a significant reduction in AIMs severity, without affecting the anti-parkinsonian action of levodopa [35]. The anti-dyskinetic effect of SSRIs is likely to be due to a combination of different mechanisms: (i) activation of presynaptic 5-HT1 receptors by serotonin, which may reduce dopamine release, as seen for selective 5-HT1 agonists; (ii) blockade of dopamine reuptake by serotonergic neurons, which may reduce swings in synaptic dopamine levels; and (iii) activation of postsynaptic serotonin receptors. In fact, the activation of postsynaptic 5-HT1 receptors by selective agonists has been shown to provide anti-dyskinetic effect in parkinsonian rats. In support of a possible postsynaptic effect of SSRIs, a significant 47 % reduction of apomorphine-induced dyskinesias was observed in patients treated with fluoxetine, without modification of parkinsonian motor disability [72]. In contrast, short-term paroxetine treatment did not affect dyskinesia induced by intravenous levodopa [73]. Clinical investigations are needed to further explore the use of SSRIs as anti-dyskinetic agents.

Conclusions

An overwhelming body of experimental evidence suggests that the serotonin system is implicated in the appearance of LID in animal models of PD. Moreover, accumulating clinical results appear to support a key role of this system also in PD patients. Dampening the release from the serotonin neurons has been shown to be a promising approach in preclinical models. Concerns have been raised about the clinical feasibility of this approach, as not only dyskinesia but also the therapeutic effect of levodopa may depend on dopamine release from the serotonin neurons in advanced stages of disease. However, few compounds have been tested in patients so far, and most of them also presented antagonist activity for the dopamine receptors, which may have played a role in the observed worsening of the levodopa therapeutic efficacy. Thus, new clinical trials employing more selective serotonin 5-HT1 receptor agonists are warranted. The effect of SSRIs should also be further investigated.

References

1. Benarroch EE. Serotonergic modulation of basal ganglia circuits: complexity and therapeutic opportunities. Neurology. 2009;73:880–6.
2. Berger M, Gray JA, Roth BL. The expanded biology of serotonin. Annu Rev Med. 2009;60:355–66.
3. Di Giovanni G, Di Matteo V, Esposito E. Serotonin-dopamine interaction: experimental evidence and therapeutic relevance. Preface. Prog Brain Res. 2008;172:ix.
4. Lavoie B, Parent A. Immunohistochemical study of the serotonergic innervation of the basal ganglia in the squirrel monkey. J Comp Neurol. 1990;299:1–16.
5. Fox SH, Chuang R, Brotchie JM. Serotonin and Parkinson's disease: on movement, mood, and madness. Mov Disord. 2009;24:1255–66.
6. Scatton B, Javoy-Agid F, Rouquier L, Dubois B, Agid Y. Reduction of cortical dopamine, noradrenaline, serotonin and their metabolites in Parkinson's disease. Brain Res. 1983;275:321–8.
7. Kish SJ, Tong J, Hornykiewicz O, Rajput A, Chang LJ, Guttman M, Furukawa Y. Preferential loss of serotonin markers in caudate versus putamen in Parkinson's disease. Brain. 2008;131:120–31.
8. Boileau I, Warsh JJ, Guttman M, Saint-Cyr JA, McCluskey T, Rusjan P, Houle S, Wilson AA, Meyer JH, Kish SJ. Elevated serotonin transporter binding in depressed patients with Parkinson's disease: a preliminary PET study with [11C]DASB. Mov Disord. 2008;23:1776–80.
9. Jellinger KA. Pathology of Parkinson's disease. Changes other than the nigrostriatal pathway. Mol Chem Neuropathol. 1991;14:153–97.
10. Peroutka SJ. 5-HT receptors: past, present and future. Trends Neurosci. 1995;18:68–9.
11. Barnes NM, Sharp T. A review of central 5-HT receptors and their function. Neuropharmacology. 1999;38:1083–152.
12. Riad M, Garcia S, Watkins KC, Jodoin N, Doucet E, Langlois X, el Mestikawy S, Hamon M, Descarries L. Somatodendritic localization of 5-HT1A and preterminal axonal localization of 5-HT1B serotonin receptors in adult rat brain. J Comp Neurol. 2000;417:181–94.

13. Pompeiano M, Palacios JM, Mengod G. Distribution and cellular localization of mRNA coding for 5-HT1A receptor in the rat brain: correlation with receptor binding. J Neurosci. 1992;12:440–53.
14. Antonelli T, Fuxe K, Tomasini MC, Bartoszyk GD, Seyfried CA, Tanganelli S, Ferraro L. Effects of sarizotan on the corticostriatal glutamate pathways. Synapse. 2005;58:193–9.
15. Ceci A, Baschirotto A, Borsini F. The inhibitory effect of 8-OH-DPAT on the firing activity of dorsal raphe serotonergic neurons in rats is attenuated by lesion of the frontal cortex. Neuropharmacology. 1994;33:709–13.
16. Casanovas JM, Hervas I, Artigas F. Postsynaptic 5-HT1A receptors control 5-HT release in the rat medial prefrontal cortex. Neuroreport. 1999;10:1441–5.
17. Hajos M, Hajos-Korcsok E, Sharp T. Role of the medial prefrontal cortex in 5-HT1A receptor-induced inhibition of 5-HT neuronal activity in the rat. Br J Pharmacol. 1999;126:1741–50.
18. Mathur BN, Capik NA, Alvarez VA, Lovinger DM. Serotonin induces long-term depression at corticostriatal synapses. J Neurosci. 2011;31:7402–11.
19. Mathur BN, Lovinger DM. Serotonergic action on dorsal striatal function. Parkinsonism Relat Disord. 2012;18 Suppl 1:S129–31.
20. Rylander D. The serotonin system: a potential target for anti-dyskinetic treatments and biomarker discovery. Parkinsonism Relat Disord. 2012;18 Suppl 1:S126–8.
21. Pazos A, Probst A, Palacios JM. Serotonin receptors in the human brain–III. Autoradiographic mapping of serotonin-1 receptors. Neuroscience. 1987;21:97–122.
22. Bishop C, Tessmer JL, Ullrich T, Rice KC, Walker PD. Serotonin 5-HT2A receptors underlie increased motor behaviors induced in dopamine-depleted rats by intrastriatal 5-HT2A/2C agonism. J Pharmacol Exp Ther. 2004;310:687–94.
23. Alex KD, Yavanian GJ, McFarlane HG, Pluto CP, Pehek EA. Modulation of dopamine release by striatal 5-HT2C receptors. Synapse. 2005;55:242–51.
24. Di Matteo V, De Blasi A, Di Giulio C, Esposito E. Role of 5-HT(2C) receptors in the control of central dopamine function. Trends Pharmacol Sci. 2001;22:229–32.
25. Benloucif S, Keegan MJ, Galloway MP. Serotonin-facilitated dopamine release in vivo: pharmacological characterization. J Pharmacol Exp Ther. 1993;265:373–7.
26. Jenner P, Sheehy M, Marsden CD. Noradrenaline and 5-hydroxytryptamine modulation of brain dopamine function: implications for the treatment of Parkinson's disease. Br J Clin Pharmacol. 1983;15 Suppl 2:277S–89.
27. Arai R, Karasawa N, Geffard M, Nagatsu T, Nagatsu I. Immunohistochemical evidence that central serotonin neurons produce dopamine from exogenous levodopa in the rat, with reference to the involvement of aromatic L-amino acid decarboxylase. Brain Res. 1994;667:295–9.
28. Arai R, Karasawa N, Geffard M, Nagatsu I. levodopa is converted to dopamine in serotonergic fibers of the striatum of the rat: a double-labeling immunofluorescence study. Neurosci Lett. 1995;195:195–8.
29. Tanaka H, Kannari K, Maeda T, Tomiyama M, Suda T, Matsunaga M. Role of serotonergic neurons in levodopa-derived extracellular dopamine in the striatum of 6-OHDA-lesioned rats. Neuroreport. 1999;10:631–4.
30. Carta M, Carlsson T, Kirik D, Bjorklund A. Dopamine released from 5-HT terminals is the cause of levodopa-induced dyskinesia in parkinsonian rats. Brain. 2007;130:1819–33.
31. Eskow KL, Dupre KB, Barnum CJ, Dickinson SO, Park JY, Bishop C. The role of the dorsal raphe nucleus in the development, expression, and treatment of levodopa-induced dyskinesia in hemiparkinsonian rats. Synapse. 2009;63:610–20.
32. Nahimi A, Holtzermann M, Landau AM, Simonsen M, Jakobsen S, Alstrup AK, Vang K, Moller A, Wegener G, Gjedde A, Doudet DJ. Serotonergic modulation of receptor occupancy in rats treated with levodopa after unilateral 6-OHDA lesioning. J Neurochem. 2012;120:806–17.
33. Rylander D, Parent M, O'Sullivan SS, Dovero S, Lees AJ, Bezard E, Descarries L, Cenci MA. Maladaptive plasticity of serotonin axon terminals in levodopa-induced dyskinesia. Ann Neurol. 2010;68:619–28.

34. Mamounas LA, Altar CA, Blue ME, Kaplan DR, Tessarollo L, Lyons WE. BDNF promotes the regenerative sprouting, but not survival, of injured serotonergic axons in the adult rat brain. J Neurosci. 2000;20:771–82.
35. Bishop C, George JA, Buchta W, Goldenberg AA, Mohamed M, Dickinson SO, Eissa S, Eskow Jaunarajs KL. Serotonin transporter inhibition attenuates levodopa-induced dyskinesia without compromising levodopa efficacy in hemi-parkinsonian rats. Eur J Neurosci. 2012;36:2839–48.
36. Tronci E, Lisci C, Stancampiano R, Fidalgo C, Collu M, Devoto P, Carta M. 5-Hydroxy-tryptophan for the treatment of levodopa-induced dyskinesia in the rat Parkinson's disease model. Neurobiol Dis. 2013;60:108–14.
37. Conti MM, Ostock CY, Lindenbach D, Goldenberg AA, Kampton E, Dell'isola R, Katzman AC, Bishop C. Effects of prolonged selective serotonin reuptake inhibition on the development and expression of levodopa-induced dyskinesia in hemi-parkinsonian rats. Neuropharmacology. 2014;77:1–8.
38. de la Fuente-Fernandez R, Sossi V, Huang Z, Furtado S, Lu JQ, Calne DB, Ruth TJ, Stoessl AJ. Levodopa-induced changes in synaptic dopamine levels increase with progression of Parkinson's disease: implications for dyskinesias. Brain. 2004;127:2747–54.
39. Antonini A, Odin P, Opiano L, Tomantschger V, Pacchetti C, Pickut B, Gasser UE, Calandrella D, Mancini F, Zibetti M, Minafra B, Bertaina I, De Deyn P, Cras C, Wolf E, Spielberger S, Poewe W. Effect and safety of duodenal levodopa infusion in advanced Parkinson's disease: a retrospective multicenter outcome assessment in patient routine care. J Neural Transm. 2013;120:1553–8.
40. Bibbiani F, Oh JD, Chase TN. Serotonin 5-HT1A agonist improves motor complications in rodent and primate parkinsonian models. Neurology. 2001;57:1829–34.
41. Bara-Jimenez W, Bibbiani F, Morris MJ, Dimitrova T, Sherzai A, Mouradian MM, Chase TN. Effects of serotonin 5-HT1A agonist in advanced Parkinson's disease. Mov Disord. 2005;20:932–6.
42. Eskow KL, Gupta V, Alam S, Park JY, Bishop C. The partial 5-HT(1A) agonist buspirone reduces the expression and development of levodopa-induced dyskinesia in rats and improves levodopa efficacy. Pharmacol Biochem Behav. 2007;87:306–14.
43. Bezard E, Munoz A, Tronci E, Pioli EY, Li Q, Porras G, Bjorklund A, Carta M. Anti-dyskinetic effect of anpirtoline in animal models of levodopa-induced dyskinesia. Neurosci Res. 2013;77:242–6.
44. Bezard E, Tronci E, Pioli EY, Li Q, Porras G, Bjorklund A, Carta M. Study of the antidyskinetic effect of eltoprazine in animal models of levodopa-induced dyskinesia. Mov Disord. 2013;28:1088–96.
45. Goetz CG, Damier P, Hicking C, Laska E, Muller T, Olanow CW, Rascol O, Russ H. Sarizotan as a treatment for dyskinesias in Parkinson's disease: a double-blind placebo-controlled trial. Mov Disord. 2007;22:179–86.
46. Carey RJ, Depalma G, Damianopoulos E, Muller CP, Huston JP. The 5-HT1A receptor and behavioral stimulation in the rat: effects of 8-OHDPAT on spontaneous and cocaine-induced behavior. Psychopharmacology (Berl). 2004;177:46–54.
47. Iravani MM, Tayarani-Binazir K, Chu WB, Jackson MJ, Jenner P. In 1-methyl-4-phenyl-1,2,3,6-tetrahydropyridine-treated primates, the selective 5-hydroxytryptamine 1a agonist (R)-(+)-8-OHDPAT inhibits levodopa-induced dyskinesia but only with\ increased motor disability. J Pharmacol Exp Ther. 2006;319:1225–34.
48. Bonifati V, Fabrizio E, Cipriani R, Vanacore N, Meco G. Buspirone in levodopa-induced dyskinesias. Clin Neuropharmacol. 1994;17:73–82.
49. Hammerstad JP, Carter J, Nutt JG, Casten GC, Shrotriya RC, Alms DR, Temple D. Buspirone in Parkinson's disease. Clin Neuropharmacol. 1986;9:556–60.
50. Kleedorfer B, Lees AJ, Stern GM. Buspirone in the treatment of levodopa induced dyskinesias. J Neurol Neurosurg Psychiatry. 1991;54:376–7.

51. Zhang X, Andren PE, Greengard P, Svenningsson P. Evidence for a role of the 5-HT1B receptor and its adaptor protein, p11, in levodopa treatment of an animal model of Parkinsonism. Proc Natl Acad Sci U S A. 2008;105:2163–8.
52. Jaunarajs KL, Dupre KB, Steiniger A, Klioueva A, Moore A, Kelly C, Bishop C. Serotonin 1B receptor stimulation reduces D1 receptor agonist-induced dyskinesia. Neuroreport. 2009;20:1265–9.
53. Bishop C, Krolewski DM, Eskow KL, Barnum CJ, Dupre KB, Deak T, Walker PD. Contribution of the striatum to the effects of 5-HT1A receptor stimulation in levodopa-treated hemiparkinsonian rats. J Neurosci Res. 2009;87:1645–58.
54. Dupre KB, Eskow KL, Barnum CJ, Bishop C. Striatal 5-HT1A receptor stimulation reduces D1 receptor-induced dyskinesia and improves movement in the hemiparkinsonian rat. Neuropharmacology. 2008;55:1321–8.
55. Dupre KB, Ostock CY, Eskow Jaunarajs KL, Button T, Savage LM, Wolf W, Bishop C. Local modulation of striatal glutamate efflux by serotonin 1A receptor stimulation in dyskinetic, hemiparkinsonian rats. Exp Neurol. 2011;229:288–99.
56. Munoz A, Li Q, Gardoni F, Marcello E, Qin C, Carlsson T, Kirik D, Di Luca M, Bjorklund A, Bezard E, Carta M. Combined 5-HT1A and 5-HT1B receptor agonists for the treatment of levodopa-induced dyskinesia. Brain. 2008;131:3380–94.
57. Lindgren HS, Andersson DR, Lagerkvist S, Nissbrandt H, Cenci MA. levodopa-induced dopamine efflux in the striatum and the substantia nigra in a rat model of Parkinson's disease: temporal and quantitative relationship to the expression of dyskinesia. J Neurochem. 2010;112:1465–76.
58. Miczek KA, Mos J, Olivier B. Brain 5-HT and inhibition of aggressive behavior in animals: 5-HIAA and receptor subtypes. Psychopharmacol Bull. 1989;25:399–403.
59. Mos J, Olivier B, Poth M, van Aken H. The effects of intraventricular administration of eltoprazine, 1-(3-trifluoromethylphenyl)piperazine hydrochloride and 8-hydroxy-2-(di-n-propylamino)tetralin on resident intruder aggression in the rat. Eur J Pharmacol. 1992;212:295–8.
60. Tronci E and Carta M. 5-HT1 receptor agonists for the treatment of levodopa-induced dyskinesia: from animal models to clinical investigation. Basal Ganglia. 2013;3:9–13.
61. Olanow CW, Damier P, Goetz CG, Mueller T, Nutt J, Rascol O, Serbanescu A, Deckers F, Russ H. Multicenter, open-label, trial of sarizotan in Parkinson disease patients with levodopa-induced dyskinesias (the SPLENDID Study). Clin Neuropharmacol. 2004;27:58–62.
62. Rijnders HJ, Slangen JL. The discriminative stimulus properties of buspirone involve dopamine-2 receptor antagonist activity. Psychopharmacology (Berl). 1993;111:55–61.
63. Riahi G, Morissette M, Parent M, Di Paolo T. Brain 5-HT(2A) receptors in MPTP monkeys and levodopa-induced dyskinesias. Eur J Neurosci. 2011;33:1823–31.
64. Oh JD, Bibbiani F, Chase TN. Quetiapine attenuates levodopa-induced motor complications in rodent and primate parkinsonian models. Exp Neurol. 2002;177:557–64.
65. Roberts C. ACP-103, a 5-HT2A receptor inverse agonist. Curr Opin Investig Drugs. 2006;7:653–60.
66. Meco G, Stirpe P, Edito F, Purcaro C, Valente M, Bernardi S, Vanacore N. Aripiprazole in Levodopa-induced dyskinesias: a one-year open-label pilot study. J Neural Transm. 2009;116:881–4.
67. Vanover KE, Betz AJ, Weber SM, Bibbiani F, Kielaite A, Weiner DM, Davis RE, Chase TN, Salamone JD. A 5-HT2A receptor inverse agonist, ACP-103, reduces tremor in a rat model and levodopa-induced dyskinesias in a monkey model. Pharmacol Biochem Behav. 2008;90:540–4.
68. Maertens de Noordhout A, Delwaide PJ. Open pilot trial of ritanserin in parkinsonism. Clin Neuropharmacol. 1986;9:480–4.
69. Meco G, Frascarelli M, Pratesi L, Linfante I, Rocchi L, Formisano R. Headache in Parkinson's disease. Headache. 1988;28:26–9.

70. Taylor JL, Bishop C, Ullrich T, Rice KC, Walker PD. Serotonin 2A receptor antagonist treatment reduces dopamine D1 receptor-mediated rotational behavior but not levodopa-induced abnormal involuntary movements in the unilateral dopamine-depleted rat. Neuropharmacology. 2006;50:761–8.
71. Kuan WL, Zhao JW, Barker RA. The role of anxiety in the development of levodopa-induced dyskinesias in an animal model of Parkinson's disease, and the effect of chronic treatment with the selective serotonin reuptake inhibitor citalopram. Psychopharmacology (Berl). 2008;197:279–93.
72. Durif F, Vidailhet M, Bonnet AM, Blin J, Agid Y. Levodopa-induced dyskinesias are improved by fluoxetine. Neurology. 1995;45:1855–8.
73. Chung KA, Carlson NE, Nutt JG. Short-term paroxetine treatment does not alter the motor response to levodopa in PD. Neurology. 2005;64:1797–8.

Chapter 12
The Opioid System in Levodopa-Induced Dyskinesia

Tom H. Johnston, Paula Ravenscroft, and Michael P. Hill

Abstract A wealth of evidence underlies the pivotal role of opioidergic neurotransmission in both normal and pathological basal ganglia function. Accompanying the development and expression of levodopa-induced dyskinesia (LID), following long-term dopamine replacement therapy in Parkinson's disease (PD), are myriad changes in both opioid receptor levels as well as the stoichiometry and processing of endogenous opioid peptides. Notably, in both patients and animal models of PD, dopamine denervation and chronic levodopa therapy are associated with an enhancement of basal ganglia opioid transmission. Whether this and other alterations are wholly causative or compensatory remains to be fully elucidated. Nevertheless, such observations have formed the basis for utilizing a variety of small molecules and other potential therapies to modulate the opioid system for the treatment of motor complications in PD. This chapter will provide an overview of the opioid system and describe both preclinical and clinical studies concerning the role of opioids in LID. New insights such as the role of receptor dimerization and potential role for biased ligands are also reviewed.

Keywords Opioid • PPE-B • Levodopa • Dyskinesia • Motor complications

The Endogenous Opioid System

The first description of a related group of endogenous opioid peptides, the enkephalins, and their structures was reported by Hughes and colleagues [1]. Other endogenous opioids were quickly discovered including dynorphin and β-endorphin [2–5]. All these peptides share the enkephalin sequence (Tyr-Gly-Gly-Phe-Leu or

T.H. Johnston PhD (✉) • P. Ravenscroft, PhD
Division of Brain, Imaging and Behaviour – Systems Neuroscience, Toronto Western Research Institute, Krembil Discovery Tower, 60 Leonard Avenue, Room 8KD-402, Toronto, ON M5T 2S8, Canada
e-mail: tom.johnston@uhnresearch.ca

M.P. Hill, PhD
Toronto Western Research Institute, Krembil Discovery Tower, 60 Leonard Avenue, Room 8KD-402, Toronto, ON M5T 2S8, Canada

S.H. Fox, J.M. Brotchie (eds.), *Levodopa-Induced Dyskinesia in Parkinson's Disease*, DOI 10.1007/978-1-4471-6503-3_12

Tyr-Gly-Gly-Phe-Met) at the N-terminus, with differing extensions at the C-terminus. Subsequently, another set of endogenous peptides with opioidergic properties, the endomorphins, were discovered and were found to be distinct from the classic opioid peptides in that they were highly μ-opioid receptor selective and did not contain the enkephalin sequence [6]. A further opioid-like peptide that has similarities to dynorphin A [7] was discovered independently by two groups and termed orphanin FQ by one [8] and nociceptin by the other [9].

Opioid Peptide Precursors

Enkephalins are pentapeptides, produced from the proteolysis of the polypeptide precursor molecule preproenkephalin-A (PPE-A), which contains six copies of methionine-enkephalin (met-enkephalin) and one copy of leucine-enkephalin (leu-enkephalin) [10, 11]. Dynorphins are produced from the proteolysis of preproenkephalin B (PPE-B, also known as preprodynorphin), which also contains sequences for leu-enkephalin, α-neoendorphin, and β-neoendorphin [12–15]. In addition, pre-proopiomelanocortin contains sequences for β-endorphin [16]. Nociceptin/Orphanin FQ (N/OFQ) is derived from prepronociceptin and is found almost exclusively in the CNS [17]. No precursor for the endomorphins has been found to date.

Opioid Receptors

The existence of opioid receptors was first demonstrated in 1973 by a number of groups [18–20], and around 20 years later molecular sequencing and cloning studies confirmed the existence of three distinct classes; μ-, δ-, and κ- [21–23]. Each class displays at least two pharmacological subtypes in vivo (reviewed in [24]). Opioid receptors, as well as other G-protein-coupled receptors, exist in oligomeric complexes [25, 26]. Homo-oligomerization of μ-, δ-, and κ-opioid receptors has been demonstrated, suggesting a critical functional role for receptor–receptor interactions. Of more interest perhaps is the phenomenon of hetero-oligomerization. Hetero-dimerization between δ- and κ-opioid receptors was the first opioid receptor complex identified and was shown to display novel pharmacology [27]. Hetero-dimerization between δ- and μ-opioid receptors has been extensively documented [28, 29], and recently a selective ligand was identified for this complex [30]. It has also been shown that the endogenous opioids endomorphin-1 and leu-enkephalin have a higher affinity for the μ–δ complex than either receptor alone [31]. Thus, specifically targeting opioid hetero-oligomers may open up new opportunities for therapeutic development. To add to the complexity, at least 31 splice variants of the μ-opioid receptor have been isolated from mice, 16 from rats, and 19 from humans (reviewed in:[32]). The endogenous opioid peptides have different degrees of selectivity for opioid receptors. Leu-enkephalin and

Table 12.1 Mammalian endogenous opioid ligands and receptors

Precursor	Endogenous peptide	Opioid receptor selectivity
Preproenkephalin-A	Met-enkephalin	δ-opioid receptor
	Leu-enkephalin	δ-opioid receptor
Preproenkephalin-B	Dynorphin A	κ-opioid receptor
	Dynorphin B	κ-opioid receptor
	Leu-enkephalin	δ-opioid receptor
	α-neoendorphin	κ-opioid receptor
	β-neoendorphin	κ-opioid receptor
Preproopioimelanocortin	β-endorphin	μ-opioid receptor
No precursor discovered	Endomorphin-1	μ-opioid receptor
	Endomorphin-2	μ-opioid receptor
Prepronociceptin	N/OFQ	N/OFQ receptor

met-enkephalin are predominantly endogenous ligands of the δ-opioid receptor [33], β-endorphin, and the endomorphins predominantly bind to the μ-opioid receptor [6, 34, 35], whereas the dynorphins and neoendorphins act mainly via the κ-opioid receptor [36]. N/OFQ binds to the nociceptin receptor (also referred to as the orphanin FQ receptor) [17]. The nociceptin receptor has a high degree of homology to other opioid receptors, but the classic endogenous opioids exhibit little or no affinity for it [37]. A summary of key endogenous opioids and their preferred target receptor is described in Table 12.1.

Opioid receptors are distributed throughout the structures of the basal ganglia and the distribution is relatively conserved across species. The distribution of opioid receptor mRNA and opioid receptor binding sites in the rat CNS has been reviewed by Mansour et al. [38]. There is a high correlation between μ-opioid receptor mRNA expression and binding in the clusters and patches of the striatum and nucleus accumbens and pallidal complex [38, 39], suggesting local receptor synthesis. δ-Opioid receptor mRNA and binding is also highly correlated in the striatum, nucleus accumbens, and globus pallidus [39, 40]. While mRNA and binding of κ-opioid receptors are highly correlated in the striatum and nucleus accumbens, there are differences in the substantia nigra pars compacta and ventral tegmental area which may be due to receptor transport [39]. The subthalamic nucleus also contains high levels of κ-opioid receptors [41].

Changes to the Opioid System in PD and LID

Many studies in both rodent [42–63] and nonhuman primate [64–72] models of Parkinson's disease have consistently demonstrated increased striatal PPE-A mRNA expression compared to normal control animals. Similarly, studies in PD patients show the same pattern of increase in striatal PPE-A mRNA expression [73–75].

Conversely, rodent [46, 47, 49, 51, 52, 60, 76–78] and nonhuman primate [66–68, 79] models of PD show reduced striatal PPE-B mRNA expression. It has also been demonstrated that prepronociceptin is increased in the substantia nigra in rodent models of PD [80–83]. Following repeated D_1/D_2 dopamine receptor stimulation in rodents, striatal PPE-A mRNA is either further elevated compared to that in the parkinsonian state or remains the same, while PPE-B mRNA levels are elevated [46–48, 50, 52, 57, 59, 63, 76, 78, 84–90]. The same pattern is seen in nonhuman primate models of PD [65–68, 70, 91–94] and in PD patients [73–75]. In contrast to the number of studies looking at precursor expression, relatively few studies have looked at the expression of the opioid peptides being produced from these precursors. However, a recent study in MPTP-lesioned nonhuman primates demonstrated significant elevations in met- and leu-enkephalin in the putamen and external segment of the globus pallidus [66]. Elevated levels of N/OFQ have also been demonstrated in the CSF of PD patients [95]. Recent studies have looked at the levels of opioid peptides in rat [87, 96] and nonhuman primate [66] models of LID. The rodent studies used the sensitive technique of matrix-assisted laser desorption ionization (MALDI) imaging mass spectrometry (IMS) that allows comprehensive detection of multiple molecular species in a single tissue section. They found elevated levels of the PPE-B derived peptides, dynorphin B and α-neoendorphin, in the dorsolateral striatum [87] and substantia nigra [96] of severely dyskinetic rats compared with mildly dyskinetic or non-dyskinetic rats. A similar elevation in the PPE-B derived peptide, dynorphin A, has been seen in dyskinetic nonhuman primates [66]. Changes in receptor levels and receptor signaling in PD and dyskinetic states have been extensively reviewed by Huot and colleagues [97] and are summarized in Table 12.2.

Potential of Selective vs. Nonselective Classical Opioid Receptor Therapies

The plethora of changes in basal ganglia function such as underactivity of output nuclei (i.e., internal segment of the globus pallidus and substantia nigra pars reticulata) likely represent key mechanisms underlying dyskinesia in PD (reviewed in [98]).

Table 12.2 Changes to opioid receptor levels and receptor signaling in PD and dyskinetic states

μ-receptor levels are reduced in the striatum and GPi of dyskinetic NHPs killed in the ON state
μ-receptor-mediated signaling is overactive in the striatum and GPi of MPTP-lesioned NHPs killed in the ON state
δ-receptor levels are unchanged in the striatum of dyskinetic NHPs killed in the ON state
δ-receptor-mediated signaling is overactive in the striatum of MPTP-lesioned NHPs killed in the ON state
κ-receptor levels are reduced in the GPe and GPi of dyskinetic NHPs killed in the ON state
κ-receptor-mediated signaling is overactive in the caudate nucleus and motor cortex of MPTP-lesioned NHPs killed in the ON state

Adapted from: Huot et al. [97]

Concomitantly, a substantial preclinical literature has implicated aberrant opioid transmission in the expression of LID. As discussed, these include the classic observations of increases in the synthesis of basal ganglia PPE-B and associated opioid peptides in animal models of LID and in *postmortem* tissue from dyskinetic PD patients [46, 74, 75, 91, 99, 100]. Additionally, functional imaging studies using positron emission tomography showed that PD patients with LID displayed heightened opioid transmission [101]. Such observations have lent support to the hypothesis that antagonism of opioid transmission, in the first instance using the approach of nonselective receptor subtype blockade, might be associated with anti-dyskinetic actions. Indeed, the nonselective opioid receptor antagonist naloxone was shown to alleviate established abnormal involuntary movements (AIMs), a correlate of dyskinesia seen in PD patients, in the 6-OHDA-lesioned rat model of PD [102], but was alternately proven both ineffective [103] and effective [104] at reducing LID in the MPTP-lesioned macaque. Responses to another nonselective opioid antagonist, naltrexone, were also ambivalent showing an alleviation of established LID evident in the MPTP-lesioned marmoset [92] while lacking effect on, or even exacerbating, LID in the macaque [105–108]. The lackluster display of benefit attributed to use of non-subtype-selective blockade was borne out in clinical studies examining both naloxone [109] and naltrexone [110, 111] in which either a total absence of, or only modest, anti-dyskinetic benefit was revealed. Such lack of clear anti-dyskinetic actions possibly reflects the interaction of non-subtype-selective ligands with multiple opioid receptors, which might provide competing pro- and anti-dyskinetic effect. Strikingly, recent advanced mass spectrometry studies have for the first time afforded insight into the exact nature of the PPE-B-derived opioid peptide species generated in the dyskinetic state that could shed light on why generalized blockade of all opioid actions likely represents too blunt a therapeutic strategy. Thus, the aforementioned MALDI-TOF-based imaging in nigral and striatal tissues of levodopa-treated, 6-OHDA-lesioned rats has revealed a strong positive correlation between AIMs severity and levels of the PPE-B derived peptides, dynorphin B and α-neoendorphin. Of note, these dyskinesia-associated peptides were not those with the highest affinity to κ-opioid receptors, but also activate μ- and δ-opioid receptors [87, 96]. Such data provide compelling evidence that enhanced activation of non-κ-opioid receptors by select peptide products of PPE-B may contribute to the development of dyskinesia after chronic levodopa therapy. In keeping with heightened activity of μ-opioid function, the selective μ-opioid receptor antagonists cyprodime and ADL5510 both alleviated LID in the MPTP-lesioned nonhuman primate, without affecting levodopa anti-parkinsonian efficacy [92, 105]. Likewise, and in agreement with hyperactive δ-opioid receptor-mediated transmission in dyskinesia, the δ-opioid receptor antagonist naltrindole reduced LID in the MPTP-lesioned nonhuman primate, also without affecting levodopa anti-parkinsonian efficacy [92]. While earlier studies remain pertinent in suggesting overactivity of κ-opioid-mediated signaling in LID (based on elevated PPE-B-heightened expression), as being a key element in the generation of LID, the more recent MALDI data described perhaps temper this assertion. Indeed, behavioral studies in which κ-opioid receptors were blocked with nor-binaltorphimine (*nor*-BNI) showed no reduction in LID in the MPTP-lesioned nonhuman primate [92]. Conversely, stimulation of κ-opioid receptors with U50,488 reduced established AIMs in the 6-OHDA-lesioned rat and dyskinesia in the MPTP-lesioned squirrel monkey, though at the expense of impairing levodopa anti-parkinsonian action [112]. In addition, TRK-820, a selective κ-opioid receptor agonist, effectively ameliorated levodopa-induced AIMs in the 6-OHDA-lesioned rat, an effect which was blocked by prior treatment with *nor*-BNI [113]. Thus, the lack of clear efficacy of non-subtype selective opioid receptor antagonists may reflect that any anti-dyskinetic benefit conferred by blocking μ- and δ-opioid receptors may be offset by blockade of κ-opioid receptors. If the failure of non-subtype selective opioid antagonists to alleviate LID in clinical trials is indeed due to their blockade of κ-opioid receptors, and if the anti-dyskinetic efficacy of the non-subtype selective opioid agonists is primarily mediated via an agonist effect at κ-opioid receptors, then selective stimulation of these receptors may yet represent a promising anti-dyskinetic target. Any κ-opioid agonist-based strategy will of course have considerable challenges to overcome in dealing with the dysphorogenicity implicit with this class of compound [114].

The optimal balance of opioid receptor stimulation and blockade with which to achieve peak anti-dyskinetic effect has yet to be fully elucidated. Indeed, an alternative explanation for the rise in levels of PPE-B and derived opioid peptides observed in striatal tissue from dyskinetic patients or animal models, is as a compensatory response to prolonged levodopa therapy and onset of dyskinesia rather than a direct causative event (reviewed in: [115]). Evidence to support this alternate hypothesis is no less compelling. Thus, nonselective stimulation of opioid receptors with meperidine alleviated LID in the MPTP-lesioned nonhuman primate [103] while not affecting the anti-parkinsonian action of levodopa. Likewise, morphine alleviated established LID as well as dyskinesia elicited by selective stimulation of either D_1 or D_2 dopamine receptors in the MPTP-lesioned nonhuman primate [116] and showed efficacy in a small open-label clinical report [117]. Given the anti-dyskinetic actions afforded by selective κ-opioid agonists, it is fair to assume that at least part of the benefit exhibited by morphine and meperidine as nonselective agonists is exerted via activity at this receptor. The obvious challenges of advancing an opioid agonist for this indication aside, the potential benefit of a selective μ-opioid agonist would be interesting to see although similar mood-altering side effects to those observed with κ-opioid selective agonists have been reported for μ-opioid agonist-based approaches [118]. It can however be said with certainty that stimulation of the δ-opioid receptor is unlikely to underlie such effects. In the untreated parkinsonian state (prior to advent of dopamine replacement therapy), δ-opioid receptor agonists can provide robust anti-parkinsonian effects. In both haloperidol-administered rats and MPTP-lesioned nonhuman primates, the selective δ-opioid receptor agonist, SNC80, reversed all parkinsonian deficits [119]. While δ-opioid receptor agonists have considerable potential as anti-parkinsonian agents, their use in more advanced patients, where dyskinesia has already emerged, may be limited. Thus, in MPTP-lesioned nonhuman primates with established levodopa-induced dyskinesia, δ-opioid agonists elicit dyskinesia even as monotherapy and exacerbate dyskinesia if given in combination with levodopa [120].

Recent work has examined both agonist and antagonist effects at the nociceptin/orphanin FQ (N/OFQ) receptor as a strategy to alleviate LID. Systemic administration of J-113397, an N/OFQ receptor antagonist, enhanced the anti-parkinsonian actions of low-dose levodopa in 6-OHDA-lesioned rats [121] but worsened established LID in the MPTP-lesioned nonhuman primate [122]. Conversely, both native N/OFQ peptide and a synthetic N/OFQ agonist, Ro 65–6570, given via intracerebroventricular injection reduced established AIMs in the 6-OHDA-lesioned rat and, given systemically to MPTP-lesioned nonhuman primates, Ro 65–6570 significantly reduced established LID without compromising the anti-parkinsonian benefit of levodopa [123].

New Avenues in Therapeutic Development

Given the multiplicity of effects, both positive and negative, offered by modulation of the opioid system in PD and LID, a combination approach that simultaneously yields the best of multiple strategies may be of value. The concept of opioid ligands with "agonist/antagonist" properties is hardly new, having been described several decades ago for pentazocine and related compounds [124]. Recent work has identified single chemical entities with dual δ-opioid receptor agonist and μ-opioid antagonist (DAMA) characteristics [125]. Such compounds, with inherent anti-parkinsonian actions (mediated via actions at the δ-opioid receptor) also, via μ-opioid receptor blockade, have the potential to suppress both the development of dyskinesia in early disease, when given as monotherapy, and its expression, when combined with levodopa, in later stages of the disease. A similar strategy centered on dual-actions of a single compound at the κ- and μ-opioid receptors has also recently been promulgated [126]. Specifically, nalbuphine, a semisynthetic opioid used clinically as an analgesic, with activity as both a weak μ-opioid receptor antagonist and a κ-opioid receptor partial agonist, has been shown to ameliorate established dyskinesia in MPTP-lesioned nonhuman primates without worsening parkinsonian symptoms. This strategy offers the combined anti-dyskinetic potential of both μ-opioid antagonism and κ-opioid agonism reportedly using sub-analgesic doses of nalbuphine below those at which any propensity for side effects relating to κ-opioid receptor stimulation (dysphoria and loss of anti-parkinsonian action) might compromise therapeutic benefit.

Avoidance of such psychotomimetic effects and other opioid agonist-related complications such as the development of tolerance with chronic administration [127] would represent a considerable advance on current therapies. Possible solutions to these issues may be forthcoming with some profound advances in the fundamental understanding of GPCR biology witnessed over the last decade. Chief among these may be the recognition that in addition to the classical repertoire of agonists and antagonists which activate or inactivate the entirety of a receptor's signaling network, "biased" ligands can selectively activate a subset of the signaling pathways available to that receptor by a particular ligand [128, 129]. For instance, a newly developed κ-opioid receptor agonist, 6′-guanidinonaltrindole, was shown to display bias toward the activation of signaling through non-β-arrestin

2 pathways [130, 131], a mechanism associated with the loss of dysphoria [132]. Such preferential signaling that avoids β-arrestin engagement is also being explored in the context of enhancing δ-opioid receptor agonist strategies in an attempt to limit seizure activity associated with "classical" agonists such as SNC80 [133]. Separation of seizure and locomotor effects may however be challenging. For example, some of the newer δ-opioid receptor agonists such as ADL5747 and ADL5859, while showing a lack of seizure activity at doses associated with analgesia, also show a total lack of SNC80-like anti-parkinsonian effect [134]. Future therapeutic development may also take into consideration discovery of yet another class of new opioid receptor ligands, those, as discussed earlier, with the capacity to selectively bind to, and activate, heteromeric opioid receptors, rather than either receptor expressed in monomeric form [30].

Conclusion

It is clear that opioidergic neurotransmission is significantly altered in Parkinson's disease and levodopa-induced dyskinesia. Current data suggest that the elevated level of opioids derived from PPE-B plays the most prominent role in dyskinesia expression following long-term dopamine replacement therapy. The most promising target explored to date appears to be the μ-opioid receptor where selective antagonists afford anti-dyskinetic activity without compromising the anti-parkinsonian benefit of levodopa. However, new avenues that utilize compounds with combined opioid agonist and antagonist properties, biased ligands, in particular κ-opioid agonists lacking β-arrestin activity or those selective for opioid receptor heterodimers, may provide novel strategies in the treatment of levodopa-induced dyskinesia.

References

1. Hughes J, Smith TW, Kosterlitz HW, Fothergill LA, Morgan BA, Morris HR. Identification of two related pentapeptides from the brain with potent opiate agonist activity. Nature. 1975;258(5536):577–80.
2. Birdsall NJ, Hulme EC. C fragment of lipotropin has a high affinity for brain opiate receptors. Nature. 1976;260(5554):793–5.
3. Cox BM, Opheim KE, Teschemacher H, Goldstein A. A peptide-like substance from pituitary that acts like morphine. 2. Purification and properties. Life Sci. 1975;16(12):1777–82.
4. Goldstein A. Opioid peptides endorphins in pituitary and brain. Science. 1976;193(4258):1081–6.
5. Li CH, Chung D, Doneen BA. Isolation, characterization and opiate activity of beta-endorphin from human pituitary glands. Biochem Biophys Res Commun. 1976;72(4):1542–7.
6. Zadina JE, Hackler L, Ge LJ, Kastin AJ. A potent and selective endogenous agonist for the mu-opiate receptor. Nature. 1997;386(6624):499–502.
7. Meunier JC. Nociceptin/orphanin FQ and the opioid receptor-like ORL1 receptor. Eur J Pharmacol. 1997;340(1):1–15.

8. Reinscheid RK, Nothacker HP, Bourson A, Ardati A, Henningsen RA, Bunzow JR, et al. Orphanin FQ: a neuropeptide that activates an opioidlike G protein-coupled receptor. Science. 1995;270(5237):792–4.
9. Meunier JC, Mollereau C, Toll L, Suaudeau C, Moisand C, Alvinerie P, et al. Isolation and structure of the endogenous agonist of opioid receptor-like ORL1 receptor. Nature. 1995;377(6549):532–5.
10. Rosen H, Douglass J, Herbert E. Isolation and characterization of the rat proenkephalin gene. J Biol Chem. 1984;259(22):14309–13.
11. Noda M, Teranishi Y, Takahashi H, Toyosato M, Notake M, Nakanishi S, et al. Isolation and structural organization of the human preproenkephalin gene. Nature. 1982;297(5865):431–4.
12. Hughes J. Isolation of an endogenous compound from the brain with pharmacological properties similar to morphine. Brain Res. 1975;88(2):295–308.
13. Civelli O, Douglass J, Goldstein A, Herbert E. Sequence and expression of the rat prodynorphin gene. Proc Natl Acad Sci U S A. 1985;82(12):4291–5.
14. Horikawa S, Takai T, Toyosato M, Takahashi H, Noda M, Kakidani H, et al. Isolation and structural organization of the human preproenkephalin B gene. Nature. 1983;306(5943):611–4.
15. Seizinger BR, Grimm C, Hollt V, Herz A. Evidence for a selective processing of proenkephalin B into different opioid peptide forms in particular regions of rat brain and pituitary. J Neurochem. 1984;42(2):447–57.
16. Chretien M, Benjannet S, Gossard F, Gianoulakis C, Crine P, Lis M, et al. From beta-lipotropin to beta-endorphin and 'pro-opio-melanocortin'. Can J Biochem. 1979;57(9):1111–21.
17. Mollereau C, Simons MJ, Soularue P, Liners F, Vassart G, Meunier JC, et al. Structure, tissue distribution, and chromosomal localization of the prepronociceptin gene. Proc Natl Acad Sci U S A. 1996;93(16):8666–70.
18. Pert CB, Snyder SH. Opiate receptor: demonstration in nervous tissue. Science. 1973;179(4077):1011–4.
19. Simon EJ, Hiller JM, Edelman I. Stereospecific binding of the potent narcotic analgesic (3H) Etorphine to rat-brain homogenate. Proc Natl Acad Sci U S A. 1973;70(7):1947–9.
20. Terenius L. Stereospecific interaction between narcotic analgesics and a synaptic plasm a membrane fraction of rat cerebral cortex. Acta Pharmacol Toxicol (Copenh). 1973;32(3):317–20.
21. Evans CJ, Keith Jr DE, Morrison H, Magendzo K, Edwards RH. Cloning of a delta opioid receptor by functional expression. Science. 1992;258(5090):1952–5.
22. Yasuda K, Raynor K, Kong H, Breder CD, Takeda J, Reisine T, et al. Cloning and functional comparison of kappa and delta opioid receptors from mouse brain. Proc Natl Acad Sci U S A. 1993;90(14):6736–40.
23. Zastawny RL, George SR, Nguyen T, Cheng R, Tsatsos J, Briones-Urbina R, et al. Cloning, characterization, and distribution of a mu-opioid receptor in rat brain. J Neurochem. 1994;62(6):2099–105.
24. Corbett AD, Henderson G, McKnight AT, Paterson SJ. 75 years of opioid research: the exciting but vain quest for the Holy Grail. Br J Pharmacol. 2006;147 Suppl 1:S153–62.
25. Bouvier M. Oligomerization of G-protein-coupled transmitter receptors. Nat Rev Neurosci. 2001;2(4):274–86.
26. Gomes I, Jordan BA, Gupta A, Rios C, Trapaidze N, Devi LA. G protein coupled receptor dimerization: implications in modulating receptor function. J Mol Med. 2001;79(5–6):226–42.
27. Jordan BA, Devi LA. G-protein-coupled receptor heterodimerization modulates receptor function. Nature. 1999;399(6737):697–700.
28. Traynor JR, Elliott J. delta-Opioid receptor subtypes and cross-talk with mu-receptors. Trends Pharmacol Sci. 1993;14(3):84–6.
29. Manglik A, Kruse AC, Kobilka TS, Thian FS, Mathiesen JM, Sunahara RK, et al. Crystal structure of the micro-opioid receptor bound to a morphinan antagonist. Nature. 2012;485(7398):321–6.

30. Gomes I, Fujita W, Gupta A, Saldanha SA, Negri A, Pinello CE, et al. Identification of a mu-delta opioid receptor heteromer-biased agonist with antinociceptive activity. Proc Natl Acad Sci U S A. 2013;110(29):12072–7.
31. George SR, Fan T, Xie Z, Tse R, Tam V, Varghese G, et al. Oligomerization of mu- and delta-opioid receptors. Generation of novel functional properties. J Biol Chem. 2000;275(34):26128–35.
32. Pasternak GW. Opioids and their receptors: are we there yet? Neuropharmacology. 2014;76(Pt B):198–203.
33. Quock RM, Burkey TH, Varga E, Hosohata Y, Hosohata K, Cowell SM, et al. The delta-opioid receptor: molecular pharmacology, signal transduction, and the determination of drug efficacy. Pharmacol Rev. 1999;51(3):503–32.
34. Wang JB, Imai Y, Eppler CM, Gregor P, Spivak CE, Uhl GR. mu opiate receptor: cDNA cloning and expression. Proc Natl Acad Sci U S A. 1993;90(21):10230–4.
35. Fichna J, Janecka A, Costentin J, Do Rego JC. The endomorphin system and its evolving neurophysiological role. Pharmacol Rev. 2007;59(1):88–123.
36. Chavkin C, James IF, Goldstein A. Dynorphin is a specific endogenous ligand of the kappa opioid receptor. Science. 1982;215(4531):413–5.
37. Henderson G, McKnight AT. The orphan opioid receptor and its endogenous ligand–nociceptin/orphanin FQ. Trends Pharmacol Sci. 1997;18(8):293–300.
38. Mansour A, Fox CA, Burke S, Akil H, Watson SJ. Immunohistochemical localization of the cloned mu opioid receptor in the rat CNS. J Chem Neuroanat. 1995;8(4):283–305.
39. Mansour A, Fox CA, Burke S, Meng F, Thompson RC, Akil H, et al. Mu, delta, and kappa opioid receptor mRNA expression in the rat CNS: an in situ hybridization study. J Comp Neurol. 1994;350(3):412–38.
40. Mansour A, Thompson RC, Akil H, Watson SJ. Delta opioid receptor mRNA distribution in the brain: comparison to delta receptor binding and proenkephalin mRNA. J Chem Neuroanat. 1993;6(6):351–62.
41. Zhu J, Chen C, Xue JC, Kunapuli S, DeRiel JK, Liu-Chen LY. Cloning of a human kappa opioid receptor from the brain. Life Sci. 1995;56(9):L201–7.
42. Angulo JA. Involvement of dopamine D1 and D2 receptors in the regulation of proenkephalin mRNA abundance in the striatum and accumbens of the rat brain. J Neurochem. 1992;58(3):1104–9.
43. Angulo JA, Cadet JL, Woolley CS, Suber F, McEwen BS. Effect of chronic typical and atypical neuroleptic treatment on proenkephalin mRNA levels in the striatum and nucleus accumbens of the rat. J Neurochem. 1990;54(6):1889–94.
44. Angulo JA, Christoph GR, Manning RW, Burkhart BA, Davis LG. Reduction of dopamine receptor activity differentially alters striatal neuropeptide mRNA levels. Adv Exp Med Biol. 1987;221:385–91.
45. Angulo JA, Davis LG, Burkhart BA, Christoph GR. Reduction of striatal dopaminergic neurotransmission elevates striatal proenkephalin mRNA. Eur J Pharmacol. 1986;130(3):341–3.
46. Duty S, Brotchie JM. Enhancement of the behavioral response to apomorphine administration following repeated treatment in the 6-hydroxydopamine-lesioned rat is temporally correlated with a rise in striatal preproenkephalin-B, but not preproenkephalin-A, gene expression. Exp Neurol. 1997;144(2):423–32.
47. Duty S, Henry B, Crossman AR, Brotchie JM. Topographical organization of opioid peptide precursor gene expression following repeated apomorphine treatment in the 6-hydroxydopamine-lesioned rat. Exp Neurol. 1998;150(2):223–34.
48. Engber TM, Susel Z, Kuo S, Gerfen CR, Chase TN. Levodopa replacement therapy alters enzyme activities in striatum and neuropeptide content in striatal output regions of 6-hydroxydopamine lesioned rats. Brain Res. 1991;552(1):113–8.
49. Gerfen CR, McGinty JF, Young 3rd WS. Dopamine differentially regulates dynorphin, substance P, and enkephalin expression in striatal neurons: in situ hybridization histochemical analysis. J Neurosci. 1991;11(4):1016–31.

50. Granata R, Wenning GK, Jolkkonen J, Jenner P, Marsden CD. Effect of repeated administration of dopamine agonists on striatal neuropeptide mRNA expression in rats with a unilateral nigral 6-hydroxydopamine lesion. J Neural Transm. 1996;103(3):249–60.
51. Harrison MB, Kumar S, Hubbard CA, Trugman JM. Early changes in neuropeptide mRNA expression in the striatum following reserpine treatment. Exp Neurol. 2001;167(2):321–8.
52. Henry B, Crossman AR, Brotchie JM. Effect of repeated L-DOPA, bromocriptine, or lisuride administration on preproenkephalin-A and preproenkephalin-B mRNA levels in the striatum of the 6-hydroxydopamine-lesioned rat. Exp Neurol. 1999;155(2):204–20.
53. Hong JS, Yang HY, Gillin JC, Costa E. Effects of long-term administration of antipsychotic drugs on enkephalinergic neurons. Adv Biochem Psychopharmacol. 1980;24:223–32.
54. Jaber M, Fournier MC, Bloch B. Reserpine treatment stimulates enkephalin and D2 dopamine receptor gene expression in the rat striatum. Brain Res Mol Brain Res. 1992;15(3–4):189–94.
55. Jaber M, Normand E, Bloch B. Effect of reserpine treatment on enkephalin mRNA level in the rat striatum: an in situ hybridization study. Brain Res Mol Brain Res. 1995;32(1):156–60.
56. Jaber M, Tison F, Fournier MC, Bloch B. Differential influence of haloperidol and sulpiride on dopamine receptors and peptide mRNA levels in the rat striatum and pituitary. Brain Res Mol Brain Res. 1994;23(1–2):14–20.
57. Salin P, Dziewczapolski G, Gershanik OS, Nieoullon A, Raisman-Vozari R. Differential regional effects of long-term L-DOPA treatment on preproenkephalin and preprotachykinin gene expression in the striatum of 6-hydroxydopamine-lesioned rat. Brain Res Mol Brain Res. 1997;47(1–2):311–21.
58. Tang F, Costa E, Schwartz JP. Increase of proenkephalin mRNA and enkephalin content of rat striatum after daily injection of haloperidol for 2 to 3 weeks. Proc Natl Acad Sci U S A. 1983;80(12):3841–4.
59. Taylor MD, De Ceballos ML, Rose S, Jenner P, Marsden CD. Effects of a unilateral 6-hydroxydopamine lesion and prolonged L-3,4-dihydroxyphenylalanine treatment on peptidergic systems in rat basal ganglia. Eur J Pharmacol. 1992;219(2):183–92.
60. Voorn P, Docter GJ, Jongen-Relo AL, Jonker AJ. Rostrocaudal subregional differences in the response of enkephalin, dynorphin and substance P synthesis in rat nucleus accumbens to dopamine depletion. Eur J Neurosci. 1994;6(3):486–96.
61. Voorn P, Roest G, Groenewegen HJ. Increase of enkephalin and decrease of substance P immunoreactivity in the dorsal and ventral striatum of the rat after midbrain 6-hydroxydopamine lesions. Brain Res. 1987;412(2):391–6.
62. Young 3rd WS, Bonner TI, Brann MR. Mesencephalic dopamine neurons regulate the expression of neuropeptide mRNAs in the rat forebrain. Proc Natl Acad Sci U S A. 1986;83(24):9827–31.
63. Zeng BY, Jolkkonen J, Jenner P, Marsden CD. Chronic L-DOPA treatment differentially regulates gene expression of glutamate decarboxylase, preproenkephalin and preprotachykinin in the striatum of 6-hydroxydopamine-lesioned rat. Neuroscience. 1995;66(1):19–28.
64. Asselin MC, Soghomonian JJ, Cote PY, Parent A. Striatal changes in preproenkephalin mRNA levels in parkinsonian monkeys. Neuroreport. 1994;5(16):2137–40.
65. Bezard E, Ravenscroft P, Gross CE, Crossman AR, Brotchie JM. Upregulation of striatal preproenkephalin gene expression occurs before the appearance of parkinsonian signs in 1-methyl-4-phenyl- 1,2,3,6-tetrahydropyridine monkeys. Neurobiol Dis. 2001;8(2):343–50.
66. Bourdenx M, Nilsson A, Wadensten H, Falth M, Li Q, Crossman AR, et al. Abnormal structure-specific peptide transmission and processing in a primate model of Parkinson's disease and l-DOPA-induced dyskinesia. Neurobiol Dis. 2014;62:307–12.
67. Herrero MT, Augood SJ, Hirsch EC, Javoy-Agid F, Luquin MR, Agid Y, et al. Effects of L-DOPA on preproenkephalin and preprotachykinin gene expression in the MPTP-treated monkey striatum. Neuroscience. 1995;68(4):1189–98.

68. Jolkkonen J, Jenner P, Marsden CD. L-DOPA reverses altered gene expression of substance P but not enkephalin in the caudate-putamen of common marmosets treated with MPTP. Brain Res Mol Brain Res. 1995;32(2):297–307.
69. Lavoie B, Parent A, Bedard PJ. Effects of dopamine denervation on striatal peptide expression in parkinsonian monkeys. Can J Neurol Sci. 1991;18(3 Suppl):373–5.
70. Morissette M, Goulet M, Soghomonian JJ, Blanchet PJ, Calon F, Bedard PJ, et al. Preproenkephalin mRNA expression in the caudate-putamen of MPTP monkeys after chronic treatment with the D2 agonist U91356A in continuous or intermittent mode of administration: comparison with L-DOPA therapy. Brain Res Mol Brain Res. 1997;49(1–2):55–62.
71. Morissette M, Grondin R, Goulet M, Bedard PJ, Di Paolo T. Differential regulation of striatal preproenkephalin and preprotachykinin mRNA levels in MPTP-lesioned monkeys chronically treated with dopamine D1 or D2 receptor agonists. J Neurochem. 1999;72(2):682–92.
72. Quik M, Police S, Langston JW, Di Monte DA. Increases in striatal preproenkephalin gene expression are associated with nigrostriatal damage but not L-DOPA-induced dyskinesias in the squirrel monkey. Neuroscience. 2002;113(1):213–20.
73. Calon F, Birdi S, Rajput AH, Hornykiewicz O, Bedard PJ, Di Paolo T. Increase of preproenkephalin mRNA levels in the putamen of Parkinson disease patients with levodopa-induced dyskinesias. J Neuropathol Exp Neurol. 2002;61(2):186–96.
74. Henry B, Duty S, Fox SH, Crossman AR, Brotchie JM. Increased striatal pre-proenkephalin B expression is associated with dyskinesia in Parkinson's disease. Exp Neurol. 2003;183(2):458–68.
75. Nisbet AP, Foster OJ, Kingsbury A, Eve DJ, Daniel SE, Marsden CD, et al. Preproenkephalin and preprotachykinin messenger RNA expression in normal human basal ganglia and in Parkinson's disease. Neuroscience. 1995;66(2):361–76.
76. Gerfen CR, Engber TM, Mahan LC, Susel Z, Chase TN, Monsma Jr FJ, et al. D1 and D2 dopamine receptor-regulated gene expression of striatonigral and striatopallidal neurons. Science. 1990;250(4986):1429–32.
77. Jiang HK, McGinty JF, Hong JS. Differential modulation of striatonigral dynorphin and enkephalin by dopamine receptor subtypes. Brain Res. 1990;507(1):57–64.
78. Li SJ, Jiang HK, Stachowiak MS, Hudson PM, Owyang V, Nanry K, et al. Influence of nigrostriatal dopaminergic tone on the biosynthesis of dynorphin and enkephalin in rat striatum. Brain Res Mol Brain Res. 1990;8(3):219–25.
79. Parent A, Asselin MC, Cote PY. Dopaminergic regulation of peptide gene expression in the striatum of normal and parkinsonian monkeys. Adv Neurol. 1996;69:73–7.
80. Gouty S, Brown JM, Rosenberger J, Cox BM. MPTP treatment increases expression of pre-pro-nociceptin/orphanin FQ mRNA in a subset of substantia nigra reticulata neurons. Neuroscience. 2010;169(1):269–78.
81. Marti M, Mela F, Fantin M, Zucchini S, Brown JM, Witta J, et al. Blockade of nociceptin/orphanin FQ transmission attenuates symptoms and neurodegeneration associated with Parkinson's disease. J Neurosci. 2005;25(42):9591–601.
82. Norton CS, Neal CR, Kumar S, Akil H, Watson SJ. Nociceptin/orphanin FQ and opioid receptor-like receptor mRNA expression in dopamine systems. J Comp Neurol. 2002;444(4):358–68.
83. Di Benedetto M, Cavina C, D'Addario C, Leoni G, Candeletti S, Cox BM, et al. Alterations of N/OFQ and NOP receptor gene expression in the substantia nigra and caudate putamen of MPP+ and 6-OHDA lesioned rats. Neuropharmacology. 2009;56(4):761–7.
84. Andersson M, Hilbertson A, Cenci MA. Striatal fosB expression is causally linked with l-DOPA-induced abnormal involuntary movements and the associated upregulation of striatal prodynorphin mRNA in a rat model of Parkinson's disease. Neurobiol Dis. 1999;6(6):461–74.
85. Bronstein DM, Ye H, Pennypacker KR, Hudson PM, Hong JS. Role of a 35 kDa fos-related antigen (FRA) in the long-term induction of striatal dynorphin expression in the 6-hydroxydopamine lesioned rat. Brain Res Mol Brain Res. 1994;23(3):191–203.

86. Cenci MA, Lee CS, Bjorklund A. L-DOPA-induced dyskinesia in the rat is associated with striatal overexpression of prodynorphin- and glutamic acid decarboxylase mRNA. Eur J Neurosci. 1998;10(8):2694–706.
87. Hanrieder J, Ljungdahl A, Falth M, Mammo SE, Bergquist J, Andersson M. L-DOPA-induced dyskinesia is associated with regional increase of striatal dynorphin peptides as elucidated by imaging mass spectrometry. Mol Cell Proteomics. 2011;10(10):M111 009308.
88. Lundblad M, Picconi B, Lindgren H, Cenci MA. A model of L-DOPA-induced dyskinesia in 6-hydroxydopamine lesioned mice: relation to motor and cellular parameters of nigrostriatal function. Neurobiol Dis. 2004;16(1):110–23.
89. van de Witte SV, Drukarch B, Stoof JC, Voorn P. Priming with L-DOPA differently affects dynorphin and substance P mRNA levels in the striatum of 6-hydroxydopamine-lesioned rats after challenge with dopamine D1-receptor agonist. Brain Res Mol Brain Res. 1998;61(1–2):219–23.
90. Winkler C, Kirik D, Bjorklund A, Cenci MA. L-DOPA-induced dyskinesia in the intrastriatal 6-hydroxydopamine model of parkinson's disease: relation to motor and cellular parameters of nigrostriatal function. Neurobiol Dis. 2002;10(2):165–86.
91. Aubert I, Guigoni C, Li Q, Dovero S, Bioulac BH, Gross CE, et al. Enhanced preproenkephalin-B-derived opioid transmission in striatum and subthalamic nucleus converges upon globus pallidus internalis in L-3,4-dihydroxyphenylalanine-induced dyskinesia. Biol Psychiatry. 2007;61(7):836–44.
92. Henry B, Fox SH, Crossman AR, Brotchie JM. Mu- and delta-opioid receptor antagonists reduce levodopa-induced dyskinesia in the MPTP-lesioned primate model of Parkinson's disease. Exp Neurol. 2001;171(1):139–46.
93. Tamim MK, Samadi P, Morissette M, Gregoire L, Ouattara B, Levesque D, et al. Effect of non-dopaminergic drug treatment on Levodopa induced dyskinesias in MPTP monkeys: common implication of striatal neuropeptides. Neuropharmacology. 2010;58(1):286–96.
94. Tel BC, Zeng BY, Cannizzaro C, Pearce RK, Rose S, Jenner P. Alterations in striatal neuropeptide mRNA produced by repeated administration of L-DOPA, ropinirole or bromocriptine correlate with dyskinesia induction in MPTP-treated common marmosets. Neuroscience. 2002;115(4):1047–58.
95. Marti M, Sarubbo S, Latini F, Cavallo M, Eleopra R, Biguzzi S, et al. Brain interstitial nociceptin/orphanin FQ levels are elevated in Parkinson's disease. Mov Disord. 2010;25(11):1723–32.
96. Ljungdahl A, Hanrieder J, Falth M, Bergquist J, Andersson M. Imaging mass spectrometry reveals elevated nigral levels of dynorphin neuropeptides in L-DOPA-induced dyskinesia in rat model of Parkinson's disease. PLoS One. 2011;6(9):e25653.
97. Huot P, Johnston TH, Koprich JB, Fox SH, Brotchie JM. The pharmacology of L-DOPA-induced dyskinesia in Parkinson's disease. Pharmacol Rev. 2013;65(1):171–222.
98. Bezard E, Brotchie JM, Gross CE. Pathophysiology of levodopa-induced dyskinesia: potential for new therapies. Nat Rev Neurosci. 2001;2(8):577–88.
99. Maneuf YP, Mitchell IJ, Crossman AR, Brotchie JM. On the role of enkephalin cotransmission in the GABAergic striatal efferents to the globus pallidus. Exp Neurol. 1994;125(1):65–71.
100. de Ceballos ML, Fernandez A, Jenner P, Marsden CD. Parallel alterations in Met-enkephalin and substance P levels in medial globus pallidus in Parkinson's disease patients. Neurosci Lett. 1993;160(2):163–6.
101. Piccini P, Weeks RA, Brooks DJ. Alterations in opioid receptor binding in Parkinson's disease patients with levodopa-induced dyskinesias. Ann Neurol. 1997;42(5):720–6.
102. Lundblad M, Andersson M, Winkler C, Kirik D, Wierup N, Cenci MA. Pharmacological validation of behavioural measures of akinesia and dyskinesia in a rat model of Parkinson's disease. Eur J Neurosci. 2002;15(1):120–32.
103. Gomez-Mancilla B, Bedard PJ. Effect of nondopaminergic drugs on L-dopa-induced dyskinesias in MPTP-treated monkeys. Clin Neuropharmacol. 1993;16(5):418–27.

104. Klintenberg R, Svenningsson P, Gunne L, Andren PE. Naloxone reduces levodopa-induced dyskinesias and apomorphine-induced rotations in primate models of parkinsonism. J Neural Transm. 2002;109(10):1295–307.
105. Koprich JB, Fox SH, Johnston TH, Goodman A, Le Bourdonnec B, Dolle RE, et al. The selective mu-opioid receptor antagonist adl5510 reduces levodopa-induced dyskinesia without affecting antiparkinsonian action in mptp-lesioned macaque model of Parkinson's disease. Mov Disord. 2011;26(7):1225–33.
106. Samadi P, Gregoire L, Bedard PJ. Opioid antagonists increase the dyskinetic response to dopaminergic agents in parkinsonian monkeys: interaction between dopamine and opioid systems. Neuropharmacology. 2003;45(7):954–63.
107. Samadi P, Gregoire L, Hadj Tahar A, Di Paolo T, Rouillard C, Bedard PJ. Naltrexone in the short-term decreases antiparkinsonian response to l-Dopa and in the long-term increases dyskinesias in drug-naive parkinsonian monkeys. Neuropharmacology. 2005;49(2):165–73.
108. Samadi P, Gregoire L, Rouillard C, Bedard PJ. Dyskinesias occur in response to saline and naltrexone alone after priming with combination of dopaminergic agents and naltrexone in the MPTP parkinsonian monkeys. Neurobiol Dis. 2005;19(1–2):266–72.
109. Fox S, Silverdale M, Kellett M, Davies R, Steiger M, Fletcher N, et al. Non-subtype-selective opioid receptor antagonism in treatment of levodopa-induced motor complications in Parkinson's disease. Mov Disord. 2004;19(5):554–60.
110. Manson AJ, Katzenschlager R, Hobart J, Lees AJ. High dose naltrexone for dyskinesias induced by levodopa. J Neurol Neurosurg Psychiatry. 2001;70(4):554–6.
111. Rascol O, Fabre N, Blin O, Poulik J, Sabatini U, Senard JM, et al. Naltrexone, an opiate antagonist, fails to modify motor symptoms in patients with Parkinson's disease. Mov Disord. 1994;9(4):437–40.
112. Cox H, Togasaki DM, Chen L, Langston JW, Di Monte DA, Quik M. The selective kappa-opioid receptor agonist U50,488 reduces L-dopa-induced dyskinesias but worsens parkinsonism in MPTP-treated primates. Exp Neurol. 2007;205(1):101–7.
113. Ikeda K, Yoshikawa S, Kurokawa T, Yuzawa N, Nakao K, Mochizuki H. TRK-820, a selective kappa opioid receptor agonist, could effectively ameliorate L-DOPA-induced dyskinesia symptoms in a rat model of Parkinson's disease. Eur J Pharmacol. 2009;620(1–3):42–8.
114. Land BB, Bruchas MR, Lemos JC, Xu M, Melief EJ, Chavkin C. The dysphoric component of stress is encoded by activation of the dynorphin kappa-opioid system. J Neurosci. 2008;28(2):407–14.
115. Samadi P, Bedard PJ, Rouillard C. Opioids and motor complications in Parkinson's disease. Trends Pharmacol Sci. 2006;27(10):512–7.
116. Samadi P, Gregoire L, Bedard PJ. The opioid agonist morphine decreases the dyskinetic response to dopaminergic agents in parkinsonian monkeys. Neurobiol Dis. 2004;16(1):246–53.
117. Berg D, Becker G, Reiners K. Reduction of dyskinesia and induction of akinesia induced by morphine in two parkinsonian patients with severe sciatica. J Neural Transm. 1999;106(7–8):725–8.
118. Wagner KJ, Valet M, Kochs EF, Kriner M, Tolle TR, Sprenger T. The mu-opioid receptor agonist remifentanil induces acute dysphoria irrespective of its analgesic properties. J Psychopharmacol. 2010;24(3):355–61.
119. Hille CJ, Fox SH, Maneuf YP, Crossman AR, Brotchie JM. Antiparkinsonian action of a delta opioid agonist in rodent and primate models of Parkinson's disease. Exp Neurol. 2001;172(1):189–98.
120. Johnston TH, Fox SH, Gomez-Ramirez J, Lee J, Brotchie JM, editors. Anti-parkinsonian effects of delta opioid receptor stimulation are accompanied by dystonia in MPTP-lesioned non-human primates previously treated with L-DOPA. 8th International Congress of Parkinson's Disease and Movement Disorders. Rome: Movement Disorder Society; 2004.
121. Marti M, Trapella C, Viaro R, Morari M. The nociceptin/orphanin FQ receptor antagonist J-113397 and L-DOPA additively attenuate experimental parkinsonism through overinhibition of the nigrothalamic pathway. J Neurosci. 2007;27(6):1297–307.

122. Visanji NP, de Bie RM, Johnston TH, McCreary AC, Brotchie JM, Fox SH. The nociceptin/orphanin FQ (NOP) receptor antagonist J-113397 enhances the effects of levodopa in the MPTP-lesioned nonhuman primate model of Parkinson's disease. Mov Disord. 2008;23(13):1922–5.
123. Marti M, Rodi D, Li Q, Guerrini R, Fasano S, Morella I, et al. Nociceptin/orphanin FQ receptor agonists attenuate L-DOPA-induced dyskinesias. J Neurosci. 2012;32(46):16106–19.
124. Martin WR. History and development of mixed opioid agonists, partial agonists and antagonists. Br J Clin Pharmacol. 1979;7 Suppl 3:273S–9.
125. Chang KJ, Pendergast W, Holt J. Method of treating parkinson's disease with diarylmethylpiperazine compounds exhibiting delta receptor agonist activity. Google Patents; 2007.
126. Braithwaite S, Mouradian M, Voronkov M. Method of treating dyskinesia. Google Patents; 2012.
127. Dumas EO, Pollack GM. Opioid tolerance development: a pharmacokinetic/pharmacodynamic perspective. AAPS J. 2008;10(4):537–51.
128. Kenakin T. Functional selectivity through protean and biased agonism: who steers the ship? Mol Pharmacol. 2007;72(6):1393–401.
129. Kenakin T, Christopoulos A. Signalling bias in new drug discovery: detection, quantification and therapeutic impact. Nat Rev Drug Discov. 2013;12(3):205–16.
130. Schmid CL, Streicher JM, Groer CE, Munro TA, Zhou L, Bohn LM. Functional selectivity of 6′-guanidinonaltrindole (6′-GNTI) at kappa-opioid receptors in striatal neurons. J Biol Chem. 2013;288(31):22387–98.
131. Zhou L, Lovell KM, Frankowski KJ, Slauson SR, Phillips AM, Streicher JM, et al. Development of functionally selective, small molecule agonists at kappa opioid receptors. J Biol Chem. 2013;288(51):36703–16.
132. Bruchas MR, Land BB, Aita M, Xu M, Barot SK, Li S, et al. Stress-induced p38 mitogen-activated protein kinase activation mediates kappa-opioid-dependent dysphoria. J Neurosci. 2007;27(43):11614–23.
133. Bausch SB, Garland JP, Yamada J. The delta opioid receptor agonist, SNC80, has complex, dose-dependent effects on pilocarpine-induced seizures in Sprague-Dawley rats. Brain Res. 2005;1045(1–2):38–44.
134. Nozaki C, Le Bourdonnec B, Reiss D, Windh RT, Little PJ, Dolle RE, et al. delta-Opioid mechanisms for ADL5747 and ADL5859 effects in mice: analgesia, locomotion, and receptor internalization. J Pharmacol Exp Ther. 2012;342(3):799–807.

Chapter 13
Glutamate Receptors and Levodopa-Induced Dyskinesia

Barbara Picconi and Paolo Calabresi

Abstract Levodopa is considered the therapy of choice for Parkinson's disease (PD) treatment. After the early phases of the disease, in which levodopa treatment is highly effective against parkinsonian symptoms, uncontrolled motor fluctuations and abnormal movements named levodopa-induced dyskinesia (LID) appears. An efficient anti-parkinsonian/antidyskinetic therapy has not so far been developed. Altered glutamatergic transmission is one of the main pathophysiological features of LID within basal ganglia circuit. Experimental evidence shows that the trafficking and the localization of the glutamate ionotropic (NMDA and AMPA) and metabotropic receptors in the synaptic cleft appear to have a relevant role in the pathogenesis of LID. Glutamate receptors have therefore been considered as potential targets for a novel pharmacological intervention in PD and LID treatment. Here we report an overview from the main preclinical studies in experimental models of PD and LID to the most recent clinical trials in PD patients describing the pros and cons of the use of glutamate receptor agents.

Keywords Levodopa-induced dyskinesia • Parkinson's disease • NMDA receptors • AMPA receptors • mGluR receptors

Molecular Mechanisms Underlying LID Induction

The mechanisms underlying the induction of levodopa-induced dyskinesia (LID) could have both central and peripheral origins; one of the possible explanations is that the nonphysiological, pulsatile stimulation of the postsynaptic

B. Picconi
Laboratorio di Neurofisiologia, Fondazione Santa Lucia, I.R.C.C.S. via Del Fosso di Fiorano 64, Rome 00143, Italy

P. Calabresi (✉)
Laboratorio di Neurofisiologia, Fondazione Santa Lucia, I.R.C.C.S. via Del Fosso di Fiorano 64, Rome 00143, Italy

Clinica Neurologica, Facoltà di Medicina e Chirurgia, Università degli Studi di Perugia, Ospedale S. Maria della Misericordia, S. Andrea delle Fratte, Perugia 06156, Italy
e-mail: paolo.calabresi@unipg.it

S.H. Fox, J.M. Brotchie (eds.), *Levodopa-Induced Dyskinesia in Parkinson's Disease*, DOI 10.1007/978-1-4471-6503-3_13

dopamine (DA) receptors can ultimately alter the physiological circuits within the basal ganglia causing an abnormal motor output that results in altered movements [1]. At the molecular and cellular levels, the functional interaction between dopaminergic and glutamatergic pathways within the striatum represents a key element in the pathophysiology of both parkinsonian symptoms and LID development [2–4].

Experimental models of LID in rodents help to understand the effect of specific receptors and molecular signaling pathways underlying the development and expression of dyskinetic movements [5]. Moreover, parkinsonian nonhuman primates (NHP) provide a useful model as they share several clinical features of PD and LID with patients [6]. Nevertheless, clinical studies designed to assess risk factors for the development of dyskinesia in patients with PD provide the most useful information toward understanding the cellular mechanisms of this disabling disorder [7].

A combination of pre and postsynaptic maladaptive changes is needed for parkinsonian animals to develop dyskinesia [7, 8]. During the progressive loss of nigrostriatal neurons, dopaminergic sprouting and reduced DA uptake contribute to preserve intrastriatal DA levels [9] and result in an alteration of glutamatergic control [10–13]. Both dopaminergic decrease and uncontrolled replacement, by the use of levodopa, might contribute to the secondary alterations in the glutamatergic transmission within the basal ganglia circuits. Presynaptic adaptive changes together with the changes occurring in postsynaptic DA receptors, such as alteration in the receptors sensitivity and density, lead to an altered substrate in which levodopa exerts its actions. Thus, initially, levodopa is converted into DA, stored in synaptic vesicles and released by surviving DA-releasing terminals. However, when degeneration advances, DA catabolism and uptake are reduced and decarboxylation of levodopa to DA and release occur in non-dopaminergic cells causing a failure in the buffering of DA levels [14–16].

Consequently, during chronic levodopa treatment, dopaminergic control on glutamatergic transmission is progressively deregulated resulting in altered synaptic localization and function of glutamate receptors.

Glutamatergic Transmission in PD and LID

Several experimental studies using PD animal models show the amelioration of parkinsonian motor symptoms when glutamatergic transmission is normalized [17]. Accordingly, in parkinsonian rats a dramatic increase of corticostriatal glutamatergic activity has been shown [18–20]. Targeting specifically this glutamatergic hyperactivity could be beneficial for the treatment of PD and LID [18, 20–23]. This increased glutamatergic transmission influences both pre- and postsynaptic events. The postsynaptic events influencing the trafficking and the localization of the glutamate ionotropic and metabotropic receptors in the synaptic cleft appear to have a relevant role in the pathogenesis of LID [24].

Metabotropic Glutamate Receptors, mGluR Families

Among the various strategies investigated in the last years to either reduce or prevent LID, the use of drugs targeting metabotropic glutamate receptors (mGluRs) revealed promising preclinical and clinical aspects. In particular, various mGluRs interact with key molecular steps implicated in the pathophysiology of LID: DA receptor activation and release, modulation of A2a adenosine receptors, regulation of N-methyl-D-aspartate (NMDA) receptor function, as well as glutamate and GABA release. Metabotropic glutamate receptors are characterized by the coupling of their subunits to second messenger systems through G-proteins and belong to the large family of G-protein-coupled receptors (GPCRs) [25]. The family of mGluR receptors is characterized by an extracellular N-terminal domain, a heptahelical domain, and an intracellular C-terminal tail variable in length. At least 8 mGluR subtypes are known, each with its splicing variants; these are classified into 3 distinct mGluR groups: mGluR group I, including mGluR1 and 5; mGluR group II, including mGluR2 and 3; and mGluR group III, including mGluR4, 6, 7, and 8. Group I mGluRs, positively coupled to phospholipase C (PLC), increases intracellular ion calcium (Ca^{2+}) levels by releasing it from intracellular stores, stimulates IP_3/Ca^{2+}/PKC pathway, potentiates L-type voltage-dependent Ca^{2+} channels (VDCCs), and inhibits K^+ conductances. Groups II and III mGluRs, negatively coupled to adenylyl cyclase (AC), exert an inhibitory action on voltage-dependent calcium channels (VDCCs). Their stimulation causes a presynaptic inhibitory action on the release of several neurotransmitters, including glutamate [17]. Many selective mGluRs agonists and antagonists are available, and ligands for specific mGluR subtypes could be potentially therapeutically efficient for PD/LID symptoms, but the strong binding at the highly conserved glutamate site in the N-terminus exerted by these drugs can produce adverse effects related to the generalized activation of the receptor. Moreover, seminal electrophysiological studies have shown that the classical antagonists of group I mGluRs reduce DA-dependent striatal synaptic plasticity (both long-term depression, LTD, and long-term potentiation, LTP) raising the possibility that their clinical use might induce detrimental effects on striatal-dependent motor and cognitive activity [26–28]. Many efforts have been devoted into the synthesis of new allosteric modulators of the different mGluR classes, in the attempt to fine-tune activity at the receptor. For example, the possible therapeutic use of negative allosteric modulators of group I or positive allosteric modulators of groups II and III receptors has been proposed [24, 29].

Ionotropic Glutamate *N*-Methyl-D-Aspartate (NMDA) Receptors

In the last decades, N-methyl-D-aspartate (NMDA) receptors have emerged as one of the key elements of the glutamatergic synaptic transmission in both physiological and pathological conditions [4, 30–32]. From different experimental approaches, the pathophysiological picture that has developed posits that the strength of corticostriatal NMDA-mediated glutamatergic signals might be dynamically regulated during PD progression. In fact, bidirectional changes of corticostriatal synaptic plasticity are critically controlled by the different degrees of nigral denervation and by the differential assembly of striatal NMDA receptor subunits [4, 33, 34].

NMDA receptor consists in a heteromeric molecule structured by the assembling between three different subunits, GluN1, GluN2 and GluN3, each of which contain several splicing variants: GluN1 subunit presents eight variants, GluN2 subunit presents four variants GluN2A-D, and GluN3 have two different forms. The functional receptor is composed by two GluN1 subunits and two GluN2 subunits [35]. The subunits GluN1 are needed for the functionality of the receptor and for the binding of the co-agonist glycine. The selectivity of the NMDA receptor for the Mg^{+} block and for Ca^{2+} permeability is dependent on a specific residue located in the pore loop of these subunits. GluN2B subunits receive the binding of the agonist glutamate and have a critical role in several pharmacological properties of the receptor complex such as the sensitivity to the Zn^{2+}, the conductance and kinetics of the single channel. The functional properties and the trafficking mechanisms of the NMDA receptor complex depend on the phosphorylation state of the single subunits [36]. Altered NMDA receptor composition and trafficking *in* and *out* of the postsynaptic density (PSD) compartment reveal an interesting role in many pathological conditions, and it have been extensively studied in the last decades [2, 30, 37]. In the PSD, NMDA receptors are clustered with several scaffolding cytoskeletal and signaling proteins (membrane-associated guanylyl kinases, MAGUK proteins) in close contact with the large pools of subunits retained within the endoplasmic reticulum (ER) [38, 39]. This accumulation of NMDA receptors in the postsynaptic compartment ensures a rapid response to neurotransmitter release and provides a molecular mechanism for linking the transmembrane ion flux to the signaling machinery responsible for specific second messenger pathways. Among the protein complexes governing the response of the signaling cascade, α-calcium-calmodulin-dependent protein kinase II (α-CaMKII) is directly linked to the GluN2A and GluN2B subunits [40, 41] and competes in NR2A binding with PSD-95 [38]. Interestingly, CaMKII- and tyrosine-dependent phosphorylation of NMDA receptors has been shown altered in experimental model of PD [13, 42]. In the striatum, as well as in other brain areas, one form of synaptic plasticity, LTP, which is lost in parkinsonian animals, requires NMDA receptors activation [43–46]. Interestingly, the NMDA receptor complex, and in particular its subunit composition, is profoundly altered in experimental PD [47, 48].

Ionotropic Alpha-Amino-3-hydroxy-5-methyl-4-Isoxazolepropionic Acid (AMPA) Receptors

Ionotropic alpha-amino-3-hydroxy-5-methyl-4-isoxazolepropionic acid (AMPA) receptors are tetramer receptors composed by a combination of four subunits GluA1, GluA2, GluA3, and GluA4 [49]. All the AMPA receptor subunits are composed of an extracellular N-terminal domain (NTD), two S1 and S2 ligand-binding domains, four membrane spanning domains (M1–M4), and two C terminus regions. The subunits GluA1 to GluA4 can form both homo- and heteromers receptor complex. Upon binding with glutamate in the PSD compartment, synaptic AMPA receptors induce membrane depolarization and consequently remove Mg^{+} block from NMDA receptor channel, reducing the threshold to induce long-term increases of the synaptic responses. AMPA receptor-dependent depolarization also opens L-type Calcium (Ca^{2+}) channels and leads to activation of CRE elements that, binding to specific promoter regions, are responsible for gene transcription. Evidence showing a possible role of the AMPA receptor in PD and dyskinesia is provided by the high levels of radioligand binding activity for this receptor found in NHP and in the putamen of parkinsonian patients displaying LID compared to non-dyskinetic subjects [50, 51].

Although, a recent paper by Lee and coworkers [52] suggests a putative role of AMPA receptors in parkinsonian state, there is still no general consensus on the mechanism underlying dysregulation of AMPA receptor subcellular distribution in PD or their pathological changes in subunit composition. No changes in AMPA GluR2, GluR3, and GluR4 mRNA levels in the globus pallidus or GluR1-4 levels in the striatum in PD patients and in 6-hydroxydopamine (6-OHDA)-lesioned rats, compared with their respective controls, have been found [53]; however, GluR1 immunoreactivity was found to be increased in the caudate and putamen of MPTP-lesioned NHPs [54]. Recently, Ba and colleagues found a significant decrease in the abundance of both serine-831-phosphorylated GluR1 and total GluR1 in striatal neuronal membrane from parkinsonian rats. Chronic levodopa treatment induced an upregulation of the serine-831-phosphorylation state of GluR1 confined to parvalbumin-positive interneurons where GluR1 subunits are exclusively expressed [55]. GluR1 levels in a triton-insoluble synaptic fraction, representing the isolated PSD compartment, presumably mainly of striatal projection neurons, were not altered in 6-OHDA-lesioned rats [13]. Although controversial, these data indicate that modified AMPA receptor-mediated transmission in the basal ganglia network could play a critical contribution to the motor symptoms of PD.

Targeting mGluR Receptors in LIDs: Evidence from Preclinical to Clinical Studies

Preclinical and clinical studies focused on metabotropic receptors has found increased binding for the mGluR5 receptors in dyskinetic MPTP-treated NHP and in parkinsonian, dyskinetic patients [56, 57]. The use of mGluR5 antagonists has been proposed as a good antidyskinetic approach in both rodent and NHP models of PD [29, 58]. Recent studies analyzed cellular and plastic changes occurring during the reduction of LID following the administration of the high selective mGluR5 antagonist, 3-[(2-methyl-1,3-thiazol-4-yl)ethynyl]pyridine (MTEP) [59–61]. Moreover, this drug has shown an ability to prevent the gene induction of striatal pro-dynorphin and pre-proenkephalin, dyskinesiogenic molecular markers [62, 63]. Two different mGluR5 antagonists significantly reduced LID induction without interfering with the physiological motor activities and resulted in the attenuation of known molecular markers of LID [59, 61, 64]. Recently a new allosteric modulator of mGluR5, fenobam, has been studied in PD and LID [65]. Fenobam was previously utilized as non-benzodiazepine, anxiolytic compound [66]. Using rodent and NHP PD models the study confirmed that acute administration of fenobam attenuates LID in both experimental models. This paper confirms that the chronic administration of fenobam decreased the development of peak-dose LID without compromising the anti-parkinsonian effect of levodopa and prolonged its effect [65].

Despite the data supporting the possible antidyskinetic use of mGluR5 antagonists, their translational to the clinic presents difficulties. Recently, the development of tolerance following the chronic administration of mGluR antagonists has been reported [67]. Additionally, a possible new antidyskinetic mGluR5 antagonist, AFQ056, has been described [68]. AFQ056 reduced LID following an acute administration, while the anti-parkinsonian efficacy of levodopa treatment was maintained at a high dose of AFQ056 and increased at suboptimal doses without apparent side effects. Thus, it was suggested that this novel mGluR5 antagonist might have the potential to be the first drug to be approved for the treatment of LID in PD.

Recently two randomized, double-blind, placebo-controlled studies showed an antidyskinetic effect of AFQ056 at 2 weeks follow-up in a small group of PD patients with LID [69].

In the first study, the authors included PD patients with moderate to severe LID (study 1) and in the second study those with severe LID (study 2) on stable dopaminergic therapy. Treatment with 25–150 mg AFQ056 or placebo was administered twice daily for 16 days. Although encouraging antidyskinetic effects have been obtained in AFQ056-treated patients, compared to placebo, serious adverse events were reported in both studies. Regarding tolerability, the most common side effects were represented by dizziness, hallucination and illusions, fatigue, nasopharingitis, diarrhea, and insomnia. When analyzing the incidence of adverse effects by the final actual dose, there was a possible dose-dependent increase of illusions.

Overall these studies support a potential antidyskinetic effect of higher doses of AFQ056. However, this improvement was not associated with a reduction of patient disability considering some of the severe side effects reported. Further studies involving larger group of patients and longer follow-up period are needed to confirm the possible antidyskinetic effect and to explore strategies to lower aversive effects of this compound.

Targeting NMDA Receptor Subunits in PD and LID: Evidence from Preclinical to Clinical Studies

NMDA receptor represented for several years the main target for the development of antidyskinetic drugs. In animal models of dyskinesia the NMDA receptor antagonist amantadine has been studied [70–72]. Although, some promising antidyskinetic effects, the beneficial response of different NMDA receptor antagonists appears weak. Recent studies have investigated the specific role of the single subunits of the NMDA receptor complex. The first candidate studied dealing with dyskinesia was the GluN2B subunit. Findings, both in parkinsonian and dyskinetic animals, showed a possible important pathogenetic role of this subunit [50, 73–76]. High striatal levels of tyrosine phosphorylation of the GluN2B subunit have been observed in dyskinetic animals [75, 77], causing a disruption of the interaction between the GluN2B-containing NMDARs and the endocytic complex and leading to the altered stabilization and efficacy of receptors on the cell surface [78, 79]. The pharmacological antagonism of GluN2B has been extensively studied in LID with contrasting results. Dyskinetic movements in NHP were significantly reduced after the application of Co101244 and CI-1041, two selective NR2B antagonists [80–83]. Conversely the use of another GluN2B antagonist CP-101,606 exacerbated LID in dyskinetic rats and marmoset [84, 85]. Moreover, although the use of the GluN2B antagonists prevented L-DOPA-induced wearing-off fluctuations [86], it failed to ameliorate LID [61].

Recently, several studies analyzed the possible role of an altered trafficking of GluN2A and/or GluN2B subunits between synaptic and extra-synaptic membranes and in and out the site on the surface of the plasma membrane [2, 3]. Decreased levels of GluN1 and GluN2B subunits have been measured in striatal membranes of parkinsonian animals, while the abundance of GluN2A was unchanged [47, 48]. Further studies, performed utilizing the 6-OHDA parkinsonian rat model, showed alterations in striatal synaptic plasticity such as LTD, LTP, and changes in the ratio between GluN2A/N2B subunits [13, 33, 87]. In particular, the NR2B subunit was found to be specifically reduced in the PSD compartment from advanced parkinsonian rats when compared to control rats and without alterations of GluN1 and GluN2A [13, 33, 87]. Interestingly, MPTP parkinsonian NHP presented massive changes in the total homogenate levels of striatal NMDA receptor proteins, such as decreased levels of GluN1 and GluN2B [12]. Moreover, in parkinsonian rats the alteration of NMDA receptor subunit localization in the PSD compartment is accompanied by a decreased recruitment of PSD-95 to GluN2A–N2B subunits. These events are paralleled by an increased activation of the pool of α-CaMKII associated with the NMDA receptor complex [13]. Moreover, decreased synaptic membrane localization and increased vesicular localization of PSD-95 and SAP97, members of the PSD-MAGUK family, have been reported [88]. This different redistribution of MAGUK components in the synaptic site could account for dysregulation of NMDA receptors at synapses of PD animals. While, in advanced PD, LTP is completely lost and this synaptic alteration is coupled to a specific reduction of the GluN2B subunits in the PSD compartment [87], the picture found in the early phases of the disease is quite different. A partial dopaminergic depletion strongly alters the LTP maintenance, and this synaptic alteration is also accompanied by a dramatic increase in the GluN2A NMDA receptor subunits in the striatal synapses, suggesting the presence of a profound rearrangement of the receptor complex composition [89]. These profound differences in NMDA receptors in the postsynaptic compartment of partially versus fully lesioned rats suggest that GluN2-type regulatory subunits are sensitive to plastic changes induced by the differential degree of

DA denervation. The GluN2A subunit might represent a major player in early phases of PD, and it seems to be sensitive to distinct degrees of DA denervation; thus, it may represent an adequate target for early therapeutic intervention. Overall, these results from dyskinetic animal models highlight that an altered ratio between GluN2A- and GluN2B-containing receptors may play an important pathophysiological role and reducing the GluN2A subunit localization to normalize the ratio GluN2A/N2B at the synapse may represent a mechanistic target for therapy [90].

To date, the only antidyskinetic therapeutic approach considered relatively efficacious and useful in the clinical practice is represented by the treatment with amantadine, a nonselective NMDA receptor antagonist. A series of double-blind, placebo-controlled studies conducted by Verhagen Metman and colleagues demonstrate the possible therapeutic use of NMDA antagonists such as amantadine and dextromethorphan on LID and motor fluctuations in PD [91–94]. In four separate trials, the use of three different NMDA receptors antagonists adjuvant to levodopa therapy reduced LID and motor fluctuations [92]. Treatment with amantadine reduced LID severity in PD patients compared to placebo group, without altering the anti-parkinsonian effect of levodopa [94–96]. These findings suggest that amantadine, given as adjuvant to levodopa, can markedly improve motor response complications and support the view that hyperfunction of NMDA receptors contributes to the pathogenesis of LID.

Contrasting data have been collected regarding another NMDA antagonist, memantine. Idiopathic PD patients with motor fluctuations and LID were randomized to be studied in a cross-over design for the NMDA antagonist memantine [97]. Unfortunately, this study showed no drug effect on dyskinesia. One other study showed that memantine can improve severe LID resistant to other pharmacologic interventions [98]. In a recent clinical trial, that studied for a longer time memantine, beneficial drug effects on parkinsonian symptoms and motor complications have been seen [99].

Recent results confirmed a relevant role of glutamate receptor supersensitivity in the putamen of dyskinetic condition following long-term levodopa therapy in PD. Radioligand binding studies conducted in PD patients experienced LID showed increased levels of binding sites to the GluN1/N2B-containing receptor in putamen nucleus compared to control subjects, whereas binding remained unchanged in the caudate nucleus [51].

Recently, Nutt and colleagues [100] analyzed in a randomized, double-blind, placebo-controlled clinical trial: the effect of CP-101,606 (GluN2B subunit selective NMDAR antagonist) versus placebo on the motor fluctuations and LID. This study showed a significant therapeutic effect of CP-101,606 on LID severity but without any improvement of parkinsonian symptoms. Moreover, the treatment with this GluN2B antagonist was associated with a dose-related dissociation and amnesia. Further studies will be necessary to find possible antidyskinetic drugs acting on the modulation of single GluN2 subunits with the aim to avoid adverse cognitive effects.

Targeting AMPA Receptors in PD and LID: Evidence from Preclinical Studies

Corticostriatal synapses containing ionotropic glutamate receptors such as AMPA and NMDA receptors have a relevant role in the capability to trigger both forms of striatal synaptic plasticity, LTD and LTP, in physiological and pathological conditions [43]. In particular, while LTP is NMDA dependent, LTD is dependent on the cooperation between dopaminergic signaling and AMPA receptors activation. Both LTD and LTP are lost in parkinsonian rats and are differentially altered in dyskinetic condition [33, 101]. Abnormal PKA signaling via high levels of phospho-Thr34–DARPP-32 have been found in the striatum of dyskinetic rats, and this has been associated with a loss of bidirectional synaptic plasticity [33]. The phosphorylation of DARPP-32 at Thr34 does not affect per se AMPA GluR1 subunit phosphorylation another target of PKA kinases. Recently, Santini and colleagues studied PKA/DARPP-32 signaling and their effects on NMDA and AMPA subunits in a model of LID both in mouse and NHP [102, 103]. These works confirmed that LID were associated with hyperphosphorylation of DARPP-32 at Thr34 and consequently of GluR1 AMPA receptor at Ser845. Increasing evidence suggests that antagonism of AMPA receptors improves LID in MPTP NHP [70]. The highly potent and selective-specific antagonist ligand for AMPA receptors, [(3)H]Ro 48–8587, reveals high levels of AMPA receptors in dyskinetic NHP, but only in specific subregional areas of the striatum [104]. An altered trafficking of AMPA receptors subunits GluR2 and GluR2/3 has been recently implicated in the development of LID in a model of dyskinetic NHP [105].

A strong support to the role of both NMDA and AMPA receptor trafficking alteration in the development of LID has arisen from the work by Bagetta and colleagues [106]. This study demonstrates that parkinsonian symptoms and LID were associated with an altered NMDA/AMPA receptor ratio and, in particular, a switch of AMPAR subunit composition within glutamatergic synapses. In particular, Bagetta and colleagues demonstrated an increased index of rectification (IR) of AMPA current in striatal medium spiny neurons recorded from dyskinetic animals, suggesting a possible abnormal insertion of GluR2-lacking AMPARs at corticostriatal synapses. In fact, this rise of this parameter indicates that, in these conditions, a switch toward a nonlinear voltage–current relationship occurs within glutamatergic synapses and stems from a major availability of Ca^{2+}-permeable AMPA receptors (i.e., Ca^{2+}-permeable GluR2-lacking AMPARs). This result is in agreement with previous data demonstrating that dopaminergic lesion and dyskinetic condition are associated with increased activity of Ca^{2+}-permeable AMPAR due to hyperphosphorylation of GluR1 subunit [103].

Together, these data show an interesting potential role of AMPA receptor activity and trafficking in the development of PD and levodopa-induced complications. However, to date, no relevant clinical study showing benefit of AMPA antagonists has been reported.

References

1. Rascol O, Goetz C, Koller W, Poewe W, Sampaio C. Treatment interventions for Parkinson's disease: an evidence based assessment. Lancet. 2002;359:1589–98.
2. Gardoni F, Ghiglieri V, Luca M, Calabresi P. Assemblies of glutamate receptor subunits with post-synaptic density proteins and their alterations in Parkinson's disease. Prog Brain Res. 2010;183:169–82.
3. Ghiglieri V, Bagetta V, Pendolino V, Picconi B, Calabresi P. Corticostriatal plastic changes in experimental L-DOPA-Induced dyskinesia. Parkinsons Dis. 2012;2012:358176.
4. Sgambato-Faure V, Cenci MA. Glutamatergic mechanisms in the dyskinesias induced by pharmacological dopamine replacement and deep brain stimulation for the treatment of Parkinson's disease. Prog Neurobiol. 2012;96:69–86.
5. Cenci MA. L-DOPA-induced dyskinesia: cellular mechanisms and approaches to treatment. Parkinsonism Relat Disord. 2007;13:S263–7.
6. Jenner P. Molecular mechanisms of L-DOPA-induced dyskinesia. Nat Rev Neurosci. 2008;9:665–77.
7. Calabresi P, Filippo MD, Ghiglieri V, Tambasco N, Picconi B. Levodopa-induced dyskinesias in patients with Parkinson's disease: filling the bench-to-bedside gap. Lancet Neurol. 2010;9:1106–17.
8. Cenci MA, Konradi C. Maladaptive striatal plasticity in L-DOPA-induced dyskinesia. Prog Brain Res. 2010;183:209–33.
9. Lee J, Zhu WM, Stanic D, Finkelstein DI, Horne MH, Henderson J, et al. Sprouting of dopamine terminals and altered dopamine release and uptake in Parkinsonian dyskinaesia. Brain. 2008;131:1574–87.
10. Ahmed I, Bose SK, Pavese N, Ramlackhansingh A, Turkheimer F, Hotton G, et al. Glutamate NMDA receptor dysregulation in Parkinson's disease with dyskinesias. Brain. 2011;134:979–86.
11. Calabresi P, Mercuri NB, Sancesario G, Bernardi G. Electrophysiology of dopamine-denervated striatal neurons. Implications for Parkinson's disease. Brain. 1993;116:433–52.
12. Hallett PJ, Dunah AW, Ravenscroft P, Zhou S, Bezard E, Crossman AR, et al. Alterations of striatal NMDA receptor subunits associated with the development of dyskinesia in the MPTP-lesioned primate model of Parkinson's disease. Neuropharmacology. 2005;48:503–16.
13. Picconi B, Gardoni F, Centonze D, Mauceri D, Cenci MA, Bernardi G, et al. Abnormal Ca2+– calmodulin-dependent protein kinase II function mediates synaptic and motor deficits in experimental parkinsonism. J Neurosci. 2004;24:5283–91.
14. Carta M, Carlsson T, Kirik D, Bjorklund A. Dopamine released from 5-HT terminals is the cause of L-DOPA-induced dyskinesia in parkinsonian rats. Brain. 2007;130:1819–33.
15. Picconi B, Ghiglieri V, Calabresi P. L-3,4-dihydroxyphenylalanine-induced sprouting of serotonin axon terminals: a useful biomarker for dyskinesias? Ann Neurol. 2010;68:578–80.
16. Rylander D, Parent M, O'Sullivan SS, Dovero S, Lees AJ, Bezard E, et al. Maladaptive plasticity of serotonin axon terminals in levodopa-induced dyskinesia. Ann Neurol. 2010;68:619–28.
17. Gubellini P, Pisani A, Centonze D, Bernardi G, Calabresi P. Metabotropic glutamate receptors and striatal synaptic plasticity: implications for neurological diseases. Prog Neurobiol. 2004;74:271–300.
18. Calabresi P, Pisani A, Mercuri NB, Bernardi G. Heterogeneity of metabotropic glutamate receptors in the striatum: electrophysiological evidence. Eur J Neurosci. 1993;5:1370–7.
19. Greenamyre JT. Glutamatergic influences on the basal ganglia. Clin Neuropharmacol. 2001;24:65–70.
20. Gubellini P, Picconi B, Bari M, Battista N, Calabresi P, Centonze D, et al. Experimental parkinsonism alters endocannabinoid degradation: implications for striatal glutamatergic transmission. J Neurosci. 2002;22:6900–7.

21. Blandini F, Armentero MT. New pharmacological avenues for the treatment of L-DOPA-induced dyskinesias in Parkinson's disease: targeting glutamate and adenosine receptors. Expert Opin Investig Drugs. 2012;21:153–68.
22. Maccarrone M, Gubellini P, Bari M, Picconi B, Battista N, Centonze D, et al. Levodopa treatment reverses endocannabinoid system abnormalities in experimental parkinsonism. J Neurochem. 2003;85:1018–25.
23. Tang K, Low MJ, Grandy DK, Lovinger DM. Dopamine-dependent synaptic plasticity in striatum during in vivo development. Proc Natl Acad Sci U S A. 2001;98:1255–60.
24. Duty S. Targeting glutamate receptors to tackle the pathogenesis, clinical symptoms and levodopa-induced dyskinesia associated with Parkinson's disease. CNS Drugs. 2012;26:1017–32.
25. Schoepp D, Bockaert J, Sladeczek F. Pharmacological and functional characteristics of metabotropic excitatory amino acid receptors. Trends Pharmacol Sci. 1990;11:508–15.
26. Gubellini P, Saulle E, Centonze D, Bonsi P, Pisani A, Bernardi G, et al. Selective involvement of mGlu1 receptors in corticostriatal LTD. Neuropharmacology. 2001;40:839–46.
27. Gubellini P, Saulle E, Centonze D, Costa C, Tropepi D, Bernardi G, et al. Corticostriatal LTP requires combined mGluR1 and mGluR5 activation. Neuropharmacology. 2003;44:8–16.
28. Sung KW, Choi S, Lovinger DM. Activation of group I mGluRs is necessary for induction of long-term depression at striatal synapses. J Neurophysiol. 2001;86:2405–12.
29. Marino MJ, Conn PJ. Glutamate-based therapeutic approaches: allosteric modulators of metabotropic glutamate receptors. Curr Opin Pharmacol. 2006;6:98–102.
30. Gardoni F, Di Luca M. New targets for pharmacological intervention in the glutamatergic synapse. Eur J Pharmacol. 2006;545:2–10.
31. Picconi B, Ghiglieri V, Bagetta V, Barone I, Sgobio C, Calabresi P. Striatal synaptic changes in experimental parkinsonism: role of NMDA receptor trafficking in PSD. Parkinsonism Relat Disord. 2008;14:S145–9.
32. Picconi B, Piccoli G, Calabresi P. Synaptic dysfunction in Parkinson's disease. Adv Exp Med Biol. 2012;970:553–72.
33. Picconi B, Centonze D, Hakansson K, Bernardi G, Greengard P, Fisone G, et al. Loss of bidirectional striatal synaptic plasticity in L-DOPA-induced dyskinesia. Nat Neurosci. 2003;6:501–6.
34. Vastagh C, Gardoni F, Bagetta V, Stanic J, Zianni E, Giampa C, et al. N-methyl-D-aspartate (NMDA) receptor composition modulates dendritic spine morphology in striatal medium spiny neurons. J Biol Chem. 2012;287:18103–14.
35. Kutsuwada T, Kashiwabuchi N, Mori H, Sakimura K, Kushiya E, Araki K, et al. Molecular diversity of the NMDA receptor channel. Nature. 1992;358:36–41.
36. Chen BS, Roche KW. Regulation of NMDA receptors by phosphorylation. Neuropharmacology. 2007;53:362–8.
37. Groc L, Bard L, Choquet D. Surface trafficking of N-methyl-D-aspartate receptors: physiological and pathological perspectives. Neuroscience. 2009;158:4–18.
38. Gardoni F, Schrama LH, Kamal A, Gispen WH, Cattabeni F, Di Luca M. Hippocampal synaptic plasticity involves competition between Ca2+/calmodulin-dependent protein kinase II and postsynaptic density 95 for binding to the NR2A subunit of the NMDA receptor. J Neurosci. 2001;21:1501–9.
39. Kennedy MB. Signal-processing machines at the postsynaptic density. Science. 2000;290:750–4.
40. Gardoni F, Caputi A, Cimino M, Pastorino L, Cattabeni F, Di Luca M. Calcium/calmodulin-dependent protein kinase II is associated with NR2A/B subunits of NMDA receptor in postsynaptic densities. J Neurochem. 1998;71:1733–41.
41. Strack S, McNeill RB, Colbran RJ. Mechanism and regulation of calcium/calmodulin-dependent protein kinase II targeting to the NR2B subunit of the N-methyl-D-aspartate receptor. J Biol Chem. 2000;275:23798–806.

42. Oh JD, Vaughan CL, Chase TN. Effect of dopamine denervation and dopamine agonist administration on serine phosphorylation of striatal NMDA receptor subunits. Brain Res. 1999;821:433–42.
43. Calabresi P, Picconi B, Tozzi A, Di Filippo M. Dopamine-mediated regulation of corticostriatal synaptic plasticity. Trends Neurosci. 2007;30:211–9.
44. Calabresi P, Maj R, Pisani A, Mercuri NB, Bernardi G. Long-term synaptic depression in the striatum: physiological and pharmacological characterization. J Neurosci. 1992;12:4224–33.
45. Collingridge GL, Bliss TV. Memories of NMDA receptors and LTP. Trends Neurosci. 1995;18:54–6.
46. Malenka RC, Bear MF. LTP and LTD: an embarrassment of riches. Neuron. 2004;44:5–21.
47. Ulas J, Cotman CW. Dopaminergic denervation of striatum results in elevated expression of NR2A subunit. Neuroreport. 1996;7:1789–93.
48. Dunah AW, Standaert DG. Dopamine D1 receptor-dependent trafficking of striatal NMDA glutamate receptors to the postsynaptic membrane. J Neurosci. 2001;21:5546–58.
49. Traynelis SF, Wollmuth LP, McBain CJ, Menniti FS, Vance KM, Ogden KK, et al. Glutamate receptor ion channels: structure, regulation, and function. Pharmacol Rev. 2010;62:405–96.
50. Calon F, Morissette M, Ghribi O, Goulet M, Grondin R, Blanchet PJ, et al. Alteration of glutamate receptors in the striatum of dyskinetic 1-methyl-4-phenyl-1,2,3,6-tetrahydropyridine-treated monkeys following dopamine agonist treatment. Prog Neuropsychopharmacol Biol Psychiatry. 2002;26:127–38.
51. Calon F, Rajput AH, Hornykiewicz O, Bedard PJ, Di Paolo T. Levodopa-induced motor complications are associated with alterations of glutamate receptors in Parkinson's disease. Neurobiol Dis. 2003;14:404–16.
52. Lee CY, Lee CH, Shih CC, Liou HH. Paraquat inhibits postsynaptic AMPA receptors on dopaminergic neurons in the substantia nigra pars compacta. Biochem Pharmacol. 2008;76:1155–64.
53. Bernard V, Gardiol A, Faucheux B, Bloch B, Agid Y, Hirsch EC. Expression of glutamate receptors in the human and rat basal ganglia: effect of the dopaminergic denervation on AMPA receptor gene expression in the striatopallidal complex in Parkinson's disease and rat with 6-OHDA lesion. J Comp Neurol. 1996;368:553–68.
54. Betarbet R, Porter RH, Greenamyre JT. GluR1 glutamate receptor subunit is regulated differentially in the primate basal ganglia following nigrostriatal dopamine denervation. J Neurochem. 2000;74:1166–74.
55. Ba M, Kong M, Yang H, Ma G, Lu G, Chen S, et al. Changes in subcellular distribution and phosphorylation of GluR1 in lesioned striatum of 6-hydroxydopamine-lesioned and l-dopa-treated rats. Neurochem Res. 2006;31:1337–47.
56. Ouattara B, Gregoire L, Morissette M, Gasparini F, Vranesic I, Bilbe G, et al. Metabotropic glutamate receptor type 5 in levodopa-induced motor complications. Neurobiol Aging. 2011;32:1286–95.
57. Samadi P, Gregoire L, Morissette M, Calon F, Hadj Tahar A, Dridi M, et al. mGluR5 metabotropic glutamate receptors and dyskinesias in MPTP monkeys. Neurobiol Aging. 2008;29:1040–51.
58. Conn PJ, Battaglia G, Marino MJ, Nicoletti F. Metabotropic glutamate receptors in the basal ganglia motor circuit. Nat Rev Neurosci. 2005;6:787–98.
59. Levandis G, Bazzini E, Armentero MT, Nappi G, Blandini F. Systemic administration of an mGluR5 antagonist, but not unilateral subthalamic lesion, counteracts l-DOPA-induced dyskinesias in a rodent model of Parkinson's disease. Neurobiol Dis. 2008;29:161–8.
60. Mela F, Marti M, Dekundy A, Danysz W, Morari M, Cenci MA. Antagonism of metabotropic glutamate receptor type 5 attenuates l-DOPA-induced dyskinesia and its molecular and neurochemical correlates in a rat model of Parkinson's disease. J Neurochem. 2007;101:483–97.
61. Rylander D, Recchia A, Mela F, Dekundy A, Danysz W, Cenci MA. Pharmacological modulation of glutamate transmission in a rat model of L-DOPA-induced dyskinesia: effects on motor behavior and striatal nuclear signaling. J Pharmacol Exp Ther. 2009;330:227–35.

62. Aubert I, Guigoni C, Li Q, Dovero S, Bioulac BH, Gross CE, et al. Enhanced preproenkephalin-B-derived opioid transmission in striatum and subthalamic nucleus converges upon globus pallidus internalis in L-3,4-dihydroxyphenylalanine-induced dyskinesia. Biol Psychiatry. 2007;61:836–44.
63. Cenci MA, Lee CS, Bjorklund A. L-DOPA-induced dyskinesia in the rat is associated with striatal overexpression of prodynorphin- and glutamic acid decarboxylase mRNA. Eur J Neurosci. 1998;10:2694–706.
64. Morin N, Gregoire L, Morissette M, Desrayaud S, Gomez-Mancilla B, Gasparini F, et al. MPEP, an mGlu5 receptor antagonist, reduces the development of L-DOPA-induced motor complications in de novo parkinsonian monkeys: biochemical correlates. Neuropharmacology. 2013;66:355–64.
65. Rylander D, Iderberg H, Li Q, Dekundy A, Zhang J, Li H, et al. A mGluR5 antagonist under clinical development improves L-DOPA-induced dyskinesia in parkinsonian rats and monkeys. Neurobiol Dis. 2010;39:352–61.
66. Pecknold JC, McClure DJ, Appeltauer L, Wrzesinski L, Allan T. Treatment of anxiety using fenobam (a nonbenzodiazepine) in a double-blind standard (diazepam) placebo-controlled study. J Clin Psychopharmacol. 1982;2:129–33.
67. Gravius A, Dekundy A, Nagel J, More L, Pietraszek M, Danysz W. Investigation on tolerance development to subchronic blockade of mGluR5 in models of learning, anxiety, and levodopa-induced dyskinesia in rats. J Neural Transm. 2008;115:1609–19.
68. Gregoire L, Morin N, Ouattara B, Gasparini F, Bilbe G, Johns D, et al. The acute antiparkinsonian and antidyskinetic effect of AFQ056, a novel metabotropic glutamate receptor type 5 antagonist, in L-Dopa-treated parkinsonian monkeys. Parkinsonism Relat Disord. 2011;17:270–6.
69. Berg D, Godau J, Trenkwalder C, Eggert K, Csoti I, Storch A, et al. AFQ056 treatment of levodopa-induced dyskinesias: results of 2 randomized controlled trials. Mov Disord. 2011;26:1243–50.
70. Bibbiani F, Oh JD, Kielaite A, Collins MA, Smith C, Chase TN. Combined blockade of AMPA and NMDA glutamate receptors reduces levodopa-induced motor complications in animal models of PD. Exp Neurol. 2005;196:422–9.
71. Dekundy A, Lundblad M, Danysz W, Cenci MA. Modulation of L-DOPA-induced abnormal involuntary movements by clinically tested compounds: further validation of the rat dyskinesia model. Behav Brain Res. 2007;179:76–89.
72. Lundblad M, Andersson M, Winkler C, Kirik D, Wierup N, Cenci MA. Pharmacological validation of behavioural measures of akinesia and dyskinesia in a rat model of Parkinson's disease. Eur J Neurosci. 2002;15:120–32.
73. Chase TN, Oh JD. Striatal dopamine- and glutamate-mediated dysregulation in experimental parkinsonism. Trends Neurosci. 2000;23:S86–91.
74. Dunah AW, Wang Y, Yasuda RP, Kameyama K, Huganir RL, Wolfe BB, et al. Alterations in subunit expression, composition, and phosphorylation of striatal N-methyl-D-aspartate glutamate receptors in a rat 6-hydroxydopamine model of Parkinson's disease. Mol Pharmacol. 2000;57:342–52.
75. Hurley MJ, Jackson MJ, Smith LA, Rose S, Jenner P. Immunoautoradiographic analysis of NMDA receptor subunits and associated postsynaptic density proteins in the brain of dyskinetic MPTP-treated common marmosets. Eur J Neurosci. 2005;21:3240–50.
76. Oh JD, Russell DS, Vaughan CL, Chase TN. Enhanced tyrosine phosphorylation of striatal NMDA receptor subunits: effect of dopaminergic denervation and L-DOPA administration. Brain Res. 1998;813:150–9.
77. Quintana A, Melon C, Kerkerian-Le Goff L, Salin P, Savasta M, Sgambato-Faure V. Forelimb dyskinesia mediated by high-frequency stimulation of the subthalamic nucleus is linked to rapid activation of the NR2B subunit of N-methyl-D-aspartate receptors. Eur J Neurosci. 2010;32:423–34.
78. Dunah AW, Sirianni AC, Fienberg AA, Bastia E, Schwarzschild MA, Standaert DG. Dopamine D1-dependent trafficking of striatal N-methyl-D-aspartate glutamate receptors requires Fyn protein tyrosine kinase but not DARPP-32. Mol Pharmacol. 2004;65:121–9.

79. Lau CG, Zukin RS. NMDA receptor trafficking in synaptic plasticity and neuropsychiatric disorders. Nat Rev Neurosci. 2007;8:413–26.
80. Blanchet PJ, Calon F, Morissette M, Hadj Tahar A, Belanger N, Samadi P, et al. Relevance of the MPTP primate model in the study of dyskinesia priming mechanisms. Parkinsonism Relat Disord. 2004;10:297–304.
81. Blanchet PJ, Konitsiotis S, Whittemore ER, Zhou ZL, Woodward RM, Chase TN. Differing effects of N-methyl-D-aspartate receptor subtype selective antagonists on dyskinesias in levodopa-treated 1-methyl-4-phenyl-tetrahydropyridine monkeys. J Pharmacol Exp Ther. 1999;290:1034–40.
82. Hadj Tahar A, Gregoire L, Darre A, Belanger N, Meltzer L, Bedard PJ. Effect of a selective glutamate antagonist on L-dopa-induced dyskinesias in drug-naive parkinsonian monkeys. Neurobiol Dis. 2004;15:171–6.
83. Morissette M, Dridi M, Calon F, Hadj Tahar A, Meltzer LT, Bedard PJ, et al. Prevention of levodopa-induced dyskinesias by a selective NR1A/2B N-methyl-D-aspartate receptor antagonist in parkinsonian monkeys: implication of preproenkephalin. Mov Disord. 2006;21:9–17.
84. Quintana A, Sgambato-Faure V, Savasta M. Effects of L-DOPA and STN-HFS dyskinesiogenic treatments on NR2B regulation in basal ganglia in the rat model of Parkinson's disease. Neurobiol Dis. 2012;48:379–90.
85. Nash JE, Ravenscroft P, McGuire S, Crossman AR, Menniti FS, Brotchie JM. The NR2B-selective NMDA receptor antagonist CP-101,606 exacerbates L-DOPA-induced dyskinesia and provides mild potentiation of anti-parkinsonian effects of L-DOPA in the MPTP-lesioned marmoset model of Parkinson's disease. Exp Neurol. 2004;188:471–9.
86. Wessell RH, Ahmed SM, Menniti FS, Dunbar GL, Chase TN, Oh JD. NR2B selective NMDA receptor antagonist CP-101,606 prevents levodopa-induced motor response alterations in hemi-parkinsonian rats. Neuropharmacology. 2004;47:184–94.
87. Gardoni F, Picconi B, Ghiglieri V, Polli F, Bagetta V, Bernardi G, et al. A critical interaction between NR2B and MAGUK in L-DOPA induced dyskinesia. J Neurosci. 2006;26:2914–22.
88. Nash JE, Johnston TH, Collingridge GL, Garner CC, Brotchie JM. Subcellular redistribution of the synapse-associated proteins PSD-95 and SAP97 in animal models of Parkinson's disease and L-DOPA-induced dyskinesia. FASEB J. 2005;19:583–5.
89. Paille V, Picconi B, Bagetta V, Ghiglieri V, Sgobio C, Di Filippo M, et al. Distinct levels of dopamine denervation differentially alter striatal synaptic plasticity and NMDA receptor subunit composition. J Neurosci. 2010;30:14182–93.
90. Gardoni F, Sgobio C, Pendolino V, Calabresi P, Di Luca M, Picconi B. Targeting NR2A-containing NMDA receptors reduces L-DOPA-induced dyskinesias. Neurobiol Aging. 2012;33:2138–44.
91. Verhagen Metman L, Blanchet PJ, van den Munckhof P, Del Dotto P, Natte R, Chase TN. A trial of dextromethorphan in parkinsonian patients with motor response complications. Mov Disord. 1998;13:414–7.
92. Verhagen Metman L, Del Dotto P, Blanchet PJ, van den Munckhof P, Chase TN. Blockade of glutamatergic transmission as treatment for dyskinesias and motor fluctuations in Parkinson's disease. Amino Acids. 1998;14:75–82.
93. Verhagen Metman L, Del Dotto P, Natte R, van den Munckhof P, Chase TN. Dextromethorphan improves levodopa-induced dyskinesias in Parkinson's disease. Neurology. 1998;51:203–6.
94. Verhagen Metman L, Del Dotto P, van den Munckhof P, Fang J, Mouradian MM, Chase TN. Amantadine as treatment for dyskinesias and motor fluctuations in Parkinson's disease. Neurology. 1998;50:1323–6.
95. Snow BJ, Macdonald L, McAuley D, Wallis W. The effect of amantadine on levodopa-induced dyskinesias in Parkinson's disease: a double-blind, placebo-controlled study. Clin Neuropharmacol. 2000;23:82–5.

96. Del Dotto P, Pavese N, Gambaccini G, Bernardini S, Metman LV, Chase TN, et al. Intravenous amantadine improves levodopa-induced dyskinesias: an acute double-blind placebo-controlled study. Mov Disord. 2001;16:515–20.
97. Merello M, Nouzeilles MI, Cammarota A, Leiguarda R. Effect of memantine (NMDA antagonist) on Parkinson's disease: a double-blind crossover randomized study. Clin Neuropharmacol. 1999;22:273–6.
98. Hanagasi HA, Kaptanoglu G, Sahin HA, Emre M. The use of NMDA antagonist memantine in drug-resistant dyskinesias resulting from L-dopa. Mov Disord. 2000;15:1016–7.
99. Varanese S, Howard J, Di Rocco A. NMDA antagonist memantine improves levodopa-induced dyskinesias and "on-off" phenomena in Parkinson's disease. Mov Disord. 2010;25:508–10.
100. Nutt JG, Gunzler SA, Kirchhoff T, Hogarth P, Weaver JL, Krams M, et al. Effects of a NR2B selective NMDA glutamate antagonist, CP-101,606, on dyskinesia and Parkinsonism. Mov Disord. 2008;23:1860–6.
101. Picconi B, Bagetta V, Ghiglieri V, Paille V, Di Filippo M, Pendolino V, et al. Inhibition of phosphodiesterases rescues striatal long-term depression and reduces levodopa-induced dyskinesia. Brain. 2011;134:375–87.
102. Santini E, Sgambato-Faure V, Li Q, Savasta M, Dovero S, Fisone G, et al. Distinct changes in cAMP and extracellular signal-regulated protein kinase signalling in L-DOPA-induced dyskinesia. PLoS One. 2010;5:e12322.
103. Santini E, Valjent E, Usiello A, Carta M, Borgkvist A, Girault JA, et al. Critical involvement of cAMP/DARPP-32 and extracellular signal-regulated protein kinase signaling in L-DOPA-induced dyskinesia. J Neurosci. 2007;27:6995–7005.
104. Ouattara B, Hoyer D, Gregoire L, Morissette M, Gasparini F, Gomez-Mancilla B, et al. Changes of AMPA receptors in MPTP monkeys with levodopa-induced dyskinesias. Neuroscience. 2010;167:1160–7.
105. Silverdale MA, Kobylecki C, Hallett PJ, Li Q, Dunah AW, Ravenscroft P, et al. Synaptic recruitment of AMPA glutamate receptor subunits in levodopa-induced dyskinesia in the MPTP-lesioned nonhuman primate. Synapse. 2010;64:177–80.
106. Bagetta V, Sgobio C, Pendolino V, Del Papa G, Tozzi A, Ghiglieri V, et al. Rebalance of striatal NMDA/AMPA receptor ratio underlies the reduced emergence of dyskinesia during D2-like dopamine agonist treatment in experimental Parkinson's disease. J Neurosci. 2012;32:17921–31.

Chapter 14
Cannabinoids and Levodopa-Induced Dyskinesia

Andrea Giuffrida and Alex Martinez

Abstract The endocannabinoid system modulates the release of excitatory and inhibitory neurotransmitters in several brain areas implicated in motor control. Cannabinoid and dopamine receptors are highly abundant and often co-expressed in the basal ganglia circuitry, and the cross talk between these two systems regulates short- and long-term synaptic plasticity in the striatum. Dysregulation of the endocannabinoid system has been reported in animal models of Parkinson's disease and parkinsonian patients and is exacerbated in dyskinetic states, following chronic levodopa administration.

This chapter reviews recent investigations on the relationships between endocannabinoids and other neurotransmitter/neuromodulator systems in the basal ganglia, with the intent to underline their relevance for the pathophysiology of levodopa-induced dyskinesia and discuss new pharmacological approaches for their treatment.

Keywords Endocannabinoid • Anandamide • CB1 • Dopamine • Parkinson's disease • Levodopa • Striatum

A. Giuffrida (✉)
Department of Pharmacology,
University of Texas Health Science Center at San Antonio,
7703 Floyd Curl Drive,
San Antonio, TX 78229, USA
e-mail: giuffrida@uthscsa.edu

A. Martinez
Department of Biology, Trinity University, San Antonio, TX USA

Department of Pharmacology,
University of Texas Health Science Center at San Antonio, 7703 Floyd Curl Drive,
San Antonio, TX 78229, USA

S.H. Fox, J.M. Brotchie (eds.), *Levodopa-Induced Dyskinesia in Parkinson's Disease*, DOI 10.1007/978-1-4471-6503-3_14

Introduction

Although levodopa remains the gold standard for the treatment of motor symptoms in Parkinson's disease (PD), its long-term use leads to the development of abnormal involuntary movement, collectively termed dyskinesia, in as many as 90–95 % of PD patients receiving treatment [1–3].

The molecular mechanisms associated with LID development are not fully understood, but several factors, including neurotransmitter abnormalities, pulsatile stimulation of dopamine receptors, and maladaptive plasticity within the striatum, are known to play a role [4].

To date, the only FDA-approved drug for the treatment of dyskinesia is the NMDA antagonist amantadine [5, 6]. This drug, however, has a short therapeutic time window, is poorly tolerated, and can worsen dyskinesia upon discontinuation or induce psychiatric complications [5, 7]. Thus, there is an urgent need to develop alternative antidyskinetic therapies targeting non-dopaminergic systems to avoid possible interferences with the antiparkinsonian effects of L-DOPA.

In the last decade, several studies have pointed to the endocannabinoid system as an important modulator of synaptic transmission and plasticity in the basal ganglia circuitry. As this system regulates dopamine-induced motor activation and is required for the coordination and fine-tuning of movement [8, 9], it represents a potential pharmacological target for the treatment of motor disorders. Indeed, both exogenous and endogenous cannabinoids show antiparkinsonian and antidyskinetic activity in animal models of PD and patients.

In this chapter, we will review the most relevant studies on the role played by the endocannabinoid system in LID, discuss the complex interactions between endocannabinoids and several neurotransmitters regulating basal ganglia function [10, 11], and provide a conceptual frame to address some conflicting findings reported in the literature.

The Endocannabinoid System

The endocannabinoid system consists of a family of lipid signaling molecules (endocannabinoids) released on demand from membrane lipid precursors, the enzymes responsible for their synthesis and degradation and distinct metabotropic (cannabinoid), ionotropic, and nuclear receptors activated by these ligands [12, 13].

Among the multiple endocannabinoids identified so far [14], arachidonoyl ethanolamine (anandamide) [15, 16] and 2-arachidonoyl glycerol (2-AG) [17] represent the two most studied examples.

Anandamide is synthesized in a Ca^{++}-dependent manner from N-arachidonoyl phosphatidylethanolamine by phospholipase D (PLD) [18, 19] or via alternative pathways, such as those initiated by alpha-beta-hydrolase 4 [20]. 2-AG is produced by diacylglycerol lipases (DAGLα and β) acting on membrane acyl arachidonoyl glycerols [21, 22]. As in the case of anandamide, multiple biosynthetic pathways have been reported for 2-AG, which can also derive from the hydrolysis of phosphatidic acid or lysophospholipids [23, 24].

The biological actions of anandamide are terminated by facilitated diffusion into cells via a carrier-mediated transport [25], followed by enzymatic hydrolysis via a fatty acid amide hydrolase (FAAH) [26–28]. To date, there is no consensus on the existence of an endocannabinoid transporter [29], and its molecular identity has not been yet identified. Anandamide can be also metabolized by lipoxygenases [30] and cyclooxygenases (such as COX-2) [31–33]. In particular, the COX-2 metabolic pathway may become physiologically relevant under conditions promoting endocannabinoid or COX-2 upregulation, as in the course of neurodegenerative processes [34].

Concerning 2-AG, although this lipid can be metabolized by FAAH and cyclooxygenases [35, 36], in the brain it is mainly hydrolyzed by a monoacylglycerolipase (MAGL), which is localized in presynaptic elements [37]. Interestingly, pharmacological blockade of FAAH by URB597 may decrease brain 2-AG in vitro via a mechanism involving TRPV1 activation and DAGL inhibition [38, 39]. This decrease, however, has not been confirmed in vivo by other groups [40–43], suggesting that it might be limited to specific brain areas.

The endocannabinoids can activate $G_{i/o}$ protein-coupled cannabinoid receptors (CB1 and CB2), some members of the transient receptor potential (TRP) family, as well as nuclear peroxisome proliferator-activated receptors (PPAR) [44]. Endocannabinoids can also serve as allosteric modulators or bind to other metabotropic receptors, including GPR55 [45–47] – a cloned orphan receptor activated by the CB1 antagonists rimonabant and AM251 [48] – and GPR18 [49]. However, the physiological roles of these receptors remain unknown, and neither anandamide nor 2-AG has shown consistent pharmacological effects following GPR55 stimulation [50].

In rodents and humans, CB1 receptors are highly expressed in the peripheral and central nervous system (CNS) [51, 52], whereas CB2 receptors are mainly restricted to immunocompetent cells, lymphoid organs, and microglia [44, 53, 54]. The "segregation" of CB2 to the immune system has been challenged by recent studies showing their presence in neurons and glial cells throughout the brain, including the substantia nigra pars reticulata (SNpr) and the striatum [55–58]. Also, CB2 receptors are upregulated in activated microglia and astrocytes in response to neurotoxic insults and neuroinflammatory events [59–62].

Within the basal ganglia, CB1 receptors are generally expressed on presynaptic elements, including GABAergic striatofugal neurons [63, 64], striatal parvalbumin-positive interneurons [65, 66], glutamatergic terminals from the cortex [67] and the subthalamic nucleus [68], and serotonergic afferents [69, 70] (see Fig. 14.1). It is now well established that activation of presynaptic CB1 receptors by retrogradely mobilized endocannabinoids inhibits the release of several neurotransmitters involved in basal ganglia function [71, 72].

Endocannabinoids and CB1 receptors have been implicated in three main forms of plasticity at striatal synapses: (1) short-term depolarization-induced suppression

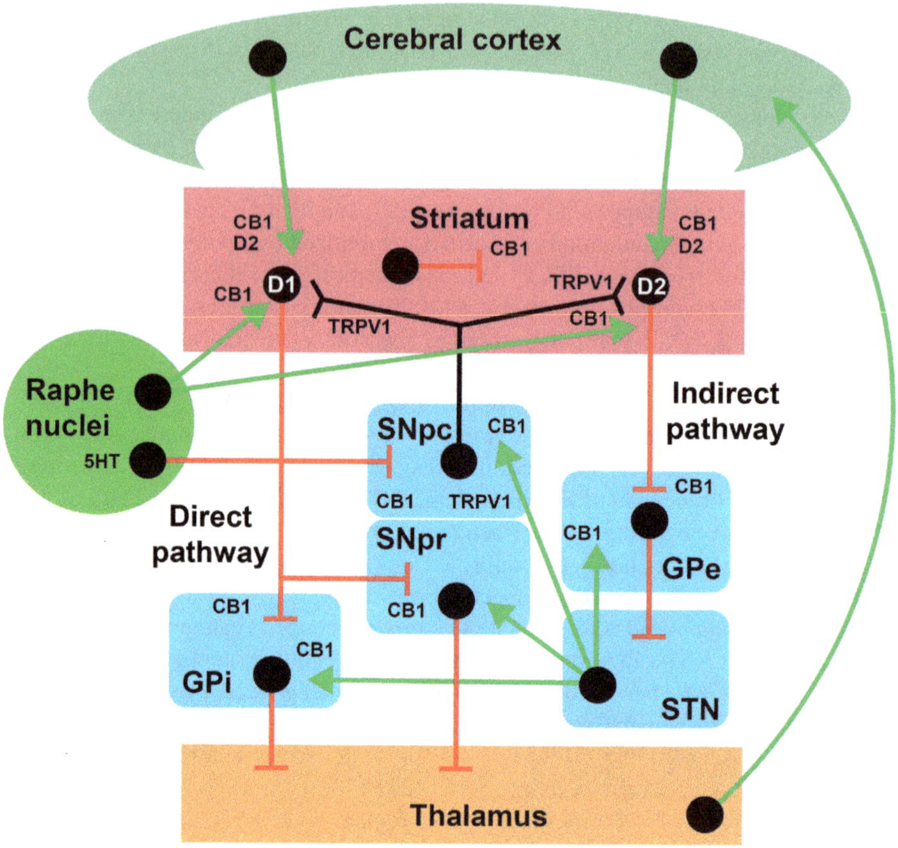

Fig. 14.1 Schematic illustration of the basal ganglia motor circuit showing striatofugal "direct" and "indirect" projections to the output nuclei and afferent projections from the cortex and raphe nuclei to the striatum. *GPi* globus pallidus pars interna, *GPe* globus pallidus pars externa, *SNpc* substantia nigra pars compacta, *SNpr* substantia nigra pars reticulata, *STN* subthalamic nucleus. Glutamatergic (*green*), GABAergic (*red*), and serotonergic projections and expression of different receptor subtypes are also indicated

of excitation (DSE) or inhibition (DSI), (2) short-term depression dependent by activation of postsynaptic Gq-coupled receptors, and (3) long-term depression (LTD) (for review, see [73]). Also, concomitant activation of CB1 and other metabotropic receptors can promote the coupling of CB1 to G isoforms other than $G_{i/o}$ [74, 75] or the formation of heterodimers with D2 and mu-opioid receptors [76, 77], leading to different downstream signaling pathways than those traditionally activated by cannabinoids.

Studies on cannabinoid agonists administered to CB1 knockout mice support the existence of non-CB1/CB2 receptors regulating synaptic transmission throughout the body (for review, see [52]).

As previously mentioned, some exogenous and endogenous cannabinoids can target at least five distinct TRP channels [78]. In particular, anandamide can bind to TRPV1 receptors [79], which are expressed in the striatum, globus pallidus, and the substantia nigra pars compacta (SNpc) [80–82]. As anandamide affinity for TRPV1 is quite low, it is not clear whether this lipid might serve as an endovanilloid ligand under physiological conditions [83, 84]. Nevertheless, blockade of FAAH activity has been shown to enhance anandamide potency at TRPV1 receptors in vitro [85]. In addition, there is evidence for a cross talk between CB1 and TRPV1 receptors, as CB1 stimulation can alter the phosphorylation state of TRPV1 and consequently its function [86].

Some cannabinoid compounds, including anandamide, noladin ether, virodhamine, and WIN55,212-2, can also bind different subtypes of PPAR receptors and enhance the expression of their target genes [87]. In particular, anandamide has been shown to activate □□□□ PPARα [88] and PPARγ □□□□□ [89]. These receptors, which are known to increase insulin sensitivity and modulate glucose and lipid metabolism, are also expressed in neuronal and glial cells of the basal ganglia [90, 91]. Although their role in the CNS is still largely unexplored, recent studies indicate that PPARα □□□□ and □□□□□ PPARγ agonists have antioxidant and neuroprotective activity in animal models of PD [92–95], Alzheimer's disease [96, 97], cerebral ischemia [98], and traumatic brain injury [99, 100], and they can reverse haloperidol-induced oral dyskinesia in rats [101].

Pharmacological Effects

Effects on PD Motor Symptoms

In general, systemic administration of exogenous cannabinoids, or enhancement of endocannabinoid tone via pharmacological blockade of their catabolic enzymes or reuptake, decreases locomotor activity in a CB1-dependent manner [28, 102–104]. In line with these observations, CB1 knockout mice exhibit motor abnormalities [105, 106] and suppression of cocaine-induced hyperlocomotion [107]. However, some of the cannabinoid-induced motor effects are not elicited via activation of CB1 receptors. For instance, anandamide produces catalepsy in both CB1 knockout mice and wild-type controls [108], and elevation of endocannabinoid tone in these animals produces hypokinesia via a TRPV1-mediated mechanism [109]. Also, pharmacological blockade of TRPV1 receptors in 6-OHDA rats has been shown to unmask the antidyskinetic effects of the FAAH inhibitor URB597 [103] (see below). These observations suggest that, under conditions in which anandamide reaches supraphysiological concentrations and consequently activates TRPV1, these channels can influence motor behaviors presumably by affecting the firing rate of nigrostriatal neurons and dopamine transmission [110].

In the context of PD, several research groups have found increased CB1 mRNA and receptor binding in the striatum of animal models [111, 112] and PD patients [113]. Numerous studies have also shown abnormal endocannabinoid levels, although there is no consensus on the direction of endocannabinoid fluctuations. While some reports indicate an increase of endocannabinoid levels in the basal ganglia of dopamine-depleted rodents [114–116], other studies showed decreased or unaltered endocannabinoid tone [103, 117, 118]. These discrepancies may be attributable to species-specific differences among PD models or to the physiological state of the animals at the time of the experiments, which is known to affect endocannabinoid release. Interestingly, the administration of levodopa to 6-OHDA rats failed to elevate anandamide levels [103, 117] and further increased CB1 expression in the striatum [119], suggesting that levodopa is not able to correct the endocannabinoid dysfunction associated with dopamine denervation. This dysfunction likely causes the disruption of the plasticity observed at corticostriatal synapses in PD models [118–121]. In this regard, elevation of endocannabinoid tone has been shown to rescue striatal LTD and to alleviate motor deficits associated with the nigrostriatal lesion [118]. Although these data point to a deficit (rather than an enhancement) of endocannabinoid mobilization in PD, improvement of motor symptoms has been achieved not only with administration of cannabinoid agonists but also with CB1 receptor antagonists in either rodents [122–124] or nonhuman primates [125] (Table 14.1). Explaining these paradoxical findings is challenging, although the answer may lie in the multiple site of actions engaged by cannabinoid drugs when administered systemically. Indeed, while increased endocannabinoid transmission may alleviate PD symptoms by reducing striatal glutamate release [71, 115], on the other hand, activation of CB1 on striatofugal terminals of the "indirect" pathway may lead to increased GABAergic drive to the external globus pallidus (GPe), which may amplify the inhibitory output of the basal ganglia and consequently contribute to PD symptoms. Therefore, in this case, CB1 antagonism may produce antiparkinsonian effects by limiting GABA release from striatopallidal projections. Finally, other studies have hypothesized that CB1 antagonists elicit antiparkinsonian effects only in animals with severe nigrostriatal lesions [123, 126], which may differentially affect endocannabinoid production and CB1 expression in the striatum and GPe of these animals versus those with less severe lesions.

Effects on LID

As endocannabinoids counteract dopamine-mediated hyperactivity [103, 136, 137] and given the fact that increased corticostriatal glutamate transmission contributes to dyskinesias [138, 139], stimulation of CB1 receptors should alleviate dyskinetic symptoms by (1) reducing levodopa-induced sensitization of dopamine receptors, (2) normalizing aberrant glutamate release, and (3) rebalancing maladaptive plasticity in

Table 14.1 Pharmacological effects of cannabinoid agents on PD motor symptoms and dyskinesia

Cannabinoid agent	Pharmacology	PD motor symptoms	Dyskinesia
THC (Cannabis)	CB1/CB2 agonist	Alleviate motor deficits in PD models [126] or no effect [122, 127]	No effect in PD patients [127]
Nabilone	CB1/CB2 agonist	Improve levodopa antiparkinsonian action [128]	Antidyskinetic in PD models [128]. Can reduce dyskinesia in PD patients [129]
URB597	FAAH inhibitor	Alleviate motor deficits in PD models [118]	Antidyskinetic in PD models in the presence of a TRPV1 blocker [103]
WIN55,212-2	CB1/CB2 agonist	Induce hypokinesia in rodents [103, 130]	Antidyskinetic in PD models [103, 117, 131]
HU-210	CB1/CB2 agonist	Impair motor function [132]	Alleviate some AIM subtypes [132]
SR141716A (rimonabant)	CB1 antagonist	Alleviate motor deficits in PD models [123–125] or no effect [122, 133]	Antidyskinetic in PD models [125] or no effect [117]. Can precipitate AIMs in non-dyskinetic animals [132]
AM251	CB1 antagonist	Alleviate motor deficits in PD models [123]	No effect [103, 132]
CE	CB1 antagonist	Enhance antiparkinsonian action of levodopa [134]	No effect [134]
Oleylethanolamide (OEA)	TRPV1 antagonist	No effect [135]	Antidyskinetic in PD models [135]

the denervated striatum. In support of this hypothesis, several groups have shown cannabinoid-mediated improvement of levodopa-induced abnormal involuntary movements (AIMs) in rodent models and nonhuman primates [103, 118, 128, 131, 132] and PD patients [129] (Table 14.1).

The antidyskinetic effects of cannabinoid agonists do not seem to result from a generalized motor suppression, as they were obtained using doses that did not produce hypomotility or catalepsy [103, 126]. Nevertheless, as in the case of PD motor deficits, significant antidyskinetic effects [125, 130], or no effects [134], were also observed with CB1 antagonists (Table 14.1). The rationale for blocking CB1 receptors as a pharmacological approach to treat dyskinesia is based on the observations that endocannabinoid transmission is elevated in dyskinetic animals [140] and PD patients [113] and that genetic deletion of CB1 receptors prevents the development of severe abnormal movements in mice [140]. However, neither striatal endocannabinoid levels nor CB1 upregulation has been correlated to LID expression or severity [125, 141].

Overall, these discrepancies reveal some limitations in generalizing cannabinoid effects across different animal models and may be ascribed to the multiple sites of action of cannabinoid agents (see above), which complicate the translation of these findings into new pharmacotherapies.

So far, studies carried out in PD patients have been inconclusive. While a randomized, double-blind, placebo-controlled pilot study by Sieradzan et al. [129] has shown an antidyskinetic action of the cannabinoid agonist nabilone in PD patients [129], other reports have not confirmed any beneficial effects of either cannabinoid agonists [127] or antagonists [133] on LID. However, the study of Carroll et al. [127] evaluated the effects of oral cannabis, which has a highly variable pharmacokinetics and a more complex pharmacological profile than synthetic cannabinoid agonists. In addition, the assessment of dyskinesia was based on patient self-reported questionnaires, which are often inaccurate in identifying symptoms [142]. On the other hand, the dose of the CB1 antagonist rimonabant used in the study of Mesnage et al. [133] was significantly lower than that used by van der Stelt and coworkers [125]. Thus, new and larger-scale clinical studies are necessary to confirm the antidyskinetic properties of cannabinoid agents in humans.

Pharmacological blockade of FAAH, which elevates anandamide and other acylethanolamides in those brain areas where they are actively synthesized, did not reduce levodopa-induced AIMs in 6-OHDA rats [103]. These findings suggest that increasing anandamide tone is not sufficient to alleviate dyskinesia, possibly because of the concomitant stimulation of CB1 and TRPV1 receptors, which exert opposite effects within the basal ganglia circuitry. In support of this hypothesis, coadministration of the FAAH inhibitor URB597 and the TRPV1 antagonist capsazepine produced a significant antidyskinetic effect in 6-OHDA rats [103, 131]. In addition, a recent study by Gonzalez-Aparicio [143] has shown that oleylethanolamide (OEA), a structural analog of anandamide that does not bind to CB1 but has antagonistic activity at TRPV1 receptors, can reduce levodopa-induced AIM via a TRPV1-mediated mechanism [135]. These observations differ from those reported by Lee et al. [143], showing that the administration of either URB597 or the TRPV1 agonist capsaicin alone reduced levodopa-induced hyperactivity in reserpine-treated rats [143]. However, it is important to note that hyperactivity in reserpine-treated rodents has not been validated as an appropriate measure of dyskinesia [144].

Although TRPV1 blockade seems necessary to unmask the antidyskinetic effect of URB597, the beneficial action of this drug is only partially mediated by CB1 receptors, since pretreatment with the CB1 antagonist AM251 did not fully reverse the combined effect of URB597 and capsazepine (CPZ) [103]. Interestingly, administration of the nonselective PPAR antagonist BADGE completely blocked the URB597+CPZ antidyskinetic effect (unpublished observations), suggesting a PPAR-dependent mechanism. Whether the involvement of PPAR in this response reflects a direct action of anandamide, or of other lipid signaling molecules elevated by FAAH blockade, on these nuclear receptors is still unclear. Nevertheless, a recent study has shown that PPARα □□□□□ and □□□□□ PPARγ agonists administered individually or in combination with antipsychotics can alleviate haloperidol-induced oral dyskinesias [101].

Endocannabinoid Modulation of Basal Ganglia Circuitry: Pathophysiology and Implications for LID

According to the classical model of basal ganglia organization (see Fig. 14.1), striatal MSN receive excitatory glutamatergic projections from the cerebral cortex. MSN are in turn modulated by nigrostriatal dopaminergic afferents that exert excitatory or inhibitory effects on "direct" and "indirect" striatofugal pathways via dopamine D1 and D2 receptors, respectively.

Although CB1 are not present on dopaminergic neurons [145], they co-localize with D1/D2-like receptors in the dorsal striatum and indirectly affect dopamine output by modulating neurotransmitter release from projecting inhibitory and excitatory terminals via stimulation of CB1 receptors [64, 67, 68, 72, 146, 147]. The overall effect of cannabinoids on dopamine release in the caudate-putamen remains controversial, as some studies have shown a decrease [72], an increase [148], or no effect at all [149, 150]. Anandamide- and endocannabinoid-enhancing drugs, such as FAAH inhibitors, can also modulate nigrostriatal dopamine transmission by acting at TRPV1 [109, 110, 151, 152] or PPAR receptors [153].

Stimulation of dopamine D1- and D2-like receptors has been shown to affect striatal endocannabinoids in opposite ways: for example, while D1 agonists tend to decrease anandamide [154], D2-like agonists increase it [103, 117, 136, 155]. These effects may depend on the ability of D1 and D2 agonists to enhance or diminish excitatory postsynaptic currents in striatal MSN, respectively, and suggest a dopamine-mediated control of endocannabinoid mobilization [156]. Indeed, studies have shown that LTD at corticostriatal synapses is regulated by D2 receptors [118, 157]. Although the precise site of this modulation is still the subject of debate, it appears to be restricted to glutamatergic projections onto MSN of the indirect pathway [118, 158] and to be mediated by anandamide or 2-AG, depending on the frequency of stimulation applied to the glutamatergic afferents [159–161].

Endocannabinoids, in particular anandamide, also mediate synaptic depression at GABAergic afferents onto striatal MSN [155, 162–164] to produce disinhibition of MSN activation.

Interestingly, endocannabinoid-mediated LTD at corticostriatal synapses is profoundly compromised after striatal dopamine denervation [118] or blockade of D2 receptors [157, 165, 166] and completely lost in dyskinetic – but not in non-dyskinetic – parkinsonian rats treated with levodopa [167].

In line with these observations, behavioral studies indicate that the anandamide elevation observed after administration of dopaminergic agonists may serve as an inhibitory feedback signal to offset dopamine-induced hyperactivity [136, 137, 168]. Thus, abnormalities in dopamine and endocannabinoid-mediated plasticity may disrupt this feedback mechanism and lead to motor disturbances, particularly upon long-term activation of dopamine receptors.

Recent studies have added a further level of complexity, showing a competitive interaction between dopamine D2 and adenosine A_{2A} receptors in the induction of endocannabinoid-mediated plasticity, such that D2 receptor activation promotes

LTD, whereas A_{2A} activation promotes LTP [158, 169]. Also, coadministration of A_{2A} and CB1 agonists has been shown to partially inhibit the CB1-dependent decrease of glutamate transmission [170]. The presence of A2A receptors on glutamatergic terminals projecting onto MNS spines [171] suggests that these might be the anatomical substrate for these complex interactions.

CB1 receptors are also expressed on serotonergic raphe-striatal fibers [69] (Fig. 14.1), which are able to (1) convert levodopa into dopamine and release it as a "false neurotransmitter," thus contributing to LID development [172]; (2) influence nigrostriatal dopamine release [173]; and consequently (3) affect the dopamine-mediated and CB1-dependent control of glutamate release [174]. Therefore, we could speculate that cannabinoid agents may exert their antidyskinetic effects by dampening the ectopic dopamine release from serotonergic terminals and/or by controlling dopamine transmission indirectly via inhibition of 5-HT release [175, 176].

Molecular Mechanisms

Overactivity of D1-positive striatofugal neurons of the direct pathway has been long known to be involved in LID [177–179]. Dopamine denervation leads to a high-affinity D1 receptor state in 6-OHDA rats [180, 181], and D1 agonist-induced GTPγS binding has been correlated with LID severity in MPTP-treated primates [182]. D1 overactivity is also accompanied by dysregulation of the cAMP/protein kinase A (PKA) signaling cascade and increased signaling of the dopamine- and cAMP-regulated phosphoprotein-32 kDa (DARPP-32), a key integrator of dopaminergic and glutamatergic inputs in the striatum [131, 183, 184].

Administration of the cannabinoid agonist WIN55,212-2 has been shown to alleviate levodopa-induced AIM in 6-OHDA rats and to reverse the concomitant overactivity of striatal PKA [131]. In keeping with these observations, blockade of PKA signaling has been proven as an effective strategy to reduce AIM expression [185, 186], possibly by preventing PKA-mediated cytoskeleton modifications, which may contribute to the long-term aberrant plasticity underlying striatal dysfunction in dyskinesia [185, 187]. The reduction of PKA activity elicited by cannabinoids may result from the direct activation of CB1 receptors, which are negatively linked to adenylyl cyclase and co-localized on D1-positive striatal neurons [69].

PKA-induced phosphorylation at the threonine (Thr)-34 site converts DARPP-32 into an inhibitor of protein phosphatase-1 (PP1) [188]. Although DARPP-32 phosphorylation appears to be required for the expression of CB1-mediated motor effects, such as catalepsy [189], WIN55,212-2 administration to dyskinetic rats produced a dephosphorylation of DARPP-32 at Thr-34 that was only partially reversed by the CB1 antagonist AM251 even at doses that fully blocked WIN55,212-2 antidyskinetic effect [131]. This discrepancy may depend on different biochemical or functional aspects underlying the behaviors measured in these studies (catalepsy versus AIM) and/or, as previously mentioned, on species-specific differences among animal models. Interestingly, Polissidis et al. [190] have shown that WIN55,212-2 can produce opposite effects on striatal Thr-34 phosphorylation across different rat strains [190].

Concluding Remarks

Experimental evidence indicates that systemic administration of cannabinoids reverses the aberrant levodopa-induced overactivity of downstream signaling that may lead to long-term maladaptive changes in striatal plasticity. However, both direct (or indirect) cannabinoid agonists and antagonists have shown antidyskinetic actions in preclinical models, and experimental evidence for their efficacy in clinical settings is still limited. Given the modulatory action played by the endocannabinoid system in the basal ganglia, understanding its dysfunction in PD and reconciling conflicting data may have important implications for the pathophysiology and treatment of levodopa-associated motor complications.

In addition, the therapeutic potentials of modulating endocannabinoid levels or targeting non-CB receptors activated by endocannabinoids, such as TRP channels and PPAR receptors, have not been fully explored. These approaches may offer more effective and specific pharmacological actions than those observed with traditional cannabinoid agents. Furthermore, as some of these drugs have shown anti-inflammatory and neuroprotective properties in the CNS [191], their application in PD therapy appears particularly appealing, as they may delay/halt the progressive neurodegenerative process occurring in this pathology.

Funding Sources NS 050401–07 and M.J. Fox Foundation (to A.G.); 1F31NS073411-03 (to A.M.)

References

1. Ahlskog JE, Muenter MD. Frequency of levodopa-related dyskinesias and motor fluctuations as estimated from the cumulative literature. Mov Disord. 2001;16:448–58.
2. Fabbrini G, Brotchie JM, Grandas F, Nomoto M, Goetz CG. Levodopa-induced dyskinesias. Mov Disord. 2007;22:1379–89.
3. Hely MA, Morris JG, Reid WG, Trafficante R. Sydney Multicenter Study of Parkinson's disease: non-L-dopa-responsive problems dominate at 15 years. Mov Disord. 2005;20: 190–9.
4. Huot P, Johnston TH, Koprich JB, Fox SH, Brotchie JM. The pharmacology of L-DOPA-induced dyskinesia in Parkinson's disease. Pharmacol Rev. 2013;65:171–222.
5. Verhagen Metman L, et al. Amantadine as treatment for dyskinesias and motor fluctuations in Parkinson's disease. Neurology. 1998;50:1323–6.
6. Del Dotto P, et al. Intravenous amantadine improves levodopa-induced dyskinesias: an acute double-blind placebo-controlled study. Mov Disord. 2001;16:515–20.
7. Sawada H, et al. Amantadine for dyskinesias in Parkinson's disease: a randomized controlled trial. PLoS One. 2010;5:e15298.

8. Benarroch E. Endocannabinoids in basal ganglia circuits: implications for Parkinson disease. Neurology. 2007;69:306–9.
9. Morera-Herreras T, Miguelez C, Aristieta A, Ruiz-Ortega JA, Ugedo L. Endocannabinoid modulation of dopaminergic motor circuits. Front Pharmacol. 2012;3:110.
10. Pacher P, Batkai S, Kunos G. The endocannabinoid system as an emerging target of pharmacotherapy. Pharmacol Rev. 2006;58:389–462.
11. Chevaleyre V, Heifets BD, Kaeser PS, Sudhof TC, Castillo PE. Endocannabinoid-mediated long-term plasticity requires cAMP/PKA signaling and RIM1alpha. Neuron. 2007;54:801–12.
12. Kano M, Ohno-Shosaku T, Hashimotodani Y, Uchigashima M, Watanabe M. Endocannabinoid-mediated control of synaptic transmission. Physiol Rev. 2009;89:309–80.
13. O'Sullivan SE. Cannabinoids go nuclear: evidence for activation of peroxisome proliferator-activated receptors. Br J Pharmacol. 2007;152:576–82.
14. Lopez-Moreno JA, Gonzalez-Cuevas G, Moreno G, Navarro M. The pharmacology of the endocannabinoid system: functional and structural interactions with other neurotransmitter systems and their repercussions in behavioral addiction. Addict Biol. 2008;13:160–87.
15. Devane WA, et al. Isolation and structure of a brain constituent that binds to the cannabinoid receptor. Science. 1992;258:1946–9.
16. Di Marzo V, et al. Formation and inactivation of endogenous cannabinoid anandamide in central neurons. Nature. 1994;372:686–91.
17. Mechoulam R, et al. Identification of an endogenous 2-monoglyceride, present in canine gut, that binds to cannabinoid receptors. Biochem Pharmacol. 1995;50:83–90.
18. Cadas H, Gaillet S, Beltramo M, Venance L, Piomelli D. Biosynthesis of an endogenous cannabinoid precursor in neurons and its control by calcium and cAMP. J Neurosci. 1996;16:3934–42.
19. Sugiura T, et al. Enzymatic synthesis of anandamide, an endogenous cannabinoid receptor ligand, through N-acylphosphatidylethanolamine pathway in testis: involvement of Ca(2+)-dependent transacylase and phosphodiesterase activities. Biochem Biophys Res Commun. 1996;218:113–7.
20. Simon GM, Cravatt BF. Endocannabinoid biosynthesis proceeding through glycerophospho-N-acyl ethanolamine and a role for alpha/beta-hydrolase 4 in this pathway. J Biol Chem. 2006;281:26465–72.
21. Bisogno T, et al. Cloning of the first sn1-DAG lipases points to the spatial and temporal regulation of endocannabinoid signaling in the brain. J Cell Biol. 2003;163:463–8.
22. Basavarajappa BS. Neuropharmacology of the endocannabinoid signaling system-molecular mechanisms, biological actions and synaptic plasticity. Curr Neuropharmacol. 2007;5:81–97.
23. Carrier EJ, et al. Cultured rat microglial cells synthesize the endocannabinoid 2-arachidonylglycerol, which increases proliferation via a CB2 receptor-dependent mechanism. Mol Pharmacol. 2004;65:999–1007.
24. Sugiura T, et al. 2-Arachidonoylglycerol: a possible endogenous cannabinoid receptor ligand in brain. Biochem Biophys Res Commun. 1995;215:89–97.
25. Beltramo M, et al. Functional role of high-affinity anandamide transport, as revealed by selective inhibition. Science. 1997;277:1094–7.
26. Cravatt BF, et al. Molecular characterization of an enzyme that degrades neuromodulatory fatty-acid amides. Nature. 1996;384:83–7.
27. Wei BQ, Mikkelsen TS, McKinney MK, Lander ES, Cravatt BF. A second fatty acid amide hydrolase with variable distribution among placental mammals. J Biol Chem. 2006;281:36569–78.
28. Giuffrida A, McMahon LR. In vivo pharmacology of endocannabinoids and their metabolic inhibitors: therapeutic implications in Parkinson's disease and abuse liability. Prostaglandins Other Lipid Mediat. 2010;91:90–103.
29. Hillard CJ, Jarrahian A. Cellular accumulation of anandamide: consensus and controversy. Br J Pharmacol. 2003;140:802–8.

30. Hampson AJ, et al. Anandamide hydroxylation by brain lipoxygenase:metabolite structures and potencies at the cannabinoid receptor. Biochim Biophys Acta. 1995;1259:173–9.
31. Yu M, Ives D, Ramesha CS. Synthesis of prostaglandin E2 ethanolamide from anandamide by cyclooxygenase-2. J Biol Chem. 1997;272:21181–6.
32. Kim J, Alger BE. Inhibition of cyclooxygenase-2 potentiates retrograde endocannabinoid effects in hippocampus. Nat Neurosci. 2004;7:697–8.
33. Fowler CJ. The contribution of cyclooxygenase-2 to endocannabinoid metabolism and action. Br J Pharmacol. 2007;152:594–601.
34. Vila M, et al. The role of glial cells in Parkinson's disease. Curr Opin Neurol. 2001;14: 483–9.
35. Goparaju SK, Ueda N, Yamaguchi H, Yamamoto S. Anandamide amidohydrolase reacting with 2-arachidonoylglycerol, another cannabinoid receptor ligand. FEBS Lett. 1998;422: 69–73.
36. Kozak KR, et al. Metabolism of the endocannabinoids, 2-arachidonylglycerol and anandamide, into prostaglandin, thromboxane, and prostacyclin glycerol esters and ethanolamides. J Biol Chem. 2002;277:44877–85.
37. Dinh TP, et al. Brain monoglyceride lipase participating in endocannabinoid inactivation. Proc Natl Acad Sci U S A. 2002;99:10819–24.
38. Justinova Z, et al. Fatty acid amide hydrolase inhibition heightens anandamide signaling without producing reinforcing effects in primates. Biol Psychiatry. 2008;64:930–7.
39. Maccarrone M, et al. Anandamide inhibits metabolism and physiological actions of 2-arachidonoylglycerol in the striatum. Nat Neurosci. 2008;11:152–9.
40. Kathuria S, et al. Modulation of anxiety through blockade of anandamide hydrolysis. Nat Med. 2003;9:76–81.
41. Gobbi G, et al. Antidepressant-like activity and modulation of brain monoaminergic transmission by blockade of anandamide hydrolysis. Proc Natl Acad Sci U S A. 2005;102:18620–5.
42. Maione S, et al. Elevation of endocannabinoid levels in the ventrolateral periaqueductal grey through inhibition of fatty acid amide hydrolase affects descending nociceptive pathways via both cannabinoid receptor type 1 and transient receptor potential vanilloid type-1 receptors. J Pharmacol Exp Ther. 2006;316:969–82.
43. Seillier A, Advani T, Cassano T, Hensler JG, Giuffrida A. Inhibition of fatty-acid amide hydrolase and CB1 receptor antagonism differentially affect behavioural responses in normal and PCP-treated rats. Int J Neuropsychopharmacol. 2010;13:373–86.
44. Console-Bram L, Marcu J, Abood ME. Cannabinoid receptors: nomenclature and pharmacological principles. Prog Neuropsychopharmacol Biol Psychiatry. 2012;38:4–15.
45. Begg M, et al. Evidence for novel cannabinoid receptors. Pharmacol Ther. 2005;106: 133–45.
46. Sharir H, et al. The endocannabinoids anandamide and virodhamine modulate the activity of the candidate cannabinoid receptor GPR55. J Neuroimmune Pharmacol. 2012;7:856–65.
47. Henstridge CM, et al. Minireview: recent developments in the physiology and pathology of the lysophosphatidylinositol-sensitive receptor GPR55. Mol Endocrinol. 2011;25:1835–48.
48. Johns DG, et al. The novel endocannabinoid receptor GPR55 is activated by atypical cannabinoids but does not mediate their vasodilator effects. Br J Pharmacol. 2007;152:825–31.
49. McHugh D, et al. N-arachidonoyl glycine, an abundant endogenous lipid, potently drives directed cellular migration through GPR18, the putative abnormal cannabidiol receptor. BMC Neurosci. 2010;11:44.
50. Yin H, et al. Lipid G protein-coupled receptor ligand identification using beta-arrestin PathHunter assay. J Biol Chem. 2009;284:12328–38.
51. Glass M, Dragunow M, Faull RL. Cannabinoid receptors in the human brain: a detailed anatomical and quantitative autoradiographic study in the fetal, neonatal and adult human brain. Neuroscience. 1997;77:299–318.
52. Mackie K, Stella N. Cannabinoid receptors and endocannabinoids: evidence for new players. AAPS J. 2006;8:E298–306.

53. Walter L, Stella N. Cannabinoids and neuroinflammation. Br J Pharmacol. 2004;141:775–85.
54. Ramirez BG, Blazquez C, Gomez del Pulgar T, Guzman M, de Ceballos ML. Prevention of Alzheimer's disease pathology by cannabinoids: neuroprotection mediated by blockade of microglial activation. J Neurosci. 2005;25:1904–13.
55. Van Sickle MD, et al. Identification and functional characterization of brainstem cannabinoid CB2 receptors. Science. 2005;310:329–32.
56. Gong JP, et al. Cannabinoid CB2 receptors: immunohistochemical localization in rat brain. Brain Res. 2006;1071:10–23.
57. Suarez J, et al. Early maternal deprivation induces gender-dependent changes on the expression of hippocampal CB(1) and CB(2) cannabinoid receptors of neonatal rats. Hippocampus. 2009;19:623–32.
58. Onaivi ES, Ishiguro H, Gu S, Liu QR. CNS effects of CB2 cannabinoid receptors: beyond neuro-immuno-cannabinoid activity. J Psychopharmacol. 2012;26:92–103.
59. Benito C, et al. Cannabinoid CB2 receptors and fatty acid amide hydrolase are selectively overexpressed in neuritic plaque-associated glia in Alzheimer's disease brains. J Neurosci. 2003;23:11136–41.
60. Maresz K, Carrier EJ, Ponomarev ED, Hillard CJ, Dittel BN. Modulation of the cannabinoid CB2 receptor in microglial cells in response to inflammatory stimuli. J Neurochem. 2005;95: 437–45.
61. Price DA, et al. WIN55,212-2, a cannabinoid receptor agonist, protects against nigrostriatal cell loss in the 1-methyl-4-phenyl-1,2,3,6-tetrahydropyridine mouse model of Parkinson's disease. Eur J Neurosci. 2009;29:2177–86.
62. Palazuelos J, et al. Microglial CB2 cannabinoid receptors are neuroprotective in Huntington's disease excitotoxicity. Brain. 2009;132:3152–64.
63. Wallmichrath I, Szabo B. Cannabinoids inhibit striatonigral GABAergic neurotransmission in the mouse. Neuroscience. 2002;113:671–82.
64. Martin AB, et al. Expression and function of CB1 receptor in the rat striatum: localization and effects on D1 and D2 dopamine receptor-mediated motor behaviors. Neuropsychopharmacology. 2008;33:1667–79.
65. Fusco FR, et al. Immunolocalization of CB1 receptor in rat striatal neurons: a confocal microscopy study. Synapse. 2004;53:159–67.
66. Uchigashima M, et al. Subcellular arrangement of molecules for 2-arachidonoyl-glycerol-mediated retrograde signaling and its physiological contribution to synaptic modulation in the striatum. J Neurosci. 2007;27:3663–76.
67. Gerdeman G, Lovinger DM. CB1 cannabinoid receptor inhibits synaptic release of glutamate in rat dorsolateral striatum. J Neurophysiol. 2001;85:468–71.
68. Morera-Herreras T, Ruiz-Ortega JA, Gomez-Urquijo S, Ugedo L. Involvement of subthalamic nucleus in the stimulatory effect of Delta(9)-tetrahydrocannabinol on dopaminergic neurons. Neuroscience. 2008;151:817–23.
69. Hermann H, Marsicano G, Lutz B. Coexpression of the cannabinoid receptor type 1 with dopamine and serotonin receptors in distinct neuronal subpopulations of the adult mouse forebrain. Neuroscience. 2002;109:451–60.
70. Haring M, Marsicano G, Lutz B, Monory K. Identification of the cannabinoid receptor type 1 in serotonergic cells of raphe nuclei in mice. Neuroscience. 2007;146:1212–9.
71. Gerdeman GL, Ronesi J, Lovinger DM. Postsynaptic endocannabinoid release is critical to long-term depression in the striatum. Nat Neurosci. 2002;5:446–51.
72. Sidlo Z, Reggio PH, Rice ME. Inhibition of striatal dopamine release by CB1 receptor activation requires nonsynaptic communication involving GABA, H2O2, and KATP channels. Neurochem Int. 2008;52:80–8.
73. Mathur BN, Lovinger DM. Endocannabinoid-dopamine interactions in striatal synaptic plasticity. Front Pharmacol. 2012;3:66.
74. Glass M, Felder CC. Concurrent stimulation of cannabinoid CB1 and dopamine D2 receptors augments cAMP accumulation in striatal neurons: evidence for a Gs linkage to the CB1 receptor. J Neurosci. 1997;17:5327–33.

75. Mukhopadhyay S, Howlett AC. Chemically distinct ligands promote differential CB1 cannabinoid receptor-Gi protein interactions. Mol Pharmacol. 2005;67:2016–24.
76. Kearn CS, Blake-Palmer K, Daniel E, Mackie K, Glass M. Concurrent stimulation of cannabinoid CB1 and dopamine D2 receptors enhances heterodimer formation: a mechanism for receptor cross-talk? Mol Pharmacol. 2005;67:1697–704.
77. Hojo M, et al. mu-Opioid receptor forms a functional heterodimer with cannabinoid CB1 receptor: electrophysiological and FRET assay analysis. J Pharmacol Sci. 2008;108:308–19.
78. Akopian AN, Ruparel NB, Jeske NA, Patwardhan A, Hargreaves KM. Role of ionotropic cannabinoid receptors in peripheral antinociception and antihyperalgesia. Trends Pharmacol Sci. 2008;30:79–84.
79. Caterina MJ, et al. The capsaicin receptor: a heat-activated ion channel in the pain pathway. Nature. 1997;389:816–24.
80. Mezey E, et al. Distribution of mRNA for vanilloid receptor subtype 1 (VR1), and VR1-like immunoreactivity, in the central nervous system of the rat and human. Proc Natl Acad Sci U S A. 2000;97:3655–60.
81. Micale V, et al. Anxiolytic effects in mice of a dual blocker of fatty acid amide hydrolase and transient receptor potential vanilloid type-1 channels. Neuropsychopharmacology. 2009;34:593–606.
82. Cristino L, et al. Immunohistochemical localization of cannabinoid type 1 and vanilloid transient receptor potential vanilloid type 1 receptors in the mouse brain. Neuroscience. 2006;139:1405–15.
83. Szolcsanyi J. Are cannabinoids endogenous ligands for the VR1 capsaicin receptor? Trends Pharmacol Sci. 2000;21:41–2.
84. Zygmunt PM, Julius I, Di Marzo I, Hogestatt ED. Anandamide – the other side of the coin. Trends Pharmacol Sci. 2000;21:43–4.
85. De Petrocellis L, et al. The activity of anandamide at vanilloid VR1 receptors requires facilitated transport across the cell membrane and is limited by intracellular metabolism. J Biol Chem. 2001;276:12856–63.
86. Starowicz K, Nigam S, Di Marzo V. Biochemistry and pharmacology of endovanilloids. Pharmacol Ther. 2007;114:13–33.
87. Sun Y, Bennett A. Cannabinoids: a new group of agonists of PPARs. PPAR Res. 2007;2007:23513.
88. Sun Y, et al. Cannabinoid activation of PPAR alpha; a novel neuroprotective mechanism. Br J Pharmacol. 2007;152:734–43.
89. Bouaboula M, et al. Anandamide induced PPARgamma transcriptional activation and 3 T3-L1 preadipocyte differentiation. Eur J Pharmacol. 2005;517:174–81.
90. Moreno S, Farioli-Vecchioli S, Ceru MP. Immunolocalization of peroxisome proliferator-activated receptors and retinoid X receptors in the adult rat CNS. Neuroscience. 2004;123:131–45.
91. Cimini A, et al. Expression of peroxisome proliferator-activated receptors (PPARs) and retinoic acid receptors (RXRs) in rat cortical neurons. Neuroscience. 2005;130:325–37.
92. Carta AR, et al. Rosiglitazone decreases peroxisome proliferator receptor-gamma levels in microglia and inhibits TNF-alpha production: new evidences on neuroprotection in a progressive Parkinson's disease model. Neuroscience. 2011;194:250–61.
93. Carroll CB, Zeissler ML, Hanemann CO, Zajicek JP. Delta(9)-tetrahydrocannabinol (Delta(9)-THC) exerts a direct neuroprotective effect in a human cell culture model of Parkinson's disease. Neuropathol Appl Neurobiol. 2012;38:535–47.
94. Swanson CR, et al. The PPAR-gamma agonist pioglitazone modulates inflammation and induces neuroprotection in parkinsonian monkeys. J Neuroinflammation. 2011;8:91.
95. Galan-Rodriguez B, et al. Oleoylethanolamide exerts partial and dose-dependent neuroprotection of substantia nigra dopamine neurons. Neuropharmacology. 2009;56:653–64.
96. Xiang GQ, et al. PPARgamma agonist pioglitazone improves scopolamine-induced memory impairment in mice. J Pharm Pharmacol. 2012;64:589–96.

97. Dhikav V, Anand K. Potential predictors of hippocampal atrophy in Alzheimer's disease. Drugs Aging. 2011;28:1–11.
98. Medhi B, Aggarwal R, Chakrabarti A. Neuroprotective effect of pioglitazone on acute phase changes induced by partial global cerebral ischemia in mice. Indian J Exp Biol. 2010; 48:793–9.
99. Besson VC, Chen XR, Plotkine M, Marchand-Verrecchia C. Fenofibrate, a peroxisome proliferator-activated receptor alpha agonist, exerts neuroprotective effects in traumatic brain injury. Neurosci Lett. 2005;388:7–12.
100. Thal SC, et al. Pioglitazone reduces secondary brain damage after experimental brain trauma by PPAR-gamma-independent mechanisms. J Neurotrauma. 2011;28:983–93.
101. Grover S, Kumar P, Singh K, Vikram V, Budhiraja RD. Possible beneficial effect of peroxisome proliferator-activated receptor (PPAR)–alpha and gamma agonist against a rat model of oral dyskinesia. Pharmacol Biochem Behav. 2013;111:17–23.
102. Compton DR, Martin BR. The effect of the enzyme inhibitor phenylmethylsulfonyl fluoride on the pharmacological effect of anandamide in the mouse model of cannabimimetic activity. J Pharmacol Exp Ther. 1997;283:1138–43.
103. Morgese MG, Cassano T, Cuomo V, Giuffrida A. Anti-dyskinetic effects of cannabinoids in a rat model of Parkinson's disease: role of CB(1) and TRPV1 receptors. Exp Neurol. 2007;208:110–9.
104. Giuffrida A, Seillier A. New insights on endocannabinoid transmission in psychomotor disorders. Prog Neuropsychopharmacol Biol Psychiatry. 2012;38:51–8.
105. Ledent C, et al. Unresponsiveness to cannabinoids and reduced addictive effects of opiates in CB1 receptor knockout mice. Science. 1999;283:401–4.
106. Zimmer A, Zimmer AM, Hohmann AG, Herkenham M, Bonner TI. Increased mortality, hypoactivity, and hypoalgesia in cannabinoid CB1 receptor knockout mice. Proc Natl Acad Sci U S A. 1999;96:5780–5.
107. Li X, et al. Attenuation of basal and cocaine-enhanced locomotion and nucleus accumbens dopamine in cannabinoid CB1-receptor-knockout mice. Psychopharmacology (Berl). 2009;204:1–11.
108. McMahon LR, Koek W. Differences in the relative potency of SR 141716A and AM 251 as antagonists of various in vivo effects of cannabinoid agonists in C57BL/6 J mice. Eur J Pharmacol. 2007;569:70–6.
109. Tzavara ET, et al. Endocannabinoids activate transient receptor potential vanilloid 1 receptors to reduce hyperdopaminergia-related hyperactivity: therapeutic implications. Biol Psychiatry. 2006;59:508–15.
110. Marinelli S, et al. Presynaptic facilitation of glutamatergic synapses to dopaminergic neurons of the rat substantia nigra by endogenous stimulation of vanilloid receptors. J Neurosci. 2003;23:3136–44.
111. Romero J, et al. Unilateral 6-hydroxydopamine lesions of nigrostriatal dopaminergic neurons increased CB1 receptor mRNA levels in the caudate-putamen. Life Sci. 2000;66:485–94.
112. Lastres-Becker I, et al. Increased cannabinoid CB1 receptor binding and activation of GTP-binding proteins in the basal ganglia of patients with Parkinson's syndrome and of MPTP-treated marmosets. Eur J Neurosci. 2001;14:1827–32.
113. Van Laere K, et al. Regional changes in type 1 cannabinoid receptor availability in Parkinson's disease in vivo. Neurobiol Aging. 2012;33(620):e621–8.
114. Di Marzo V, Hill MP, Bisogno T, Crossman AR, Brotchie JM. Enhanced levels of endogenous cannabinoids in the globus pallidus are associated with a reduction in movement in an animal model of Parkinson's disease. FASEB J. 2000;14:1432–8.
115. Gubellini P, et al. Experimental parkinsonism alters endocannabinoid degradation: implications for striatal glutamatergic transmission. J Neurosci. 2002;22:6900–7.
116. Maccarrone M, et al. Levodopa treatment reverses endocannabinoid system abnormalities in experimental parkinsonism. J Neurochem. 2003;85:1018–25.
117. Ferrer B, Asbrock N, Kathuria S, Piomelli D, Giuffrida A. Effects of levodopa on endocannabinoid levels in rat basal ganglia: implications for the treatment of levodopa-induced dyskinesias. Eur J Neurosci. 2003;18:1607–14.

118. Kreitzer AC, Malenka RC. Endocannabinoid-mediated rescue of striatal LTD and motor deficits in Parkinson's disease models. Nature. 2007;445:643–7.
119. Zeng BY, et al. Chronic L-DOPA treatment increases striatal cannabinoid CB1 receptor mRNA expression in 6-hydroxydopamine-lesioned rats. Neurosci Lett. 1999;276:71–4.
120. Calabresi P, Giacomini P, Centonze D, Bernardi G. Levodopa-induced dyskinesia: a pathological form of striatal synaptic plasticity? Ann Neurol. 2001;47:60–8.
121. Picconi B, et al. Pathological synaptic plasticity in the striatum: implications for Parkinson's disease. Neurotoxicology. 2005;26:779–83.
122. Meschler JP, Howlett AC, Madras BK. Cannabinoid receptor agonist and antagonist effects on motor function in normal and 1-methyl-4-phenyl-1,2,5,6-tetrahydropyridine (MPTP)-treated non-human primates. Psychopharmacology (Berl). 2001;156:79–85.
123. Fernandez-Espejo E, et al. Cannabinoid CB1 antagonists possess antiparkinsonian efficacy only in rats with severe nigral lesion in experimental parkinsonism. Neurobiol Dis. 2005;18:591–601.
124. Kelsey JE, Harris O, Cassin J. The CB(1) antagonist rimonabant is adjunctively therapeutic as well as monotherapeutic in an animal model of Parkinson's disease. Behav Brain Res. 2009;203:304–7.
125. van der Stelt M, et al. A role for endocannabinoids in the generation of parkinsonism and levodopa-induced dyskinesia in MPTP-lesioned non-human primate models of Parkinson's disease. FASEB J. 2005;19:1140–2.
126. van Vliet SA, Vanwersch RA, Jongsma MJ, Olivier B, Philippens IH. Therapeutic effects of Delta9-THC and modafinil in a marmoset Parkinson model. Eur Neuropsychopharmacol. 2008;18:383–9.
127. Carroll CB, et al. Cannabis for dyskinesia in Parkinson disease. Neurology. 2004;63: 1245–50.
128. Fox SH, Henry B, Hill M, Crossman A, Brotchie J. Stimulation of cannabinoid receptors reduces levodopa-induced dyskinesia in the MPTP-lesioned nonhuman primate model of Parkinson's disease. Mov Disord. 2002;17:1180–7.
129. Sieradzan KA, et al. Cannabinoids reduce levodopa-induced dyskinesia in Parkinson's disease: a pilot study. Neurology. 2001;57:2108–11.
130. Segovia G, Mora F, Crossman AR, Brotchie JM. Effects of CB1 cannabinoid receptor modulating compounds on the hyperkinesia induced by high-dose levodopa in the reserpine-treated rat model of Parkinson's disease. Mov Disord. 2003;18:138–49.
131. Martinez A, Macheda T, Morgese MG, Trabace L, Giuffrida A. The cannabinoid agonist WIN55212-2 decreases L-DOPA-induced PKA activation and dyskinetic behavior in 6-OHDA-treated rats. Neurosci Res. 2012;72:236–42.
132. Walsh S, Gorman AM, Finn DP, Dowd E. The effects of cannabinoid drugs on abnormal involuntary movements in dyskinetic and non-dyskinetic 6-hydroxydopamine lesioned rats. Brain Res. 2010;1363:40–8.
133. Mesnage V, et al. Neurokinin B, neurotensin, and cannabinoid receptor antagonists and Parkinson disease. Clin Neuropharmacol. 2004;27:108–10.
134. Cao X, et al. Blockade of cannabinoid type 1 receptors augments the antiparkinsonian action of levodopa without affecting dyskinesias in 1-methyl-4-phenyl-1,2,3,6-tetrahydropyridine-treated rhesus monkeys. J Pharmacol Exp Ther. 2007;323:318–26.
135. Gonzalez-Aparicio R, Moratalla R. Oleoylethanolamide reduces L-DOPA-induced dyskinesia via TRPV1 receptor in a mouse model of Parkinson s disease. Neurobiol Dis. 2013;62C:416–25.
136. Giuffrida A, et al. Dopamine activation of endogenous cannabinoid signaling in dorsal striatum. Nat Neurosci. 1999;2:358–63.
137. Masserano JM, Karoum F, Wyatt RJ. SR 141716A, a CB1 cannabinoid receptor antagonist, potentiates the locomotor stimulant effects of amphetamine and apomorphine. Behav Pharmacol. 1999;10:429–32.
138. Bido S, Marti M, Morari M. Amantadine attenuates levodopa-induced dyskinesia in mice and rats preventing the accompanying rise in nigral GABA levels. J Neurochem. 2011;118: 1043–55.

139. Dupre KB, et al. Local modulation of striatal glutamate efflux by serotonin 1A receptor stimulation in dyskinetic, hemiparkinsonian rats. Exp Neurol. 2011;229:288–99.
140. Perez-Rial S, et al. Increased vulnerability to 6-hydroxydopamine lesion and reduced development of dyskinesias in mice lacking CB1 cannabinoid receptors. Neurobiol Aging. 2011;32:631–45.
141. Van Laere K. In vivo imaging of the endocannabinoid system: a novel window to a central modulatory mechanism in humans. Eur J Nucl Med Mol Imaging. 2007;34:1719–26.
142. Vitale C, et al. Unawareness of dyskinesias in Parkinson's and Huntington's diseases. Neurol Sci. 2001;22:105–6.
143. Lee J, Di Marzo V, Brotchie JM. A role for vanilloid receptor 1 (TRPV1) and endocannabinnoid signalling in the regulation of spontaneous and L-DOPA induced locomotion in normal and reserpine-treated rats. Neuropharmacology. 2006;51:557–65.
144. Cenci MA, Whishaw IQ, Schallert T. Animal models of neurological deficits: how relevant is the rat? Nat Rev Neurosci. 2002;3:574–9.
145. Matyas F, et al. Subcellular localization of type 1 cannabinoid receptors in the rat basal ganglia. Neuroscience. 2006;137:337–61.
146. Szabo B, Wallmichrath I, Mathonia P, Pfreundtner C. Cannabinoids inhibit excitatory neurotransmission in the substantia nigra pars reticulata. Neuroscience. 2000;97:89–97.
147. Julian MD, et al. Neuroanatomical relationship between type 1 cannabinoid receptors and dopaminergic systems in the rat basal ganglia. Neuroscience. 2003;119:309–18.
148. Malone DT, Taylor DA. Modulation by fluoxetine of striatal dopamine release following Delta9-tetrahydrocannabinol: a microdialysis study in conscious rats. Br J Pharmacol. 1999;128:21–6.
149. Szabo B, Muller T, Koch H. Effects of cannabinoids on dopamine release in the corpus striatum and the nucleus accumbens in vitro. J Neurochem. 1999;73:1084–9.
150. Kofalvi A, et al. Involvement of cannabinoid receptors in the regulation of neurotransmitter release in the rodent striatum: a combined immunochemical and pharmacological analysis. J Neurosci. 2005;25:2874–84.
151. de Lago E, de Miguel R, Lastres-Becker I, Ramos JA, Fernandez-Ruiz J. Involvement of vanilloid-like receptors in the effects of anandamide on motor behavior and nigrostriatal dopaminergic activity: in vivo and in vitro evidence. Brain Res. 2004;1007:152–9.
152. Marinelli S, et al. N-arachidonoyl-dopamine tunes synaptic transmission onto dopaminergic neurons by activating both cannabinoid and vanilloid receptors. Neuropsychopharmacology. 2007;32:298–308.
153. Melis M, et al. Endogenous fatty acid ethanolamides suppress nicotine-induced activation of mesolimbic dopamine neurons through nuclear receptors. J Neurosci. 2008;28:13985–94.
154. Patel S, Rademacher DJ, Hillard CJ. Differential regulation of the endocannabinoids anandamide and 2-arachidonylglycerol within the limbic forebrain by dopamine receptor activity. J Pharmacol Exp Ther. 2003;306:880–8.
155. Centonze D, et al. A critical interaction between dopamine D2 receptors and endocannabinoids mediates the effects of cocaine on striatal gabaergic Transmission. Neuropsychopharmacology. 2004;29:1488–97.
156. Andre VM, et al. Dopamine modulation of excitatory currents in the striatum is dictated by the expression of D1 or D2 receptors and modified by endocannabinoids. Eur J Neurosci. 2010;31:14–28.
157. Tang K, Low MJ, Grandy DK, Lovinger DM. Dopamine-dependent synaptic plasticity in striatum during in vivo development. Proc Natl Acad Sci U S A. 2001;98:1255–60.
158. Shen W, Flajolet M, Greengard P, Surmeier DJ. Dichotomous dopaminergic control of striatal synaptic plasticity. Science. 2008;321:848–51.
159. Ade KK, Lovinger DM. Anandamide regulates postnatal development of long-term synaptic plasticity in the rat dorsolateral striatum. J Neurosci. 2007;27:2403–9.
160. Adermark L, Lovinger DM. Frequency-dependent inversion of net striatal output by endocannabinoid-dependent plasticity at different synaptic inputs. J Neurosci. 2009;29: 1375–80.

161. Lerner TN, Kreitzer AC. RGS4 is required for dopaminergic control of striatal LTD and susceptibility to parkinsonian motor deficits. Neuron. 2012;73:347–59.
162. Szabo B, Dorner L, Pfreundtner C, Norenberg W, Starke K. Inhibition of GABAergic inhibitory postsynaptic currents by cannabinoids in rat corpus striatum. Neuroscience. 1998;85: 395–403.
163. Narushima M, Uchigashima M, Hashimoto K, Watanabe M, Kano M. Depolarization-induced suppression of inhibition mediated by endocannabinoids at synapses from fast-spiking interneurons to medium spiny neurons in the striatum. Eur J Neurosci. 2006;24: 2246–52.
164. Adermark L, Talani G, Lovinger DM. Endocannabinoid-dependent plasticity at GABAergic and glutamatergic synapses in the striatum is regulated by synaptic activity. Eur J Neurosci. 2009;29:32–41.
165. Kreitzer AC, Malenka RC. Dopamine modulation of state-dependent endocannabinoid release and long-term depression in the striatum. J Neurosci. 2005;25:10537–45.
166. Wang Z, et al. Dopaminergic control of corticostriatal long-term synaptic depression in medium spiny neurons is mediated by cholinergic interneurons. Neuron. 2006;50:443–52.
167. Picconi B, et al. Inhibition of phosphodiesterases rescues striatal long-term depression and reduces levodopa-induced dyskinesia. Brain. 2011;134:375–87.
168. Beltramo M, et al. Reversal of dopamine D(2) receptor responses by an anandamide transport inhibitor. J Neurosci. 2000;20:3401–7.
169. Tozzi A, et al. The distinct role of medium spiny neurons and cholinergic interneurons in the D/AA receptor interaction in the striatum: implications for Parkinson's disease. J Neurosci. 2011;31:1850–62.
170. Martire A, et al. Pre-synaptic adenosine A2A receptors control cannabinoid CB1 receptor-mediated inhibition of striatal glutamatergic neurotransmission. J Neurochem. 2011;116: 273–80.
171. Quiroz C, et al. Key modulatory role of presynaptic adenosine A2A receptors in cortical neurotransmission to the striatal direct pathway. ScientificWorldJournal. 2009;9:1321–44.
172. Carta M, Carlsson T, Kirik D, Bjorklund A. Dopamine released from 5-HT terminals is the cause of L-DOPA-induced dyskinesia in parkinsonian rats. Brain. 2007;130:1819–33.
173. Abdallah L, et al. Impact of serotonin 2C receptor null mutation on physiology and behavior associated with nigrostriatal dopamine pathway function. J Neurosci. 2009;29:8156–65.
174. Mathur BN, Capik NA, Alvarez VA, Lovinger DM. Serotonin induces long-term depression at corticostriatal synapses. J Neurosci. 2011;31:7402–11.
175. Nakazi M, Bauer U, Nickel T, Kathmann M, Schlicker E. Inhibition of serotonin release in the mouse brain via presynaptic cannabinoid CB1 receptors. Naunyn Schmiedebergs Arch Pharmacol. 2000;361:19–24.
176. Haring M, Grieb M, Monory K, Lutz B, Moreira FA. Cannabinoid CB(1) receptor in the modulation of stress coping behavior in mice: the role of serotonin and different forebrain neuronal subpopulations. Neuropharmacology. 2013;65:83–9.
177. Bezard E, Brotchie JM, Gross CE. Pathophysiology of levodopa-induced dyskinesia: potential for new therapies. Nat Rev Neurosci. 2001;2:577–88.
178. DeLong MR, Wichmann T. Circuits and circuit disorders of the basal ganglia. Arch Neurol. 2007;64:20–4.
179. Koprich JB, Johnston TH, Huot P, Fox SH, Brotchie JM. New insights into the organization of the basal ganglia. Curr Neurol Neurosci Rep. 2009;9:298–304.
180. Pifl C, Reither H, Hornykiewicz O. Functional sensitization of striatal dopamine D1 receptors in the 6-hydroxydopamine-lesioned rat. Brain Res. 1992;572:87–93. doi:0006-8993(92)90455-I [pii].
181. Corvol JC, et al. Persistent increase in olfactory type G-protein alpha subunit levels may underlie D1 receptor functional hypersensitivity in Parkinson disease. J Neurosci. 2004;24:7007–14.
182. Aubert I, et al. Increased D1 dopamine receptor signaling in levodopa-induced dyskinesia. Ann Neurol. 2005;57:17–26.

183. Oh JD, Del Dotto P, Chase TN. Protein kinase A inhibitor attenuates levodopa-induced motor response alterations in the hemi-parkinsonian rat. Neurosci Lett. 1997;228:5–8.
184. Santini E, Valjent E, Fisone G. Parkinson's disease: levodopa-induced dyskinesia and signal transduction. FEBS J. 2008;275:1392–9.
185. Lebel M, Chagniel L, Bureau G, Cyr M. Striatal inhibition of PKA prevents levodopa-induced behavioural and molecular changes in the hemiparkinsonian rat. Neurobiol Dis. 2010;38:59–67.
186. Santini E, et al. Critical involvement of cAMP/DARPP-32 and extracellular signal-regulated protein kinase signaling in L-DOPA-induced dyskinesia. J Neurosci. 2007;27:6995–7005.
187. Cyr M, et al. Sustained elevation of extracellular dopamine causes motor dysfunction and selective degeneration of striatal GABAergic neurons. Proc Natl Acad Sci U S A. 2003;100: 11035–40.
188. Hemmings Jr HC, Greengard P, Tung HY, Cohen P. DARPP-32, a dopamine-regulated neuronal phosphoprotein, is a potent inhibitor of protein phosphatase-1. Nature. 1984;310:503–5.
189. Andersson M, et al. Cannabinoid action depends on phosphorylation of dopamine- and cAMP-regulated phosphoprotein of 32 kDa at the protein kinase A site in striatal projection neurons. J Neurosci. 2005;25:8432–8.
190. Polissidis A, et al. Individual differences in the effects of cannabinoids on motor activity, dopaminergic activity and DARPP-32 phosphorylation in distinct regions of the brain. Int J Neuropsychopharmacol. 2010;13:1175–91.
191. Carta AR. PPAR-gamma: therapeutic prospects in Parkinson's disease. Curr Drug Targets. 2013;14:743–51.

Chapter 15
The Role of the Noradrenergic System and Its Receptors in Levodopa-Induced Dyskinesia

Corinne Y. Ostock and Christopher Bishop

Abstract Chronic dopamine replacement therapy with levodopa in Parkinson's disease (PD) is often complicated by the emergence of deleterious motor sequelae including levodopa-induced dyskinesia (LID). The mechanism(s) underlying the pathogenesis of LID remain speculative; however, accumulating evidence has highlighted a role for the noradrenergic system. In this chapter, we evaluate the body of research that has implicated the NA system in the development and treatment of LID and discuss the following: (1) changes in the noradrenergic system originating in the locus coeruleus in the parkinsonian brain, (2) the use of experimental models with noradrenergic lesions for the investigation of LID, and (3) the efficacy of targeting noradrenergic receptors for the treatment of LID.

Keywords Levodopa-induced dyskinesia • Noradrenaline • Locus coeruleus • α(alpha) adrenergic receptor • β(beta) adrenergic receptor

Introduction

The Noradrenergic System

The Locus Coeruleus

Two main ascending systems are responsible for noradrenergic innervation of the central nervous system (CNS). The first and primary source of noradrenaline (NA) within the CNS is the locus coeruleus (LC) located along the fourth ventricle in the pons. Noradrenergic neurons of the LC display extensive axonal branching

C.Y. Ostock • C. Bishop, PhD (✉)
Behavioral Neuroscience Program, Department of Psychology, Binghamton University, 4400 Vestal Parkway East, Binghamton, NY 13902-6000, USA
e-mail: costock1@binghamton.edu; cbishop@binghamton.edu

S.H. Fox, J.M. Brotchie (eds.), *Levodopa-Induced Dyskinesia in Parkinson's Disease*, DOI 10.1007/978-1-4471-6503-3_15

allowing single NA neurons to have widespread cerebral innervation [1, 2]. Large fusiform cells originating in the LC project to the hippocampus and cortex, while medium-sized multipolar cells project to the cortex, spinal cord, and cerebellum [1, 3]. The LC also projects to areas implicated in PD including the substantia nigra pars compacta (SNc) [4] and striatum [5]. Under normal conditions, melanin-containing noradrenergic neurons within the rostral and dorsal planes of the LC innervate the forebrain, while noradrenergic neurons located within the ventral and caudal planes of the LC project to the spinal cord and cerebellum [6]. The medullary noradrenergic system is the second source of NA within the CNS and is comprised of scattered groupings of noradrenergic neurons in the ventrolateral reticular formation and nucleus of the solitary tract that innervate the hypothalamus and regulate endocrine, feeding, and sexual behavior [7–9]. Neurotransmitter release from NA neurons occurs through typical synapses or volume transmission via axonal varicosities [10].

Noradrenergic Receptors

NA exerts potent effects on CNS neurotransmission via interactions with noradrenergic receptors. Two classes of metabotropic noradrenergic receptors are found throughout the central and peripheral nervous system: alpha (α(alpha)1, α(alpha)2) and beta (β(beta)1, β(beta)2, β(beta)3) receptors. The excitatory effects of NA are mediated through α(alpha)1- and β(beta)- receptors within the CNS. Gq-coupled (alpha)1-adrenoceptors (α(alpha)1R) are comprised of three distinct subtypes: α(alpha)1_A, α(alpha)1_B, and α(alpha)$1_{D,}$ which when stimulated activate phospholipase C activity. α(alpha)1R are found in regions associated with PD including the LC, cortex, striatum, and subthalamic nucleus (STN) (Fig. 15.1) [11–14]. Gs-coupled β(beta)1- and β(beta)2- adrenergic receptors (β(beta)R) are found postsynaptically throughout the cortex, olfactory bulbs, septum, hippocampus, striatum, and thalamus [15, 16], while β(beta)3R are found mainly in the periphery. Stimulation of these receptors activates adenylyl cyclase (AC) activity. The inhibitory effects of NA are mediated by Gi-coupled α(alpha)2-adrenoceptors (α(alpha)2R) negatively coupled to AC. This receptor class is comprised of three distinct subtypes: α(alpha)2_A, α(alpha)2_B, and α(alpha)2_C, respectively. α(alpha)2_B receptors are found mainly in the periphery. α(alpha)2_ARs are widely distributed throughout the brain including in the hippocampus, amygdala, LC, cerebral cortex, and striatum, while α(alpha)2_CRs are found in the olfactory bulbs, basal ganglia, and most densely in the striatum (Fig. 15.1) [17–19], an area only sparsely innervated by NA neurons and containing low NA content. α(alpha)2Rs exist presynaptically as autoreceptors on dendrites of NA neurons where they tonically regulate noradrenergic efflux. α(alpha)1-, 2-, and β(beta)- receptors are found postsynaptically as heteroreceptors on targets of NA neurons where they regulate the release of other neurotransmitters. Stimulation and blockade of these receptors have anatomically distinct effects on motor behavior that will be discussed in more detail later.

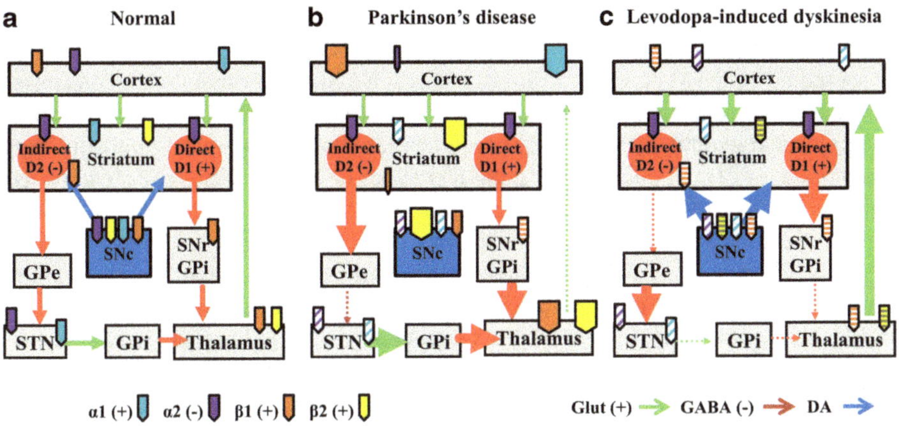

Fig. 15.1 Changes in basal ganglia signaling following DA loss in PD and following exogenous levodopa administration in LID. (**a**) DA signaling in the normal brain. (**b**) DA loss in PD causes an imbalance between the "direct" and "indirect" striatal output pathways. PD is associated with overactivity of the "indirect" striatal output neurons and underactivity in "direct" striatal output neurons. (**c**) In LID, exogenous levodopa treatment shifts the imbalance between the "direct" and "indirect" pathway in favor of the "direct" striatal output pathway. Noradrenergic receptors are represented as colored boxes: **α**(alpha)1R- *blue*, **α**(alpha)2R- *purple*, **β**(beta)1R- *orange*, **β**(beta)2R- *yellow*. Changes in receptor expression are expressed in size (compared to the normal brain). *Smaller boxes* represent decreased receptor expression and *larger boxes* represent increased receptor expression [13–18, 35–41]. *Striped boxes* are added when receptor expression changes are unknown

The Role of Noradrenaline in Parkinson's Disease

Neuroanatomical Alterations in the Noradrenergic System

Anatomical studies of patients with PD have revealed a disease-specific pattern of noradrenergic cell loss from rostral to caudal within the LC as PD severity progresses [20]. LC noradrenergic cell loss is often greater than 80 % in the parkinsonian brain [21]. This far exceeds that seen in normal aging where moderate (~25–50 %) reductions in LC cell number and NA brain concentrations are common [6, 22]. The remaining melanin-containing neurons of the LC display morphological changes in the PD brain including reduced dendritic arborizations and dendritic length, loss of synaptic spines, and shrinkage or swelling of the soma [23–26]. Other markers for NA are also altered in postmortem LC tissue including NA transporter (NAT), the NA rate-limiting enzyme tyrosine hydroxylase (TH), and the NA-synthesizing enzyme dopamine (DA)-β-hydroxylase (DBH) [27]. LC NA degeneration in PD is also associated with reduced NA concentrations and altered NA functioning throughout the CNS. For example, there are marked reductions in postmortem NA tissue concentrations in the cortex, cerebellum, motor thalamus, and hypothalamus of PD patients [22, 28, 29]. Cerebrospinal fluid (CSF) measures of NA and DBH concentration in PD patients are inconsistent. Moderate reductions in CSF concentrations of DBH (50–60 %) [30, 31] and NA (~60 %) [32] have been observed in some PD patients, while no change in CSF NA or metabolite concentration has been reported in others displaying comparable parkinsonian disabilities [33]. Cortical axonal TH and DBH immunoreactivity is diminished in postmortem brain tissue of PD patients [34]. Although changes in TH may also reflect pathological DA cell loss, cortical DBH reductions are likely a direct result of LC degeneration because the cortex receives sole NA innervation from LC.

Noradrenergic receptor expression is also altered in the parkinsonian brain. Excitatory α(alpha)1R and β(beta)1R densities are increased in the cerebral cortex of PD patients [13, 35] (Fig. 15.1). This effect has also been mirrored in experimental models of PD as cortical α(alpha)1R expression is enhanced in rodents with DA and NA lesions [36], and β(beta)1R density is increased in the cortex, thalamus, hippocampus, and amygdala of hemiparkinsonian rats [37]. Striatal β(beta)1R expression is reduced following DA lesions in an experimental model of PD [38] and shows a trend for reductions in PD humans patients [39] suggesting that these receptors may reside presynaptically on nigrostriatal terminals. Striatal β(beta)2R expression is increased in DA-lesioned rats and human PD brains, while enhanced nigral β(beta)2R expression is found in DA-lesioned rats but not human PD brains [37, 39]. Inhibitory α(alpha)2Rs are reduced in the cerebral cortex of PD patients [13]. This is in opposition to experimental work showing that chronic reserpine treatment to deplete CNS DA, NA, and 5-HT led to an upregulation in α(alpha)2R density in the cerebral cortex [40]. In rats, unilateral DA depletion increased α(alpha)2_A mRNA expression within the intact LC, and levodopa restored LC α(alpha)2_AR to control levels in these animals [41]. In the same animals, striatal α(alpha)2_CR expression remained stable following DA lesion and levodopa treatment (Fig. 15.1). Despite this study, how noradrenergic receptor expression and function change in the parkinsonian brain as a result of levodopa treatment remains critical but largely unanswered questions.

Behavioral Outcome of Noradrenergic Depletion in Parkinson's Disease

Correlative post- and antemortem studies have attributed reductions in biochemical markers in the NA system to motor symptoms such as "freezing," akinesia, postural instability, and tremor as well as reduced treatment efficacy [10, 42, 43]. Experimentally, DBH knock-out mice that lack NA display parkinsonian motor deficits and spontaneous dyskinesia even when striatal DA content is normal [44]; however, it is not clear if NA loss in early neurodevelopment affects motor symptoms differently than NA loss induced in adulthood. Parkinsonian motor deficits induced in mice and monkeys by administering the DA neurotoxin 1-methyl-4-phenyl-1,2,3,6-tetrahydropyridine (MPTP) can be exacerbated by additional noradrenergic LC degeneration [45–47]. Interestingly, MPTP treatment in mice that depleted striatal DA content by 80 % did not produce PD-like motor deficits when the NA system remained intact [44].

Experimental Models of Noradrenergic Loss in Levodopa-Induced Dyskinesia

Several animal models have been developed that recapitulate nigrostriatal DA cell loss observed in PD including striatal, SNc, or medial forebrain bundle (MFB) lesions with 6-hydroxydopamine (6-OHDA) or peripheral injection of MPTP. Unfortunately, the majority of animal models used to study levodopa-induced dyskinesia (LID) neither display nor account for NA loss. In fact, the NAT blocker desipramine is often given prior to 6-OHDA infusion to prevent noradrenergic cell loss. Although evidence suggests that additional NA lesions exacerbate primary PD symptoms, only four studies have directly examined the effect of additional noradrenergic lesions on LID severity. These models have thus far produced contradictory behavioral effects on LID severity and duration. The strengths and limitations of the lesion models discussed below highlight the necessity of more accurately modeling the progression and treatment of PD.

The DSP-4 Model

The NA neurotoxin [N-(2-chloroethyl)-N-ethyl-2-bromobenzylamine] (DSP-4) has been widely employed due, in part, to the ease with which the NA lesion can be created. DSP-4 is a systemically administered, blood-brain barrier penetrating molecule that causes rapid but transient reductions in brain NA tissue content [48, 49] and cortical NAT [50]. Following uptake by NAT, DSP-4 accumulates in the neuron and induces degeneration of NA terminals via alkylation of vital neuronal structures [51]. DSP-4 has gained more use in PD research since it preferentially targets noradrenergic neurons originating in the LC, leaving medullary NA neurons intact [48, 52]; however, a number of issues limit its widespread utility in experimental PD and LID studies.

First, the extent of DSP-4-induced LC cell loss remains debatable. It has been suggested that DSP-4 preferentially affects NA terminals, not cell bodies. Szot et al. [53] reported that DSP-4 administration alone does not result in LC cell loss in rats, though NA tissue concentrations, terminal NAT-binding sites, and α(alpha)2R binding sites were reduced in a number of forebrain regions. However, week prior to 6-OHDA, others have reported that DSP-4 treatment suppressed TH-positive cell staining in the LC for up to 2 months [54, 55] and reduced NA tissue concentrations in multiple forebrain regions [56, 57]. In addition to such variability, other limitations of DSP-4 treatment include damage to serotonergic neurons [58, 59] and noradrenergic system recovery over time [49, 53]. There is also evidence of species and strain differences in the susceptibility of different populations of NA neurons to DSP-4 [59].

Only two studies have evaluated the effect of DSP-4 treatment on LID. The order of DA and NA lesions differed between these groups but produced similar behavioral outcomes. DSP-4 administered 7 days prior to 6-OHDA DA lesions [55] or 4 weeks after 6-OHDA DA lesion [60] did not alter the severity or duration of LID compared to rats with DA lesions alone. However, DSP-4 treatment altered the trajectory of changes in levodopa-induced rotational and locomotor behavior in rodents over repeated levodopa treatments. Daily levodopa treatment produces a sensitization-desensitization of contralateral rotations in 6-OHDA DA lesioned rats where rotational activity increases for the first 2 weeks then decreases by the third week of consecutive levodopa treatment [54, 61]. DSP-4 pretreatment in these animals prevented the desensitization response such that dual DA + NA-lesioned rats continued to rotate at high levels by the third week of testing, while rats with just DA lesions rotated significantly less. In acute tests, rats with additional DSP-4-induced NA lesions rotated more than those without NA lesions during the first 20 min of testing, an effect that vanished during the last 40 min of testing [57]. Furthermore, rats with additional DSP-4 lesions displayed increased activity in locomotor chambers following acute levodopa treatment compared to rats with just DA lesions [56]. These findings are in contrast to more general effects of levodopa. For example, MPTP-treated mice given DSP-4 displayed less levodopa-induced locomotor activity [47, 62] following levodopa treatment compared to those without NA lesions.

The 6-OHDA Model

6-OHDA infusions into the striatum, SNc, or MFB have been employed to model DA and/or NA loss in a number of animal models. 6-OHDA is transported into the cell via NAT or the DA transporter (DAT). Once in the cell, 6-OHDA interferes with mitochondrial function leading to metabolic deficits and eventually cell death [63, 64]. 6-OHDA has also been used to destroy NA neurons by local infusion into the LC or its fiber tracts [45, 65, 66].

NA lesions with 6-OHDA in hemiparkinsonian rats have been shown to modify the severity and duration of LID; however, only two studies have systematically examined this, and the direction of these effects is in opposition. Barnum et al. [65] and Fulceri et al. [66] administered 6-OHDA unilaterally into the MFB of rats with or without pretreatment of the NAT blocker desipramine to protect NA neurons in order to compare rats with purely dopaminergic lesions from those with dual DA and NA lesions. Dual lesions resulted in marked reductions in hippocampal (80 %) and moderate (64 %) [65] to marked (~75 %) [66] reductions in striatal NA content compared to rats with an intact noradrenergic system. In one study, 6-OHDA-induced NA loss exacerbated the severity and prolonged the duration of LID [66]. In these rats, NA loss was positively correlated with LID severity. This is in stark contrast to the work by Barnum et al. [65] which showed that 6-OHDA-induced NA loss attenuated LID severity. In this work, dual lesioned rats were less dyskinetic than just DA-lesioned rats on the first day of levodopa treatment, an effect that disappeared by the third day of chronic levodopa treatment. Furthermore, rats with additional NA loss required higher doses of levodopa to elicit dyskinesia compared to those with intact NA systems. Levodopa-induced contralateral rotations were also reduced in these dual lesioned rats [65].

The Ibotenic Acid Model

Direct LC infusions of ibotenic acid have also been reported to alter LID in 6-OHDA DA-lesioned rats that had previously been rendered dyskinetic [60]. Although ibotenic acid lesions are not neurochemically selective, they were reported to produce moderate (~42 %) reductions in cresyl violet-stained LC cell numbers compared to animals with DA alone lesions. Such LC lesions did not impact the severity of peak LID or levodopa-induced locomotor behavior compared to previous test days. Instead, LID duration was extended by approximately 40 min within a single testing period.

Considerations

Several features of these studies may explain behavioral discrepancies that have complicated the interpretation of NA lesion models in LID. First, lesion order and timing are probable sources of variance since NA loss precedes DA loss in the human condition [21]; unfortunately this has not been consistently recapitulated in experimental models. Second, the behavioral consequence of bilateral NA loss with a unilateral DA lesion remains unknown. Finally, whether NA-lesion-induced changes in LID are due to the loss of the transmitter itself or noradrenergic machinery like NAT has not been elucidated. Evidence shows that NA binds to DA receptors within the basal ganglia [67], and striatal infusion of exogenous NA elicits LID in hemiparkinsonian rats [68]. Therefore, it has been argued that lesion-induced reductions in NA binding at DA receptors may manifest as a dampened behavioral response to levodopa [47, 62, 65]. Alternatively, evidence shows that NAT can take up DA [69, 70] and may play a more integral role in clearing levodopa-derived DA following loss of striatal DAT in PD [71]. Under these circumstances, NA lesions that destroy terminal NAT would be expected to augment LID severity or duration [60, 66]. To date, no studies have directly examined the mechanism(s) through which NA modifies levodopa-induced behaviors. A concerted effort toward systematic investigation of these issues is clearly warranted in order to determine the exact contribution of the LC and NA in LID.

Pharmacological Targeting of Noradrenergic Receptors in Levodopa-Induced Dyskinesia

According to classic models of basal ganglia neurocircuitry, under normal conditions nigrostriatal DA stimulates DA receptors on GABAergic output neurons of the "direct" and "indirect" signaling pathways to induce motor movements (Fig. 15.1a). SNc DA degeneration in PD is accompanied by marked reductions in striatal DA content which alters the activity of these GABAergic striatal output neurons.

As discussed in Chap. 7, PD is associated with reduced activity in the "direct" striatonigral/striatopallidal pathway and enhanced activity of the "indirect" striatopallidal pathway resulting in disinhibition of the SNr and STN (Fig. 15.1b). Nigral and STN disinhibition increases thalamic inhibition, the net result of which is disrupted motor movement for the PD patient. DA replacement with levodopa while initially correcting for PD disturbances eventually results in imbalanced basal ganglia circuitry in favor of the direct pathway (Fig 15.1c) where SNr inhibition is increased and thalamic inhibition is decreased. At the same, there is underactivity of the indirect pathway which enhances STN inhibition and reduces thalamic inhibition. Behaviorally, this manifests in excessive movements such as LID. Noradrenergic compounds may improve LID expression by restoring proper basal ganglia neurotransmission.

α(alpha)1-Adrenoceptors

Recently, α(alpha)1Rs have become of interest in LID because they influence DA neurotransmission within the CNS. Direct stimulation of α(alpha)1R within the medial prefrontal cortex enhances local DA efflux in healthy rodent brains, and antagonism at this receptor can block NA-induced DA release [72]. These effects are also seen in subcortical structures directly implicated in PD and LID. For example, α(alpha)1R stimulation has excitatory effects on DA neurons within the SNc [73], while intrastriatal α(alpha)1R blockade reduces striatal DA efflux in intact rodents [74].

There is a small but growing body of literature implicating this receptor in dyskinesia. As shown in Table 15.1, to date three studies have examined the α(alpha)1R in respect to LID. Administration of the selective α(alpha)1R antagonist 2-[[b-(−4-hydroxyphenyl)ethyl]aminomethyl]-1-tetralone (HEAT) dose dependently reduced LID in 6-OHDA-lesioned rats without producing sedative locomotor effects [75]. In contrast, pretreatment with the α(alpha)1R antagonist prazosin did not influence LID in 6-OHDA-lesioned rats [76] or MPTP-treated nonhuman primates [77] but did reduce levodopa-induced motor activity in nonhuman primates [77]. Prazosin has also been shown to attenuate amphetamine-induced ipsilateral circling behavior in a rodent model of PD [78, 79]. Collectively, this suggests that α(alpha)1R blockade may blunt DA-mediated hyperactivity.

The neuroanatomical site of action for these effects has not been determined. Direct infusion of the α(alpha)1R agonist cirazoline into the striatum via reverse-microdialysis alone failed to produce dyskinesia in parkinsonian rats [75]. This does not rule out striatal α(alpha)1R as a point of articulation for α(alpha)1R antagonists because concomitant DA receptor stimulation is likely necessary to induce LID. However, extra-striatal α(alpha)1Rs may also contribute to the anti-LID effects of α(alpha)1R antagonists. A key basal ganglia intermediate, the STN, receives noradrenergic innervation [80, 81] and contains postsynaptic α(alpha)1R that have been shown to influence locomotor behavior in a rodent model of PD [12]. To date, this receptor population has not been directly investigated for its role in LID.

Table 15.1 Review of the literature on behavioral effects of pharmacological targeting of noradrenergic receptor subtypes on dopaminergic- and levodopa-mediated behaviors

Receptor target		Species	Levodopa-induced recovery in PD symptoms	Levodopa-induced dyskinesia	Dopaminergic-induced hyperactivity/rotations
α1-	Antagonist	Human	–	–	–
		Primate	No change [77]	No change [77]	Decrease [77]
		Rodent	–	Decrease [75] No change [76]	Decrease [78, 79]
α2-	Agonist	Human	–	–	–
		Primate	Decrease [85]	Decrease [85]	–
		Rodent	Decrease [84]	Decrease [84]	Decrease [78, 79, 86, 87] No change [86]
	Antagonist	Human	No change [97–99]	Decrease [99] No change [97[a], 98]	–
		Primate	No change [85] Increase [94–96, 102]	Decrease [85, 94, 95] No change [102[b]]	Decrease [100]
		Rodent	Decrease [84] No change [93] Increase [101]	Decrease [65[c], 75, 84, 92, 93, 101]	Increase [78, 79, 86, 87]
β1/2-	Antagonist	Human	No change [109]	Decrease [109]	–
		Primate	Decrease [85]	Decrease [85]	–
		Rodent	No change [84]	Decrease [65, 75, 84, 110]	Decrease [110]
β2-	Antagonist	Human	Increase [112, 113]	–	–
		Primate	–	–	–
		Rodent	–	–	–

Symbols denote studies in which mixed results were found with further clarification

[a]In humans, α(alpha)2R antagonism reduced LID in patients with severe but not moderate dyskinesia

[b]In primates, α(alpha)2R antagonism did not alter disabling dyskinesia but worsened non-disabling dyskinesia associated with prolonged levodopa "on-time"

[c]In rodents, α(alpha)2R antagonism was associated with an early reduction, but later potentiation, of LID severity

α(alpha)2-Adrenergic Receptors

Inhibitory α(alpha)2R are the most extensively studied adrenoceptor in LID because they are abundantly expressed in the basal ganglia and have been shown to modify DA synthesis and turnover [82]. For example, knockout mice lacking the $\alpha 2(\text{alpha})_C$R display reduced striatal DA turnover, while mice with overexpression of this receptor display increased DA metabolism [83]. Both agonists and antagonists for this receptor show promise as adjunctive treatment for the reduction of LID.

α(alpha)2R Agonism

Since the early 1990s, several α(alpha)2R agonists have been examined for antidyskinetic properties. Acute treatment with the α(alpha)2R agonist clonidine effectively blocked LID in 6-OHDA-lesioned rats [84] and MPTP-treated nonhuman primates [85] (Table 15.1). Unfortunately, dyskinesia reversal is often accompanied by the return of parkinsonian motor deficits. Levodopa-, apomorphine-, and methylphenidate-induced rotations are all attenuated following α(alpha)2R stimulation [78, 79, 86, 87] (Table 15.1). Thus, the relatively small therapeutic window of current α(alpha)2R agonists may hinder clinical use because the antidyskinetic actions of α(alpha)2R agonists may be confounded by a global suppression of motor movements.

Evidence suggests that striatal α(alpha)2R may underlie the antidyskinetic properties of α(alpha)2R agonists. Over 90 % of GABAergic striatal medium spiny neurons express Gi-coupled-α(alpha)2_CR [88] which when stimulated reduce AC activity [89, 90]. Striatal α(alpha)2R exert simultaneous, opposing effects on striatal signaling pathway activity that may account for the antidyskinetic effects of α(alpha)2R agonists. As shown in Fig. 15.1, LID is associated with overactivity of striatal projection neurons in the "direct" signaling pathway and underactivity of striatal projection neurons in the "indirect" signaling pathway. α2R stimulation inhibits DA D1R signaling pathway activity in striatonigral neurons of the "direct" pathway and enhances DA D2R signaling pathway activity in striatopallidal neurons of the "indirect" pathway [91]. Thus, α(alpha)2R stimulation may reduce LID by decreasing the activity and firing rate of inhibitory GABAergic projections to the hypoactive SNr, which could restore thalamic inhibition and alleviate LID symptoms.

α(alpha)2R Antagonism

α(alpha)2R antagonists have gained more widespread attention as potential levodopa adjuncts because they may possess dual antidyskinetic and antiparkinsonian actions. Many α(alpha)2R antagonists including idazoxan, yohimbine, rauwlscine, and fipamezole reduce the severity or duration of LID in rodents [65, 75, 84, 92, 93] and nonhuman primates [85, 94–96] (Table 15.1). However, these compounds show mixed efficacy in humans. For example, in a phase IIb clinical trial, high doses of fipamezole reduced dyskinesia without hindering levodopa's antiparkinsonian benefits (measured using the Unified Parkinson's Disease Rating Scale; UPDRS) in a population of American PD patients but did not change LID severity in a population of somewhat dissimilar Indian PD patients [97]. Idazoxan treatment did not alter LID severity in PD patients [98]; however, in a Phase IIa pilot study, a moderate dose of idazoxan mildly reduced LID severity [99]. Notably, the high dose of idazoxan actually increased LID severity in some patients [99], a finding supported by recent experimental evidence where idazoxan reduced LID and levodopa-induced rotations during the first hour of testing but extended the duration of LID and rotations during the last hour of testing [65, 100]. Furthermore, α(alpha)2R blockade with the specific antagonist atipamezole has been shown to augment levodopa-induced rotational behavior in hemiparkinsonian rats [79, 86] (Table 15.1). Interestingly, α(alpha)2R antagonists prevent severe/disabling LID but do not completely abolish levodopa-induced behaviors. This feature may underlie the ability of α(alpha)2R antagonists to maintain levodopa's promotor benefits (Table 15.1). In fact, both idazoxan and fipamezole have the capacity to extend the duration of levodopa's antiparkinsonian effects [94, 95, 101, 102]. Only one study described a worsening of parkinsonian symptoms where yohimbine treatment diminished time spent on the rotorod [84].

Several mechanisms discussed herein could potentially account for the antidyskinetic and antiparkinsonian effects of this class of compound. Recent experimental evidence has shown that idazoxan treatment simultaneously reduces levodopa-induced striatal DA efflux and LID [92]. This is surprising because intrastriatal infusions of idazoxan actually enhance striatal DA efflux in DA-lesioned rats [103], and systemic administration of the selective α(alpha)2R antagonists RX821002 and RX811059, both derivatives of idazoxan, increases striatal DA overflow [104]. Simultaneous blockade of α(alpha)2R at several points within the basal ganglia circuitry could contribute to the purported coincident antidyskinetic and antiparkinsonian effects. Striatal α(alpha)2R blockade may facilitate the antiparkinsonian effects of α(alpha)2R antagonists via enhancement of DA D1R-mediated AC activity and signaling. Extrastriatal α(alpha)2R likely underlie the antidyskinetic effects of α(alpha)2R antagonists. Blockade of α(alpha)2R on hypoactive SNr neurons in LID (Fig. 15.1c) increases GABA release [105] from neurons that innervate the thalamus. Thus, SNr α(alpha)2R antagonism may restore thalamic inhibition to ultimately reduce LID.

β(beta)-Adrenergic Receptors

β(beta)R are found in high concentrations in the striatum [106] rendering them an intriguing but underexplored target for the control of LID. As shown in Table 15.1, β(beta)R compounds show promise as adjuncts for reducing LID and parkinsonian signs. Peripherally active β(beta)2R antagonists (i.e., beta-blockers) reduce postural tremor in MPTP-treated nonhuman primates [107] and essential tremor in humans [108], while centrally active compounds modify LID expression. In a small open-label study in humans, the pan β(beta)1/2R antagonist propranolol was reported to blunt choreic and ballistic dyskinetic movements with little efficacy for reducing dystonia [109]. Propranolol's antidyskinetic effects have been supported in rodent [65, 75, 84, 110] and nonhuman primate [85] models of PD as well. There is one nonhuman primate study suggesting that doses of propranolol that reduce LID may also alter levodopa's promotor benefits [85], although others, including clinical work, report that no change in levodopa-mediated behaviors [84, 109, 110].

Striatal β(beta)R may mediate the antidyskinetic effects of β(beta)R antagonists. Direct infusion of propranolol into the lesioned striatum reduced LID in hemiparkinsonian rats [110]. In vivo and in vitro work shows that β(beta)R stimulation enhanced, while blockade reduced striatal DA efflux [111]. This could explain how β(beta)R antagonists reduce LID but at higher doses may also potentially interfere with levodopa's antiparkinsonian effects. In support, co-treatment of the centrally active β(beta)2R agonist albuterol with levodopa improved parkinsonian motor deficits as measured by the UPDRS and finger-tapping tests [112, 113]. β(beta)2R agonism also extended levodopa "on-time" in PD patients [114], perhaps by enhancing striatal DA levels. Alternatively, striatal β(beta)R antagonism may reduce LID by blunting aberrant DA receptor signaling pathway activity [91, 110]. The extent to which β(beta)Rs are co-localized with DA D1R or D2R expressing neurons is unknown, but recent evidence suggests that β(beta)1R are found on DA D1R expressing neurons [115]. Striatal preprodynorphin (PPD) gene expression used as a marker of D1R-mediated activity of the direct pathway [116] is increased following levodopa treatment [117]. A low dose of propranolol that moderately reduced LID also blocked striatal PPD mRNA expression in levodopa-treated hemiparkinsonian rats [110].

Considerations

As is often the case, several noradrenergic compounds have proceeded to clinical trials despite the lack of knowledge surrounding how they work. Key issues must be addressed in order to fully elucidate the role of noradrenergic adjuncts in LID. First, several α(alpha)2R antagonists including yohimbine, idazoxan, rauwolscine and to a lesser extent fipamezole and atipamezole have pharmacological effects at non-noradrenergic receptors. Chief among these are agonism of serotonin 1A receptors and antagonism at DA D2R [95, 118–121], both of which have repeatedly been shown to reduce LID [122–126]. Thus, off-target action should be an important consideration when testing α(alpha)2R antagonists. Second, levodopa methyl ester, DA, and the DA metabolite 3-MT show moderate to high binding affinities for the α(alpha)2R [90, 127, 128]. Therefore, it is important to determine if α(alpha)2R blockade reduces LID by preventing the action of levodopa-derived DA and/or NE at α(alpha)2R within the basal ganglia. Finally, PD is often associated with reduced autonomic nervous system activity and some PD patients experience orthostatic hypotension and bradycardia [129, 130]. Peripheral β(beta)R blockade could exacerbate these symptoms and may explain why these compounds have not seen widespread use.

Conclusions

A large body of work shows that the noradrenergic system originating in the LC is profoundly affected in PD. Despite this knowledge, the impact of such loss on levodopa treatment efficacy and side effects remains speculative. Significant points of noradrenergic articulation in the basal ganglia circuitry suggest that α(alpha) and β(beta) receptor compounds may be viable adjuncts to levodopa but the mechanisms through which they exert their effects remain underexplored. However, it has become increasingly apparent that LID is a multifaceted disorder with complex interactions between many neurotransmitter systems. As such, drug development in LID needs to move beyond the antiquated notion that targeting a discreet receptor population is sufficient to ablate dyskinesia symptoms. The future of pharmacological treatments for LID likely lies in systematically targeting multiple noradrenergic and/or other neurotransmitters receptor classes simultaneously in order to normalize aberrant neurotransmission in the basal ganglia responsible for LID.

References

1. Loughlin SE, Foote SL, Grzanna R. Efferent projections of nucleus locus coeruleus: morphologic subpopulations have different efferent targets. Neuroscience. 1986;18:307–19.
2. Room P, Postema F, Korf J. Divergent axon collaterals of rat locus coeruleus neurons: demonstration by a fluorescent double labeling technique. Brain Res. 1981;221:219–30.
3. Loughlin SE, Foote SL, Fallon JH. Locus coeruleus projections to cortex: topography, morphology and collateralization. Brain Res Bull. 1982;9:287–94.
4. Berridge CW, Waterhouse BD. The locus coeruleus–noradrenergic system: modulation of behavioral state and state-dependent cognitive processes. Brain Res Rev. 2003;42:33–84.
5. Mason ST, Fibiger HC. Regional topography within noradrenergic locus coeruleus as revealed by retrograde transport of horseradish peroxidase. J Comp Neurol. 1979;187:703–24.
6. Mann DM. The locus coeruleus and its possible role in ageing and degenerative disease of the human central nervous system. Mech Ageing Dev. 1983;23:73–94.
7. Gaspar P. Anatomy of the noradrenergic pathways in the primate brain: alterations in Parkinson's disease. In: Briley M, Marien M, editors. Noradrenergic mechanisms in Parkinson's disease. Boca Raton: CRC Press; 1994. p. 73–88.
8. Margules DL, Lewis MJ, Dragovich JA, Margules AS. Hypothalamic norepinephrine: circadian rhythms and the control of feeding behavior. Science. 1972;178:640–3.
9. Pfaus JG. Pathways of sexual desire. J Sex Med. 2009;6:1506–33.
10. Gesi M, Soldani P, Giorgi FS, Santinami A, Bonaccorsi I, et al. The role of the locus coeruleus in the development of Parkinson's disease. Neurosci Biobehav Rev. 2000;24:655–68.
11. Arcos D, Sierra A, Nunez A, Flores G, Aceves J, Arias-Montano JA. Noradrenaline increases the firing rate of a subpopulation of rat subthalamic neurons through the activation of alpha1-adrenoceptors. Neuropharmacology. 2003;45:1070–9.
12. Belujon P, Bezard E, Taupignon A, Bioulac B, Benazzouz A. Noradrenergic modulation of subthalamic nucleus activity: behavioral and electrophysiological evidence in intact and 6-hydroxydopamine-lesioned rats. J Neurosci. 2007;27:9595–606.
13. Cash R, Ruberg M, Raisman R, Agid Y. Adrenergic receptors in Parkinson's disease. Brain Res. 1984;322:269–75.
14. Rommelfanger KS, Mitrano DA, Smith Y, Weinshenker D. Light and electron microscopic localization of alpha-1 adrenergic receptor immunoreactivity in the rat striatum and ventral midbrain. Neuroscience. 2009;158:1530–40.
15. Atlas D, Melamed E. Direct mapping of beta-adrenergic receptors in the rat central nervous system by a novel fluorescent beta-blocker. Brain Res. 1978;150:377–85.
16. Nicholas AP, Pieribone VA, Hökfelt T. Cellular localization of messenger RNA for beta-1 and beta-2 adrenergic receptors in rat brain: an in situ hybridization study. Neuroscience. 1993;56:1023–39.

17. Rosin DL, Talley EM, Lee A, Stornetta RL, Gaylinn BD, Guyenet PG, Lynch KR. Distribution of alpha 2C-adrenergic receptor-like immunoreactivity in the rat central nervous system. J Comp Neurol. 1996;372:135–65.
18. MacDonald E, Scheinin M. Distribution and pharmacology of alpha 2-adrenoceptors in the central nervous system. J Physiol Pharmacol. 1995;46:241–58.
19. Zhang W, Ordway GA. The alpha2C-adrenoceptor modulates GABA release in mouse striatum. Brain Res Mol Brain Res. 2003;112:24–32.
20. German DC, Manaye KF, White CL, Woodward DJ, McIntire DD, Smith WK, et al. Disease-specific patterns of locus coeruleus cell loss. Ann Neurol. 1992;32:667–76.
21. Zarow C, Lyness SA, Mortimer JA, Chui HC. Neuronal loss is greater in the locus coeruleus than nucleus basalis and substantia nigra in Alzheimer and Parkinson diseases. Arch Neurol. 2003;60:337–41.
22. Hornykiewicz O, Kish SJ. Biochemical pathophysiology of Parkinson's disease. Adv Neurol. 1986;45:19–34.
23. Patt S, Gerhard L. A golgi study of human locus coeruleus in normal brains and in Parkinson's disease. Neuropathol Appl Neurobiol. 1993;19:519–23.
24. Baloyannis SJ, Costa V, Baloyannis IS. Morphological alterations of the synapses in the locus coeruleus in Parkinson's disease. J Neurol Sci. 2006;248:35–41.
25. Bertrand E, Lechowicz W, Szpak GA, Dymecki J. Qualitative and quantitative analysis of locus coeruleus neuron in Parkinson's disease. Folia Neuropathol. 1997;35:80–6.
26. Braak H, Braak E. Pathoanatomy of Parkinson's disease. J Neurol. 2000;247 Suppl 2:II/3–10.
27. McMillan PJ, White SS, Franklin A, Greenup JL, Leverenz JB, Raskind MA, Szot P. Differential response of the central noradrenergic nervous system to the loss of locus coeruleus neurons in Parkinson's disease and Alzheimer's disease. Brain Res. 2011;1373:240–52.
28. Shannak K, Rajput A, Rozdilsky B, Kish S, Gilbert J, Hornykiewicz O. Noradrenaline, dopamine and serotonin levels and metabolism in the human hypothalamus: observations in Parkinson's disease and normal subjects. Brain Res. 1994;639:33–41.
29. Pifl C, Kish SJ, Hornykiewicz O. Thalamic noradrenaline in Parkinson's disease: deficits suggest role in motor and non-motor symptoms. Mov Disord. 2012;27:1618–24.
30. Hurst JH, LeWitt PA, Burns RS, Foster NL, Lovenberg W. CSF dopamine-beta-hydroxylase activity in Parkinson's disease. Neurology. 1985;3:565–8.
31. O'Connor DT, Cervenka JH, Stone RA, Levine GL, Parmer RJ, Franco-Bourland RE, et al. Dopamine beta-hydroxylase immunoreactivity in human cerebrospinal fluid: properties, relationship to central noradrenergic neuronal activity and variation in Parkinson's disease and congenital dopamine beta-hydroxylase deficiency. Clin Sci. 1994;86:149–58.
32. Eldrup E, Mogensen P, Jacobsen J, Pakkenberg H, Christensen NJ. CSF and plasma concentrations of free norepinephrine, dopamine, 3,4-dihydroxyphenylacetic acid (DOPAC), 3,4-dihydroxyphenylalanine (DOPA), and epinephrine in Parkinson's disease. Acta Neurol Scand. 1995;92:116–21.
33. Chia LG, Cheng FC, Kuo JS. Monoamines and their metabolites in plasma and lumbar cerebrospinal fluid of Chinese patients with Parkinson's disease. J Neurol Sci. 1993;116:125–34.
34. Gaspar P, Duyckaerts C, Alvarez C, Javoy-Agid F, Berger B. Alterations of dopaminergic and noradrenergic innervations in motor cortex in Parkinson's disease. Ann Neurol. 1991;30: 365–74.
35. Cash R, Raisman R, Lanfumey L, Ploska A, Agid Y. Cellular localization of adrenergic receptors in rat and human brain. Brain Res. 1986;370:127–35.
36. Caccia C, Achilli G, Buonamici M, Carpentieri M, Cervini MA, Maj R, et al. Receptor adaptive responsiveness in disease models: 6-OHDA lesioned and spontaneously hypertensive rats. J Recept Res. 1988;8:97–105.
37. Johnson EW, Wolfe BB, Molinoff PB. Regulation of subtypes of beta-adrenergic receptors in rat brain following treatment with 6-hydroxydopamine. J Neurosci. 1989;9:2297–305.
38. Reisine TD, Nagy JI, Beaumont K, Fibiger HC, Yamamura HI. The localization of receptor binding sites in the substantia nigra and striatum of the rat. Brain Res. 1979;177:241–52.

39. Waeber C, Rigo M, Chinaglia G, Probst A, Palacios JM. Beta-adrenergic receptor subtypes in the basal ganglia of patients with Huntington's chorea and Parkinson's disease. Synapse. 1991;8:270–80.
40. Ribas C, Miralles A, Busquets X, García-Sevilla JA. Brain alpha(2)-adrenoceptors in monoamine-depleted rats: increased receptor density, G coupling proteins, receptor turnover and receptor mRNA. Br J Pharmacol. 2001;132:1467–76.
41. Alachkar A, Brotchie JM, Jones OT. Changes in the mRNA levels of alpha(2A) and alpha (2C) adrenergic receptors in rat models of Parkinson's disease and L-DOPA-induced dyskinesia. J Mol Neurosci. 2001;46:145–52.
42. Delaville C, De Deurwaerdère P, Benazzouz A. Noradrenaline and Parkinson's disease. Front Syst Neurosci. 2011;5:1–12.
43. Grimbergen YA, Langston JW, Roos RA, Bloem BR. Postural instability in Parkinson's disease: the adrenergic hypothesis and the locus coeruleus. Expert Rev Neurother. 2009;9: 279–90.
44. Rommelfanger KS, Edwards GL, Freeman KG, Liles LC, Miller GW, Weinshenker D. Norepinephrine loss produces more profound motor deficits than MPTP treatment in mice. Proc Natl Acad Sci. 2007;104:13804–9.
45. Mavridis M, Degryse AD, Lategan AJ, Marien MR, Colpaert FC. Effects of locus coeruleus lesions on parkinsonian signs, striatal dopamine and substantia nigra cell loss after 1-methyl-4-phenyl-1,2,3,6-tetrahydropyridine in monkeys: a possible role for the locus coeruleus in the progression of Parkinson's disease. Neuroscience. 1991;41:507–23.
46. Srinivasan J, Schmidt WJ. Potentiation of parkinsonian symptoms by depletion of locus coeruleus noradrenaline in 6-hydroxydopamine-induced partial degeneration of substantia nigra in rats. Eur J Neurosci. 2003;17:2586–92.
47. Archer T, Fredriksson A. Influence of noradrenaline denervation on MPTP-induced deficits in mice. J Neural Transm. 2006;113:1119–29.
48. Fritschy JM, Grzanna R. Immunohistochemical analysis of the neurotoxic effects of DSP-4 identifies two populations of noradrenergic axon terminals. Neuroscience. 1989;30:181–97.
49. Wolfman C, Abó V, Calvo D, Medina J, Dajas F, Silveira R. Recovery of central noradrenergic neurons one year after the administration of the neurotoxin DSP4. Neurochem Int. 1994;25:395–400.
50. Cheetham SC, Viggers JA, Butler SA, Prow MR, Heal DJ. [3H]nisoxetin-- radioligand for noradrenaline reuptake sites: correlation with inhibition of [3H]noradrenaline uptake and effect of DSP-4 lesioning and antidepressant treatments. Neuropharmacology. 1996;35:63–70.
51. Ross SB, Johansson JG, Lindberg B, Dahlbom R. Cyclizing compounds. I. Tertiary N-(2-bromobenzyl-N-haloalkylamines) with adrenergic blocking action. Acta Pharm Suec. 1973;10:29–42.
52. Fritschy JM, Grzanna R. Selective effects of DSP-4 on locus coeruleus axons: are there pharmacologically different types of noradrenergic axons in the central nervous system? Prog Brain Res. 1991;88:257–68.
53. Szot P, Miguelez C, White SS, Franklin A, Sikkema C, Wilkinson CW, et al. A comprehensive analysis of the effect of DSP4 on the locus coeruleus noradrenergic system in the rat. Neuroscience. 2010;166:279–91.
54. Marin C, Aguilar E, Bonastre M. Effect of locus coeruleus denervation on levodopa-induced motor fluctuations in hemiparkinsonian rats. J Neural Transm. 2008;115:1133–9.
55. Pérez V, Marin C, Rubio A, Aguilar E, Barbanoj M, Kulisevsky J. Effect of the additional noradrenergic neurodegeneration to 6-OHDA-lesioned rats in levodopa-induced dyskinesias and in cognitive disturbances. J Neural Transm. 2009;116:1257–66.
56. Srinivasan J, Schmidt WJ. Behavioral and neurochemical effects of noradrenergic depletions with N-(2-chloroethyl)-N-ethyl-2-bromobenzylamine in 6-hydroxydopamine-induced rat model of Parkinson's disease. Behav Brain Res. 2004;151:191–9.
57. Pérez V, Sosti V, Rubio A, Barbanoj M, Rodríguez-Alvarez J, Kulisevsky J. Modulation of the motor response to dopaminergic drugs in a parkinsonian model of combined dopaminergic and noradrenergic degeneration. Eur J Pharmacol. 2007;576:83–90.

58. Hallman H, Sundström E, Jonsson G. Effects of the noradrenaline neurotoxin DSP 4 on monoamine neurons and their transmitter turnover in rat CNS. J Neural Transm. 1984;60:89–102.
59. Fornai F, Bassi L, Torracca MT, Alessandrì MG, Scalori V, Corsini GU. Region- and neurotransmitter-dependent species and strain differences in DSP-4-induced monoamine depletion in rodents. Neurodegeneration. 1996;5:241–9.
60. Miguelez C, Aristieta A, Cenci MA, Ugedo L. The locus coeruleus is directly implicated in L-DOPA induced dyskinesia in Parkinsonian rats: An electrophysiological and behavioural study. PLoS One. 2011;6:1–12.
61. Pérez V, Sosti V, Rubio A, Barbanoj M, Gich I, Rodríguez-Alvarez J, et al. Noradrenergic modulation of the motor response induced by long-term levodopa administration in Parkinsonian rats. J Neural Transm. 2009;116:867–74.
62. Nishi K, Kondo T, Narabayashi H. Destruction of norepinephrine terminals in 1-methyl-4-phenyl-1,2,3,6-tetrahydropyridine (MPTP)-treated mice reduces locomotor activity induced by L-dopa. Neurosci Lett. 1991;123:244–7.
63. Blum D, Torch S, Lambeng N, Nissou M, Benabid AL, Sadoul R, et al. Molecular pathways involved in the neurotoxicity of 6-OHDA, dopamine and MPTP: contribution to the apoptotic theory in Parkinson's disease. Prog Neurobiol. 2001;65:135–72.
64. Deumens R, Blokland A, Prickaerts J. Modeling Parkinson's disease in rats: an evaluation of 6-OHDA lesions of the nigrostriatal pathway. Exp Neurol. 2002;175:303–17.
65. Barnum CJ, Bhide N, Lindenbach D, Surrena MA, Goldenberg AA, Tignor S, et al. Effects of noradrenergic denervation on L-DOPA-induced dyskinesia and its treatment by α- and β-adrenergic receptor antagonists in hemiparkinsonian rats. Pharmacol Biochem Behav. 2012;100:607–15.
66. Fulceri F, Biagioni F, Ferrucci M, Lazzeri G, Bartalucci A, Galli V, et al. Abnormal involuntary movements (AIMs) following pulsatile dopaminergic stimulation: severe deterioration and morphological correlates following the loss of locus coeruleus neurons. Brain Res. 2007;1135:219–29.
67. Pinnock RD. Sensitivity of compacta neurones in the rat substantia nigra slice to dopamine agonists. Eur J Pharmacol. 1983;96:269–76.
68. Buck K, Ferger B. Comparison of intrastriatal administration of noradrenaline and l-DOPA on dyskinetic movements: a bilateral reverse in vivo microdialysis study in 6-hydroxydopamine-lesioned rats. Neuroscience. 2009;159:16–20.
69. Raiteri M, Del Carmine R, Bertollini A, Levi G. Effect of sympathomimetic amines on the synaptosomal transport of noradrenaline, dopamine and 5-hydroxytryptamine. Eur J Pharmacol. 1977;41:133–43.
70. Moron JA, Brockington A, Wise RA, Rocha BA, Hope BT. Dopamine uptake through the norepinephrine transporter in brain regions with low levels of the dopamine transporter: evidence from knock-out mouse lines. J Neurosci. 2002;22:389–95.
71. Arai A, Tomiyama M, Kannari K, Kimura T, Suzuki C, Watanabe M, et al. Reuptake of L-DOPA-derived extracellular DA in the striatum of a rodent model of Parkinson's disease via norepinephrine transporter. Synapse. 2008;62:632–5.
72. Pan WH, Yang SY, Lin SK. Neurochemical interaction between dopaminergic and noradrenergic neurons in the medial prefrontal cortex. Synapse. 2004;53:44–52.
73. Grenhoff J, North RA, Johnson SW. Alpha 1-adrenergic effects on dopamine neurons recorded intracellularly in the rat midbrain slice. Eur J Neurosci. 1995;7:1707–13.
74. Sommermeyer H, Frielingsdorf J, Knorr A. Effects of prazosin on the dopaminergic neurotransmission in rat brain. Eur J Pharmacol. 1995;276:267–70.
75. Buck K, Ferger B. The selective alpha1 adrenoceptor antagonist HEAT reduces L-DOPA-induced dyskinesia in a rat model of Parkinson's disease. Synapse. 2010;64:117–26.
76. Paquette MA, Foley K, Brudney EG, Meshul CK, Johnson SW, Berger SP. The sigma-1 antagonist BMY-14802 inhibits L-DOPA-induced abnormal involuntary movements by a WAY-100635-sensitive mechanism. Psychopharmacology (Berl). 2009;204:743–54.

77. Visanji NP, Fox SH, Johnston TH, Millan MJ, Brotchie JM. Alpha1-adrenoceptors mediate dihydroxyphenylalanine-induced activity in 1-methyl-4-phenyl-1,2,3,6-tetrahydropyridine-lesioned macaques. J Pharmacol Exp Ther. 2009;328:276–83.
78. Mavridis M, Colpaert FC, Millan MJ. Differential modulation of (+)-amphetamine-induced rotation in unilateral substantia nigra-lesioned rats by alpha 1 as compared to alpha 2 agonists and antagonists. Brain Res. 1991;562:216–24.
79. Haapalinna A, Leino T, Heinonen E. The alpha 2-adrenoceptor antagonist atipamezole potentiates anti-Parkinsonian effects and can reduce the adverse cardiovascular effects of dopaminergic drugs in rats. Naunyn Schmiedebergs Arch Pharmacol. 2003;368:342–51.
80. Parent A, Hazrati LN. Functional anatomy of the basal ganglia. II. The place of subthalamic nucleus and external pallidum in basal ganglia circuitry. Brain Res Brain Res Rev. 1995;20:128–54.
81. Canteras NS, Shammah-Lagnado SJ, Silva BA, Ricardo JA. Afferent connections of the subthalamic nucleus: a combined retrograde and anterograde horseradish peroxidase study in the rat. Brain Res. 1990;1990(513):43–59.
82. Anden N, Grabowska M. Pharmacological evidence for a stimulation of dopamine neurons by noradrenaline neurons in the brain. Eur J Pharmacol. 1976;39:275–82.
83. Sallinen J, Link RE, Haapalinna A, Viitamaa T, Kulatunga M, Sjoholm B, et al. Genetic alteration of alpha 2C-adrenoceptor expression in mice: influence on locomotor hypothermic, and neurochemical effects of dexmedetomidine, a subtypenonselective alpha 2-adrenoceptor agonist. Mol Pharmacol. 1997;51:36–46.
84. Dekundy A, Lundblad M, Danysz W, Cenci MA. Modulation of L-DOPA-induced abnormal involuntary movements by clinically tested compounds: further validation of the rat dyskinesia model. Behav Brain Res. 2007;179:76–89.
85. Gomez-Mandilla B, Bedard PJ. Effect of nondopaminergic drugs on L-dopa induced dyskinesias in MPTP-treated monkeys. Clin Neuropharmacol. 1993;16:418–27.
86. Huotari M, Kukkonen K, Liikka N, Potasev T, Raasmaja A, Männistö PT. Effects of histamine H(3)-ligands on the levodopa-induced turning behavior of hemiparkinsonian rats. Parkinsonism Relat Disord. 2000;6:159–64.
87. Chopin P, Colpaert FC, Marien M. Effects of alpha-2 adrenoceptor agonists and antagonists on circling behavior in rats with unilateral 6-hydroxydopamine lesions of the nigrostriatal pathway. J Pharmacol Exp Ther. 1999;288:798–804.
88. Holmberg M, Scheinin M, Kurose H, Miettinen R. Adrenergic alpha2C-receptors reside in rat striatal GABAergic projection neurons: comparison of radioligand binding and immunohistochemistry. Neuroscience. 1999;93:1323–33.
89. Lu L, Ordway GA. Alpha2C-adrenoceptors mediate inhibition of forskolin-stimulated cAMP production in rat striatum. Brain Res Mol Brain Res. 1997;52:228–34.
90. Zhang W, Klimek V, Farley JT, Zhu MY, Ordway GA. alpha2C adrenoceptors inhibit adenylyl cyclase in mouse striatum: potential activation by dopamine. J Pharmacol Exp Ther. 1999;289:1286–92.
91. Hara M, Fukui R, Hieda E, Kuroiwa M, Bateup HS, Kano T, et al. Role of adrenoceptors in the regulation of dopamine/DARPP-32 signaling in neostriatal neurons. J Neurochem. 2010;4:1046–59.
92. Buck K, Voehringer P, Ferger B. The alpha(2) adrenoceptor antagonist idazoxan alleviates L-DOPA-induced dyskinesia by reduction of striatal dopamine levels: an in vivo microdialysis study in 6-hydroxydopamine-lesioned rats. J Neurochem. 2010;112:444–52.
93. Lundblad M, Andersson M, Winkler C, Kirik D, Wierup N, Cenci MA. Pharmacological validation of behavioural measures of akinesia and dyskinesia in a rat model of Parkinson's disease. Eur J Neurosci. 2002;15:120–32.
94. Henry B, Fox SH, Peggs D, Crossman AR, Brotchie JM. The alpha2-adrenergic receptor antagonist idazoxan reduces dyskinesia and enhances anti-parkinsonian actions of L-dopa in the MPTP-lesioned primate model of Parkinson's disease. Mov Disord. 1999;14:744–53.

95. Savola JM, Hill M, Engstrom M, Merivuori H, Wurster S, McGuire SG, et al. Fipamezole (JP-1730) is a potent alpha2 adrenergic receptor antagonist that reduces levodopa-induced dyskinesia in the MPTP-lesioned primate model of Parkinson's disease. Mov Disord. 2003;18:872–83.
96. Grondin R, Hadj Tahar A, Doan VD, Ladure P, Bédard PJ. Noradrenoceptor antagonism with idazoxan improves L-dopa-induced dyskinesias in MPTP monkeys. Naunyn Schmiedebergs Arch Pharmacol. 2000;361:181–6.
97. Lewitt PA, Hauser RA, Lu M, Nicholas AP, Weiner W, Coppard N, et al. Randomized clinical trial of fipamezole for dyskinesia in Parkinson disease (FJORD study). Neurology. 2012;79:163–9.
98. Manson AJ, Iakovidou E, Lees AJ. Idazoxan is ineffective for levodopa-induced dyskinesias in Parkinson's disease. Mov Disord. 2000;15:336–7.
99. Rascol O, Arnulf I, Peyro-Saint Paul H, Brefel-Courbon C, Vidailhet M, Thalamas C, et al. Idazoxan, an alpha-2 antagonist, and L-DOPA-induced dyskinesias in patients with Parkinson's disease. Mov Disord. 2001;16:708–13.
100. Domino EF, Ni L, Colpaert F, Marien M. Effects of (+/−)-idazoxan alone and in combination with L-DOPA methyl ester in MPTP-induced hemiparkinsonian monkeys. Receptors Channels. 2003;9:335–8.
101. Fox SH, Henry B, Hill MP, Peggs D, Crossman AR, Brotchie JM. Neural mechanisms underlying peak-dose dyskinesia induced by levodopa and apomorphine are distinct: evidence from the effects of the alpha(2) adrenoceptor antagonist idazoxan. Mov Disord. 2001;16:642–50.
102. Johnston TH, Fox SH, Piggott MJ, Savola JM, Brotchie JM. The α_2 adrenergic antagonist fipamezole improves quality of levodopa action in Parkinsonian primates. Mov Disord. 2010;25:2084–93.
103. Hertel P, Nomikos GG, Svensson TH. Idazoxan preferentially increases dopamine output in the rat medial prefrontal cortex at the nerve terminal level. Eur J Pharmacol. 1999;371:153–8.
104. Hudson AL, Robinson ES, Lalies MD, Tyacke RJ, Jackson HC, Nutt DJ. In vitro and in vivo approaches to the characterization of the alpha2-adrenoceptor. J Auton Pharmacol. 1999;19:311–20.
105. Alachkar A, Brotchie J, Jones OT. alpha2-Adrenoceptor-mediated modulation of the release of GABA and noradrenaline in the rat substantia nigra pars reticulata. Neurosci Lett. 2006;395:138–42.
106. Rainbow TC, Parsons B, Wolfe BB. Quantitative autoradiography of beta 1- and beta 2-adrenergic receptors in rat brain. Proc Natl Acad Sci U S A. 1984;81:1585–9.
107. Kuno S, Mizuta E, Nishida J, Takechi M. Therapeutic effects of arotinolol, a beta-adrenergic blocker, on tremor in MPTP-induced parkinsonian monkeys. Clin Neuropharmacol. 1992;15:381–6.
108. Lyons KE, Pahwa R. Pharmacotherapy of essential tremor: an overview of existing and upcoming agents. CNS Drugs. 2008;22:1037–45.
109. Carpentier AF, Bonnet AM, Vidailhet M, Agid Y. Improvement of levodopa-induced dyskinesia by propranolol in Parkinson's disease. Neurology. 1996;46:1548–51.
110. Lindenbach D, Ostock CY, Eskow Jaunarajs KL, Dupre KB, Barnum CJ, Bhide N, et al. Behavioral and cellular modulation of L-DOPA-induced dyskinesia by beta-adrenoceptor blockade in the 6-hydroxydopamine-lesioned rat. J Pharmacol Exp Ther. 2011;337:755–65.
111. Reisine TD, Chesselt MF, Lubetzki C, Cheramy A, Glowinski J. A role for striatal beta-adrenergic receptors in the regulation of dopamine release. Brain Res. 1982;241:123–30.
112. Alexander GM, Schwartzman RJ, Nukes TA, Grothusen JR, Hooker MD. Beta 2-adrenergic agonist as adjunct therapy to levodopa in Parkinson's disease. Neurology. 1994;44:1511–3.
113. Uc EY, Lambert CP, Harik SI, Rodnitzky RL, Evans WJ. Albuterol improves response to levodopa and increases skeletal muscle mass in patients with fluctuating Parkinson disease. Clin Neuropharmacol. 2003;26:207–12.

114. Hishida R, Kurahashi K, Narita S, Baba T, Matsunaga M. "Wearing-off" and beta 2-adrenoceptor agonist in Parkinson's disease. Lancet. 1992;339:870.
115. Meitzen J, Luoma JI, Stern CM, Mermelstein PG. β1-Adrenergic receptors activate two distinct signaling pathways in striatal neurons. J Neurochem. 2011;116:984–95.
116. Gerfen CR, Engber TM, Mahan LC, Susel Z, Chase TN, Monsma Jr FJ, et al. D1 and D2 dopamine receptor-regulated gene expression of striatonigral and striatopallidal neurons. Science. 1990;250:1429–32.
117. Cenci MA, Lee CS, Bjorklund A. L-DOPA-induced dyskinesia in the rat is associated with striatal overexpression of prodynorphin- and glutamic acid decarboxylase mRNA. Eur J Neurosci. 1998;10:2694–706.
118. Millan MJ, Newman-Tancredi A, Audinot V, Cussac D, Lejeune F, Nicolas JP, et al. Agonist and antagonist actions of yohimbine as compared to fluparoxan at alpha(2)-adrenergic receptors (AR)s, serotonin (5-HT)(1A), 5-HT(1B), 5-HT(1D) and dopamine D(2) and D(3) receptors. Significance for the modulation of frontocortical monoaminergic transmission and depressive states. Synapse. 2000;35:79–95.
119. Newman-Tancredi A, Nicolas JP, Audinot V, Gavaudan S, Verrièle L, Touzard M, et al. Actions of alpha2 adrenoceptor ligands at alpha2A and 5-HT1A receptors: the antagonist, atipamezole, and the agonist, dexmedetomidine, are highly selective for alpha2A adrenoceptors. Naunyn Schmiedebergs Arch Pharmacol. 1998;358:197–206.
120. De Vos H, Convents A, De Keyser J, De Backer JP, Van Megen IJ, Ebinger G, et al. Autoradiographic distribution of alpha 2 adrenoceptors, NAIBS, and 5-HT1A receptors in human brain using [3H]idazoxan and [3H]rauwolscine. Brain Res. 1991;566:13–20.
121. Pertovaara A, Haapalinna A, Sirviö J, Virtanen R. Pharmacological properties, central nervous system effects, and potential therapeutic applications of atipamezole, a selective alpha2-adrenoceptor antagonist. CNS Drug Rev. 2005;11:273–88.
122. Eskow KL, Gupta V, Alam S, Park JY, Bishop C. The partial 5-HT(1A) agonist buspirone reduces the expression and development of l-DOPA-induced dyskinesia in rats and improves l-DOPA efficacy. Pharmacol Biochem Behav. 2007;87:306–14.
123. Bishop C, Krolewski DM, Eskow KL, Barnum CJ, Dupre KB, Deak T, et al. Contribution of the striatum to the effects of 5-HT1A receptor stimulation in L-DOPA-treated hemiparkinsonian rats. J Neurosci Res. 2009;87:1645–58.
124. Huot P, Brotchie JM. 5-HT(1A) receptor stimulation and L-DOPA-induced dyskinesia in Parkinson's disease: bridging the gap between serotonergic and glutamatergic mechanisms. Exp Neurol. 2011;231:195–8.
125. Taylor JL, Bishop C, Walker PD. Dopamine D1 and D2 receptor contributions to L-DOPA-induced dyskinesia in the dopamine-depleted rat. Pharmacol Biochem Behav. 2005;81:887–93.
126. Koprich JB, Huot P, Fox SH, Jarvie K, Lang AE, Seeman P, et al. The effects of fast-off-D2 receptor antagonism on L-DOPA-induced dyskinesia and psychosis in parkinsonian macaques. Prog Neuropsychopharmacol Biol Psychiatry. 2013;43:151–6.
127. Alachkar A, Brotchie JM, Jones OT. Binding of dopamine and 3-methoxytyramine as l-DOPA metabolites to human alpha(2)-adrenergic and dopaminergic receptors. Neurosci Res. 2010;67:245–9.
128. Antkiewicz-Michaluk L, Ossowska K, Romańska I, Michaluk J, Vetulani J. 3-Methoxytyramine, an extraneuronal dopamine metabolite plays a physiological role in the brain as an inhibitory regulator of catecholaminergic activity. Eur J Pharmacol. 2008;599:32–5.
129. Stuebner E, Vichayanrat E, Low DA, Mathias CJ, Isenmann S, Haensch CA. Twenty-four hour non-invasive ambulatory blood pressure and heart rate monitoring in Parkinson's disease. Front Neurol. 2013;4(49):1–14.
130. Oka H, Morita M, Onouchi K, Yoshioka M, Mochio S, Inoue K. Cardiovascular autonomic dysfunction in dementia with Lewy bodies and Parkinson's disease. J Neurol Sci. 2007;254:72–7.

Chapter 16
Involvement of the Cholinergic System in Levodopa-Induced Dyskinesia

Maryka Quik, Tanuja Bordia, Danhui Zhang, and Xiomara Perez

Abstract Although levodopa-induced dyskinesia (LID) in Parkinson's disease arises because of aberrant dopaminergic transmission, extensive evidence indicates that nondopaminergic drugs may be useful in the suppression of these abnormal involuntary movements. Here, we review a compelling literature which suggests that drugs that act on the nicotinic cholinergic system are beneficial in reducing LID. Nicotine treatment decreased LID in several parkinsonian animal models including mice, rats, and monkeys using treatment modes that readily extend to human use (patch or oral administration). Nicotine decreased LID when given either before or several months after the start of levodopa treatment, with no tolerance to its beneficial effect during the course of the study (30 weeks). Work with nicotinic acetylcholine receptor (nAChR) null mutant mice shows that nicotine exerts its antidyskinetic effects by acting at nAChRs, with the $\alpha4\beta2$, $\alpha6\beta2$, and $\alpha7$ receptor subtypes all contributing to the occurrence of LID. An involvement of multiple subtypes in LID is also supported by studies with drugs targeting select nAChR populations. Notably, nicotine and nAChR drugs did not worsen parkinsonism in any animal model. The mechanisms whereby nicotine and nAChR drugs reduce LID may involve long-term nAChR downregulation and/or desensitization followed by a decline in striatal dopamine release. In addition to its ability to reduce LID, nicotine also protects against nigrostriatal damage and has cognitive-enhancing and antidepressant effects. These combined properties suggest that nicotine and nAChR drugs may be of benefit in the management of LID in Parkinson's disease.

Keywords Levodopa • Dyskinesia • Nicotine • nAChRs • Parkinson's disease

M. Quik (✉) • T. Bordia • D. Zhang • X. Perez
Center for Health Sciences, SRI International,
333 Ravenswood Ave, Menlo Park, CA 94025, USA
e-mail: maryka.quik@sri.com

S.H. Fox, J.M. Brotchie (eds.), *Levodopa-Induced Dyskinesia in Parkinson's Disease*, DOI 10.1007/978-1-4471-6503-3_16

Introduction

Current treatments for Parkinson's disease primarily consist of symptomatic management, with the most effective therapy involving dopamine replacement with the dopamine precursor levodopa. Unfortunately, long-term levodopa use is associated with complications such as on-off phenomena, wearing off, and levodopa-induced dyskinesia (LID), which represent substantial barriers to effective disease management [1, 2]. Aberrant dopaminergic function plays a key role in the generation of LID. However, other central nervous system (CNS) neurotransmitter systems are also involved in LID, including the cholinergic one. Here, we first provide a brief overview of nicotinic acetylcholine receptors (nAChRs), followed by preclinical studies showing that nicotine reduces LID in several parkinsonian animal models. We next focus on the nAChR subtypes that mediate LID using genetically modified mice and selective nAChR drugs. Lastly, we discuss the role of muscarinic acetylcholine receptors (mAChRs). Overall, the data suggest that nicotine and drugs directed to CNS nAChRs may be most useful for the treatment of the dyskinesia that arises with levodopa therapy.

nAChR Subtypes in the Nervous System

Converging evidence indicates that nicotine exerts its antidyskinetic effect by interacting at CNS nAChRs, for which the endogenous neurotransmitter is acetylcholine. nAChRs are ligand-gated ion channels composed of five membrane-spanning subunits around a central channel [3, 4]. Various types of nAChRs have been identified throughout the body. Some consist of only α subunits (α7), whereas other are composed of a combination of α (α1–α6) and β (β1–β4) subunits, with the α subunit containing the acetylcholine or agonist recognition site [3, 4]. The predominant nAChR subtype in skeletal muscle is composed of $\alpha1\beta1\gamma\delta$ subunits, while the $\alpha3\beta4^*$ and α7 subtypes are present in the peripheral nervous system (the asterisk indicates the possible presence of other subunits in the receptor). By contrast, the most abundant and widespread CNS nAChR populations are the heteromeric $\beta2^*$ and homomeric α7 receptors (Fig. 16.1), with only minimal expression of $\alpha3\beta4^*$ and no $\alpha1\beta1\gamma\delta$ nAChRs. The observation that nAChRs in the brain vary from those in the peripheral autonomic nervous system and skeletal muscle is of note as it allows for the development of targeted drug therapies for CNS disorders. The primary $\beta2^*$ nAChRs in the brain are the $\alpha4\beta2^*$ receptors, which are widely expressed in numerous brain regions, and the $\alpha6\beta2^*$ nAChRs, which exhibit a more restricted distribution primarily to catecholaminergic neurons (Fig. 16.1). Homomeric α7 nAChRs are widely localized throughout the brain, although expression is relatively sparse in the basal ganglia. nAChRs containing the α2, α3, and/or α5 subunits are also present in the brain but to a much lesser extent [3, 4, 7].

In summary, the predominant nAChR subtypes in regions such as the cortex, hippocampus, thalamus, and cerebellum are the $\alpha4\beta2^*$ and α7 nAChRs, while the

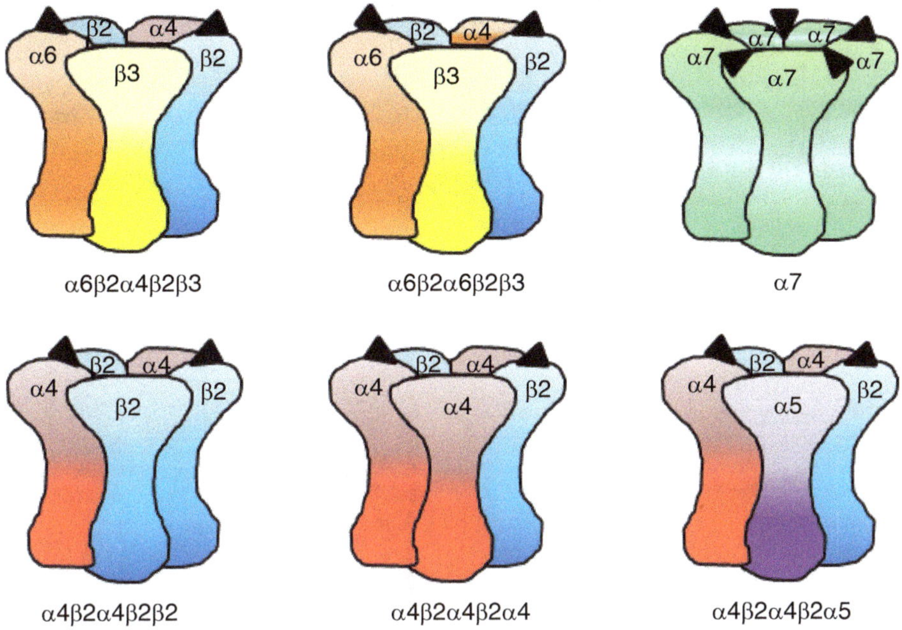

Fig. 16.1 Subunit composition of nAChR subtypes in the nigrostriatal dopaminergic pathway. Nicotine, like acetylcholine, causes its effects in the brain by acting at nAChRs. These ligand-gated ion channels are pentamers composed of different combinations of **α** (**α**2–**α**6) and **β** (**β**2–**β**4) subunits. Work using genetically modified mice, nAChR subtype-specific antibodies and drugs, as well as nAChR subunit chimeras and concatamers has shown that the primary nAChRs functionally active in the nigrostriatal pathway are the **α6β2***, **α4β2***, and **α**7 nAChRs. The **α6β2*** nAChR subtype consists of the high-affinity **α6β2α4β2β**3 and the lower-affinity **α6β2α6β2β**3 nAChRs [5]. **α4β2*** nAChRs include the **α4β2α4β2α**5 and **α4β**2 subtypes, with the latter existing in two different stoichiometries, the higher-affinity **α4β2α4β2β**2 conformation and the lower-affinity **α4β2α4β2α**4 conformation [6]. **α6β2α4β2β**3, **α6β2α6β2β**3, **α4β2α5β**2, **α4β2α4β2β**2, and **α4β2α4β2α**4 are all present in dopaminergic terminals [7]. GABAergic interneurons and medium spiny neurons express **α4β2α4β2β**2 and **α4β2α4β2α**4, whereas **α**7 nAChRs are thought to be exclusively localized to glutamatergic terminals [7]. Two agonist binding sites (*triangles*) are depicted at the interface between **α** and **β**2 subunits in heteromeric receptors, while the homomeric **α**7 receptors have five binding sites [4]

primary ones in the basal ganglia are the α4β2* and α6β2* subtypes with α7 nAChRs less densely expressed [3, 4, 7].

Nicotine Administration Reduces LID in Parkinsonian Animal Models

Parkinsonian animal models have been used as a first approach to investigate the potential of nicotine and nAChR drugs to alleviate LID. Although the available models recapitulate many features of Parkinson's disease, they each have their

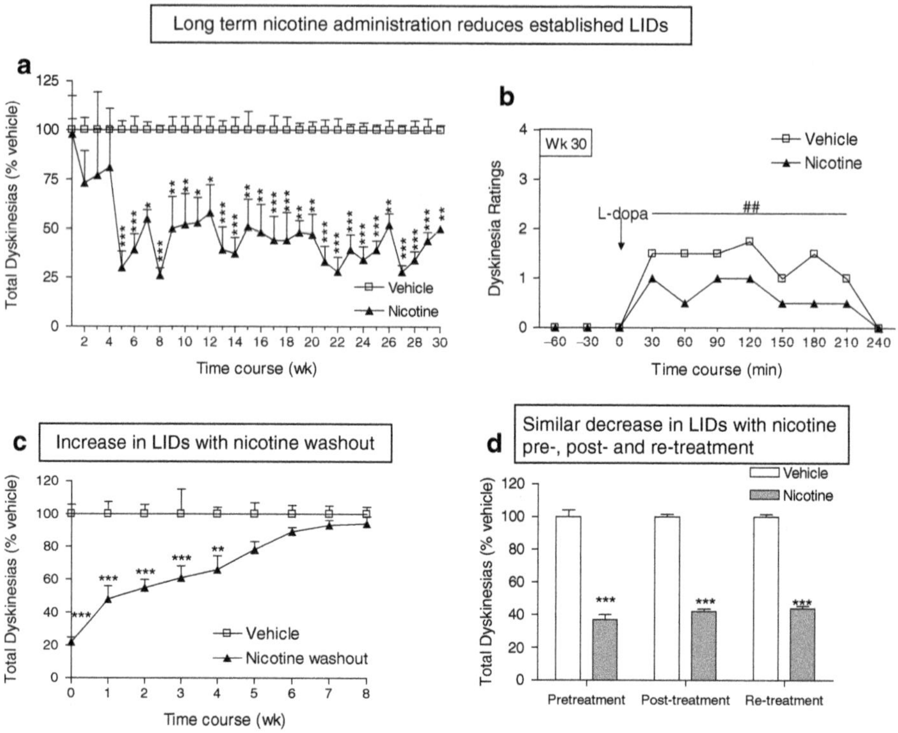

Fig. 16.2 Nicotine treatment reduces LID in nonhuman primates. In panel (**a**), MPTP-lesioned monkeys were gavaged with levodopa (10 mg/kg)/carbidopa (2.5 mg/kg) twice daily for 5 days a week. Eight weeks later, they were given nicotine (300 μg/ml in 50 % diluted Gatorade in the drinking water). The total dyskinesia scores (expressed as % vehicle) were averaged over 2–3 days during the 4 h period following the afternoon dose of levodopa. Values are the mean ± SEM of 5–6 monkeys. Panel (**b**) shows a decline in the daily time course of LID with nicotine treatment at week 30, with similar results during the entire course of the study. The symbols depict the median of 5–6 monkeys. Panel (**c**) shows that the nicotine-mediated improvement in LID is lost by 6 weeks of nicotine washout (mean ± SEM of 5–6 monkeys). Panel (**d**) compares the effect of nicotine on the total dyskinesia scores in animals pretreated for 2 weeks with nicotine, treated with nicotine 8 weeks after the start of levodopa treatment, and re-treated with nicotine after a 6-week washout period. The *bars* represent the mean ± SEM of 15–30 weeks of nicotine treatment. Significance of difference of drug treatment from vehicle, $*P<0.05$, $**P<0.01$, and $***P<0.001$ using two-way ANOVA followed by a Bonferroni post hoc test. Significance of difference from vehicle using a Mann-Whitney test, $\#\#P<0.01$ (Data taken in modified from Refs. [14, 15])

limitations. For this reason, nicotine and nAChR drugs have been tested in several models, including parkinsonian monkeys, rats, and mice (Table 16.1).

MPTP-lesioned, (1-methyl-4-phenyl-1,2,3,6-tetrahydropyridine) nonhuman primates (NHPs) offer the advantage that they develop parkinsonian motor symptoms that closely resemble those in Parkinson's disease. Moreover, the abnormal choreiform and dystonic movements that arise in NHPs with levodopa treatment are similar to those observed clinically. Thus, levodopa-treated parkinsonian NHPs provide a good model for pharmacological investigations [20–22]. Long-term administration of nicotine alleviates both peak and total LID by ~60 % in this animal model for the entire length of the studies (30 weeks) (Fig. 16.2a, b) [13–15]. The antidyskinetic effect of nicotine persists for several weeks after discontinuation (Fig. 16.2c). Interestingly, nicotine readministration after a 10-week washout period led to an immediate decline in LID (re-treatment group) (Fig. 16.2d). Thus, prior nicotine exposure appears to exert a priming effect, that is, it modulates receptor-mediated responsiveness such that the effect to subsequent exposure occurs more readily. Nicotine treatment reduced LID to a similar extent whether administered 2 months before (pretreatment) or after (posttreatment) levodopa (Fig. 16.2d) [13, 15]. Thus, nicotine can be used prophylactically to minimize the occurrence of LID and can also reduce existing LID. No tolerance developed to continued daily nicotine treatment (Fig. 16.2a), a key concern as patients generally require lifelong treatment with levodopa [13, 15]. The effect of nicotine was optimal in NHPs with moderate nigrostriatal damage [14], suggesting that nicotine administration would most effectively reduce LID in patients with moderate Parkinson's disease.

The ability of nicotine to reduce LID was also evaluated in rodents with a unilateral nigrostriatal lesion, as they more readily allow for the testing of varying treatment paradigms and delivery modes, as well as for work to investigate mechanisms. Long-term nicotine administration reduced LID in rats and mice when provided in the drinking water, via slow-release minipumps or by systemic injection (Table 16.1) [8–11, 16]. The findings that the nicotine-mediated improvement in LID is observed

Table 16.1 Nicotine and nAChR drugs decrease LID in different parkinsonian animal models

nAChR drug	nAChR subtype	Mode of treatment	Animal model	Moderate lesion	Severe lesion	References
				% decline in LID		
Nicotine	Multiple	Drinking water	Rats	50–70	–	[8]
		Injection		–	30	[9]
		Minipump		50–70	–	[8]
		Drinking water	Mice	50–60	20	[10–12]
		Drinking water	Monkeys	50–70	0	[13–15]
Varenicline	Multiple	Injection	Rats	40	0	[16]
		Oral	Monkeys	45	–	[17]
A-85380	β2*	Injection	Rats	50	20	[16]
Sazetidine	β2*	Injection	Rats	–	23	[18]
TC-2696	β2*	Injection	Rats	–	30	[18]
TI-10165	β2*	Injection	Rats	–	32	[18]
TC-8831	β2*	Injection	Rats	–	24	[18]
		Oral	Monkeys	50	–	[17, 19]
TC-10600	β2*	Injection	Rats	–	32	[18]
ABT-089	β2*	Oral	Monkeys	35	–	[17]

Rats and mice were lesioned by unilateral injection of 6-OHDA into the medial forebrain bundle, while monkeys were given MPTP via subcutaneous injection using administration paradigms that resulted in moderate and more severe nigrostriatal damage [8–17]. Levodopa was administered once or twice daily 5 days per week for weeks to months [8–17]. All declines in LID shown are significant ($P<0.05$) from corresponding vehicle-treated groups

– Not done

Table 16.2 nAChR subunit deletion modulates expression of LID

nAChR subunit knockout	Baseline LID in knockout	Nicotine reduces LID in knockout	nAChR subtype involved in LID	Reference
β2	Decreased	No	Yes	[10]
α6	Decreased	No	Yes	[11]
α4	No change	No	Yes	[12]
α7	Increased	Yes	Yes	[12]

across species and with varying modes of administration support the idea that nicotine treatment may be useful in Parkinson's disease patients. In fact, a small clinical trial showed that 4 months of oral nicotine administration reduced various components of LID in patients with moderate Parkinson's disease (http://www.neuraltus.com/pages/news_rel12_03_10.html).

α4β2*, α6β2*, and α7 nAChR Subtypes Modulate LID

The knowledge of the nAChR subtypes involved in LID is important for the development of selective pharmacotherapies to reduce their occurrence with a minimum of side effects. The use of nAChR knockout mice has proved very useful in this regard. β2 nAChR subunit knockout mice, which lack both α4β2* and α6β2* nAChRs, had reduced baseline LID levels. Moreover, nicotine did not reduce LID in β2 nAChR subunit null mutant mice (Table 16.2). These two findings suggest that nAChRs expressing the β2 subunit are important for both the generation of LID and the antidyskinetic effect of nicotine. Subsequent work showed that α6 nAChR subunit deletion led to similar results as β2 deletion, indicating α6β2* nAChRs play a critical role [10, 11] (Table 16.2). α4 nAChR subunit deletion also prevented the antidyskinetic effect of nicotine, although it did not result in a reduction in baseline LID scores (Table 16.2). The observation that β2, α4, and α6 nAChR subunit deletion all affected LID suggests that α6β2* and α4β2* nAChRs are important for the expression of LID. As well or alternatively, the finding that both these subtypes mediate LID may also suggest an involvement of the α6α4β2* nAChR population. This idea stems from recent reports showing that the α6α4β2* nAChR mediates several of nicotine's effects on dopaminergic function including dopamine neuronal firing and endogenous dopamine release [23–26].

Additionally, LID were modified in α7 nAChR null mutant mice. However, while the lack of β2* nAChRs prevented the nicotine-mediated decrease in LID and decreased baseline LID, α7 nAChR deletion increased baseline LID [12]. These data suggest that the α7 nAChR subtype has an inhibitory impact on the development of LID. Moreover, nicotine treatment still decreased LID in α7 nAChR knockout mice. This differential regulation by α7 and β2* nAChRs may relate to the fact that these receptors exhibit unique molecular and functional properties. For instance, α7 nAChRs are much more permeable to calcium, desensitize more rapidly, and are linked to different intracellular cell signaling steps compared to β2* nAChRs [27–31].

In summary, findings with nAChR knockout mice suggest that multiple nAChR populations influence LID, including the α4β2*, α6β2*, and α7 subtypes. Since Parkinson's disease is progressive with multiple compensatory changes throughout, one possibility is that various nAChR subtype drugs may be differentially effective during the course of the disease.

CNS Selective nAChR Drugs Reduce LID

Pharmacological studies further support the idea that CNS selective nAChRs are involved in LID. Evidence for this possibility stems from studies which show that β2* nAChR agonists attenuate LID in parkinsonian rodents and NHPs (Table 16.1). A-85380 and a series of novel nicotinic receptor compounds primarily acting at β2* nAChRs all reduced LID in 6-hydroxydopamine (6-OHDA)-lesioned rats [16, 18]. In addition, one of the agonists used in the rodent studies (TC-8831) reduced LID in both macaques and squirrel monkeys, without worsening parkinsonism on or off levodopa [17, 19]. Since currently available drugs act at both α4β2* and α6β2* nAChRs, it has not been possible to evaluate a role for the individual subtypes pharmacologically. The β2* nAChR antagonist mecamylamine also decreased LID in parkinsonian rats [9], supporting the idea that the decline in LID is mediated via an interaction at β2* nAChRs.

Localization of nAChRs Involved in Nicotine's Antidyskinetic Effect

Current data suggest that nAChRs in the striatum play a key role in the antidyskinetic effect of nicotine. Evidence for this idea stems from lesion studies involving unilateral injection of the dopaminergic neurotoxin 6-OHDA in rats and mice or subcutaneous injection of MPTP in NHPs. Both nicotine and nAChR drugs most effectively reduce LID in mice, rats, or NHPs with moderate nigrostriatal damage, that is, when the dopaminergic system is still partially intact (Table 16.1) [10, 14, 16]. Nicotine administration is much less effective in severely lesioned rodents and NHPs [10, 14, 16]. These results suggest that α6β2* and/or α4β2* nAChRs on nigrostriatal dopamine terminals are critical for the nAChR-mediated decline in LID. However, since nicotine does reduce LID to some extent with severe nigrostriatal damage, nAChRs on nondopaminergic neurons may also contribute. Alternatively, α4β2* nAChRs in the thalamus, cortex, and cerebellum, regions linked to motor control and coordination, may be important. Additionally, α7 nAChRs on cortical afferents to the striatum and/or α7 receptors in other CNS regions may be involved in the antidyskinetic effect of nicotine.

In summary, α4β2* and α6β2* nAChRs on dopamine terminals of the nigrostriatal pathway contribute to the expression of LID. Other striatal α4β2* and α7 nAChRs may also be involved, as well as nAChRs in other brain regions. The idea that multiple nAChRs in various brain regions are important is consistent with studies showing that the striatal dopaminergic system is tightly interconnected with numerous other neurotransmitter systems involved in movement.

Mechanism of Action of Nicotine to Reduce Dyskinesia

nAChR Desensitization and Downregulation

An important issue is the mechanism of action of nAChR drugs to reduce LID, as this may help in the development of more targeted therapeutic approaches. Unexpectedly, previous work showed that both nAChR agonists and the antagonist mecamylamine reduced LID to a similar extent [9]. One explanation for these findings is that agonists produce their behavioral effect via nAChR desensitization. Although agonists initially activate nAChRs, this is rapidly followed by receptor desensitization or a functional blockade, similar to that observed with antagonists [28, 32, 33]. Further support for this idea stems from findings that nicotine administration results in a similar decline in LID whether given intermittently via injection or continuously via slow-release minipump, a regimen that readily promotes receptor desensitization [8, 9]. The idea is consistent with its potential mechanism of action in addiction, depression, and anxiety [28, 32, 34]. Long-term nicotine treatment also leads to the downregulation of α6β2* nAChRs (Fig. 16.3a, b) [36]. Thus, both nAChR desensitization and downregulation may underlie the nAChR-mediated improvement in LID.

Decline in Striatal nAChR-Mediated Dopamine Release

LID are thought to arise because of a loss of nigrostriatal dopamine cells, coupled with intermittent exposure to large doses of levodopa. Together these two conditions lead to large transient increases in striatal dopamine release, which consequently

Fig. 16.3 (continued) Long-term nicotine treatment also decreased nicotine-evoked ^{3}H-dopamine release from synaptosomes prepared from the unlesioned (**c**) and lesioned (**d**) striatum [35]. These changes may underlie the improvement in LID observed with long-term nicotine treatment. Levodopa treatment had no effect on nAChR expression or function. A schematic representation of the effects of lesioning and/or nicotine is shown in the *lower panels*. (**e**) Depicts nAChR-mediated dopamine release in vehicle-treated intact striatum with nicotine stimulation. (**f**) Illustrates the decrease in release in vehicle-treated lesioned striatum due to lesioning (*dotted neuron*). (**g**) Shows the decrease in stimulated release in intact striatum with nicotine treatment, which results in a decrease in **α6β2*** nAChR levels. (**h**) Shows that long-term nicotine treatment further decreases release in lesioned rats. Thus, nicotine treatment may dampen the enhanced dopamine release that arises with chronic levodopa treatment and consequently reduce LID. *X* denotes nAChR regulation. The *red dots* represent dopamine. The *arrows* signify a decrease in dopamine release under different conditions

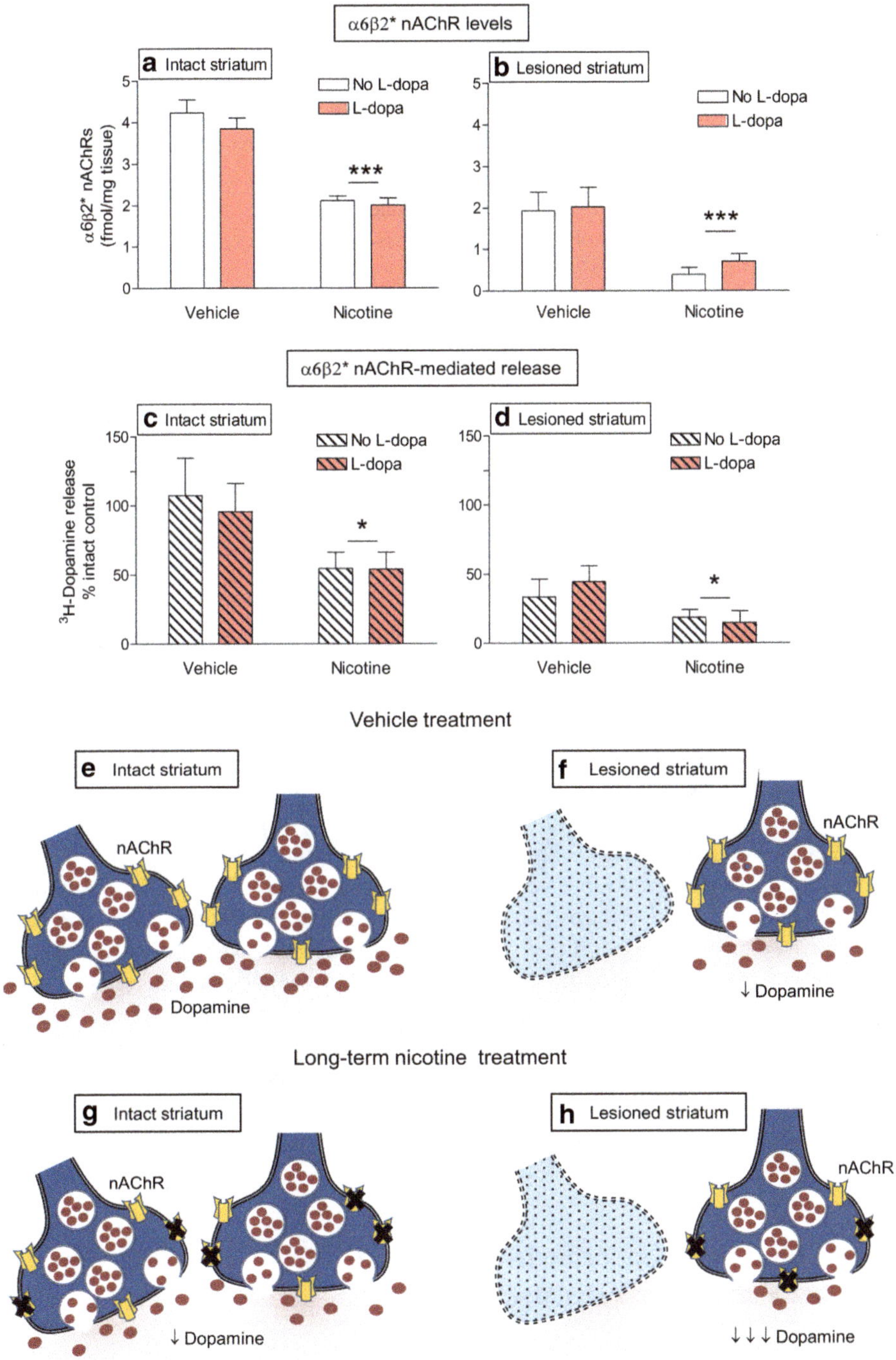

Fig. 16.3 Long-term nicotine treatment decreases **α6β2*** nAChR expression and function. Previous work had shown that long-term nicotine treatment decreased **α6β2*** nAChR binding levels in the unlesioned (**a**) and lesioned (**b**) striatum of rats with a unilateral 6-OHDA lesion [35].

result in excessive dopaminergic stimulation and LID [37–40]. As mentioned earlier, nAChRs in the striatum are present on nigrostriatal dopamine terminals and regulate dopamine release [3, 7]. Long-term nicotine treatment may act to dampen this enhanced striatal dopamine release by modulating α6β2* and α4β2* nAChR expression and function (Fig. 16.3a, d) [35]. Long-term nicotine treatment decreases striatal α6β2* nAChR levels by ~50 %, with corresponding declines in α6β2* nAChR-mediated dopamine release. α4β2* nAChR-mediated dopamine release is also decreased by ~50 % [35]. Although α4β2* nAChR levels are not decreased after long-term nicotine treatment, they are functionally desensitized, as are α6β2* nAChRs to a lesser degree [28, 41]. Altogether these data suggest that nicotine may improve LID in parkinsonian animal models by chronically desensitizing and/or downregulating nAChRs and consequently decreasing dopaminergic function (Fig. 16.3).

Nicotine and nAChR Drugs Do Not Affect Parkinson's Disease Motor Symptoms

In contrast to the consistent decline in LID in parkinsonian mice, rats, and NHPs, the effect of nicotine or nAChR agonists on motor symptoms of parkinsonism has yielded variable results with an improvement in motor function in some studies but not others [8, 10, 13, 42–44]. This inconsistency in preclinical studies extends to clinical reports and trials, which have also yielded variable improvements in Parkinson's disease motor symptoms (Table 16.3). These differential outcomes may be due to differences in nicotine dosing, duration of treatment, outcome measure evaluated, small cohort size, and/or clinical stage (early versus late Parkinson's disease). However, a more likely explanation for the variable outcome may relate to the type of trial, that is, open-label versus double-blinded (Table 16.3). The open-label studies generally yielded a positive outcome [45–50], although there was one with no improvement [51]. By contrast, nicotine treatment led to very little clinical efficacy on motor symptoms in the double-blinded trials, which also had significantly larger groups of patients [52–55]. In fact, there was improvement in only one double-blinded study that was limited to two patients [56]. Although differential treatment paradigms among the different studies cannot be ruled out, the weight of the evidence suggests that nicotine treatment does not acutely improve motor symptoms in levodopa-treated Parkinson's disease patients.

Role of Smoking/Nicotine in Protection Against Parkinson's Disease

In addition to its ability to reduce LID, one very interesting property of nicotine resides in its capacity to protect against neuronal degeneration in experimental models [31, 57–59]. Nicotine and nAChR agonists have been reported to attenuate a host of

Table 16.3 Variable improvement in Parkinson's disease motor symptoms with nicotine or nicotinic agonists

Study	Improvement in parkinsonism	Drug	# of patients	Outcome measure used	Reference or ClinicalTrials.gov Identifier
Open-label	Yes	Nicotine IV	6	Tremor[a]	[45]
	Yes	Smoking and nicotine gum	6	Tremor, rigidity, bradykinesia, posture[a]	[46]
	Yes	Nicotine IV and patch	15	Hoehn and Yahr	[47]
	Yes	Nicotine gum	8	UPDRS	[48]
	Yes	Nicotine patch	6	UPDRS	[49]
	Yes	Smoking	1	UPDRS, Hoehn and Yahr	[50]
	No	Nicotine patch	22	UPDRS	[51]
Double-blinded	No	Nicotine gum	48	UPDRS	[52]
	No	Nicotine patch	16	UPDRS	[53]
	No	Nicotine patch	32	Columbia Univ rating scale, Schwab-England	[54]
	No	SIB-1508Y oral	77	UPDRS	[55]
	Yes	Nicotine gum and patch	2	Tremor, rigidity[a]	[56]
	Data not yet available	Nicotine patch	40	UPDRS	NCT00873392
	Data not yet available	Varenicline	40	UPDRS, Berg Balance Scale	NCT01341080

[a]Scale not provided. Unified Parkinson's Disease Rating Scale (UPDRS)

toxic insults in different neuronal cells in culture. More relevant to Parkinson's disease, nicotine administration also protects against nigrostriatal damage in parkinsonian monkey, rat, and mouse models [60–62]. These basic research findings provide a molecular basis for the well-known epidemiological observation of a negative correlation between Parkinson's disease and tobacco use [63–67]. The reduced incidence of Parkinson's disease and smoking is dose and time dependent and diminishes with smoking cessation. Importantly, it is not due to a selective mortality [63–67]. Furthermore, twin studies show that Parkinson's disease is less likely to develop in the twin that smokes [68]. These combined findings all point to a true biologic effect of smoking in reducing Parkinson's disease. To our knowledge, studies have not been reported as to whether smoking modulates the expression of LID.

Although tobacco contains thousands of chemicals, anyone of which may contribute to smoking's neuroprotective effect, the animal studies described above suggest that nicotine may represent a component that plays a role, at least in part. Indeed, a clinical trial funded by the Michael J. Fox Foundation is currently recruiting participants to investigate the disease-modifying potential of transdermal nicotine in early Parkinson's disease (ClinicalTrials.gov Identifier NCT01560754).

Muscarinic Receptors

Acetylcholine not only interacts with nAChRs but also mAChRs of which there are five subtypes (M1 to M5) [69, 70]. These G-protein-coupled receptors are quite distinct structurally and functionally from nAChRs; they are also differentially distributed in the CNS with high expression in the striatum and numerous other brain areas linked to motor function [69, 70]. Interestingly, muscarinic receptor antagonists were initially a primary treatment for Parkinson's disease motor symptoms. However, levodopa and dopamine agonists have now largely superseded their use because antimuscarinic drugs have numerous side effects due to the stimulation of peripheral nervous system mAChRs (nausea, constipation, urinary retention, dry mouth) and also CNS mAChRs (sedation, confusion) [70, 71]. The effect of antimuscarinics has also been tested on LID. Atropine did not attenuate LID in levodopa-treated parkinsonian NHPs [72], although the muscarinic blocker dicyclomine did reduce LID in 6-OHDA-lesioned mice. Further work is thus required to understand the role of mAChRs in LID.

Summary and Concluding Remarks

Accumulating studies suggest that the nicotinic cholinergic system represents a target for improving LID. Long-term administration of nicotine or nAChR drugs substantially reduces LID in parkinsonian mice, rats, and nonhuman primates (Fig. 16.2, Table 16.1). The mechanism may involve nAChR desensitization and/or downregulation, with a consequent decline in striatal dopamine release (Fig. 16.3). Presynaptic α6β2* and α4β2* nAChRs on striatal dopaminergic terminals appear to be prime regulators of LID. However, α4β2* and α7 nAChRs on other striatal neurons and/or in other brain regions may also contribute. Long-term molecular and cellular alterations are involved since the antidyskinetic effect of nicotine requires several weeks to develop and to wash out (Fig. 16.2a, c). Nicotine administration appears to result in long-term alterations in synaptic plasticity or priming, as evidenced by findings that its beneficial effect is very quickly restored upon nicotine reintroduction [14].

Despite extensive preclinical studies and much clinical experience, the identification of pharmacotherapies for the management of LID in Parkinson's disease patients has proved difficult. This may be because their origin is multifactorial and involves alterations in numerous CNS neurotransmitter systems. Variability in drug responsiveness may arise because of a complex interplay between the effects of progressive dopamine denervation and chronic dopaminergic drug treatment, with the continual development of novel compensatory mechanisms. Possibly, there may be subgroups of LID that respond more effectively to certain classes of drugs than others. An alternate or concurrent possibility is that drug combinations may prove optimal in the treatment of LID as is the case for the treatment of numerous other disease states ranging from cancer to psychiatric conditions.

Altogether, the present review provides evidence for a role of nAChR drugs for the treatment of LID. Additional advantages of such drugs are that they exhibit pro-cognitive and antidepressant properties and most importantly may have disease-modifying potential. These combined characteristics suggest a compelling role for nAChR drugs in Parkinson's disease treatment.

Acknowledgments We gratefully acknowledge support for NIH grants NS59910 and NS65851.

References

1. Huot P, Johnston TH, Koprich JB, Fox SH, Brotchie JM. The pharmacology of LEVODOPA-induced dyskinesia in Parkinson's disease. Pharmacol Rev. 2013;65:171–222.
2. Poewe W, Mahlknecht P, Jankovic J. Emerging therapies for Parkinson's disease. Curr Opin Neurol. 2012;25:448–59.
3. Millar NS, Gotti C. Diversity of vertebrate nicotinic acetylcholine receptors. Neuropharmacology. 2009;56:237–46.
4. Albuquerque EX, Pereira EF, Alkondon M, Rogers SW. Mammalian nicotinic acetylcholine receptors: from structure to function. Physiol Rev. 2009;89:73–120.
5. Kuryatov A, Lindstrom J. Expression of functional human alpha6beta2beta3* acetylcholine receptors in Xenopus laevis oocytes achieved through subunit chimeras and concatamers. Mol Pharmacol. 2011;79:126–40.
6. Harpsoe K, Hald H, Timmermann DB, Jensen ML, Dyhring T, Nielsen EO, et al. Molecular determinants of subtype-selective efficacies of cytisine and the novel compound NS3861 at heteromeric nicotinic acetylcholine receptors. J Biol Chem. 2013;288:2559–70.
7. Quik M, Wonnacott S. {alpha}6{beta}2* and {alpha}4{beta}2* nicotinic acetylcholine receptors as drug targets for Parkinson's disease. Pharmacol Rev. 2011;63:938–66.
8. Bordia T, Campos C, Huang L, Quik M. Continuous and intermittent nicotine treatment reduces L-3,4-dihydroxyphenylalanine (LEVODOPA)-induced dyskinesia in a rat model of Parkinson's disease. J Pharmacol Exp Ther. 2008;327:239–47.
9. Bordia T, Campos C, McIntosh JM, Quik M. Nicotinic receptor-mediated reduction in Levodopa-induced dyskinesia may occur via desensitization. J Pharmacol Exp Ther. 2010;333:929–38.
10. Huang L, Grady SR, Quik M. Nicotine reduces levodopa-induced dyskinesia by acting at {beta}2 nicotinic receptors. J Pharmacol Exp Ther. 2011;338:932–41.
11. Quik M, Park KM, Hrachova M, Mallela A, Huang LZ, McIntosh JM, et al. Role for alpha6 nicotinic receptors in levodopa-induced dyskinesia in parkinsonian mice. Neuropharmacology. 2012;63:450–9.
12. Quik M, Campos C, Grady S. Multiple CNS nicotinic receptors mediate Levodopa-induced dyskinesia; studies with parkinsonian nicotinic receptor knockout mice. Biochem Pharmacol. 2013;86(8):1153–62.
13. Quik M, Cox H, Parameswaran N, O'Leary K, Langston JW, Di Monte D. Nicotine reduces levodopa-induced dyskinesia in lesioned monkeys. Ann Neurol. 2007;62:588–96.
14. Quik M, Mallela A, Chin M, McIntosh JM, Perez XA, Bordia T. Nicotine-mediated improvement in levodopa-induced dyskinesia in MPTP-lesioned monkeys is dependent on dopamine nerve terminal function. Neurobiol Dis. 2013;50:30–41.
15. Quik M, Mallela A, Ly J, Zhang D. Nicotine reduces established Levodopa-induced dyskinesia in a monkey model of Parkinson's disease. Mov Disord. 2013;28(10):1398–406.
16. Huang LZ, Campos C, Ly J, Carroll FI, Quik M. Nicotinic receptor agonists decrease Levodopa-induced dyskinesia most effectively in moderately lesioned parkinsonian rats. Neuropharmacology. 2011;60:861–8.

17. Zhang D, Mallela A, Sohn D, Carroll FI, Bencherif M, Letchworth S, Quik M. Nicotinic receptor agonists reduce L-dopa-induced dyskinesias in a monkey model of Parkinson's disease. J Pharmacol Exp Ther 2013;347:225–235. PMCID 3781407.
18. Quik M, Campos C, Bordia T, Strachan JP, Zhang J, McIntosh JM, et al. alpha4beta2 nicotinic receptors play a role in the nAChR-mediated decline in levodopa-induced dyskinesia in parkinsonian rats. Neuropharmacology. 2013;71:191–203.
19. Johnston TH, Huot P, Fox SH, Koprich JB, Szeliga KT, James JW, et al. TC-8831, a nicotinic acetylcholine receptor agonist, reduces levodopa-induced dyskinesia in the MPTP macaque. Neuropharmacology. 2013;73:337–47.
20. Jenner P. Functional models of Parkinson's disease: a valuable tool in the development of novel therapies. Ann Neurol. 2009;64:S16–29.
21. Duty S, Jenner P. Animal models of Parkinson's disease: a source of novel treatments and clues to the cause of the disease. Br J Pharmacol. 2011;164:1357–91.
22. Iderberg H, Francardo V, Pioli EY. Animal models of LEVODOPA-induced dyskinesia: an update on the current options. Neuroscience. 2012;211:13–27.
23. Exley R, Maubourguet N, David V, Eddine R, Evrard A, Pons S, et al. Distinct contributions of nicotinic acetylcholine receptor subunit alpha4 and subunit alpha6 to the reinforcing effects of nicotine. Proc Natl Acad Sci U S A. 2011;108:7577–82.
24. Liu L, Zhao-Shea R, McIntosh JM, Gardner PD, Tapper AR. Nicotine persistently activates ventral tegmental area dopaminergic neurons via nicotinic acetylcholine receptors containing alpha4 and alpha6 subunits. Mol Pharmacol. 2012;81:541–8.
25. Zhao-Shea R, Liu L, Soll LG, Improgo MR, Meyers EE, McIntosh JM, et al. Nicotine-mediated activation of dopaminergic neurons in distinct regions of the ventral tegmental area. Neuropsychopharmacology. 2011;36:1021–32.
26. Gotti C, Guiducci S, Tedesco V, Corbioli S, Zanetti L, Moretti M, et al. Nicotinic acetylcholine receptors in the mesolimbic pathway: primary role of ventral tegmental area alpha6beta2* receptors in mediating systemic nicotine effects on dopamine release, locomotion, and reinforcement. J Neurosci. 2010;30:5311–25.
27. Giniatullin R, Nistri A, Yakel JL. Desensitization of nicotinic ACh receptors: shaping cholinergic signaling. Trends Neurosci. 2005;28:371–8.
28. Picciotto MR, Addy NA, Mineur YS, Brunzell DH. It is not "either/or": activation and desensitization of nicotinic acetylcholine receptors both contribute to behaviors related to nicotine addiction and mood. Prog Neurobiol. 2008;84:329–42.
29. Wonnacott S, Sidhpura N, Balfour DJ. Nicotine: from molecular mechanisms to behaviour. Curr Opin Pharmacol. 2005;5:53–9.
30. Changeux JP. Allosteric receptors: from electric organ to cognition. Annu Rev Pharmacol Toxicol. 2010;50:1–38.
31. Quik M, Perez XA, Bordia T. Nicotine as a potential neuroprotective agent for Parkinson's disease. Mov Disord. 2012;27:947–57.
32. Buccafusco JJ, Beach JW, Terry AV. Desensitization of nicotinic acetylcholine receptors as a strategy for drug development. J Pharmacol Exp Ther. 2009;328:364–70.
33. Corringer PJ, Sallette J, Changeux JP. Nicotine enhances intracellular nicotinic receptor maturation: a novel mechanism of neural plasticity? J Physiol Paris. 2006;99:162–71.
34. Anderson SM, Brunzell DH. Low dose nicotine and antagonism of beta2 subunit containing nicotinic acetylcholine receptors have similar effects on affective behavior in mice. PLoS One. 2012;7:e48665.
35. Bordia T, McIntosh JM, Quik M. The nicotine-mediated decline in L-dopa-induced dyskinesias is associated with a decrease in striatal dopamine release. J Neurochem 2013;125: 291–302. PMCID 377804.
36. Lai A, Parameswaran N, Khwaja M, Whiteaker P, Lindstrom JM, Fan H, et al. Long-term nicotine treatment decreases striatal alpha6* nicotinic acetylcholine receptor sites and function in mice. Mol Pharmacol. 2005;67:1639–47.
37. Carta M, Bezard E. Contribution of pre-synaptic mechanisms to levodopa-induced dyskinesia. Neuroscience. 2011;198:245–51.

38. Cenci MA. Dopamine dysregulation of movement control in LEVODOPA-induced dyskinesia. Trends Neurosci. 2007;30:236–43.
39. Fisone G, Bezard E. Molecular mechanisms of levodopa-induced dyskinesia. Int Rev Neurobiol. 2011;98:95–122.
40. Lindgren HS, Andersson DR, Lagerkvist S, Nissbrandt H, Cenci MA. LEVODOPA-induced dopamine efflux in the striatum and the substantia nigra in a rat model of Parkinson's disease: temporal and quantitative relationship to the expression of dyskinesia. J Neurochem. 2010;112: 1465–76.
41. Grady SR, Wageman CR, Patzlaff NE, Marks MJ. Low concentrations of nicotine differentially desensitize nicotinic acetylcholine receptors that include alpha5 or alpha6 subunits and that mediate synaptosomal neurotransmitter release. Neuropharmacology. 2012;62: 1935–43.
42. Domino EF, Ni L, Zhang H. Nicotine alone and in combination with levodopa methyl ester or the D(2) agonist N-0923 in MPTP-induced chronic hemiparkinsonian monkeys. Exp Neurol. 1999;158:414–21.
43. Meshul CK, Kamel D, Moore C, Kay TS, Krentz L. Nicotine alters striatal glutamate function and decreases the apomorphine-induced contralateral rotations in 6-OHDA-lesioned rats. Exp Neurol. 2002;175:257–74.
44. Schneider JS, Van Velson M, Menzaghi F, Lloyd GK. Effects of the nicotinic acetylcholine receptor agonist SIB-1508Y on object retrieval performance in MPTP-treated monkeys: comparison with levodopa treatment. Ann Neurol. 1998;43:311–7.
45. Marshall J, Schnieden H. Effect of adrenaline, noradrenaline, atropine, and nicotine on some types of human tremor. J Neurol Neurosurg Psychiatry. 1966;29:214–8.
46. Ishikawa A, Miyatake T. Effects of smoking in patients with early-onset Parkinson's disease. J Neurol Sci. 1993;117:28–32.
47. Kelton MC, Kahn HJ, Conrath CL, Newhouse PA. The effects of nicotine on Parkinson's disease. Brain Cogn. 2000;43:274–82.
48. Mitsuoka T, Kaseda Y, Yamashita H, Kohriyama T, Kawakami H, Nakamura S, et al. Effects of nicotine chewing gum on UPDRS score and P300 in early-onset parkinsonism. Hiroshima J Med Sci. 2002;51:33–9.
49. Villafane G, Cesaro P, Rialland A, Baloul S, Azimi S, Bourdet C, et al. Chronic high dose transdermal nicotine in Parkinson's disease: an open trial. Eur J Neurol. 2007;14:1313–6.
50. Hanagasi HA, Lees A, Johnson JO, Singleton A, Emre M. Smoking-responsive juvenile-onset Parkinsonism. Mov Disord. 2007;22:115–9.
51. Lemay S, Chouinard S, Blanchet P, Masson H, Soland V, Beuter A, et al. Lack of efficacy of a nicotine transdermal treatment on motor and cognitive deficits in Parkinson's disease. Prog Neuropsychopharmacol Biol Psychiatry. 2004;28:31–9.
52. Clemens P, Baron JA, Coffey D, Reeves A. The short-term effect of nicotine chewing gum in patients with Parkinson's disease. Psychopharmacology (Berl). 1995;117:253–6.
53. Ebersbach G, Stock M, Muller J, Wenning G, Wissel J, Poewe W. Worsening of motor performance in patients with Parkinson's disease following transdermal nicotine administration. Mov Disord. 1999;14:1011–3.
54. Vieregge A, Sieberer M, Jacobs H, Hagenah JM, Vieregge P. Transdermal nicotine in PD: a randomized, double-blind, placebo-controlled study. Neurology. 2001;57:1032–5.
55. Shoulson I. Randomized placebo-controlled study of the nicotinic agonist SIB-1508Y in Parkinson disease. Neurology. 2006;66:408–10.
56. Fagerstrom KO, Pomerleau O, Giordani B, Stelson F. Nicotine may relieve symptoms of Parkinson's disease. Psychopharmacology (Berl). 1994;116:117–9.
57. Kawamata J, Suzuki S, Shimohama S. alpha7 nicotinic acetylcholine receptor mediated neuroprotection in Parkinson's disease. Curr Drug Targets. 2012;13:623–30.
58. Thiriez C, Villafane G, Grapin F, Fenelon G, Remy P, Cesaro P. Can nicotine be used medicinally in Parkinson's disease? Expert Rev Clin Pharmacol. 2011;4:429–36.
59. Picciotto MR, Zoli M. Neuroprotection via nAChRs: the role of nAChRs in neurodegenerative disorders such as Alzheimer's and Parkinson's disease. Front Biosci. 2008;13:492–504.

60. Mudo G, Belluardo N, Mauro A, Fuxe K. Acute intermittent nicotine treatment induces fibroblast growth factor-2 in the subventricular zone of the adult rat brain and enhances neuronal precursor cell proliferation. Neuroscience. 2007;145:470–83.
61. Quik M, O'Neill M, Perez XA. Nicotine neuroprotection against nigrostriatal damage: importance of the animal model. Trends Pharmacol Sci. 2007;28:229–35.
62. Shimohama S. Nicotinic receptor-mediated neuroprotection in neurodegenerative disease models. Biol Pharm Bull. 2009;32:332–6.
63. Noyce AJ, Bestwick JP, Silveira-Moriyama L, Hawkes CH, Giovannoni G, Lees AJ, et al. Meta-analysis of early nonmotor features and risk factors for Parkinson disease. Ann Neurol. 2012;72:893–901.
64. Wirdefeldt K, Adami HO, Cole P, Trichopoulos D, Mandel J. Epidemiology and etiology of Parkinson's disease: a review of the evidence. Eur J Epidemiol. 2011;26 Suppl 1:S1–58.
65. Tanner CM. Advances in environmental epidemiology. Mov Disord. 2010;25 Suppl 1:S58–62.
66. Searles Nielsen S, Gallagher LG, Lundin JI, Longstreth Jr WT, Smith-Weller T, Franklin GM, et al. Environmental tobacco smoke and Parkinson's disease. Mov Disord. 2012;27:293–6.
67. Elbaz A, Moisan F. Update in the epidemiology of Parkinson's disease. Curr Opin Neurol. 2008;21:454–60.
68. Tanner CM, Goldman SM, Aston DA, Ottman R, Ellenberg J, Mayeux R, et al. Smoking and Parkinson's disease in twins. Neurology. 2002;58:581–8.
69. Caulfield MP, Birdsall NJ. International Union of Pharmacology. XVII. Classification of muscarinic acetylcholine receptors. Pharmacol Rev. 1998;50:279–90.
70. Langmead CJ, Watson J, Reavill C. Muscarinic acetylcholine receptors as CNS drug targets. Pharmacol Ther. 2008;117:232–43.
71. Yarnall A, Rochester L, Burn DJ. The interplay of cholinergic function, attention, and falls in Parkinson's disease. Mov Disord. 2011;26:2496–503.
72. Gomez-Mancilla B, Bedard PJ. Effect of nondopaminergic drugs on Levodopa-induced dyskinesia in MPTP-treated monkeys. Clin Neuropharmacol. 1993;16:418–27.

Chapter 17
New Clinical Trials for Levodopa-Induced Dyskinesia

Susan H. Fox and Isabelle Boileau-Boire

Abstract Ongoing clinical trials are looking at new strategies for treatment of levodopa-induced dyskinesia (LID). While the pathophysiology of LID is still not completely understood, preclinical studies have provided more insights into the underlying mechanisms. To date, however, translation to human therapeutic trials has generally been disappointing. Two main therapeutic strategies are recognized: (1) agents that may *prevent* the development of dyskinesia and can be used in early PD and (2) interventions that *reduce established* dyskinesia in advanced PD.

As LID are thought to relate to chronic pulsatile stimulation of dopamine receptors, continuous dopaminergic stimulation might reduce established dyskinesia and possibly prevent or delay the appearance. Ongoing clinical trials are investigating novel dopamine preparations with a more stable delivery that might also allow reductions in oral levodopa. The effect of levodopa-sparing agents on LID is also being investigated, both in early and advanced PD. Moreover, non-dopaminergic agents are being studied as add-on therapies in established LID. Such agents are also being studied in early PD, either as monotherapy to improve parkinsonian symptoms without causing dyskinesia or as add-on treatments to prevent development of dyskinesia in levodopa-treated patients.

In this chapter, we will review recently published and ongoing Phase II–IV clinical trials for the treatment of LID. As the research field is constantly evolving, this chapter will be updated regularly through a website (LINK), which will include a database of ongoing studies and recent results from clinical trials. This is meant to be a practical tool for the clinician to follow new developments in the field of LID treatment and to have an easy access to information on ongoing trials.

Keywords Levodopa-induced dyskinesias • Parkinson's disease • Motor fluctuations • Treatment • Clinical trials

S.H. Fox (✉) • I. Boileau-Boire
Movement Disorder Clinic, University of Toronto,
Toronto Western Hospital, Toronto, ON, Canada
e-mail: sfox@uhnresearch.ca

S.H. Fox, J.M. Brotchie (eds.), *Levodopa-Induced Dyskinesia in Parkinson's Disease*, DOI 10.1007/978-1-4471-6503-3_17

Levodopa-induced dyskinesias (LID) pose a challenge to every clinician facing patients with long-standing Parkinson's disease (PD). Levodopa is currently the most effective treatment for motor symptoms of PD. However, the long-term use of levodopa is almost invariably associated with the development of motor complications such as wearing-off and dyskinesias. Currently, modifications in the levodopa regimen and amantadine remain the mainstay of the medical treatment for LID (see Chap. 5). However, reduction of levodopa formulations is frequently at the expense of an increase in parkinsonian symptoms, and amantadine can be associated with side effects and its efficacy may be limited. Some studies have shown a modest improvement in LID with clozapine [1], but its administration is limited by the risk of agranulocytosis and the need for frequent laboratory monitoring. Deep brain stimulation or continuous administration of levodopa or dopamine agonist can also be used in patients with significant motor fluctuations and dyskinesias, but inclusion criteria are very restricted and not every patient is amenable to these interventions.

Many clinical studies are trying to find new therapeutic targets for LID. While the pathophysiology of LID is still not completely understood, ongoing preclinical studies have provided more insights on the biochemical mechanisms of this motor complication. Translation to human clinical studies has, however, been generally disappointing so far. Some molecules seem to have positive results in small studies, and further research is needed to establish the real efficacy of these new molecules.

In this chapter, we will review ongoing Phase II–IV clinical trials for the treatment of LID. As the research field is constantly moving, this chapter will soon be out-of-date and therefore a website will be updated regularly with new studies and recent results from clinical trials. This is meant to be a useful tool for the clinician to keep tract of new developments in the field of LID treatment and to provide, in a centralized fashion, information on ongoing studies. We have divided the trials into two main therapeutic strategies:

1. Studies investigating agents that may *prevent* the development of dyskinesia and can be used in early PD
2. Studies for interventions that *reduce established* dyskinesia in advanced PD

Prevention of Dyskinesia in Early PD Patients (Table 17.1)

One focus of research is directed at *preventing* the appearance of LID in early PD patients. Several approaches have been adopted, and the background pathophysiology underlying these targets has been extensively covered in earlier chapters. One strategy is to provide more continuous dopaminergic stimulation, to minimize pulsatile dopamine receptor stimulation using longer-acting dopaminergics such as dopamine agonists. Another approach is to use add-on, "levodopa-sparing" agents, such as dopamine agonists or monoamine oxidase-B (MAO-B) inhibitors, to reduce levodopa dosage. Some studies are also investigating non-dopaminergic agents for

Table 17.1 Prevention of dyskinesia in early PD patients (Ongoing studies as of September 3rd, 2013)

Agent	Class of molecule	Study type	Dose	Duration	Number of pts	Inclusion criteria	Primary outcomes	Secondary outcomes	Measure for dyskinesias
Amantadine (PREMANDYSK study)	NMDA glutamate receptor antagonist	Phase 2	200 mg/day	18 months	202 pts	Early idiopathic PD by UKPDBB criteria	Ratio of patients with dyskinesia	Ratio of patients with dyskinesia after wash-out	'NS'
NCT01538329		Double-blind, placebo-controlled randomized study				<3 years of Dx		Ratio of patients with motor fluctuations	
Study start date: March 2012						<1 year of levodopa		Time to onset of dyskinesias	
						Age >35 years			
						Lack of motor complications of levodopa therapy			
						Informed consent			
						Health Insurance Coverage			
						Stable antiparkinsonian treatment for at least 2 months			
						Possible to maintain this treatment unchanged during the study period (except the dose of L-dopa which can be adjusted during the study after the third month of Phase 1)			

(continued)

Table 17.1 (continued)

Agent	Class of molecule	Study type	Dose	Duration	Number of pts	Inclusion criteria	Primary outcomes	Secondary outcomes	Measure for dyskinesias
Amantadine (vs. dopamine agonist as initial treatment)	NMDA glutamate receptor antagonist	Observational, prospective case–control study	NS	10 years	500 pts	PD	Dyskinesia onset	UPDRS	NS
NCT01338662						Age 30–60 years		Dyskinesia severity	
Study start date: May 2011						Hoehn and Yahr <3			
Omega-3 fatty acids	Anti-oxidant	Phase 1	NS	1.5 years	40 pts	Idiopathic PD	Adverse events (periodic safety lab information, monthly telephone calls, therapeutic level monitoring, and assessment of dietary intakes)	Dyskinesias (during levodopa administration)	Measurements with a force plate
NCT01563913		Double-blind, placebo-controlled randomized study				No levodopa treatment or prior exposure to levodopa			Clinical scale (NS)
Study start date: October 2012									

use in early PD that can improve PD symptoms either as monotherapy without causing dyskinesia or as add-on treatments that could prevent development of dyskinesias in levodopa-treated patients.

Dopamine Agonists

Dopamine agonists, as monotherapy in early PD, have been shown to delay the development of dyskinesia; however this effect is generally not maintained once levodopa is started. Although at present there are no new active trials in this category; a brief review is outlined below. The dopamine D2 agonists ropinirole [2, 3], pramipexole [4, 5], and bromocriptine [6, 7] all showed less dyskinesia in the short term (2–5 years), but this benefit was generally lost after 10–14 years follow-up, once levodopa was added in. The ability to reduce LID development may be partly due to the longer half-life compared to levodopa, as well as the ability to use lower levodopa doses. However, the longer duration of action is not likely the main effect on reducing dyskinesia as early use of the dopamine agonist cabergoline, with the very long half-life of 63–68 h [8], appears to have similar long-term effects as the shorter-acting agonists [9, 10].

The new longer-acting preparations of ropinirole and pramipexole have been investigated as monotherapy in early PD, although to date dyskinesia development has not been the primary outcome. The slightly longer duration of action may be of advantage, but studies have not yet shown this benefit. Ropinirole prolonged release (PR), as monotherapy in early PD, was significantly more effective when compared to ropinirole immediate release (IR) [11]. The results were probably driven by higher doses of ropinirole PR; however despite the higher dosage, tolerance was similar in both groups. A randomized controlled trial (RCT) comparing add-on ropinirole PR versus additional levodopa in early PD patients already on levodopa showed significant difference in the proportion of patients who developed dyskinesia in the ropinirole PR group (3 % with ropinirole PR versus 17 % with additional levodopa) [12]. Pramipexole extended release (ER) has been studied as monotherapy in early PD in two double-blind, placebo- and active-comparator-controlled study [13, 14]. The difference in changes in the Unified Parkinson's Disease Rating Scale (UPDRS) parts II + III composite score was statistically significant for pramipexole ER when compared to placebo and non-inferior when compared to pramipexole IR. Dyskinesias were as frequent in both treatment groups (16.5 % with pramipexole ER vs. 18.3 % with pramipexole IR) but significantly more than with placebo (7.9 %).

Rotigotine, a transdermal patch of dopamine agonist, has also been studied in early PD. The continuous release of the dopamine agonists would be expected to have a positive effect on preventing LID, assuming the concept of continuous dopamine receptor stimulation (CDS); however, to date, studies have not been undertaken to investigate this concept. Results of a placebo-controlled trial demonstrated significant improvement in UPDRS part III [15]. Another study in early PD has shown non-inferiority to pramipexole and superiority to placebo in terms of reduction in OFF time [16]. Dyskinesias were more frequent with both pramipexole and rotigotine when compared to placebo (15, 12, and 3 %, respectively), but no difference was found between pramipexole and rotigotine during the study period.

Amantadine

Amantadine is a nonselective N-methyl-*D*-aspartate (NMDA) glutamate receptor antagonist that is currently recommended for the treatment of LID [17]. The effect as an antidyskinetic agent has been described in previous studies (see Chap. XX). Although no preclinical study has been published on evaluating the effect of pretreatment with amantadine to prevent LID, early use of amantadine to prevent striatal glutamate-induced neuroplasticity changes appears to be a reasonable approach. A recent retrospective study investigating the effect of early treatment with amantadine on risk of developing dyskinesia failed to demonstrate a difference in time to onset of dyskinesias, or proportion of patients with dyskinesias at 5 years, in patients who first started on amantadine versus levodopa [18].

A prospective phase II RCT trial of amantadine, 200 mg/day vs. placebo, is currently investigating possible protective effects of this medication on appearance of dyskinesias (PREMANDYSK study) (Table 17.1). Patients with early PD with less than 3 years of diagnosis and less than 1 year of levodopa usage, and no motor complications, are currently recruited. The ratio of patients with dyskinesia at 18 months will serve as the primary outcome. Time to onset of dyskinesias and ratio of patients with dyskinesia after a washout period will also be analyzed as secondary outcomes.

Furthermore, an observational, prospective, open-label study comparing the onset time and severity of LID in patients initially treated with amantadine or dopamine agonists is currently recruiting (Table 17.1). Patients will initially be divided on the basis of first treatment into either amantadine or dopamine agonist groups. When patients on amantadine necessitate a more potent symptomatic treatment, they will be further subdivided into two groups depending on whether the add-on treatment was levodopa or dopamine agonists. Patients in the dopamine agonists group cannot use amantadine. Dyskinesia onset will be analyzed over a period of 10 years, and secondary outcomes will consist of UPDRS scores and severity of dyskinesias.

Omega-3 Fatty Acids

Docosahexaenoic acid (DHA) is an omega-3 fatty acid that has been shown to delay the onset and reduce the severity of LID in MPTP monkeys [19]. A recent case report in a PD patient also showed benefit from fish oil containing high-dose DHA for treatment of LID [20], with a decrease on the Unified Dyskinesia Rating Scale

(UDysRS) score from 46 to 37 with treatment. A phase I study is currently recruiting untreated PD patients (Table 17.1). Adverse events will be monitored over a period of 1.5 years. As a secondary endpoint, dyskinesia will be measured during levodopa administration, with a clinical scale and objective dyskinesia measurements on a force plate. This device, which is similar to a doormat, has been validated in a recent study [21]. Patients are instructed to stand still on the mat first quietly and then with distraction maneuvers. Measures of variability of total body center of pressure were correlated to clinical rating scales of dyskinesia.

Reducing Established Dyskinesia (Table 17.2)

Treatment of established LID can be challenging. Three approaches are reviewed:

1. Continuous dopaminergic stimulation: such trials are investigating novel dopamine preparations that also allow reductions in oral levodopa. Surgical options such as deep brain stimulation (DBS) are reviewed in this section.
2. Add-on therapies to oral levodopa to reduce wearing off without causing LID.
3. Non-dopaminergic add-on therapies, with reducing LID as primary end point.

Continuous Dopaminergic Stimulation

Chronic stimulation of dopaminergic receptors, secondary to abnormal pulsatile levodopa dosage, is thought to lead to LID by inducing altered synaptic plasticity of striatal neurons and changes in striato-pallidal transmission (see Chap. XX). CDS might therefore be associated with a reduction in established LID. Levodopa has a short plasma half-life (around 90 min) and due to multiple factors (including peripheral and central pharmacokinetic effects) is associated with a progressively shorter duration of action with continued disease duration [22]. As a consequence, PD subjects require more frequent dosing schedules that further exacerbate the risk of LID. Methods to improve the duration of action of levodopa thus include improving gastric absorption: adaptation of the drug delivery to enhance duration of action or additional enzyme inhibition to prevent breakdown. Directly acting dopamine agonists are also able to bypass gastrointestinal issues.

Levodopa-Carbidopa Intestinal Gel

Levodopa-carbidopa intestinal gel (LCIG) is a method of direct infusion of levodopa into the small intestine, with the aim of bypassing slow gastric absorption issues, and thus increasing bioavailability of levodopa. The preparation is available for use in advanced PD in several countries, based on data from prior open-label and

Table 17.2 Reducing established dyskinesia

Agent	Class of molecule	Study type	Dose	Duration	Number of pts	Inclusion criteria	Primary outcomes	Secondary outcomes	Measure for dyskinesias
Continuous dopaminergic stimulation									
Levodopa/ Carbidopa intestinal gel (Duodopa)	Dopamine precursor and peripheral dopa-decar boxylase inhibitor	Obser vational, post-marketing long-term extension study	Variable	2 years	150 pts	PD already on with treatment with Duodopa (having already concluded the naso-intestinal phase)	Change in UPDRS part IV – item 39 (proportion of waking day spent in "OFF")	Change UPDRS IV	UPDRS part IV items 32 and 33
NCT01754129						Available data on Duodopa treatment, on previous PD conventional treatments and with at least one of the scales/questionnaires under study already collected on the patient clinical chart		Change in UPDRS I and II for both in OFF and in ON phase	
Study start date: December 2012						Written informed consent		Change in PDQ 39	
						Non-professional caregiver (relative or familiar who give daily assistance to the patient) has given his/her written consent		Change in PDSS-2	
								Change in Gait and Fall Questionnaire	
								Change in QUIP-RS	
								Change in Economical and social impact of the familiar healthcare	
								Change in Relative Stress Scale questionnaire (RSS)	
								Change in concomitant diseases and therapies	
								Change in global efficacy on motor symptoms rated by neurologists vs. baseline on a three-point scale	
								Change in self-assessment patients scale regarding their judgment on Duodopa therapy	
								Change in Duodopa daily infusion dosage	

Levodopa/Carbidopa intestinal gel (Duodopa)	Dopamine precursor and peripheral dopa-decarboxylase inhibitor	Observational, post-marketing long-term extension study	Variable	12 months	60 pts	Advanced levodopa-responsive PD	UPDRS part II	Proportion of patients who continue treatment	UPDRS part IV items 32 and 33
NCT01747655						Decision to treat with Duodopa prior to approach of the patient for participation in the study		UPDRS part III	
Study start date: February 2013						PD medication stable ×4 weeks		UPDRS part IV (items 32, 33, 34, 39)	
						Oral medication >4 times daily		Non-motor symptoms scale	
						Either 2–4 h OFF or 3 h of dyskinesias		PDQ-39	
								Healthcare resource utilization	
DBS surgery	Deep Brain Stimulator Electrodes Placed Using Intraoperative Computed Tomography and Frameless Stereotaxis Versus Microelectrode Recording and Frame-based Stereotaxis	Phase 4, open-label safety/efficacy, single arm study	–	6 months	50 pts	Idiopathic PD identified by specialists as surgical candidates	Change in ON time without dyskinesia	Change in PDQ-39	Patient diaries
NCT01703598								Change in motor UPDRS	
Study start date: September 2011									

(continued)

Table 17.2 (continued)

Agent	Class of molecule	Study type	Dose	Duration	Number of pts	Inclusion criteria	Primary outcomes	Secondary outcomes	Measure for dyskinesias
DBS surgery	Deep brain stimulation using single or multiple electrodes, target site: STN	Randomized, double-blind, parallel assignment of DBS with single or multiple electrodes	–	12 months	70 pts	Idiopathic PD ≥5 years	Change on UPDRS part III OFF medication	Change in Clinical Dyskinesia Rating Scale score	Clinical Dyskinesia Rating Scale
NCT00855621						UPDRS motor scale ≥20 in OFF medication state		Change in UPDRS part II	
Study start date: March 2009						Marked fluctuations of motor symptoms and/or troublesome dyskinesias, severe tremor, intolerable side effects of dopaminergic drugs		Change in Mattis Dementia Rating Scale score	
						Failure of medical treatment to sufficiently control symptoms		Change in social adjustment scale-SR	
						L-dopa responsive symptoms (≥30 % reduction in UPDRS motor score in drug ON state compared to OFF state) OR severe L-dopa unresponsive tremor		Change in scale of caregivers quality of life	
								Change in PDQ-39	
								New or worsened psychiatric symptoms	
Early vs. late DBS surgery	Deep brain stimulation, target site: bilateral STN; early vs. late implantation in patients with motor complications	Open-label, parallel groups study	–	3–4 years	200 pts (133 late compli cation group, 67 early compli cation group)	Idiopathic PD with >30 % improvement with levodopa challenge	Change in PDQ-39	Change from MDS-UPDRS part IB and II	Patient diary
NCT01922388						Fluctuations and/or dyskinesias	Change in MDS-UPDRS part III	PET imaging	
Study start date: August 2013						18–75 years	Change in ON time without troublesome dyskinesia		
						Normal brain MRI			
						Absence of dementia or severe psychiatric diseases			
						Written informed consent			

Ropinirole Prolonged Release (PR)	Dopamine agonist	Phase 4	4, 8, 12, 16 or 20 mg/day	4 weeks	350 pts	Idiopathic PD	Patient-reported awake time OFF (h)	NS	NS
NCT01494532		Randomized, double-blind, placebo-controlled study (add-on to levodopa)				Hoehn and Yahr criteria Stages II-IV)			
Study start date: April 2012						Lack of control with L-dopa therapy			
						Stable dose of L-dopa 4 weeks prior to screening			
						Minimum of 3 h awake "off-time" per day			
						>30 years of age			
						Non-pregnant/non-breastfeeding, women of child-bearing potential must be practicing a clinically accepted method of contraception			
						Written informed consent			
						Being willing and able to comply with study procedures, including diary card completion and follow-up clinic visits			
Ropinirole Controlled Release (CR)	Dopamine agonist	Phase 3	18–24 mg/day (Ropinirole CR 16 mg/day, plus 2 mg/day or 8 mg/day or placebo)	Screening phase 4 weeks, dose increase 12 weeks, down titration 1 week, long-term 39 weeks, down titration 1–2 weeks	80 pts	Idiopathic PD	Change in UPDRS part III	Number of pts with 30 and 20 % reduction UPDRS part III	Patient diary
NCT01929317		Randomized, double-blind, placebo-controlled study (high dose vs. maintenance)				Monotherapy arm: never received levodopa, or <3 months treatment with levodopa and discontinued >4 weeks vs. levodopa adjunct: levodopa >4 weeks		Change in UPDRS part 1, 2, and 4	
Study start date: August 2013						Age >20 years		UPDRS	
						Informed consent		CGIC	
						Corrected QTc <450 ms		Change in OFF time	
						Liver function tests <1.5–2× N		Change in ON time	
								Change in ON time without troublesome dyskinesia	

(continued)

Table 17.2 (continued)

Agent	Class of molecule	Study type	Dose	Duration	Number of pts	Inclusion criteria	Primary outcomes	Secondary outcomes	Measure for dyskinesias
Pramipexole Extended Release (ER)	Dopamine agonist	Phase 4	Variable	4 months (2 months QD, 2 months BID)	200 pts	Idiopathic PD	Patient preference QD vs. BID	Motor complications (visual rating scale for OFF severity and duration, dyskinesia severity and duration	Visual rating scale for dyskinesia duration and severity, and patient preference
NCT01515774		Open-label, randomized crossover study (QD vs. BID dosing)				Age 30–80 years		Sleep problems	
Study start date: September 2011						On dopamine agonists (ropinirole or pramipexole) and are considering to change into Mirapex ER		Motor UPDRS	
						Stable PD medication >4 weeks		Hoehn and Yahr	
						Consent to the study		Side-effects (rating scale on 0–10)	
								Preference of QD vs. BID on each factor (OFF duration and sverity, dyskinesia duration and severity, ON quality, adverse events, sleep quality, convenience)	
								Patient choice (4 months)	

Rotigotine transdermal patch	Dopamine agonist	Phase 3	4 mg/24 h to 16 mg/24 h	27 weeks (4-week Screening, 7-week Titration, 12-week maintenance, 12-day De-escalation, 30-day Safety Follow-Up)	346 pts	Idiopathic PD >3 years	Change in absolute OFF time	Responders to therapy, defined as a ≥30 % decrease in absolute OFF time	Patient diary
NCT01646255		Randomized, double-blind, placebo-controlled study				written informed consent		Percent change in absolute and relative OFF time	
Study start date: July 2012						Subject/legal representative is considered reliable and capable of adhering to the		Change and percent change in absolute and relative ON time	
						Hoehn and Yahr stage 2–4 in both ON and OFF state		Change in the number of OFF periods	
						Age ≥30 years		Change in status of the subject (on/off) after wake-up	
						MMSE score of ≥25		Change in UPDRS Part III during ON periods	
						Stable dose of L-dopa of at least 200 mg/day, in at least two doses, for >28 days			
						Not adequately controlled on a L-dopa dose which was judged by the treating physician to be optimal			
						Willing and able to accurately complete a subject diary			
						Able to differentiate between the ON and OFF states			
						≥2.5 h/day spent in the "off" state			
						If the subject is receiving an anticholinergic agent, MAOB inhibitor and/or an NMDA antagonist: stable dose for >28 days, and maintained on that dose for the duration of the study			
						Stable dose of all anti-Parkinsonian medications >20 days			

(continued)

Table 17.2 (continued)

Agent	Class of molecule	Study type	Dose	Duration	Number of pts	Inclusion criteria	Primary outcomes	Secondary outcomes	Measure for dyskinesias
Add-on therapies to reduce LID									
AVP-923 (dextro methorphan – quinine)	Modulator of glutamate and 5-HT release	Phase 2	45 mg dextro methorphan + 10 mg quinine	14 days (2-week treatment, 2-week break, 2-week placebo)	16 pts	Idiopathic PD (UK-BB criteria)	Dyskinesia (UDysRS part III)	UDysRS part IV	UDysRS
NCT01767129		Randomized, double-blind, placebo-controlled, crossover study with levodopa infusion				30–80 years		Bradykinesia	Motor diaries
Study start date: July 2013						LID >25 % of the day and at least moderate (MDS-UPDRS)		MDS-UPDRS part III	
						No Amantadine, MAOI >3 weeks		MDS-UPDRS parts I, II and IV	
						Stable PD meds >1 month		UDysRS part I and II	
						SSRI allowed (stable >1 month)		PD motor diary	
Topiramate (adjunct to amantadine)	AMPA and kainate glutamate receptor antagonist	Phase 2	NS	18 weeks	55 pts	Idiopathic PD (UK-BB criteria)	Dyskinesias (UDysRS)	CGIC	UDysRS
NCT01789047		Double-blind, placebo-controlled randomized study (add-on TPX vs. add-on placebo to amantadine)				30–90 years		MDS-UPDRS	
Study start date: March 2013						Clinically pertinent dyskinesias (CGI-s >3)		Hoehn and Yahr stage	
						Stable PD meds >4 weeks			
						Amantadine >200 mg stable >4 weeks			
						Caregiver willing to participate			
						Possible to maintain PD meds for duration of study			
						No dementia, depression, psychosis			
						Subject willing to participate in all visits			

Mavoglurant (AFQ056)	mGluR5 antagonist	Phase 2	NS	Anticipated to be on average 3 years	N/A	Patients who have completed a previous AFQ056 study	Incidence of AE	Dyskinesias (change in mAIMS, LFADLDS, items 32-33-34 in UPDRS)	mAIMS
NCT01491932		Open-label extension study				Outpatients	Incidence of SAE	MMSE	LFADLS
Study start date: March 2012						Primary caregiver willing and able to assess the condition of the patient	Changes in vital signs/lab values/ECG	SCOPA-PC	UPDRS parts 32-33-34
							Changes in UPDRS part III	C-CASA	
								C-SSRS	
Famotidine	Histamine 2 (H2) receptor antagonist	Multiple “N-of-1” studies of 3 different doses of Famotidine	3 different doses: 80, 120, and 160 mg/day	2 weeks per dose, washout 1 week in between different doses		PD by UKPDS Brain Bank Criteria	Dyskinesias (UDysRS)	MDS-UPDRS parts III and IV	UDysRS
Not registered in clinicaltrials.gov						18–79 years		CGIC	LFADLDS
Study start date = September 2010						Stable levodopa-induced dyskinesia		Dyskinesias (LFADLDS)	
						Stable PD drug regimen for 1 month prior to enrollment		Adverse effects	
Famotidine	Histamine 2 (H2) receptor antagonist	Double-blind, placebo-controlled randomized crossover study	40 mg BID	3 weeks		PD >3 years	Change in dyskinesia (Rush dyskinesia scale)	Dyskinesias (UDysRS)	Rush dyskinesia scale
Not registered in clinicaltrials.gov						30–85 years			UDysRS
Study start date = October 2010						Post-menopausal female or surgically sterilized or method of contraception			
						Levodopa responsive, stable levodopa dose (3–10 doses per day)			
						Daily dyskinesia			
						Willing and able to participate			
						Informed, written consent			
						MMSE >24			

(continued)

Table 17.2 (continued)

Agent	Class of molecule	Study type	Dose	Duration	Number of pts	Inclusion criteria	Primary outcomes	Secondary outcomes	Measure for dyskinesias
AQW051	Neuronal nicotinic receptor (a7nAchR) ligand	Phase 2	High and low dose (NS)	28 days	71 pts	PD	Dyskinesias (change in the mAIMS score	Dyskinesias (LFADLDS, UPDRS – Part IV #32–33)	mAIMS
NCT01474421		Randomized, double-blind, placebo-controlled, parallel-group multiple-dose study				Dyskinesias for >3 months	UPDRS Part III	Track-PD	LFADLDS
Study start date: September 2011						Moderate-to-severe dyskinesias	Safety and tolerability (number of participants with adverse events, any clinically significant abnormalities in safety labs or electrocardiograms (ECGs), and relevant orthostatic changes in blood pressure)	CogState	UPDRS part IV items 32 + 33
Completed March 2013, no data available						L-dopa treatment for at least 3 years		Area under the curve (AUC [0–24 h]) of AQW051	

retrospective studies, which revealed benefit in reducing OFF time without an increase in dyskinesias [23–25]. The phase III double-blind, double-dummy RCT was published recently [26]. This study, conducted in 71 patients, showed reduction in OFF time of −1.91 h in the LCIG group when compared to immediate-release oral levodopa-carbidopa, without an increase in dyskinesias. The study was not designed to show a difference in LID, as the recruited patients had low baseline levels of dyskinesias. However, there was a trend toward a greater increase in levodopa total daily dose in patients treated with oral levodopa-carbidopa, which could possibly correlate with a lower risk of dyskinesias. This will have to be studied in further detail in long-term studies.

There is an ongoing large multicenter (86 sites in 16 countries) open-label pragmatic study to evaluate clinical usefulness of levodopa-carbidopa gel (Table 17.2). Advanced PD subjects with motor fluctuations despite optimal medical treatment were recruited. Interim results of 192 patients, reported after 12 weeks, showed increased daily ON time without troublesome dyskinesia was 4.6 h per day, compared to baseline [23]. The benefit was sustained among the 61 patients who completed the 54 weeks' period, with a mean increase in ON time without dyskinesias of 5.3 h per day. Further analysis on 323 subjects who reached 54 weeks has been presented in abstract only and showed increased ON time without troublesome dyskinesias, irrespective of age, PD duration, and number of PD medications at baseline [27]. Benefits were noticed as early as 4 weeks and were maintained throughout the study period (54 weeks).

A real-life observational study comparing levodopa-carbidopa gel with apomorphine has been completed and interim results reported [28]. This analysis, performed on 87 patients with advanced PD, showed similar effects on motor symptoms for both treatments, with improvements on UPDRS parts III and dyskinesia as assessed using UPDRS part IV after 6 months, and better scores on the short version of Parkinson's Disease Questionnaire (PDQ-8). Response to levodopa-carbidopa intestinal gel was superior to apomorphine in terms of non-motor symptoms, mainly sleep/fatigue and gastroenterological/urinary symptoms, while improvements in mood were greater with apomorphine. Even though a nonsignificant tendency toward greater impact on LID with levodopa-carbidopa intestinal gel was found [29], both methods of continuous dopaminergic stimulation appear to have similar efficacy on reducing dyskinesia. There was no data on changes in levodopa dosing over the course of the study. A direct "de-priming" effect on striatal neuroplasticity due to CDS however may also play a role in reduced LID expression.

Bilayered Immediate-Release and Extended-Release Carbidopa/Levodopa

A new formulation of bilayered immediate-release and extended-release carbidopa/levodopa (IPX066) has been developed with the aim of providing a longer duration of action of levodopa. A recent RCT reported decreased OFF time by 1.69 h (as measured by patients self-assessed PD diaries) compared to 0.66 h for immediate-release carbidopa/levodopa, without increasing ON time with troublesome dyskinesia [30]. Improvements in UPDRS part III scores were also better in the IPX066 group.

Gastric-Retentive, Extended-Release Formulation of Levodopa/Carbidopa

DM1992, a gastric-retentive, extended-release formulation of levodopa/carbidopa, has been studied in a phase II trial. Preliminary data show a reduction in OFF time of 1.1 h in the DM1992 group compared to immediate-release carbidopa/levodopa, based on patient diaries [31]; however, data concerning dyskinesias are not available yet.

Sustained-Release Levodopa-Carbidopa (Accordion Pill)

Another sustained-release carbidopa-levodopa preparation, the Accordion Pill (AP), was assessed in a phase II multicenter, crossover RCT [32]. When compared to the usual regimen of carbidopa-levodopa, AP decreased OFF time by 45 % without increasing troublesome dyskinesias for the smaller dose (50–375 mg BID); for the higher dose (50–500 mg BID), not only was OFF time decreased, but time with troublesome dyskinesias was reduced as well by 40 %.

Surgical Interventions

Surgical treatments such as pallidotomy and bilateral deep brain stimulation (DBS) of the subthalamic nucleus (STN) have shown benefit in improving motor symptoms in patients with PD suffering from levodopa-induced complications, and these results have been reproduced in various studies. Disabling dyskinesias still remains one of the main indications for surgical intervention. The target of choice for PD is usually bilateral STN-DBS, as a concomitant reduction in levodopa results in a marked reduction in dyskinesia by average of 69.1 % [33]. However, for some individuals with contraindications to STN-DBS, targeting the globus pallidus interna (GPi) specifically can reduce dyskinesia, without a change in levodopa dose [34]. There has been a recent interest in targeting GPi over STN for PD due to potential for less adverse effects on mood and cognition. Thus, a large RCT compared bilateral STN-DBS with GPi DBS and found there was a significant improvement in troublesome dyskinesia of about 3.0 h at 2 years for both targets but more depression and adverse effects on visual processing and other neurocognitive measures within the STN group [35, 36]. In another RCT published recently, cognitive, mood, and behavioral adverse effects were similar in patients with GPi or STN stimulation [37]. However, dyskinesias in the ON-stimulation/ON-medication state were lower in the GPi group as measured by the Clinical Dyskinesia Rating Scale (CDRS).

A recent study investigated the benefits of functional surgery if performed earlier in the disease [38]. This trial included patients with levodopa-related motor complications present for less than 3 years who were randomized to bilateral STN-DBS or best medical treatment. There was an improvement in quality of life as measured by Parkinson's Disease Questionnaire-39 (PDQ-39). Motor scores were also improved, including the UPDRS parts III and IV, and ON time without troublesome dyskinesias as measured by patient diaries. A dyskinesia assessment during levodopa challenge also showed improvement in the neurostimulation group, as measured by the Marconi dyskinesia scale.

New studies for DBS are under way to better determine the optimal targets, timing, and long-term effects of surgical treatment (Table 17.2). One ongoing trial is evaluating the effect of the method of placement of electrodes on clinical benefit. Microelectrode recording and frame-based stereotaxis will be compared to preoperative computed tomography and frameless stereotaxis. The primary outcome measure will be change in ON time without dyskinesia. Another study of STN-DBS is comparing single versus multiple electrodes, the main endpoint being the change on UPDRS part III OFF medication. LIDs will also be assessed by the CDRS. Finally, a trial is currently recruiting PD patients with early (<3 years) and late (>3 years) motor fluctuations for STN-DBS. The change in ON time without troublesome dyskinesias will be analyzed as a primary endpoint.

Add-On Therapies to Oral Levodopa to Reduce Wearing Off Without Causing LID

Many new studies for PD are focused on pharmacological approaches to treat wearing off with the additional benefit of not worsening or inducing dyskinesia.

Prolonged-Release Ropinirole

Prolonged-release (PR) ropinirole is efficacious as add-on to levodopa, with a higher proportion of patients showing >20 % reduction in OFF time compared to ropinirole immediate release (66 % versus 51 %) [11], with no significant change in time ON with troublesome dyskinesia or the Abnormal Involuntary Movement Scale (AIMS) score. Recruitment for the phase IV extension study is ongoing (Table 17.2). Another trial is analyzing maintenance dose versus increased dose of ropinirole PR, as monotherapy or levodopa adjunct (Table 17.2). Patient diaries will be used to determine the outcomes.

Extended-Release Pramipexole

Pramipexole extended release also showed reduction in OFF time when compared to placebo (−2.1 h vs. −1.4 h), while the magnitude of the effect was comparable to Pramipexole IR (−2.5 h) [14]. UPDRS parts II+III scores also improved with pramipexole ER (−11.0 points) and IR (−12.8 points) when compared to placebo (−6.1 points). There was no increase in ON time without troublesome dyskinesia with either formulation of pramipexole. An extension study evaluating once daily (QD) versus twice daily (BID) dosage is ongoing (Table 17.2).

Rotigotine

Rotigotine transdermal patch has also shown benefit on motor symptoms in multiple studies [15, 39–41]. Recently published results from two open-label extension studies confirmed safety and efficacy of the rotigotine patch after a mean of 6 years [42]. Incidence of dyskinesia was similar in patients initially randomized in the rotigotine treatment group and placebo (study 1 – PREFER) or pramipexole (study 2 – CLEOPATRA-PD) groups. A phase III study is ongoing for patients with advanced PD and motor fluctuations (Table 17.2). The primary efficacy measure is the change in OFF time, as measured by patient diaries. ON time without troublesome dyskinesias will be available from the diaries for further evaluation.

Apomorphine

Apomorphine is a short-acting dopamine agonist that can be delivered via subcutaneous pump. Studies comparing DBS and subcutaneous infusion of apomorphine revealed similar improvements in OFF time, but dyskinesia reduction was only found in the stimulation group [43, 44]. Apomorphine infusion has also been studied in patients with contraindications to DBS. OFF time was reduced by 36 % and dyskinesias remained stable [45]. Further results from another retrospective study confirmed the same findings [46], with the absence of improvement in UPDRS part IV or dyskinesia scores.

Safinamide

Safinamide is a mixed monoamine oxidase-B (MAO-B) inhibitor, which also acts as a glutamate release inhibitor and reduces dopamine reuptake. This agent has been evaluated in advanced PD with motor complications. Preliminary results of two placebo-controlled trials were presented (add-on to levodopa and add-on to dopamine agonist), showing a significant improvement in motor symptoms without increasing troublesome dyskinesias [47, 48].

Ordopidine

Dopidines are a new class of medications that stabilize dopamine transmission by acting as competitive dopamine D2 receptor antagonists with fast dissociation properties [49]. Pridopidine, a dopamine stabilizer, has been studied in Huntington's disease [50]. The study did not reach statistical significance, but trends toward motor improvements were observed. Ordopidine (ACR325), a new member of the dopidines class, has completed in a phase I trial. Phase II trials focusing on LID are planned [51].

Zonisamide

Zonisamide is a drug mainly known for its antiepileptic properties. Its mechanisms of action are diverse and include blockade of voltage-sensitive T-type calcium channels, inhibition carbonic anhydrase, inhibition of glutamate release, and modulation of GABAa receptors. Beneficial effect of zonisamide has been demonstrated in animal models of PD. The mechanisms of action on parkinsonian symptoms are not well defined, but probably include inhibition of monoamine oxidase, activation of dopamine synthesis, and modulation of a δ-opioid receptor in the basal ganglia [52]. In a recent RCT [53], time in OFF state was significantly reduced by 0.719 h in the zonisamide group compared to a decrease of 0.011 h in the placebo group. Dyskinesia rates were similar in both groups.

Add-On Therapies to Reduce LID as Primary Endpoint

Glutamatergic Pathways

As reviewed in Chap. XX, glutamatergic pathways have a role in the pathophysiology of LID. Glutamate receptors are divided into two main categories, ionotropic (NMDA, AMPA) and metabotropic (mGluR) receptors. Amantadine, a recommended treatment for LID, probably exerts its antidyskinetic effect via the blockade of NMDA receptors causing attenuation of the glutamatergic stimulation of the direct pathway. This rationale has led to multiple clinical studies.

NMDA Receptors

Amantadine

As mentioned previously, amantadine is the only medication specifically approved for treatment of LID. Even though this medication is efficient in reducing dyskinesia, use is often limited by poor tolerance. A new formulation of extended-release amantadine (ADS-5102), which is thought to have a better side effects profile, has been studied in PD patients with LID (EASED Study). Preliminary results of the phase II/III randomized, placebo-controlled, double-blind study reported significant differences between ER-amantadine-treated patients and placebo patients on the UDysRS [54] with a statistically significant dose response [55]. There was a decrease of ON time with troublesome dyskinesias of 1.8 h with the 340 mg dose when compared to placebo. Comparisons with immediate-release amantadine have not been done, however.

Long-term effects of amantadine on LID have been questioned, especially since results from a long-term study suggested that the beneficial effect of amantadine lasted only 4.3 months [56]. A recent study examining withdrawal of amantadine in treated patients demonstrated a significant increase in dyskinesias at follow-up, as measured by UPDRS items 32+33, the AIMS scores, and patient diaries [57]. There was also a higher dropout rate related to increased dyskinesias in the placebo group. Although the results of this study could be related to a withdrawal rebound effect of amantadine, this study provides new evidence to support the long-term use of amantadine in LID.

Dextromethorphan and Quinidine

Dextromethorphan is an uncompetitive NMDA receptor antagonist. Two small studies demonstrated reduction of dyskinesia scores with dextromethorphan without increasing parkinsonian symptoms [58, 59]. The combination of dextromethorphan with quinidine, a cytochrome P450 2D6 (CYP2D6) enzyme inhibitor, improves the bioavailability of the drug by decreasing its metabolism. Dextromethorphan and quinidine (AVP923) are being evaluated in a phase II, double-blind, placebo-controlled crossover study (Table 17.2). This study will focus on dyskinesia with UDysRS scores as a primary outcome and Movement Disorders Society UPDRS (MDS-UPDRS), PD Motor Diary, bradykinesia, and subscales of the UDysRS as secondary outcomes.

D-Serine

D-Serine, an NMDA receptor agonist, has demonstrated an improvement in both drug-induced parkinsonism and tardive dyskinesias in neuroleptic-treated schizophrenic patients, as rated by the Simpson-Angus Scale for Extrapyramidal symptoms (SAS) and Abnormal Involuntary Movement Scale (AIMS) scores, respectively [60]. A recent double-blind, controlled crossover study on a small number of patients with PD patients resulted in improved UPDRS score but no improvement in dyskinesias scores on the AIMS [61].

Neu-120 and Neu-240

Other molecules may be studied in phase II and III clinical trials in future years. Neu-120, a selective, uncompetitive NMDA receptor modulator, has completed phase I studies. Neu-240, another NMDA receptor modulator, is still in the preclinical stage.

AMPA Receptors

Topiramate

Topiramate is a kainate and alpha-amino-3-hydroxyl-5-methyl-4-isoxazolepropionate (AMPA) glutamate receptor antagonist mainly known for its anticonvulsant properties.

This molecule has demonstrated benefit in levodopa-treated MPTP primates with LID. A phase 2, double-blind, placebo-controlled, add-on to amantadine study is active (Table 17.2). Dyskinesias will be measured with the UDysRS, and other motor outcomes such as the MDS-UPDRS and Hoehn and Yahr stage will be analyzed.

Perampanel

Perampanel (E2007), a highly selective AMPA glutamate receptor antagonist, showed nonsignificant trends toward reduction in OFF time in a dose-ranging phase II study [62], but those findings were not reproduced in two phase III studies [63]. Perampanel also failed to reduce dyskinesias, and the perampanel PD development program was terminated.

Talampanel

Another AMPA receptor antagonist, Talampanel (LY300164) has been studied in two phase II studies, but results are not available.

mGluR5 Receptors

Mavoglurant

Mavoglurant (AFQ056) is a metabotropic glutamate receptor 5 (mGluR5) antagonist. Two previous trials reported jointly evaluated the effect of AFQ056 on PD patients with moderate-to-severe (study 1) or severe (study 2) LID [64]. Both trials found a significant improvement in dyskinesia scores measured by either the Lang-Fahn Activities of Daily Living Dyskinesia Scale (LFADLDS) (study 1) or modified Abnormal Involuntary Movement Scale (mAIMS) (study 2) after a 16-day period, without increasing parkinsonian symptoms. AFQ056 was reasonably well tolerated, main adverse events being related to reemergence of dyskinesias in the down-titration period. The most recent trial was a dose-finding study in PD patients with moderate-to-severe peak-dose dyskinesias [65]. LID were measured after 12 weeks by the mAIMS, PD dyskinesia Scale (PDYS-26), UPDRS part IV items 32 and 33 and patient diaries. AFQ056 at a dose of 200 mg/day demonstrated decreased mAIMS scores when compared to placebo, but no effect was observed at smaller doses. PDYS-26 scores, UPDRS items 32 + 33 composite score and ON time with dyskinesias were not improved with treatment. An open-label extension study has recently stopped and further development of mavoglurant appears to be on hold (Table 17.2).

Dipraglurant

Another metabotropic glutamatergic agent studied recently is Dipraglurant (ADX48621), which also acts as an mGluR5 modulator. Preliminary results from a randomized, double-blind, placebo-controlled phase 2 study were recently presented [66]. PD patients with moderate-to-severe LID were treated for 28 days. Significant differences in the mAIMS score were demonstrated at day 1, 14, but not at day 28. It is unclear if further studies are planned.

Serotoninergic Pathways

Animal models have shown that serotonin (5-HT) might be involved in the development of LID (see Chap. XX). However, to date, results of clinical trials have been disappointing.

Sarizotan

Sarizotan (EMD128130) is a 5-HT1a receptor agonist with additional dopaminergic properties. In a phase IIb, randomized placebo-controlled study in PD patients with LID, Sarizotan at a dose of 2 mg/day demonstrated a clinical benefit on the UPDRS IV indices of dyskinesia (items 32 + 33), but no improvement was noted in ON time without dyskinesias or the mAIMS score [67]. However, the beneficial effect on dyskinesias was not reproduced in two phase III trials (PADDY-I and PADDY-II trials) [68–70]. Worsening of parkinsonism was noted in addition and further development for LID has now stopped.

Piclozotan

Another 5-HT1A receptor partial agonist with partial D3 agonist activity, Piclozotan (SUN-N4057), has shown decrease in both ON time with dyskinesias and OFF time in a phase II study, suggesting potential benefit for treatment of LID [71].

Selective Serotonin Reuptake Inhibitors (SSRIs)

SSRIs are used in PD mainly for the treatment of depression and anxiety. Dyskinetic rat models have shown improvements in LID with treatment with SSRIs. A retrospective study reviewed incidence of LID in patients with SSRI treatment [72]. No difference was found in the proportion of patients who developed dyskinesias at the end of follow-up; however mean time to onset of dyskinesia was longer in SSRI-treated patients when adjusted for disease onset (6.7 vs. 5.5 years) and levodopa treatment starts (5.6 vs. 4.7 years). Prospective studies are needed before conclusions can be drawn. A single RCT using fluoxetine demonstrated mild significant reduction in dyskinesia after apomorphine infusion, without worsening parkinsonian symptoms, in seven PD patients [73]. However, no new studies have been initiated.

Adrenergic Pathways

Adrenergic pathways are also potentially involved in the development of LID (see Chap. XX). Noradrenergic neurones are particularly abundant in the striatum, and their overactivation may lead to overstimulation of the direct pathway, causing involuntary movements. Several drugs modulating these receptors have shown benefit in animal models.

Fipamezole

A recent double-blind, placebo-controlled phase II study of fipamezole (JP-1730), a selective alpha 2-adrenergic receptor antagonist, showed potential clinical benefit for treatment of dyskinesias in a prespecified subgroup analysis [74]. The scoring system used was the Levodopa-Induced Dyskinesia Scale (LIDS), a modification of the AIMS. A phase III study is apparently in the pipeline [75].

Adenosine Pathways

Adenosine A2A receptors activation in the striatum regulate dopamine and glutamate release in the brain. Autopsy series have shown a higher number of these receptors in the striatum in dyskinetic patients compared to PD patients without LID.

Preladenant

Preladenant (SCH420814), an antagonist of the adenosine A2A receptor, improved motor function in MPTP primates without increasing dyskinesias. Improvement in OFF time without increasing dyskinesia was demonstrated in a phase II, double-blind, randomized trial [76]. However, very recently, failure to produce the same results in phase III trials led to discontinuation of further studies [77].

Istradefylline

Another adenosine A2A receptor antagonist, Istradefylline (KW-6002), has shown, in multiple phase II and III studies, improvement in OFF time without increasing troublesome dyskinesias [78–80]. A recent placebo-controlled RCT demonstrated reduction in OFF time of 0.99 and 0.96 h with the 20 and 40 mg doses, respectively, versus 0.23 h in the placebo group, without significant increases in ON time with troublesome dyskinesias [81].

Tozadenant

Similar results were found in another phase II, placebo-controlled trial of Tozadenant (SYN115) presented recently. OFF time was reduced by 1.1 h and motor UPDRS improved, without worsening ON time with troublesome dyskinesias [82]. A phase III study is in the pipeline of the company, with enrollment of patients planned in 2015 [83]. This novel mechanism for treatment of parkinsonism could offer an alternative to levodopa and therefore decrease LID; however, specific studies addressing this issue have not been done yet. Nevertheless, a direct antidyskinetic effect seems unlikely.

Histamine Pathways

Famotidine

Famotidine is histamine 2 (H2) receptor antagonist mostly used in gastroesophageal reflux disease. H2 and H3 receptors are abundant in the basal ganglia and are thought to modulate the output of the direct and indirect pathways. A study consisting of multiple"N-of-1" studies of three different doses of Famotidine vs. placebo is currently recruiting pts with PD and LID (Table 17.2). The rationale behind that study is that H2 receptors stimulation is thought to decrease acetylcholine in the striatum and therefore stimulate the direct pathway, causing excessive movements. Blocking H2 receptors could therefore improve symptoms of LID. UDysRS will be analyzed in the primary outcome and other dyskinesia scales such as LFADLDS and UPDRS part IV will also be used. Another placebo-controlled crossover study is currently recruiting (Table 17.2); dyskinesia measurements will be using Rush Dyskinesia Scale and UDysRs.

Cholinergic Pathways

Enhanced striatal cholinergic activity has been associated with LID in animal models, possibly by causing hyperstimulation of the direct pathway. Blocking acetylcholine activity might therefore reduce symptoms of dyskinesias.

AQW051

AQW051, a neuronal nicotinic receptor (a7nAchR) ligand, has recently been studied in a randomized, double-blind, placebo-controlled phase II study of PD patients with moderate-to-severe LID [84]. The study has been completed recently but no data are currently available. Primary outcomes are change in mAIMS score, change in UPDRS part III, and safety and tolerability. Dyskinesias will be further addressed by the LFADLDS score and UPDRS part IV items 32–33.

Nicotine

Nicotine has been reported in case series to improve PD motor symptoms; however this has never been reproduced in randomized placebo-controlled studies. An anti-dyskinetic effect has been found in different animal models of LID [85, 86]. Nicotine patches, in a single study, improved neuroleptic-induced akathisia [87]. A recent double-blind, placebo-controlled phase I/II study of NP002 (nicotine receptor agonist) showed nonsignificant trends in improving LID on different rating scales [88]. Other phase II studies are planned but not currently under way.

Botulinum Toxin

Botulinum toxin has been studied in multiple studies for LID, but significant benefit was not achieved except for painful OFF dystonia of the foot [89]. The most recent study of botulinum toxin in LID [90] focused on cervical-predominant dyskinesias, and primary outcome consisted of the Goetz Dyskinesias Rating Scale (GDRS) while secondary outcomes were Clinician Global Impression of Change (CGIC) and items 32–34 of the UPDRS. Clinical benefit was seen in the GDRS scores for resting dyskinesias, and there was a trend for reduction in ON time with LID, but transient dysphagia and neck weakness led to premature discontinuation of the study because of safety concerns.

MAO-B Inhibitors

Safinamide

A phase II trial designed specifically to study the antidyskinetic effect of Safinamide, a MAO-B inhibitor, showed a trend toward improvement in dyskinesias as measured by the Dyskinesia Rating Scale, but did not reach statistical significance [91]. A post-hoc analysis in the subgroup of patients with severe dyskinesia demonstrated a significant decrease in LID, and it has been said that the low proportion of patients with dyskinesias might have influenced the results.

Antiepileptics

Levetiracetam

Levetiracetam, a synaptic vesicle glycoprotein (SV2A) modulator with well-known antiepileptic properties, has been studied in MPTP-lesioned primates with LID. The rationale for the treatment was effect on the pathological synchronization/desynchronization in the basal ganglia in LID patients. Positive results in animal studies led to human clinical trials, with conflicting results [92–95]. No definitive conclusion has been drawn from the results, and more clinical studies do not seem to be in the pipeline for the moment.

References

1. Durif F, Debilly B, Galitzky M, Morand D, Viallet F, Borg M, et al. Clozapine improves dyskinesias in Parkinson disease: a double-blind, placebo-controlled study. Neurology. 2004; 62(3):381–8.
2. Rascol O, Brooks DJ, Korczyn AD, De Deyn PP, Clarke CE, Lang AE. A five-year study of the incidence of dyskinesia in patients with early Parkinson's disease who were treated with ropinirole or levodopa. 056 Study Group. N Engl J Med. 2000;342(20):1484–91.
3. Hauser RA, Rascol O, Korczyn AD, Jon Stoessl A, Watts RL, Poewe W, et al. Ten-year follow-up of Parkinson's disease patients randomized to initial therapy with ropinirole or levodopa. Mov Disord. 2007;22(16):2409–17.
4. Holloway RG, Shoulson I, Fahn S, Kieburtz K, Lang A, Marek K, et al. Pramipexole vs levodopa as initial treatment for Parkinson disease: a 4-year randomized controlled trial. Arch Neurol. 2004;61(7):1044–53.
5. Holloway R, Marek K, Biglan K, Dick A, Fahn S, Julian-Baros E, et al. Long-term effect of initiating pramipexole vs levodopa in early Parkinson disease. Arch Neurol. 2009;66(5): 563–70.
6. Lees AJ, Katzenschlager R, Head J, Ben-Shlomo Y. Ten-year follow-up of three different initial treatments in de-novo PD: a randomized trial. Neurology. 2001;57(9):1687–94.
7. Katzenschlager R, Head J, Schrag A, Ben-Shlomo Y, Evans A, Lees AJ, et al. Fourteen-year final report of the randomized PDRG-UK trial comparing three initial treatments in PD. Neurology. 2008;71(7):474–80.
8. Product Monograph. PRDostinex® (cabergoline) [Internet]. Pfizer Canada Inc;2013. [Cited 2014 Jan 8]. Available from: http://www.pfizer.ca/en/our_products/products/monograph/159
9. Bracco F, Battaglia A, Chouza C, Dupont E, Gershanik O, Marti Masso JF, et al. The long-acting dopamine receptor agonist cabergoline in early Parkinson's disease: final results of a 5-year, double-blind, levodopa-controlled study. CNS Drugs. 2004;18(11):733–46.
10. Inzelberg R, Schechtman E, Nisipeanu P. Cabergoline, pramipexole and ropinirole used as monotherapy in early Parkinson's disease: an evidence-based comparison. Drugs Aging. 2003;20(11):847–55.
11. Stocchi F, Giorgi L, Hunter B, Schapira AH. PREPARED: comparison of prolonged and immediate release ropinirole in advanced Parkinson's disease. Mov Disord. 2011;26(7): 1259–65.
12. Watts RL, Lyons KE, Pahwa R, Sethi K, Stern M, Hauser RA, et al. Onset of dyskinesia with adjunct ropinirole prolonged-release or additional levodopa in early Parkinson's disease. Mov Disord. 2010;25(7):858–66.
13. Poewe W, Rascol O, Barone P, Hauser RA, Mizuno Y, Haaksma M, et al. Extended-release pramipexole in early Parkinson disease: a 33-week randomized controlled trial. Neurology. 2011;77(8):759–66.
14. Schapira AH, Barone P, Hauser RA, Mizuno Y, Rascol O, Busse M, et al. Extended-release pramipexole in advanced Parkinson disease: a randomized controlled trial. Neurology. 2011; 77(8):767–74.
15. Watts RL, Jankovic J, Waters C, Rajput A, Boroojerdi B, Rao J. Randomized, blind, controlled trial of transdermal rotigotine in early Parkinson disease. Neurology. 2007;68(4):272–6.
16. Poewe WH, Rascol O, Quinn N, Tolosa E, Oertel WH, Martignoni E, et al. Efficacy of pramipexole and transdermal rotigotine in advanced Parkinson's disease: a double-blind, double-dummy, randomised controlled trial. Lancet Neurol. 2007;6(6):513–20.

17. Fox SH, Katzenschlager R, Lim SY, Ravina B, Seppi K, Coelho M, et al. The movement disorder society evidence-based medicine review update: treatments for the motor symptoms of Parkinson's disease. Mov Disord. 2011;26 Suppl 3:S2–41.
18. Jahangirvand A, Rajput A. Early use of amantadine to prevent or delay onset of levodopa-induced dyskinesia in Parkinson's disease (abstract), 2013. MDS Conference, Sydney.
19. Samadi P, Grégoire L, Rouillard C, Bédard PJ, Di Paolo T, Lévesque D. Docosahexaenoic acid reduces levodopa-induced dyskinesias in 1-methyl-4-phenyl-1,2,3,6-tetrahydropyridine monkeys. Ann Neurol. 2006;59(2):282–8.
20. Mischley LK, McNames J. A case report of fish oil for the treatment of levodopa-induced dyskinesia (abstract), 2013. MDS Conference, Sydney.
21. Chung KA, Lobb BM, Nutt JG, McNames J, Horak F. Objective measurement of dyskinesia in Parkinson's disease using a force plate. Mov Disord. 2010;25(5):602–8.
22. Nutts J. Pharmacokinetics and pharmacodynamics of levodopa. Mov Disord. 2008;23(S3): S580–4.
23. Fernandez HH, Vanagunas A, Odin P, Espay AJ, Hauser RA, Standaert DG, et al. Levodopa-carbidopa intestinal gel in advanced Parkinson's disease open-label study: interim results. Parkinsonism Relat Disord. 2013;19(3):339–45.
24. Zibetti M, Merola A, Ricchi V, Marchisio A, Artusi CA, Rizzi L, et al. Long-term duodenal levodopa infusion in Parkinson's disease: a 3-year motor and cognitive follow-up study. J Neurol. 2013;260(1):105–14.
25. Nyholm D, Klangemo K, Johansson A. Levodopa/carbidopa intestinal gel infusion long-term therapy in advanced Parkinson's disease. Eur J Neurol. 2012;19(8):1079–85.
26. Olanow CW, Kieburtz K, Odin P, Espay AJ, Standaert DG, Fernandez HH, et al. Continuous intrajejunal infusion of levodopa-carbidopa intestinal gel for patients with advanced Parkinson's disease: a randomised, controlled, double-blind, double-dummy study. Lancet Neurol. 2014;13(2):141–9. Epub 2013 Dec 19.
27. Odin P, Espay A, Vanagunas A, Hauser R, Fernandez H, Standaert D, et al. Interim Results from an International, open-label study of levodopa-carbidopa intestinal gel in patients with advanced Parkinson's disease: efficacy analyses by subgroups (abstract). [Internet] AAN annual meeting, March 2013. [Cited 2014 Jan 8]. Available from: http://www.abstracts2view.com/aan/view.php?nu=AAN13L_P01.060
28. Reddy P, Martinez-Martin P, Todorova A, Antonini A, Odin P, Martin A, et al. The EuroInf study: a multi-centre European comparative study of apomorphine versus intrajejunal levodopa infusion in a real life cohort of Parkinson's patients (abstract). MDS Conference, Sydney.
29. P Reddy, P Martinez-Martin, P Odin, A Antonini, D Calandrella, M Pilleri, et al. The EUROINF study: a multicentre European comparative case control study of apomorphine versus intrajejunal levodopa infusion in advanced parkinson's disease (abstract). [Internet] [Cited 2014 Jan 8]. Available from: http://www.epda.eu.com/en/resources/publications-and-web-database/?entryid2=24577&cid=32343
30. Hauser RA, Hsu A, Kell S, Espay AJ, Sethi K, Stacy M, et al. Extended-release carbidopa-levodopa (IPX066) compared with immediate-release carbidopa-levodopa in patients with Parkinson's disease and motor fluctuations: a phase 3 randomised, double-blind trial. Lancet Neurol. 2013;12(4):346–56.
31. Press release. Moretti AJ. Depomed reports top line data for phase 2 study in Parkinson's disease. [Internet] Depomed, inc; 2012. [Cited 2014 Jan 8] Available from: http://investor.depomedinc.com/phoenix.zhtml?c=97276&p=irol-newsArticle&ID=1755781&highlight=
32. LeWitt P, Friedman H, Giladi N, Gurevich T, Shabtai H, Djaldetti R, et al. Sustained-release carbidopa-levodopa (accordion pill) in patients with advanced Parkinson's disease: Pharmacokinetic and clinical experience (abstract), 2013. MDS Conference, Sydney.
33. Kleiner-Fisman G, Herzog J, Fisman DN, Tamma F, Lyons KE, Pahwa R, et al. Subthalamic nucleus deep brain stimulation: summary and meta-analysis of outcomes. Mov Disord. 2006; 21 Suppl 14:S290–304.
34. Vitek JL, Bakay RA, Freeman A, Evatt M, Green J, McDonald W, et al. Randomized trial of pallidotomy versus medical therapy for Parkinson's disease. Ann Neurol. 2003;53(5):558–69.

35. Follett KA, Weaver FM, Stern M, Hur K, Harris CL, Luo P, et al. Pallidal versus subthalamic deep-brain stimulation for Parkinson's disease. N Engl J Med. 2010;362(22):2077–91.
36. Weaver FM, Follett KA, Stern M, Luo P, Harris CL, Hur K, et al. Randomized trial of deep brain stimulation for Parkinson disease: thirty-six-month outcomes. Neurology. 2012;79(1):55–65.
37. Odekerken VJ, van Laar T, Staal MJ, Mosch A, Hoffmann CF, Nijssen PC, et al. Subthalamic nucleus versus globus pallidus bilateral deep brain stimulation for advanced Parkinson's disease (NSTAPS study): a randomised controlled trial. Lancet Neurol. 2013;12(1):37–44.
38. Schuepbach WM, Rau J, Knudsen K, Volkmann J, Krack P, Timmermann L, et al. Neurostimulation for Parkinson's disease with early motor complications. N Engl J Med. 2013;368(7):610–22.
39. Giladi N, Boroojerdi B, Korczyn AD, Burn DJ, Clarke CE, Schapira AH, et al. Rotigotine transdermal patch in early Parkinson's disease: a randomized, double-blind, controlled study versus placebo and ropinirole. Mov Disord. 2007;22(16):2398–404.
40. LeWitt PA, Lyons KE, Pahwa R, SP 650 Study Group. Advanced Parkinson disease treated with rotigotine transdermal system: PREFER Study. Neurology. 2007;68(16):1262–7.
41. Elmer LW, Surmann E, Boroojerdi B, Jankovic J. Long-term safety and tolerability of rotigotine transdermal system in patients with early-stage idiopathic Parkinson's disease: a prospective, open-label extension study. Parkinsonism Relat Disord. 2012;18(5):488–93.
42. LeWitt PA, Boroojerdi B, Surmann E, Poewe W, SP716 Study Group. Rotigotine transdermal system for long-term treatment of patients with advanced Parkinson's disease: results of two open-label extension studies, CLEOPATRA-PD and PREFER. J Neural Transm. 2013;120(7):1069–81.
43. Antonini A, Isaias IU, Rodolfi G, Landi A, Natuzzi F, Siri C, et al. A 5-year prospective assessment of advanced Parkinson disease patients treated with subcutaneous apomorphine infusion or deep brain stimulation. J Neurol. 2011;258(4):579–85.
44. De Gaspari D, Siri C, Landi A, Cilia R, Bonetti A, Natuzzi F, et al. Clinical and neuropsychological follow up at 12 months in patients with complicated Parkinson's disease treated with subcutaneous apomorphine infusion or deep brain stimulation of the subthalamic nucleus. J Neurol Neurosurg Psychiatry. 2006;77(4):450–3.
45. Drapier S, Gillioz AS, Leray E, Péron J, Rouaud T, Marchand A, et al. Apomorphine infusion in advanced Parkinson's patients with subthalamic stimulation contraindications. Parkinsonism Relat Disord. 2012;18(1):40–4.
46. Rambour M, Moreau C, Devos D, Kreisler A, Mutez E, Simonin C, et al. Continuous subcutaneous infusion of apomorphine in Parkinson's disease: Retrospective analysis of a series of 81 patients (abstract), 2013. In: MDS Conference, Sydney.
47. Barone P, Fernandez H, Ferreira J, Mueller T, Saint-Hilaire M, Stacy M, et al. Safinamide as an add-on therapy to a stable dose of a single dopamine agonist: results from a randomized, placebo-controlled, 24-week multicenter trial in early idiopathic Parkinson disease (PD) patients (MOTION study) (abstract). [Internet] AAN annual meeting, March 2013. [Cited 2014 Jan 8] Available from: http://www.abstracts2view.com/aan/view.php?nu=AAN13L_P01.061&terms
48. Schapira A, Fox S, Hauser R, Jankovic J, Jost W, Kulisevsky J, et al. Safinamide add on to L-Dopa: a randomized, placebo-controlled, 24-week global trial in patients with Parkinson's disease (PD) and motor fluctuations (SETTLE) (abstract). [Internet] AAN annual meeting, March 2013. [Cited 2014 Jan 8] Available from: http://www.abstracts2view.com/aan/view.php?nu=AAN13L_P01.062&terms
49. Dyhring T, Nielsen EØ, Sonesson C, Pettersson F, Karlsson J, Svensson P, et al. The dopaminergic stabilizers pridopidine (ACR16) and (−)-OSU6162 display dopamine D(2) receptor antagonism and fast receptor dissociation properties. Eur J Pharmacol. 2010;628(1–3):19–26.
50. de Yebenes JG, Landwehrmeyer B, Squitieri F, Reilmann R, Rosser A, Barker RA, et al. Pridopidine for the treatment of motor function in patients with Huntington's disease (MermaiHD): a phase 3, randomised, double-blind, placebo-controlled trial. Lancet Neurol. 2011;10(12):1049–57.
51. Ordopidine (ACR325). [Internet] NeuroSearch. [Cited 2014 Jan 8] Available from: http://neurosearch.com/Default.aspx?ID=8268

52. Bermejo PE, Anciones B. A review of the use of zonisamide in Parkinson's disease. Ther Adv Neurol Disord. 2009;2(5):313–7.
53. Murata M, Hasegawa K, Fukasaka J, Kochi K, Kanazawa I; The Japan Zonisamide on PD Study Group. Zonisamide improves wearing-off in Parkinson's disease: a nation-wide randomized, double-blind study (abstract), 2013. MDS Conference, Sydney.
54. Pahwa R, Tanner CM, Hauser RA, Sethi KD, Isaacson SH, Truong DD, et al. Randomized trial of extended release amantadine in Parkinson's disease patients with levodopa-induced dyskinesia (EASED study) (abstract), 2013. MDS Conference, Sydney.
55. Press release. Adamas Pharmaceuticals presents positive phase 2/3 results for ADS-5102 for the treatment of levodopa-induced dyskinesia (LID) in Parkinson's disease. [Internet] Adamas Pharmaceuticals, Inc; 2003. [Cited 2014 Jan 8] Available from: www.adamaspharma.com/PR20130618.aspx.
56. Thomas A, Iacono D, Luciano AL, Armellino K, Di Iorio A, Onofrj M. Duration of amantadine benefit on dyskinesia of severe Parkinson's disease. J Neurol Neurosurg Psychiatry. 2004;75(1):141–3.
57. Ory-Magne F, Corvol J-C, Azulay J-P, Bonnet A-M, Brefel-Courbon C, Damier P, et al. Withdrawing amantadine in dyskinetic patients with Parkinson disease: the AMANDYSK trial. Neurology. 2014;82(4):300–7. Epub 2013 Dec 26.
58. Verhagen Metman L, Del Dotto P, Natté R, van den Munckhof P, Chase TN. Dextromethorphan improves levodopa-induced dyskinesias in Parkinson's disease. Neurology. 1998;51(1):203–6.
59. Verhagen Metman L, Blanchet PJ, van den Munckhof P, Del Dotto P, Natté R, Chase TN. A trial of dextromethorphan in parkinsonian patients with motor response complications. Mov Disord. 1998;13(3):414–7.
60. Heresco-Levy U, Javitt DC, Ebstein R, Vass A, Lichtenberg P, Bar G, et al. D-serine efficacy as add-on pharmacotherapy to risperidone and olanzapine for treatment-refractory schizophrenia. Biol Psychiatry. 2005;57(6):577–85.
61. Gelfin E, Kaufman Y, Korn-Lubetzki I, Bloch B, Kremer I, Javitt DC, et al. D-serine adjuvant treatment alleviates behavioural and motor symptoms in Parkinson's disease. Int J Neuropsychopharmacol. 2012;15(4):543–9.
62. Eggert K, Squillacote D, Barone P, Dodel R, Katzenschlager R, Emre M, et al. Safety and efficacy of perampanel in advanced Parkinson's disease: a randomized, placebo-controlled study. Mov Disord. 2010;25(7):896–905.
63. Lees A, Fahn S, Eggert KM, Jankovic J, Lang A, Micheli F, et al. Perampanel, an AMPA antagonist, found to have no benefit in reducing "off" time in Parkinson's disease. Mov Disord. 2012;27(2):284–8.
64. Berg D, Godau J, Trenkwalder C, Eggert K, Csoti I, Storch A, et al. AFQ056 treatment of levodopa-induced dyskinesias: results of 2 randomized controlled trials. Mov Disord. 2011;26(7):1243–50.
65. Stocchi F, Rascol O, Destee A, Hattori N, Hauser RA, Lang AE, et al. AFQ056 in Parkinson patients with levodopa-induced dyskinesia: 13-week, randomized, dose-finding study. Mov Disord. 2013;28(13):1838–46.
66. Tison F, Durif F, Corvol JC, Eggert K, Trenkwalder C, Lew M, et al. Safety, tolerability and anti-dyskinetic efficacy of dipraglurant, a novel mGluR5 Negative Allosteric Modulator (NAM) in Parkinson's disease (PD) patients with Levodopa-Induced Dyskinesia (LID) (abstract). [Internet] AAN annual meeting, March 2013. [Cited 2014 Jan 8] Available from: http://www.abstracts2view.com/aan/view.php?nu=AAN13L_S23.004&terms.
67. Goetz CG, Damier P, Hicking C, Laska E, Müller T, Olanow CW, et al. Sarizotan as a treatment for dyskinesias in Parkinson's disease: a double-blind placebo-controlled trial. Mov Disord. 2007;22(2):179–86.
68. Goetz CG, Laska E, Hicking C, Damier P, Müller T, Nutt J, et al. Placebo influences on dyskinesia in Parkinson's disease. Mov Disord. 2008;23(5):700–7.
69. Rascol O, Damier P, Goetz CG, Hicking C, Hock K, Muller T, et al. A large phase III study to evaluate the safety and efficacy of sarizotan in the treatment of l-dopa-induced-dyskinesia associated with PD: the Paddy-1 study. Mov Disord. 2006;21 Suppl 15:S492–3.

70. Müller T, Olanow CW, Nutt J, Hicking C, Laska E, Russ H. The Paddy-2 study: the evaluation of sarizotan for treatment-associated for dyskinesia in PD patients. Mov Disord. 2006;21 Suppl 15:S591.
71. Team Asubio Update. Team Asubio and Parkinson's disease. [Internet] Asubio Pharmaceuticals, Inc. [Cited 2014 Jan 8]. Available from: http://www.asubio.com/team/TeamAsubio_Update.html.
72. Mazzucchi S, Frosini D, Unti E, Del Prete E, Del Gamba C, Bonuccelli U, et al. Can serotoninergic antidepressants prevent or delay the development of L-dopa induced dyskinesias in PD patients? (abstract), 2013. MDS Conference, Sydney.
73. Durif F, Vidailhet M, Bonnet AM, Blin J, Agid Y. Levodopa-induced duskinesias are improved by fluoxetine. Neurology. 1995;45(10):1855–8.
74. Lewitt PA, Hauser RA, Lu M, Nicholas AP, Weiner W, Coppard N, et al. Randomized clinical trial of fipamezole for dyskinesia in Parkinson disease (FJORD study). Neurology. 2012;79(2):163–9.
75. Overview: Fipamezole for treatment of levodopa (L-dopa) induced Dyskinesia in Parkinson's Disease; Expected USD 700m+ peak sales first-in-class opportunity for US and EU ready to enter Phase III. [Internet] Santhera Pharmaceuticals. [Cited 2014 Jan 8] Available from: http://www.santhera.com/index.php?mid=4&vid=&lang=en.
76. Hauser RA, Cantillon M, Pourcher E, Micheli F, Mok V, Onofrj M, et al. Preladenant in patients with Parkinson's disease and motor fluctuations: a phase 2, double-blind, randomised trial. Lancet Neurol. 2011;10(3):221–9.
77. Corporate news. Merck provides update on phase III clinical program for preladenant, the company's investigational Parkinson's disease medicine. [Internet] Merck & Co., Inc; 2013. [Cited 2014 Jan 8] Available from: http://www.mercknewsroom.com/press-release/research-and-development-news/merck-provides-update-phase-iii-clinical-program-prelade.
78. LeWitt PA, Guttman M, Tetrud JW, Tuite PJ, Mori A, Chaikin P, et al. Adenosine A2A receptor antagonist istradefylline (KW-6002) reduces "off" time in Parkinson's disease: a double-blind, randomized, multicenter clinical trial (6002-US-005). Ann Neurol. 2008;63(3):295–302.
79. Hauser RA, Shulman LM, Trugman JM, Roberts JW, Mori A, Ballerini R, et al. Study of istradefylline in patients with Parkinson's disease on levodopa with motor fluctuations. Mov Disord. 2008;23(15):2177–85.
80. Stacy M, Silver D, Mendis T, Sutton J, Mori A, Chaikin P, et al. A 12-week, placebo-controlled study (6002-US-006) of istradefylline in Parkinson disease. Neurology. 2008;70(23):2233–40.
81. Mizuno Y, Kondo T, Japanese Istradefylline Study Group. Adenosine A2A receptor antagonist istradefylline reduces daily OFF time in Parkinson's disease. Mov Disord. 2013;28(8):1138–41.
82. Hauser RA, Olanow CW, Kieburtz K, Neale A, Resburg C, Maya U, et al. A phase 2, placebo-controlled, randomized, double-blind trial of tozadenant (SYN-115) in patients with Parkinson's disease with wearing-off fluctuations on levodopa (abstract), 2013. MDS Conference, Sydney.
83. Tozadenant (SYN115): a highly differentiated product for Parkinson's disease. [Internet] Biotie Therapies; 2012. [Cited 2014 Jan 8] Available from: http://www.biotie.com/en/product_and_development/development_pipeline/syn115.
84. clinicaltrials.gov. [Internet] Study NCT01474421. [Cited 2014 Jan 8] Available from: http://clinicaltrials.gov/ct2/show/NCT01474421?term=NCT01474421&rank=1.
85. Huang LZ, Campos C, Ly J, Ivy Carroll F, Quik M. Nicotinic receptor agonists decrease L-dopa-induced dyskinesias most effectively in partially lesioned parkinsonian rats. Neuropharmacology. 2011;60(6):861–8.
86. Quik M, Mallela A, Ly J, Zhang D. Nicotine reduces established levodopa-induced dyskinesias in a monkey model of Parkinson's disease. Mov Disord. 2013;28(10):1398–406.
87. Anfang MK, Pope Jr HG. Treatment of neuroleptic-induced akathisia with nicotine patches. Psychopharmacology (Berl). 1997;134(2):153–6.
88. News release. Neuraltus Pharmaceuticals reports clinical results from phase 1/2 NP002 study in the treatment of dyskinesias resulting from levodopa therapy for Parkinson's disease.

[Internet] Neuraltus Pharmaceuticals, Inc; 2010. [Cited 2014 Jan 8] Available from: http://www.neuraltus.com/pages/news_rel12_03_10.html

89. Pacchetti C, Albani G, Martignoni E, Godi L, Alfonsi E, Nappi G. "Off" painful dystonia in Parkinson's disease treated with botulinum toxin. Mov Disord. 1995;10(3):333–6.
90. Espay AJ, Vaughan JE, Shukla R, Gartner M, Sahay A, Revilla FJ, et al. Botulinum toxin type A for Levodopa-induced cervical dyskinesias in Parkinson's disease: unfavorable risk-benefit ratio. Mov Disord. 2011;26(5):913–4.
91. Susman E. News from the AAN annual meeting: Safinamide Misses in Attempt to Prevent Dyskinesia. Neurol Today. 2011;11(9):19.
92. Wong KK, Alty JE, Goy AG, Raghav S, Reutens DC, Kempster PA. A randomized, double-blind, placebo-controlled trial of levetiracetam for dyskinesia in Parkinson's disease. Mov Disord. 2011;26(8):1552–5.
93. Stathis P, Konitsiotis S, Tagaris G, Peterson D, VALID-PD Study Group. Levetiracetam for the management of levodopa-induced dyskinesias in Parkinson's disease. Mov Disord. 2011;26(2):264–70.
94. Wolz M, Löhle M, Strecker K, Schwanebeck U, Schneider C, Reichmann H, et al. Levetiracetam for levodopa-induced dyskinesia in Parkinson's disease: a randomized, double-blind, placebo-controlled trial. J Neural Transm. 2010;117(11):1279–86.
95. Lyons KE, Pahwa R. Efficacy and tolerability of levetiracetam in Parkinson disease patients with levodopa-induced dyskinesia. Clin Neuropharmacol. 2006;29(3):148–53.

Chapter 18
Preclinical Models of Levodopa-Induced Dyskinesia

Veronica Francardo and M. Angela Cenci

Abstract L-DOPA-induced dyskinesia (LID) represents one of the major limitations in the current pharmacotherapy of Parkinson's disease (PD) and affects the majority of PD patients. Animal models are the most important preclinical tool for molecular investigations of LID mechanisms and therapeutic targets.

Over the last two decades, models of LID have been developed in both nonhuman primate and rodent species, recapitulating several aspects of the human dyskinesia. This chapter will review and compare the main features of the rodent and non-primate models of LID currently available and summarize some of the main neurobiological findings obtained from these models.

Keywords L-DOPA-induced dyskinesia • 6-Hydroxydopamine • MPTP • Animal models • Abnormal involuntary movements • Parkinson's disease

Introduction

A key and still unmet objective in PD treatment is to relieve motor symptoms without inducing dyskinesia. In order to achieve this goal, animal models of PD and LID are an essential tool to investigate pathophysiological mechanisms and test new potential therapies. Several animal models of LID showing similarities to the peak-dose pattern have been extensively studied. These models have been created both in nonhuman primates, rats and mice. Nonhuman primate models were the first to be introduced to study dyskinesia and are considered very reliable for pathophysiological and pharmacological investigations due to their phenomenological similarities to the human condition. On the other hand, the time and cost-effectiveness of rodents

V. Francardo • M.A. Cenci (✉)
Basal Ganglia Pathophysiology Unit, Department of Experimental Medical Science, Lund University, BMC F11, Sölvegatan 19, Lund, 22184, Sweden
e-mail: Angela.Cenci_Nilsson@med.lu.se

S.H. Fox, J.M. Brotchie (eds.), *Levodopa-Induced Dyskinesia in Parkinson's Disease*, DOI 10.1007/978-1-4471-6503-3_18

make them very advantageous for preclinical investigations. During the past few years, rodents have become the species of choice to explore cellular and molecular mechanisms of LID.

Nonhuman Primate Models of LID

The irreversible parkinsonian condition caused by 1-methyl-4-phenyl-1,2,3,6-tetrahydropyridine (MPTP) was described for the first time in the early 1980s [1, 2] in drug abusers inadvertently taking this toxin as a contaminant derivative of a synthetic opioid narcotic. Following these observations, MPTP became the most widely used toxin to mimic PD in nonhuman primate (NHP) models [3]. The corresponding animal model exhibits both the primary motor features (in particular, rigidity and bradykinesia) and some of the cognitive and autonomic impairments that occur in human PD [4–6]. MPTP-lesioned NHP models of PD have been produced by toxin administration via different routes (such as intravenous, subcutaneous, intramuscular injection, via osmotic mini-pumps or intracarotid [7–10], although this latter model has been reported to not develop LID [11]). MPTP causes a bilateral DA degeneration, binding the dopamine transporter and inducing therefore neuronal death of dopaminergic cells through radical stress [12], with consequent DA depletion in the putamen [3, 13–16]. Moreover, in addition to the degeneration of the nigrostriatal DA pathway, also noradrenergic and serotonergic damages have been reported in some studies [17, 18].

MPTP-lesioned NHP chronically treated with L-DOPA provide excellent models to mimic LID and test potential anti-dyskinetic agents. L-DOPA can be administered either orally or via several injection routes. Subcutaneous injections are the preferred rout of administration, offering more stable plasma levels [19].

In the dyskinesia literature, four different species of NHP have been used, namely, squirrel monkey (*Saimiri sciureus*) (only in [20, 21]), marmoset (*Callithrix jacchus*), cynomolgus (*Macaca fascicularis*), and rhesus macaque (*Macaca mulatta*) (reviewed in [22]). Among these, squirrel monkeys and marmosets are particularly advantageous for their small size that confers an easy handling and housing. However, for unknown reasons, even non-lesioned squirrel monkeys have been described to develop LID following levodopa administration [23, 24], a feature that is not encountered in humans. On the contrary, MPTP-treated marmosets show parkinsonian motor symptoms reversible upon the administration of DA-mimetic compounds [13] and exhibit dyskinetic-like behaviors, including chorea and dystonia, when treated with levodopa. Some authors find that LID in the marmoset is difficult to rate due to the fast hyperkinetic movements and limited spectrum of dyskinetic behaviors exhibited by this species (reviewed in [22, 25]. In this model, the robust locomotor activity induced by levodopa provides a gross measure of the total motor activity associated with the treatment, including both antiparkinsonian and dyskinetic components. Marmosets have also been used to obtain unilateral [26, 27] or bilateral 6-OHDA lesion models of PD [28, 29]. To our knowledge, there are no studies reporting the development of LID upon levodopa administration to 6-OHDA-lesioned marmosets.

MPTP-lesioned macaques (both cynomolgus and rhesus) show stable and reproducible dyskinesia upon chronic levodopa administration. In this model, dyskinesia involves one or more parts of the body [10] and is characterized by choreic, dystonic, and ballistic movements or the combination of those [30–35]. LID presents several similarities with the dyskinesia observed in PD patients, exhibiting the same repertoire as well as interindividual differences in the patterns of dyskinetic behaviors [22, 25]. Thanks to these similarities to the human symptoms, MPTP-lesioned macaques are often referred to as the "gold-standard model" of LID, particularly by the laboratories where this model is commonly used.

Behavioral assessments in MPTP-lesioned NHP are performed post hoc on video recordings of animals freely moving in their home cages. The efficacy of levodopa in improving parkinsonian motor deficits is measured with a scale derived from the United Parkinson's Disease Rating Scale (UPDRS) used in PD patients [36, 37], which gives scores for mobility, bradykinesia, and posture. To quantify LID, several rating scales are available, such as the Abnormal Involuntary Movement Scales, the Dyskinesia Disability Scales for MPTP-treated primates, the Monkey Quality On-Time Rating, the Global Non-Human Primate Dyskinesia Rating Scale, and the St. Kitts Biomedical Primate Dyskinesia Scale (for a review, see [38]). All these scales have been used to measure the severity of dyskinesia in either pharmacological or pathophysiological studies of LID and have been compared to clinical rating scales in order to define their translational value [39]. Weaknesses of these scales often lie in the difficulty to anchor the different severity grades (moderate, mild, severe LID) to objective, unequivocal, physiological parameters. This difficulty may give rise to significant inter-rater and intra-laboratory variability. Today, the most widely used rating scale to measure LID in MPTP primates is the Dyskinesia Disability Rating Scale [40], which has been recently revised [38]. The original scale goes from 0 to 4 and assesses both severity and duration of the dyskinetic behavior: 0, no dyskinesia; 1, rare dyskinetic postures and movements; 2, moderate dyskinetic movements but not interfering with the normal behavior; 3, marked and frequent dyskinesia interfering with the normal behavior; and 4, severe and virtually continuous dyskinesia disabling to the animal and replacing the normal behavior. The revision of this scale includes the evaluation of disability under dyskinetic conditions based on the duration and the continuity of the behavior during 6 h observation time [41]. This requires examining the impact of dyskinesia on specific motor tasks that the animals can normally perform (such as walking on the floor, climbing a branch or the wall, grasping fruit). These modifications of the original rating scale thus allow one to distinguish between disabling and non-disabling dyskinesia, setting the threshold of disability over grade 2 (moderate severity) of the original rating scale. This revised scale not only adds clarity, simplicity, and standardization to the rating procedure (training videos are also included), but it also improves its clinical relevance. Indeed, it is closer to the way in which dyskinesia is rated in clinical trials, where PD patients are asked to keep diaries of "ON time with troublesome dyskinesia" versus "ON time with non-troublesome dyskinesia."

Rodent Models of LID

Abnormal Involuntary Movements in 6-OHDA-Lesioned Rats

Rodent models of PD have proven value to study neurobiological mechanisms, formulate new hypotheses, and screen new potential treatments. In the past two decades, several rodent models of PD that mimic different aspects of the human pathology have been developed [42]. However, the 6-hydroxydopamine (6-OHDA) model [43] is the only one so far used to produce a rat model of LID. The reproducibility of the lesions and the possibility to target different areas make the 6-OHDA injections a solid procedure to reproduce particular patterns of nigrostriatal degeneration [44]. Varying degrees of DA denervation can be induced by injecting the toxin in different sites along the nigrostriatal DA pathway, mimicking in this way early- or late-stage PD [45]. 6-OHDA injections in the medial forebrain bundle (MFB) lead to a virtually complete nigrostriatal lesion, with up to 100 % loss of dopaminergic terminals in the striatum, whereas intrastriatal injections lead to a less severe lesion, with a varying degree of DA denervation depending on the toxin concentration used [46]. The toxin is generally injected unilaterally, as a complete bilateral lesion causes a dramatic akinetic state, requiring intense postoperative nursing protocols. In addition, a major practical advantage offered by unilateral lesion models is that motor performance on the non-impaired side of the body can serve as a control relative to the impaired side in all tests assessing lateralized behavior, for example, rotational locomotion, sensorimotor integration, and forelimb use [47].

Levodopa-induced abnormal involuntary movements (AIMs) with dystonic and choreiform-like features were described for the first time in the rat at the end of the "1980s" [48]. The AIMs had the same time profile as peak-dose dyskinesia in PD and showed increased severity with repeated levodopa administration [49, 50]. Levodopa-induced dyskinesia in the rat presents many functional and phenomenological similarities with human peak-dose LID. Indeed, it affects virtually all muscle groups in the body, it interferes with normal motor activities when severe [47, 51], and its incidence and severity are dependent on the levodopa dose [52]. Importantly, LID in rats is modulated by clinically recognized anti-dyskinetic compounds such as amantadine and clozapine [47, 51].

Earlier studies had described levodopa-induced stereotypies, mainly consisting of augmented repetitive oral behaviors in rodents [53]. A link has been demonstrated between an overstimulation of the dopaminergic system and the appearance of these stereotypic behaviors [48]. However, in the rodent AIM scale first introduced by Cenci and coworkers [49, 50], increased manifestations of normal, rodent-specific behaviors (e.g., licking, gnawing, rearing) were not included. A recent study has indeed confirmed that levodopa-induced abnormal involuntary movements and stereotypic behaviors represent distinct entities that respond very differently to a range of compounds, including the anti-dyskinetic drug amantadine [54].

6-OHDA-lesioned rats treated with levodopa exhibit three main subtypes of dyskinesia, which involve distinct areas of the body, i.e., the forelimb, orofacial, and neck-trunk regions. These movements are qualitatively distinguishable from an augmented manifestation of typical rodent behaviors, like grooming, exploratory sniffing, licking, or gnawing. The latter are goal directed, bilateral, and superimposed on a pattern of general behavioral activation. The original rating scale of AIM severity described by Cenci and coworkers [49] scores axial, limb, orolingual, and locomotive dyskinetic subtypes (for illustrations and video clips, see [45, 46] and [55] based on the proportion of the observation time during which the dyskinetic behavior is expressed. Animals are monitored for 1 min every 20 min over the 3 h immediately post levodopa injection, and ratings are done through online observations. Each AIM subtype is scored on a severity scale from 0 to 4, as follows: 0, no dyskinesia; 1, signs of dyskinesia during half of the observation time; 2, dyskinesia present during more than half of the observation time; 3, dyskinesia present all the time but suppressible by light sensory stimuli; and 4, continuous, severe, and not suppressible dyskinesia. Anchoring a dyskinesia severity grade to the time during which dyskinetic items are present increases the objectivity of the assessments. Summing all the scores per session (or treatment period) gives a total dyskinesia score with excellent metric properties, as indicated by the striking linear correlations found between this score and levels of gene or protein expression in specific brain regions (see, e.g., [49, 56–58]). Although apparently different from the dyskinesia rating scales used in nonhuman primates, the rodent AIM scale in fact presents some important similarities. Thus, severity grades 1 and 2 of the rodent rating scale appear to correspond to the grades defined as mild and moderate, respectively, on the monkey rating scale. Indeed, in the rat, a severity score of two sets the threshold above which dyskinesia becomes disabling and interferes with the animals' performance in tests of spontaneous behavior [47] (video clips have been published in [59]).

In order to give a more articulate description of the dyskinetic behaviors, an additional rating scale assessing the amplitude of the AIMs has been recently introduced. This scale was first developed to capture different features of dyskinetic behaviors in rats having different degrees of striatal DA denervation [46]. This scale has now been tested in several laboratories and found to provide a sensitive tool to detect treatment-related improvements in LID, particularly when combined with the basic severity scale of Cenci and coworkers to obtain a "global AIM score" [18, 54, 60].

Another AIM rating scale for 6-OHDA-lesioned rats was developed by Steece-Collier and colleagues, originally with the purpose of evaluating the response to levodopa in rats with intrastriatal transplants of embryonic ventral mesencephalic tissue [61, 62]. This scale introduces qualitative components to classify the dyskinetic movements as stereotypic or dystonic and takes one additional body region into consideration, i.e., the hind limb. This more complex method has the advantage of giving a broader description of the behaviors observed, but it inevitably also makes the online scoring more difficult and time demanding. Indeed, in the original description of this method, animals were rated only at two time points following the administration of levodopa [61, 62]. The Steece-Collier scale has not been subjected to the same extent of pharmacological validation as the one proposed by Cenci et al. (1998) [47, 49, 51]. However, a recent comparison of different methods to study dyskinesia in the rat [54] showed a good internal correlation between the two scales.

Both of them were found capable of detecting reductions in levodopa-induced AIMs by compounds with proven anti-dyskinetic properties in nonhuman primates and PD patients, such as amantadine, and D1 or D2 receptor antagonists [54].

Summary of the Main Neurobiological Correlates of LID in the Rat

The occurrence of LID in the rat is highly dependent on the degree of striatal denervation [46, 63, 64]. In rats receiving 6-OHDA injections in the MFB, the degeneration of the nigrostriatal DA pathway is complete in the ipsilateral hemisphere [46], and clinically relevant doses of levodopa [52] or DA agonists [65] are sufficient to induce severe dyskinetic movements. On the contrary, the same levodopa doses induce less severe AIMs in rats sustaining partial lesions (i.e., intrastriatal lesions) [46].

The susceptibility to develop dyskinesia varies among rats, even when working with rats with complete MFB lesions [49, 56]. This mimics the clinical situation, since the susceptibility to LID varies greatly among PD patients (reviewed in [66]). This variability is particularly advantageous to perform correlation analyses between dyskinesia severity and molecular changes in the brain, as first illustrated by [49, 56]. Molecular changes correlating with dyskinesia severity are assumed to drive the development of a maladaptive plasticity that maintains the brain in a dyskinesia-prone state. Indeed, clinical observations have indicated that, once established, LID is very difficult to reverse (reviewed in [67]). The many studies performed in the rat model of LID have shaped current mainstream pathophysiological notions. According to these, the extracellular DA derived from levodopa administration [68, 69] acts on supersensitive DA receptors in the denervated striatum and strongly activates signaling cascades that mediate upregulation of immediate early genes, neurotransmitter-related genes, and other plasticity molecules (reviewed in [67]. The mitogen-activated protein kinase (MAPK), extracellular signal-regulated kinase 1/2 (ERK1/2), is the pathway downstream of the DA receptors that most strongly contributes to the events outlined above [34, 49, 56, 58, 70–72]. In addition to neurons, it is most likely that nonneuronal compartments also contribute to the development of LID. We have found that levodopa-treated rats exhibit endothelial proliferation and angiogenic activity in the basal ganglia in a manner that correlates positively to dyskinesia severity [73–75]. Recent microarray expression studies using striatal mRNA from dyskinetic rats have reported pronounced upregulation of genes involved in extracellular matrix remodeling, cell growth, and immune regulation [76]. These data support the concept that LID is associated with a comprehensive reorganization of the striatal microenvironment.

The importance of serotonergic and noradrenergic systems in the pathophysiology of LID has been investigated in several studies [18, 77–81]. In particular, the density of striatal serotonin (5-HT) fibers was shown to correlate with dyskinesia both in rats and nonhuman primate models [18, 82] and in one postmortem study in PD patients [18]. Moreover, transplants of serotonergic neurons in the striatum exacerbate LID in the rat [83]. Compounds acting on the 5-HT system are currently under clinical evaluation and represent a potential therapeutic strategy for LID (reviewed in [84]).

Contralateral Rotations

Upon levodopa administration, hemi-parkinsonian rats show contralateral turning behavior, a feature that was originally presented as an index of the drug's antiparkinsonian effect (reviewed in [45]. Contralateral rotational activity has been variably used to assess both the therapeutic effect and the dyskinesiogenic potential of dopaminergic drugs, raising interpretational difficulties [45, 85]. The dose of levodopa necessary to induce robust contralateral turning is higher than the one required for inducing LID [52]. The lower levodopa doses sufficient to induce AIMs are more clinically relevant and reflect the daily levodopa doses given to advanced PD patients [52].

Rather than focusing on the total number of turns, some studies have measured changes in the duration of rotational behavior and used them as a rodent equivalent of the wearing-off fluctuations seen in PD. Thus, by counting the number of contralateral rotations during 3 h post levodopa administration, the group led by T. Chase reported that the duration of maximal rotational behavior became approx. 20–25 % shorter over 3 weeks of levodopa treatment, resembling a "wearing-off"-like phenomenon [86–89]. This model has not been widely used by many other groups, possibly because it measures a relatively subtle phenomenon that requires high doses of levodopa to be consistently induced [52].

The rats' rotational activity can be also rated as locomotive AIMs according to the same scoring criteria as used for the axial, limb, and orolingual AIMs [49]. This AIM subtype does not provide a specific measure of dyskinesia in the rat, as locomotive AIM scores are highly induced by drugs with low dyskinesiogenic potential (e.g., bromocriptine), and they are not attenuated by amantadine nor by many other compounds that significantly reduce axial, limb, and orolingual AIMs [47, 51, 54, 77, 90]. When included in the ratings, locomotive AIMs should therefore be considered separately from the other three AIM subtypes. In line with our own experience, a recent study where AIM scales, ratings of stereotypic behaviors, and rotational counts were recorded in parallel from 6-OHDA-lesioned rats treated with levodopa concluded that these behavioral manifestations have distinct time profiles and respond differently to DA antagonist treatments [54].

The 6-OHDA-Lesioned Mouse Model of LID

The vast availability of genetically engineered mouse strains has raised a great interest toward using the mouse for studies of LID. Over the last decade, both toxin-based and genetic mouse models of PD have been used to replicate parkinsonian motor deficits and LID [22]. 6-OHDA lesions are currently the preferred model to replicate LID in the mouse [57, 58, 91–93]. Although MPTP lesions are widely used for neuroprotection studies in the mouse, they do not seem suitable for the purpose of studying LID. Indeed, the degree of DA depletion obtained after MPTP administration in the mouse seems to vary greatly depending on administration protocols and strain [94], and the behavioral and biochemical impairments are also variable [94–97]. The only study reporting LID in MPTP-lesioned mice has described the occurrence of AIMs only in old animals (10–12 months old) treated with very high doses of levodopa (200 mg/kg). Dyskinesia was described as hyperactivity (running, jumping) and stereotypic movements (scratching, gnawing) [98].

Upon levodopa treatment, 6-OHDA-lesioned mice exhibit AIMs with dystonic and hyperkinetic features similar to those characterized in the rat [93]. The rating scale used for mouse AIMs follows the same principles as that used in rats and reflects the topographic distribution, frequency, and duration of the dyskinetic behavior affecting orofacial, trunk, and limb muscles [57, 91–93]. The mouse model of LID has been pharmacologically validated, to some extent, in studies showing that the severity of AIMs can be reduced by acute administration of compounds like amantadine and buspirone, which alleviate LID in other animal models of PD [99] and in PD patients [100]. Moreover, molecular interventions that either increase or aggravate levodopa-induced AIMs in the mouse exert similar effects in nonhuman primate models of LID [34, 101]. However, for therapeutic screening purposes, the rat is to be preferred over the mouse. Indeed, AIMs are easier to quantify in rats, they have been more extensively validated, and they display a large degree of stability and reproducibility over prolonged treatment periods (see, e.g., [50, 102]).

Given the increasing availability of transgenic mice, the use of 6-OHDA-lesioned mice for the purpose of studying LID is bound to expand. It is therefore important to be open about the difficulties that the lesioning procedure in mice may entail. The first studies performing 6-OHDA lesions in the mouse highlighted the high degree of postoperative mortality both in MFB (82 %) and striatally lesioned animals (30 %) [93]. In recent years, advances have been made in optimizing the lesion procedures, the postoperative care protocols, and the routines for behavioral testing in 6-OHDA-lesioned mice [57, 103]. The quality of postsurgical care administered to the animals in the first 2–3 weeks post lesion has proven to be absolutely essential to ensure a good recovery [57]. It is important that laboratories interested in using 6-OHDA-lesioned mice acquire sufficient training and apply suitable nursing protocols on the mice in the first three postoperative weeks.

As described in the rat [46], 6-OHDA-lesioned mice with different degree of DA denervation exhibit different dyskinesia severity. Indeed, mice with a virtually complete lesion (i.e., receiving 6-OHDA injections in the MFB) have been described to exhibit severe AIMs (grades 3 and 4) upon treatment with low, therapeutic levodopa doses. At the same doses, mice with partial DA denervation (intrastriatal lesions) exhibit less severe dyskinetic behavior (grades 1 and 2), and dyskinetic movements are interrupted by normal behavior [59]. Severe AIMs (corresponding to grades 3 and 4 of the basic rating scale of Cenci et al. 1998) may provide a parallel to the disabling dyskinesia in the monkey LID rating scale, while mild AIMs (grades 1 and 2) represent dyskinesia without disabilities. Indeed, AIMs with grade 1 or 2 do not substitute the animals' normal behavior, whereas grade 3 or 4 AIMs completely replace the animals' normal behavior [59], comparable to what is observed in severely dyskinetic monkeys [41].

When setting up a mouse model of LID, an important consideration is the potential difference between strains in the sensitivity to levodopa. One study comparing the rotational behavior and the severity of AIMs following chronic levodopa treatment in pure FVB and FVB/C57BL6 mice reported a higher sensitivity to levodopa in the latter ones [104]. However, to our knowledge, no other comparisons have been done after this study, and most laboratories currently use the C57BL6 strain to generate mouse models of LID [34, 57, 58, 92, 93, 105].

Mouse and rat models present some differences that should be taken into account when choosing the rodent species for studies of LID. First of all, dyskinetic mice present faster and less articulate limb and orofacial AIMs compared to the rats, precluding in this way a reliable application of the AIM amplitude scale. Moreover, when mice have a complete MFB lesion, AIM scores rapidly reach a plateau of severity, within 2–3 days of commencing levodopa treatment. Some studies have reported a reduction of total AIM scores due to a shortened duration of the dyskinetic behavior when treatment with levodopa was given for several weeks [99]. As described in the rat model, levodopa induces contralateral rotations, which are more pronounced in dyskinetic mice than in non-dyskinetic ones [57]. Although rotations have been used as measure of dyskinesia in some studies [106], we observed that pronounced rotations may also occur in mice that do not exhibit any abnormal movement of the body [57]. Lundblad et al. (2005) reported that locomotive AIMs (a correlate of rotations) are not reduced by amantadine [99]. Fasano et al. (2010) reported that genetic inactivation of RasGRF1 (a positive modulator of ERK signaling) alleviates axial, limb, and orolingual AIMs but has no effects on locomotive AIMs [34]. For all the above reasons, rotational behavior should be scored as a separate measure from limb and orolingual AIMs [34, 57, 99].

The degree of DA denervation does not only affect the severity of the dyskinetic behaviors, but also the susceptibility to LID and the expression pattern of postsynaptic markers of dyskinesia. MFB-lesioned mice treated with levodopa develop all subtypes of AIMs to their maximal severity, whereas intrastriatally lesioned animals show a high interindividual variability, some of them being free from dyskinesia [57]. In these two different PD models, postsynaptic markers of dyskinesia are expressed in the most denervated striatal regions, providing a mirror image of TH fiber density [57]. Among them, pERK1/2 provides an early marker of aberrant neuroplasticity and postsynaptic D1 receptor supersensitivity (reviewed in [67]). In MFB-lesioned mice with a complete DA denervation throughout striatum, an acute levodopa challenge activates pERK1/2 in all striatal regions. By contrast, in intrastriatally lesioned mice, pERK1/2 is expressed only in the dorsolateral striatum, which is the most dopamine-depleted striatal region, but not in medial areas having more than 60 % of spared tyrosine hydroxylase fibers [57]. Another postsynaptic marker of dyskinesia that shows the same expression pattern of pERK1/2 is ΔFosB [57, 70], a stable transcription factor that accumulates in the brain after chronic perturbations [107]. Striatal upregulation of ΔFosB in dyskinesia has been described in rats [56], nonhuman primates [33, 34], mice [34, 57, 70, 93], and recently also in a postmortem investigation of PD patients with dyskinesia [108]. Studies targeting upstream components of the Ras-ERK-ΔFosB signaling cascade intervention in the mouse [34] as in the rat [56, 109] and in the nonhuman primate models [33] have demonstrated the important causal contribution of this pathway to LID.

An interesting observation described in the mouse 6-OHDA lesion model is the presence of dopaminergic neurons in the striatum correlating with the severity of LID [57, 92]. These cells have been described also in rats [110–112], nonhuman primates [113, 114], and PD patients [115]. Interestingly, these neurons seem to be regulated by 6-OHDA lesion and levodopa treatment, and they have been recently described to correlate both with dyskinesia severity and with the expression levels of ΔFosB [57]. However, the functional importance of these neurons is yet to be clarified. Some laboratories are currently interested in finding out whether these neurons play a role in the pathophysiology of LID [116, 117].

Other Mouse Models

Only 10 % of the human PD is caused by genetic mutations. However, the use of knockout and transgenic mice for genes involved in the human disease is very important to investigate molecular pathways of neurodegeneration and devise both biomarkers and targets for therapeutic intervention [118]. Several mouse strains have been created, overexpressing either genes involved in autosomal-dominant PD, such as α-synuclein [119] and leucine-rich repeat kinase 2 (LRRK2) [120, 121], or knockdown or knockout of genes involved in the autosomal recessive forms such as parkin [122, 123], PTEN-induced putative kinase 1 (PINK-1) [124–126], and DJ-1 [127]. However, the use of genetic models for studies of LID has been very limited, due to their lacking a severe nigrostriatal degeneration, required for the development of dyskinesia. Thus far, the only genetic model reported to exhibit pronounced nigrostriatal DA denervation and LID upon levodopa administration is the aphakia mouse (lacking Ptx3 gene, which codes for a transcription factor required for DA cell differentiation) [128]. However, the type of AIMs developed by this mouse model are different from the ones described in the toxin models, showing simultaneous fluttering movements of both forepaws and one hindpaw while rearing. This pattern was termed "three-paw dyskinesia" and can be reduced following treatment with anti-dyskinetic drugs such as amantadine and buspirone [128]. In this model, some well-established postsynaptic markers of LID are expressed in the striatum upon chronic levodopa treatment, such as immunoreactivity for FosB/ΔFosB and phospho-ERK1/2 [128, 129]. These markers were originally described in classical neurotoxin-based models in mice with 6-OHDA lesions [34, 57, 70, 93], in the rat [56, 71] and in nonhuman primate models of LID [33, 34].

Conclusions

Nonhuman primate and rodent models have been developed for preclinical research in order to study molecular and biochemical mechanisms underlying the pathophysiology of LID. Dyskinetic behaviors are seen in both rodents and nonhuman primates that sustain severe nigrostriatal lesions followed by treatment with levodopa. These movements show dystonic and hyperkinetic components seen in PD patients (face validity). These models have contributed to progress in understanding many molecular mechanisms involved in the development of dyskinesia. Moreover, they all have been pharmacologically validated for their response to compounds already used in the clinic for the treatment of dyskinesia, such as amantadine. For the sake of testing potential treatments for dyskinesia, it is essential to exclude that a reduction of AIM scores is not due to a motor depressant effect or to a potential interference of the treatment with the therapeutic effect of levodopa. This has been addressed in different ways in NHP and in rodent models (see Table 18.1). In monkeys, scores of parkinsonian motor behavior (similar to the UPDRS scale used in clinic) and locomotor behavior are used to measure parkinsonian motor deficits [2, 13, 34–36, 41, 130–134]. By contrast, these tests cannot be applied to rodents, in which, however, other behavioral tests have been used (e.g., rotarod, cylinder test, etc; see Table 18.1) [47, 49, 57, 92, 93, 99, 105, 135–138].

Table 18.1 Most common behavioral tests used to assess antiparkinsonian motor responses to dopaminergic medication

Specie	Lesion model	Test used	Seminal references
Marmoset	Chronic MPTP	Motor disability	[13, 130]
		Locomotor activity	[130]
Squirrel monkey	Chronic MPTP	Motor disability	[2]
		Locomotor activity	[131]
Macaque	Chronic MPTP	Motor disability	[34–36, 41, 132]
		Locomotor activity	[34, 35, 41, 132]
		Electromyography	[133, 134]
Rat	Intracerebral	Rotational activity	[47, 49]
	6-OHDA injection	Rotarod	[132, 135]
		Cylinder test	[47]
		Locomotor activity	[132]
		Corridor test	[136]
		Stepping test	[137]
Mouse	Intracerebral	Rotational activity	[57, 92, 93]
	6-OHDA injection	Rotarod	[92]
		Cylinder test	[57, 92, 93, 99]
		Locomotor activity	[34, 57, 92, 99]
		Corridor test	[92, 105]
		Stepping test	[105]

Thanks to their translational applicability, animal models of LID represent a valuable tool to further understand molecular pathways through which levodopa induces dyskinetic responses and to test new potential anti-dyskinetic treatments acting on these pathways.

References

1. Davis GC, Williams AC, Markey SP, Ebert MH, Caine ED, Reichert CM, et al. Chronic Parkinsonism secondary to intravenous injection of meperidine analogues. Psychiatry Res. 1979;1(3):249–54.
2. Langston JW, Forno LS, Rebert CS, Irwin I. Selective nigral toxicity after systemic administration of 1-methyl-4-phenyl-1,2,5,6-tetrahydropyrine (MPTP) in the squirrel monkey. Brain Res. 1984;292(2):390–4.
3. Burns RS, Chiueh CC, Markey SP, Ebert MH, Jacobowitz DM, Kopin IJ. A primate model of parkinsonism: selective destruction of dopaminergic neurons in the pars compacta of the substantia nigra by N-methyl-4-phenyl-1,2,3,6-tetrahydropyridine. Proc Natl Acad Sci U S A. 1983;80(14):4546–50.
4. Albanese A, Granata R, Gregori B, Piccardi MP, Colosimo C, Tonali P. Chronic administration of 1-methyl-4-phenyl-1,2,3,6-tetrahydropyridine to monkeys: behavioural, morphological and biochemical correlates. Neuroscience. 1993;55(3):823–32.
5. Chaumette T, Lebouvier T, Aubert P, Lardeux B, Qin C, Li Q, et al. Neurochemical plasticity in the enteric nervous system of a primate animal model of experimental Parkinsonism. Neurogastroenterol Motil. 2009;21(2):215–22.
6. Schneider JS. Chronic exposure to low doses of MPTP. II. Neurochemical and pathological consequences in cognitively-impaired, motor asymptomatic monkeys. Brain Res. 1990;534(1–2):25–36.
7. Bezard E, Imbert C, Deloire X, Bioulac B, Gross CE. A chronic MPTP model reproducing the slow evolution of Parkinson's disease: evolution of motor symptoms in the monkey. Brain Res. 1997;766(1–2):107–12.
8. Emborg ME, Moirano J, Schafernak KT, Moirano M, Evans M, Konecny T, et al. Basal ganglia lesions after MPTP administration in rhesus monkeys. Neurobiol Dis. 2006;23(2):281–9.
9. Belanger N, Gregoire L, Hadj Tahar A, Bedard PJ. Chronic treatment with small doses of cabergoline prevents dopa-induced dyskinesias in parkinsonian monkeys. Mov Disord. 2003;18(12):1436–41.
10. Hadj Tahar A, Gregoire L, Darre A, Belanger N, Meltzer L, Bedard PJ. Effect of a selective glutamate antagonist on L-dopa-induced dyskinesias in drug-naive parkinsonian monkeys. Neurobiol Dis. 2004;15(2):171–6.
11. Lieu CA, Deogaonkar M, Bakay RA, Subramanian T. Dyskinesias do not develop after chronic intermittent levodopa therapy in clinically hemiparkinsonian rhesus monkeys. Parkinsonism Relat Disord. 2011;17(1):34–9.
12. Smeyne RJ, Jackson-Lewis V. The MPTP model of Parkinson's disease. Brain Res Mol Brain Res. 2005;134(1):57–66.
13. Jenner P, Rupniak NM, Rose S, Kelly E, Kilpatrick G, Lees A, et al. 1-Methyl-4-phenyl-1,2,3,6-tetrahydropyridine-induced parkinsonism in the common marmoset. Neurosci Lett. 1984;50(1–3):85–90.
14. Bezard E, Dovero S, Prunier C, Ravenscroft P, Chalon S, Guilloteau D, et al. Relationship between the appearance of symptoms and the level of nigrostriatal degeneration in a progressive 1-methyl-4-phenyl-1,2,3,6-tetrahydropyridine-lesioned macaque model of Parkinson's disease. J Neurosci. 2001;21(17):6853–61.

15. Jan C, Pessiglione M, Tremblay L, Tande D, Hirsch EC, Francois C. Quantitative analysis of dopaminergic loss in relation to functional territories in MPTP-treated monkeys. Eur J Neurosci. 2003;18(7):2082–6.
16. Guigoni C, Li Q, Aubert I, Dovero S, Bioulac BH, Bloch B, et al. Involvement of sensorimotor, limbic, and associative basal ganglia domains in L-3,4-dihydroxyphenylalanine-induced dyskinesia. J Neurosci. 2005;25(8):2102–7.
17. Pifl C, Schingnitz G, Hornykiewicz O. Effect of 1-methyl-4-phenyl-1,2,3,6-tetrahydropyridine on the regional distribution of brain monoamines in the rhesus monkey. Neuroscience. 1991;44(3):591–605.
18. Rylander D, Parent M, O'Sullivan SS, Dovero S, Lees AJ, Bezard E, et al. Maladaptive plasticity of serotonin axon terminals in levodopa-induced dyskinesia. Ann Neurol. 2010;68(5):619–28.
19. Cooper DR, Marrel C, Testa B, van de Waterbeemd H, Quinn N, Jenner P, et al. L-Dopa methyl ester–a candidate for chronic systemic delivery of L-Dopa in Parkinson's disease. Clin Neuropharmacol. 1984;7(1):89–98.
20. Langston JW, Quik M, Petzinger G, Jakowec M, Di Monte DA. Investigating levodopa-induced dyskinesias in the parkinsonian primate. Ann Neurol. 2000;47(4 Suppl 1):S79–89.
21. Hsu A, Togasaki DM, Bezard E, Sokoloff P, Langston JW, Di Monte DA, et al. Effect of the D3 dopamine receptor partial agonist BP897 [N-[4-(4-(2-methoxyphenyl)piperazinyl)butyl]-2-naphthamide] on L-3,4-dihydroxyphenylalanine-induced dyskinesias and parkinsonism in squirrel monkeys. J Pharmacol Exp Ther. 2004;311(2):770–7.
22. Iderberg H, Francardo V, Pioli EY. Animal models of L-DOPA-induced dyskinesia: an update on the current options. Neuroscience. 2012;211:13–27.
23. Quik M, Police S, Langston JW, Di Monte DA. Increases in striatal preproenkephalin gene expression are associated with nigrostriatal damage but not L-DOPA-induced dyskinesias in the squirrel monkey. Neuroscience. 2002;113(1):213–20.
24. Togasaki DM, Tan L, Protell P, Di Monte DA, Quik M, Langston JW. Levodopa induces dyskinesias in normal squirrel monkeys. Ann Neurol. 2001;50(2):254–7.
25. Morin N, Jourdain VA, Di Paolo T. Modeling dyskinesia in animal models of Parkinson disease. Exp Neurol. 2014;256:105–16.
26. Pirker W, Tedroff J, Ponten H, Gunne L, Andren PE, Hurd YL. Coadministration of (−)-OSU6162 with l-DOPA normalizes preproenkephalin mRNA expression in the sensorimotor striatum of primates with unilateral 6-OHDA lesions. Exp Neurol. 2001;169(1):122–34.
27. Svenningsson P, Gunne L, Andren PE. L-DOPA produces strong induction of c-fos messenger RNA in dopamine-denervated cortical and striatal areas of the common marmoset. Neuroscience. 2000;99(3):457–68.
28. Mitchell IJ, Carroll CB. Reversal of parkinsonian symptoms in primates by antagonism of excitatory amino acid transmission: potential mechanisms of action. Neurosci Biobehav Rev. 1997;21(4):469–75.
29. Mitchell IJ, Hughes N, Carroll CB, Brotchie JM. Reversal of parkinsonian symptoms by intrastriatal and systemic manipulations of excitatory amino acid and dopamine transmission in the bilateral 6-OHDA lesioned marmoset. Behav Pharmacol. 1995;6(5–6):492–507.
30. Clarke CE, Sambrook MA, Mitchell IJ, Crossman AR. Levodopa-induced dyskinesia and response fluctuations in primates rendered parkinsonian with 1-methyl-4-phenyl-1,2,3,6-tetrahydropyridine (MPTP). J Neurol Sci. 1987;78(3):273–80.
31. Boyce S, Clarke CE, Luquin R, Peggs D, Robertson RG, Mitchell IJ, et al. Induction of chorea and dystonia in parkinsonian primates. Mov Disord. 1990;5(1):3–7.
32. Gold SJ, Hoang CV, Potts BW, Porras G, Pioli E, Kim KW, et al. RGS9-2 negatively modulates L-3,4-dihydroxyphenylalanine-induced dyskinesia in experimental Parkinson's disease. J Neurosci. 2007;27(52):14338–48.
33. Berton O, Guigoni C, Li Q, Bioulac BH, Aubert I, Gross CE, et al. Striatal overexpression of DeltaJunD resets L-DOPA-induced dyskinesia in a primate model of Parkinson disease. Biol Psychiatry. 2009;66(6):554–61.

34. Fasano S, Bezard E, D'Antoni A, Francardo V, Indrigo M, Qin L, et al. Inhibition of Ras-guanine nucleotide-releasing factor 1 (Ras-GRF1) signaling in the striatum reverts motor symptoms associated with L-dopa-induced dyskinesia. Proc Natl Acad Sci U S A. 2010;107(50):21824–9.
35. Gregoire L, Morin N, Ouattara B, Gasparini F, Bilbe G, Johns D, et al. The acute antiparkinsonian and antidyskinetic effect of AFQ056, a novel metabotropic glutamate receptor type 5 antagonist, in L-Dopa-treated parkinsonian monkeys. Parkinsonism Relat Disord. 2011;17(4):270–6.
36. Imbert C, Bezard E, Guitraud S, Boraud T, Gross CE. Comparison of eight clinical rating scales used for the assessment of MPTP-induced parkinsonism in the Macaque monkey. J Neurosci Methods. 2000;96(1):71–6.
37. Bezard E, Przedborski S. A tale on animal models of Parkinson's disease. Mov Disord. 2011;26(6):993–1002.
38. Fox SH, Johnston TH, Li Q, Brotchie J, Bezard E. A critique of available scales and presentation of the Non-human primate dyskinesia rating scale. Mov Disord. 2012;27(11):1373–8.
39. Colosimo C, Martinez-Martin P, Fabbrini G, Hauser RA, Merello M, Miyasaki J, et al. Task force report on scales to assess dyskinesia in Parkinson's disease: critique and recommendations. Mov Disord. 2010;25(9):1131–42.
40. Brotchie JM, Fox SH. Quantitative assessment of dyskinesias in subhuman primates. Mov Disord. 1999;14 Suppl 1:40–7.
41. Huot P, Johnston TH, Koprich JB, Aman A, Fox SH, Brotchie JM. L-745,870 reduces L-DOPA-induced dyskinesia in the 1-methyl-4-phenyl-1,2,3,6-tetrahydropyridine-lesioned macaque model of Parkinson's disease. J Pharmacol Exp Ther. 2012;342(2):576–85.
42. Blesa J, Phani S, Jackson-Lewis V, Przedborski S. Classic and new animal models of Parkinson's disease. J Biomed Biotechnol. 2012;2012:845618.
43. Ungerstedt U. 6-Hydroxy-dopamine induced degeneration of central monoamine neurons. Eur J Pharmacol. 1968;5(1):107–10.
44. Blandini F, Armentero MT, Martignoni E. The 6-hydroxydopamine model: news from the past. Parkinsonism Relat Disord. 2008;14 Suppl 2:S124–9.
45. Cenci MA, Whishaw IQ, Schallert T. Animal models of neurological deficits: how relevant is the rat? Nat Rev Neurosci. 2002;3(7):574–9.
46. Winkler C, Kirik D, Bjorklund A, Cenci MA. L-DOPA-induced dyskinesia in the intrastriatal 6-hydroxydopamine model of parkinson's disease: relation to motor and cellular parameters of nigrostriatal function. Neurobiol Dis. 2002;10(2):165–86.
47. Lundblad M, Andersson M, Winkler C, Kirik D, Wierup N, Cenci MA. Pharmacological validation of behavioural measures of akinesia and dyskinesia in a rat model of Parkinson's disease. Eur J Neurosci. 2002;15(1):120–32.
48. Fog R. Stereotyped and non-stereotyped behaviour in rats induced by various stimulant drugs. Psychopharmacologia. 1969;14(4):299–304.
49. Cenci MA, Lee CS, Bjorklund A. L-DOPA-induced dyskinesia in the rat is associated with striatal overexpression of prodynorphin- and glutamic acid decarboxylase mRNA. Eur J Neurosci. 1998;10(8):2694–706.
50. Lee CS, Cenci MA, Schulzer M, Bjorklund A. Embryonic ventral mesencephalic grafts improve levodopa-induced dyskinesia in a rat model of Parkinson's disease. Brain. 2000;123(Pt 7):1365–79.
51. Dekundy A, Lundblad M, Danysz W, Cenci MA. Modulation of L-DOPA-induced abnormal involuntary movements by clinically tested compounds: further validation of the rat dyskinesia model. Behav Brain Res. 2007;179(1):76–89.
52. Lindgren HS, Rylander D, Ohlin KE, Lundblad M, Cenci MA. The "motor complication syndrome" in rats with 6-OHDA lesions treated chronically with L-DOPA: relation to dose and route of administration. Behav Brain Res. 2007;177(1):150–9.
53. Ervin GN, Fink JS, Young RC, Smith GP. Different behavioral responses to L-DOPA after anterolateral or posterolateral hypothalamic injections of 6-hydroxydopamine. Brain Res. 1977;132(3):507–20.

54. Breger LS, Dunnett SB, Lane EL. Comparison of rating scales used to evaluate l-DOPA-induced dyskinesia in the 6-OHDA lesioned rat. Neurobiol Dis. 2013;50:142–50.
55. Cenci MA, Lundblad M. Utility of 6-hydroxydopamine lesioned rats in the preclinical screening of novel treatments for Parkinson's disease. In: LeDoux M, editor. Animal models of movement disorders. San Diego: Elsevier Academic Press; 2005. p. 193–208.
56. Andersson M, Hilbertson A, Cenci MA. Striatal fosB expression is causally linked with l-DOPA-induced abnormal involuntary movements and the associated upregulation of striatal prodynorphin mRNA in a rat model of Parkinson's disease. Neurobiol Dis. 1999;6(6):461–74.
57. Francardo V, Recchia A, Popovic N, Andersson D, Nissbrandt H, Cenci MA. Impact of the lesion procedure on the profiles of motor impairment and molecular responsiveness to L-DOPA in the 6-hydroxydopamine mouse model of Parkinson's disease. Neurobiol Dis. 2011;42(3):327–40.
58. Santini E, Valjent E, Usiello A, Carta M, Borgkvist A, Girault JA, et al. Critical involvement of cAMP/DARPP-32 and extracellular signal-regulated protein kinase signaling in L-DOPA-induced dyskinesia. J Neurosci. 2007;27(26):6995–7005.
59. Francardo V, Iderberg H, Lindgren H, Cenci MA. Rodent models of PD treatment-related complications. In: LeDoux M, editor. Movement disorders: genetic models. 2nd ed, (in Press).
60. Iderberg H, Rylander D, Bimpisidis Z, Cenci MA. Modulating mGluR5 and 5-HT1A/1B receptors to treat l-DOPA-induced dyskinesia: Effects of combined treatment and possible mechanisms of action. Exp Neurol. 2013;250:116–24.
61. Maries E, Kordower JH, Chu Y, Collier TJ, Sortwell CE, Olaru E, et al. Focal not widespread grafts induce novel dyskinetic behavior in parkinsonian rats. Neurobiol Dis. 2006;21(1):165–80.
62. Steece-Collier K, Collier TJ, Danielson PD, Kurlan R, Yurek DM, Sladek Jr JR. Embryonic mesencephalic grafts increase levodopa-induced forelimb hyperkinesia in parkinsonian rats. Mov Disord. 2003;18(12):1442–54.
63. Paille V, Henry V, Lescaudron L, Brachet P, Damier P. Rat model of Parkinson's disease with bilateral motor abnormalities, reversible with levodopa, and dyskinesias. Mov Disord. 2007;22(4):533–9.
64. Nadjar A, Gerfen CR, Bezard E. Priming for l-dopa-induced dyskinesia in Parkinson's disease: a feature inherent to the treatment or the disease? Prog Neurobiol. 2009;87(1):1–9.
65. Delfino MA, Stefano AV, Ferrario JE, Taravini IR, Murer MG, Gershanik OS. Behavioral sensitization to different dopamine agonists in a parkinsonian rodent model of drug-induced dyskinesias. Behav Brain Res. 2004;152(2):297–306.
66. Manson A, Stirpe P, Schrag A. Levodopa-induced-dyskinesias clinical features, incidence, risk factors, management and impact on quality of life. J Parkinsons Dis. 2012;2:189–98.
67. Cenci MA, Konradi C. Maladaptive striatal plasticity in L-DOPA-induced dyskinesia. Prog Brain Res. 2010;183:209–33.
68. Meissner W, Ravenscroft P, Reese R, Harnack D, Morgenstern R, Kupsch A, et al. Increased slow oscillatory activity in substantia nigra pars reticulata triggers abnormal involuntary movements in the 6-OHDA-lesioned rat in the presence of excessive extracellular striatal dopamine. Neurobiol Dis. 2006;22(3):586–98.
69. Lindgren HS, Andersson DR, Lagerkvist S, Nissbrandt H, Cenci MA. L-DOPA-induced dopamine efflux in the striatum and the substantia nigra in a rat model of Parkinson's disease: temporal and quantitative relationship to the expression of dyskinesia. J Neurochem. 2010;112(6):1465–76.
70. Pavon N, Martin AB, Mendialdua A, Moratalla R. ERK phosphorylation and FosB expression are associated with L-DOPA-induced dyskinesia in hemiparkinsonian mice. Biol Psychiatry. 2006;59(1):64–74.
71. Westin JE, Vercammen L, Strome EM, Konradi C, Cenci MA. Spatiotemporal pattern of striatal ERK1/2 phosphorylation in a rat model of L-DOPA-induced dyskinesia and the role of dopamine D1 receptors. Biol Psychiatry. 2007;62(7):800–10.
72. Sgambato-Faure V, Buggia V, Gilbert F, Levesque D, Benabid AL, Berger F. Coordinated and spatial upregulation of arc in striatonigral neurons correlates with L-dopa-induced behavioral sensitization in dyskinetic rats. J Neuropathol Exp Neurol. 2005;64(11):936–47.

73. Westin JE, Lindgren HS, Gardi J, Nyengaard JR, Brundin P, Mohapel P, et al. Endothelial proliferation and increased blood-brain barrier permeability in the basal ganglia in a rat model of 3,4-dihydroxyphenyl-L-alanine-induced dyskinesia. J Neurosci. 2006;26(37):9448–61.
74. Lindgren HS, Ohlin KE, Cenci MA. Differential involvement of D1 and D2 dopamine receptors in L-DOPA-induced angiogenic activity in a rat model of Parkinson's disease. Neuropsychopharmacology. 2009;34(12):2477–88.
75. Ohlin KE, Francardo V, Lindgren HS, Sillivan SE, O'Sullivan SS, Luksik AS, et al. Vascular endothelial growth factor is upregulated by L-dopa in the parkinsonian brain: implications for the development of dyskinesia. Brain. 2011;134(Pt 8):2339–57.
76. Lortet S, Lacombe E, Boulanger N, Rihet P, Nguyen C, Goff LK, et al. Striatal molecular signature of subchronic subthalamic nucleus high frequency stimulation in parkinsonian rat. PLoS One. 2013;8(4):e60447.
77. Carta M, Carlsson T, Kirik D, Bjorklund A. Dopamine released from 5-HT terminals is the cause of L-DOPA-induced dyskinesia in parkinsonian rats. Brain. 2007;130(Pt 7):1819–33.
78. Dupre KB, Ostock CY, Eskow Jaunarajs KL, Button T, Savage LM, Wolf W, et al. Local modulation of striatal glutamate efflux by serotonin 1A receptor stimulation in dyskinetic, hemiparkinsonian rats. Exp Neurol. 2011;229(2):288–99.
79. Miguelez C, Aristieta A, Cenci MA, Ugedo L. The locus coeruleus is directly implicated in L-DOPA-induced dyskinesia in parkinsonian rats: an electrophysiological and behavioural study. PLoS One. 2011;6(9):e24679.
80. Ostock CY, Dupre KB, Jaunarajs KL, Walters H, George J, Krolewski D, et al. Role of the primary motor cortex in L-Dopa-induced dyskinesia and its modulation by 5-HT1A receptor stimulation. Neuropharmacology. 2011;61(4):753–60.
81. Rylander D. The serotonin system: a potential target for anti-dyskinetic treatments and biomarker discovery. Parkinsonism Relat Disord. 2012;18 Suppl 1:S126–8.
82. Zeng BY, Iravani MM, Jackson MJ, Rose S, Parent A, Jenner P. Morphological changes in serotoninergic neurites in the striatum and globus pallidus in levodopa primed MPTP treated common marmosets with dyskinesia. Neurobiol Dis. 2010;40(3):599–607.
83. Carlsson T, Carta M, Winkler C, Bjorklund A, Kirik D. Serotonin neuron transplants exacerbate L-DOPA-induced dyskinesias in a rat model of Parkinson's disease. J Neurosci. 2007;27(30):8011–22.
84. Cenci MA, Ohlin KE, Odin P. Current options and future possibilities for the treatment of dyskinesia and motor fluctuations in Parkinson's disease. CNS Neurol Disord Drug Targets. 2011;10(6):670–84.
85. Marin C, Aguilar E, Obeso JA. Coadministration of entacapone with levodopa attenuates the severity of dyskinesias in hemiparkinsonian rats. Mov Disord. 2006;21(5):646–53.
86. Bibbiani F, Oh JD, Kielaite A, Collins MA, Smith C, Chase TN. Combined blockade of AMPA and NMDA glutamate receptors reduces levodopa-induced motor complications in animal models of PD. Exp Neurol. 2005;196(2):422–9.
87. Marin C, Aguilar E, Bonastre M, Tolosa E, Obeso JA. Early administration of entacapone prevents levodopa-induced motor fluctuations in hemiparkinsonian rats. Exp Neurol. 2005;192(1):184–93.
88. Oh JD, Bibbiani F, Chase TN. Quetiapine attenuates levodopa-induced motor complications in rodent and primate parkinsonian models. Exp Neurol. 2002;177(2):557–64.
89. Papa SM, Engber TM, Kask AM, Chase TN. Motor fluctuations in levodopa treated parkinsonian rats: relation to lesion extent and treatment duration. Brain Res. 1994;662(1–2):69–74.
90. Dupre KB, Eskow KL, Barnum CJ, Bishop C. Striatal 5-HT1A receptor stimulation reduces D1 receptor-induced dyskinesia and improves movement in the hemiparkinsonian rat. Neuropharmacology. 2008;55(8):1321–8.
91. Darmopil S, Muneton-Gomez VC, de Ceballos ML, Bernson M, Moratalla R. Tyrosine hydroxylase cells appearing in the mouse striatum after dopamine denervation are likely to be projection neurones regulated by L-DOPA. Eur J Neurosci. 2008;27(3):580–92.
92. Smith GA, Heuer A, Dunnett SB, Lane EL. Unilateral nigrostriatal 6-hydroxydopamine lesions in mice II: predicting l-DOPA-induced dyskinesia. Behav Brain Res. 2012;226(1):281–92.

93. Lundblad M, Picconi B, Lindgren H, Cenci MA. A model of L-DOPA-induced dyskinesia in 6-hydroxydopamine lesioned mice: relation to motor and cellular parameters of nigrostriatal function. Neurobiol Dis. 2004;16(1):110–23.
94. Jackson-Lewis V, Przedborski S. Protocol for the MPTP mouse model of Parkinson's disease. Nat Protoc. 2007;2(1):141–51.
95. Date I, Felten DL, Felten SY. Long-term effect of MPTP in the mouse brain in relation to aging: neurochemical and immunocytochemical analysis. Brain Res. 1990;519(1–2):266–76.
96. Mitsumoto Y, Watanabe A, Mori A, Koga N. Spontaneous regeneration of nigrostriatal dopaminergic neurons in MPTP-treated C57BL/6 mice. Biochem Biophys Res Commun. 1998;248(3):660–3.
97. Sedelis M, Hofele K, Auburger GW, Morgan S, Huston JP, Schwarting RK. MPTP susceptibility in the mouse: behavioral, neurochemical, and histological analysis of gender and strain differences. Behav Genet. 2000;30(3):171–82.
98. Nicholas AP. Levodopa-induced hyperactivity in mice treated with 1-methyl-4-phenyl-1,2,3,6-tetrahydropyridine. Mov Disord. 2007;22(1):99–104.
99. Lundblad M, Usiello A, Carta M, Hakansson K, Fisone G, Cenci MA. Pharmacological validation of a mouse model of l-DOPA-induced dyskinesia. Exp Neurol. 2005;194(1):66–75.
100. Jankovic J, Stacy M. Medical management of levodopa-associated motor complications in patients with Parkinson's disease. CNS Drugs. 2007;21(8):677–92.
101. Ahmed MR, Berthet A, Bychkov E, Porras G, Li Q, Bioulac BH, et al. Lentiviral overexpression of GRK6 alleviates L-dopa-induced dyskinesia in experimental Parkinson's disease. Sci Transl Med. 2010;2(28):28ra.
102. Westin JE, Andersson M, Lundblad M, Cenci MA. Persistent changes in striatal gene expression induced by long-term L-DOPA treatment in a rat model of Parkinson's disease. Eur J Neurosci. 2001;14(7):1171–6.
103. Thiele SL, Warre R, Nash JE. Development of a unilaterally-lesioned 6-OHDA mouse model of Parkinson's disease. J Vis Exp. 2012;(60).
104. Thiele SL, Warre R, Khademullah CS, Fahana N, Lo C, Lam D, et al. Generation of a model of L-DOPA-induced dyskinesia in two different mouse strains. J Neurosci Methods. 2011;197(2):193–208.
105. Grealish S, Mattsson B, Draxler P, Bjorklund A. Characterisation of behavioural and neurodegenerative changes induced by intranigral 6-hydroxydopamine lesions in a mouse model of Parkinson's disease. Eur J Neurosci. 2010;31(12):2266–78.
106. Fredduzzi S, Moratalla R, Monopoli A, Cuellar B, Xu K, Ongini E, et al. Persistent behavioral sensitization to chronic L-DOPA requires A2A adenosine receptors. J Neurosci. 2002;22(3):1054–62.
107. Hope BT, Nye HE, Kelz MB, Self DW, Iadarola MJ, Nakabeppu Y, et al. Induction of a long-lasting AP-1 complex composed of altered Fos-like proteins in brain by chronic cocaine and other chronic treatments. Neuron. 1994;13(5):1235–44.
108. Lindgren HS, Rylander D, Iderberg H, Andersson M, O'Sullivan SS, Williams DR, et al. Putaminal upregulation of FosB/FosB-like immunoreactivity in Parkinson's disease patients with dyskinesia. J Parkinsons Dis. 2011;177:347–57.
109. Cao X, Yasuda T, Uthayathas S, Watts RL, Mouradian MM, Mochizuki H, et al. Striatal overexpression of DeltaFosB reproduces chronic levodopa-induced involuntary movements. J Neurosci. 2010;30(21):7335–43.
110. Tashiro Y, Sugimoto T, Hattori T, Uemura Y, Nagatsu I, Kikuchi H, et al. Tyrosine hydroxylase-like immunoreactive neurons in the striatum of the rat. Neurosci Lett. 1989;97(1–2):6–10.
111. Jollivet C, Montero-Menei CN, Venier-Julienne MC, Sapin A, Benoit JP, Menei P. Striatal tyrosine hydroxylase immunoreactive neurons are induced by L-dihydroxyphenylalanine and nerve growth factor treatment in 6-hydroxydopamine lesioned rats. Neurosci Lett. 2004;362(2):79–82.
112. Meredith GE, Farrell T, Kellaghan P, Tan Y, Zahm DS, Totterdell S. Immunocytochemical characterization of catecholaminergic neurons in the rat striatum following dopamine-depleting lesions. Eur J Neurosci. 1999;11(10):3585–96.

113. Dubach M, Schmidt R, Kunkel D, Bowden DM, Martin R, German DC. Primate neostriatal neurons containing tyrosine hydroxylase: immunohistochemical evidence. Neurosci Lett. 1987;75(2):205–10.
114. Palfi S, Leventhal L, Chu Y, Ma SY, Emborg M, Bakay R, et al. Lentivirally delivered glial cell line-derived neurotrophic factor increases the number of striatal dopaminergic neurons in primate models of nigrostriatal degeneration. J Neurosci. 2002;22(12):4942–54.
115. Huot P, Parent A. Dopaminergic neurons intrinsic to the striatum. J Neurochem. 2007;101(6):1441–7.
116. DiCaudo C, Riverol M, Mundinano IC, Ordonez C, Hernandez M, Marcilla I, et al. Chronic levodopa administration followed by a washout period increased number and induced phenotypic changes in striatal dopaminergic cells in MPTP-monkeys. PLoS One. 2012;7(11):e50842.
117. Espadas I, Darmopil S, Vergano-Vera E, Ortiz O, Oliva I, Vicario-Abejon C, et al. L-DOPA-induced increase in TH-immunoreactive striatal neurons in parkinsonian mice: insights into regulation and function. Neurobiol Dis. 2012;48(3):271–81.
118. Dawson TM, Ko HS, Dawson VL. Genetic animal models of Parkinson's disease. Neuron. 2010;66(5):646–61.
119. Chesselet MF. In vivo alpha-synuclein overexpression in rodents: a useful model of Parkinson's disease? Exp Neurol. 2008;209(1):22–7.
120. Li Y, Liu W, Oo TF, Wang L, Tang Y, Jackson-Lewis V, et al. Mutant LRRK2(R1441G) BAC transgenic mice recapitulate cardinal features of Parkinson's disease. Nat Neurosci. 2009;12(7):826–8.
121. Li X, Patel JC, Wang J, Avshalumov MV, Nicholson C, Buxbaum JD, et al. Enhanced striatal dopamine transmission and motor performance with LRRK2 overexpression in mice is eliminated by familial Parkinson's disease mutation G2019S. J Neurosci. 2010;30(5):1788–97.
122. Itier JM, Ibanez P, Mena MA, Abbas N, Cohen-Salmon C, Bohme GA, et al. Parkin gene inactivation alters behaviour and dopamine neurotransmission in the mouse. Hum Mol Genet. 2003;12(18):2277–91.
123. Perez FA, Palmiter RD. Parkin-deficient mice are not a robust model of parkinsonism. Proc Natl Acad Sci U S A. 2005;102(6):2174–9.
124. Gautier CA, Kitada T, Shen J. Loss of PINK1 causes mitochondrial functional defects and increased sensitivity to oxidative stress. Proc Natl Acad Sci U S A. 2008;105(32):11364–9.
125. Gispert S, Ricciardi F, Kurz A, Azizov M, Hoepken HH, Becker D, et al. Parkinson phenotype in aged PINK1-deficient mice is accompanied by progressive mitochondrial dysfunction in absence of neurodegeneration. PLoS One. 2009;4(6):e5777.
126. Kitada T, Pisani A, Porter DR, Yamaguchi H, Tscherter A, Martella G, et al. Impaired dopamine release and synaptic plasticity in the striatum of PINK1-deficient mice. Proc Natl Acad Sci U S A. 2007;104(27):11441–6.
127. Goldberg MS, Pisani A, Haburcak M, Vortherms TA, Kitada T, Costa C, et al. Nigrostriatal dopaminergic deficits and hypokinesia caused by inactivation of the familial Parkinsonism-linked gene DJ-1. Neuron. 2005;45(4):489–96.
128. Ding Y, Restrepo J, Won L, Hwang DY, Kim KS, Kang UJ. Chronic 3,4-dihydroxyphenylalanine treatment induces dyskinesia in aphakia mice, a novel genetic model of Parkinson's disease. Neurobiol Dis. 2007;27(1):11–23.
129. Ding Y, Won L, Britt JP, Lim SA, McGehee DS, Kang UJ. Enhanced striatal cholinergic neuronal activity mediates L-DOPA-induced dyskinesia in parkinsonian mice. Proc Natl Acad Sci U S A. 2011;108(2):840–5.
130. Tayarani-Binazir KA, Jackson MJ, Rose S, Olanow CW, Jenner P. Pramipexole combined with levodopa improves motor function but reduces dyskinesia in MPTP-treated common marmosets. Mov Disord. 2010;25(3):377–84.
131. Rupniak NM, Tye SJ, Jennings CA, Loper AE, Bondi JV, Hichens M, et al. Antiparkinsonian efficacy of a novel transdermal delivery system for (+)-PHNO in MPTP-treated squirrel monkeys. Neurology. 1989;39(3):329–35.
132. Bezard E, Tronci E, Pioli EY, Li Q, Porras G, Bjorklund A, et al. Study of the antidyskinetic effect of eltoprazine in animal models of levodopa-induced dyskinesia. Mov Disord. 2013;28(8):1088–96.

133. Doudet D, Gross C, Lebrun-Grandie P, Bioulac B. MPTP primate model of Parkinson's disease: a mechanographic and electromyographic study. Brain Res. 1985;335(1):194–9.
134. Doudet DJ, Gross C, Arluison M, Bioulac B. Modifications of precentral cortex discharge and EMG activity in monkeys with MPTP-induced lesions of DA nigral neurons. Exp Brain Res. 1990;80(1):177–88.
135. Monville C, Torres EM, Dunnett SB. Comparison of incremental and accelerating protocols of the rotarod test for the assessment of motor deficits in the 6-OHDA model. J Neurosci Methods. 2006;158(2):219–23.
136. Dowd E, Monville C, Torres EM, Dunnett SB. The Corridor Task: a simple test of lateralised response selection sensitive to unilateral dopamine deafferentation and graft-derived dopamine replacement in the striatum. Brain Res Bull. 2005;68(1–2):24–30.
137. Olsson M, Nikkhah G, Bentlage C, Bjorklund A. Forelimb akinesia in the rat Parkinson model: differential effects of dopamine agonists and nigral transplants as assessed by a new stepping test. J Neurosci. 1995;15(5 Pt 2):3863–75.
138. Lundblad M, Vaudano E, Cenci MA. Cellular and behavioural effects of the adenosine A2a receptor antagonist KW-6002 in a rat model of l-DOPA-induced dyskinesia. J Neurochem. 2003;84(6):1398–410.

Chapter 19
Final Thoughts: Summary and Future Therapeutic Strategies in Levodopa-Induced Dyskinesia

Jonathan M. Brotchie and Susan H. Fox

Abstract Levodopa-induced dyskinesia represents an on-going challenge in the management of PD. Clinicians and scientists need to continue to work together in developing strategies to reduce and prevent LID. Here, we review and summarize the field to date as presented in the book and give an overview of future perspectives.

Keywords Levodopa • Non-dopaminergic • Animal models

We would like to take the opportunity to conclude this book by thanking all the authors who have contributed their time and expertise to make it a comprehensive survey of LID as it exists in 2014. As editors, it has been our pleasure to work with such esteemed, and valued, colleagues and friends in the field. We are also extremely grateful for the support and enabling role of the team at Springer who made the book possible. We would like to conclude by sharing our personal views on how we see the field moving forward. The views are solely our own and based upon experience, and perhaps prejudice, developed in our work over the last decades where we have been involved in the assessment of more than 40 potential antidyskinetic therapies in nonhuman primates and more than a dozen in clinical trials.

Throughout this book, authors have provided reviews of the state of the art with respect to clinical management (Chaps. 1, 3, 5, and 6) and understanding of the pathophysiology of LID (Chaps. 4, 7, and 8). Levodopa is still the most effective antiparkinsonian drug with least propensity for side effects and most cost-effective at improving PD patients' quality of life. The development of LID becomes part of the "cost" of this potential improvement in PD symptoms. Despite decades of study, the pharmacological properties of levodopa preclude long-term administration of the drug in a way that does not result in LID (Chaps. 9 and 10).

J.M. Brotchie (✉) • S.H. Fox
Division of Neurology, SHF Dept of Medicine, JMB Toronto Western Research Institute, Toronto Western Hospital, Toronto, ON, Canada
e-mail: j.brotchie@atuka.com

S.H. Fox, J.M. Brotchie (eds.), *Levodopa-Induced Dyskinesia in Parkinson's Disease*, DOI 10.1007/978-1-4471-6503-3_19

There is increasing understanding that for many PD subjects, putting up with a small degree of LID is better than the opposite clinical state of being off, slow, and stiff. We know that PD patients often do not appreciate that they even have LID, and so the clinical necessity to treat such movements may be driven by family and the physician, rather than the patient themselves. However, there still exists a significant number of PD patients, for whom LID is a bothersome symptom, including young-onset patients who will need a lifetime of levodopa therapy and an increasing proportion of patients who do not tolerate levodopa-sparing agents in particular dopamine agonists. The early use of surgical options (e.g., bilateral STN DBS) is still only for a select group. There thus remain a significant number of PD patients requiring management of LID. Targeting certain individuals based on genetic propensity to develop LID (Chap. 4) may be a future option, to rationalize and optimize such therapies.

We have come to understand that there exists a panoply of neurotransmitter systems, including mu-opioid, alpha-adrenergic, 5-HT-1A, −1B and 2A serotonergic, nicotinic cholinergic, CB1 cannabinoid, and both inotropic and metabotropic glutamate receptors, that have been validated as potential antidyskinetic therapeutic targets (Chaps. 11, 12, 13, 14, 15, and 16). Such validation has been delivered in rodent and nonhuman primate models (Chap. 18) and in many cases, in proof-of-concept Phase II clinical trials. Potential therapeutics acting at these targets have in common that their anticipated mode of utilization would most likely be as adjunctive therapy to levodopa. That is, they show potential to reduce the expression of dyskinesia once it has been established, without reducing the antiparkinsonian benefits of levodopa. In this scenario, they are analogous to, and potentially an extension of, the current use of amantadine. However, these new targets offer hope to provide benefit to those patients who currently do not benefit, or receive nonoptimal benefit, from combination of dopamine replacement therapy and amantadine. A major challenge in delivering this promise appears to be successful translation from demonstration of efficacy in nonhuman primates and Phase II clinical studies into success at Phase III trials and ultimately regulatory approval for clinical use. Thus, the nonhuman primate models of LID, based upon MPTP administration and chronic levodopa therapy, have proved extremely reliable in defining compounds and target drug exposure levels that show efficacy in Phase II. Moreover, the availability of the intravenous levodopa Phase IIa trial paradigm, pioneered by Dr Chase in the 1990s and early twenty-first century, allows rapidly demonstration of clinical proof of concept of approaches with efficacy in nonhuman primate, though admittedly in a non-real-world situation, in a small number of patients. Beyond Phase II, success seems more difficult to attain; in Phase III, we only have trials showing no significant efficacy, compared to placebo, to report.

The reasons behind the lack of success beyond Phase II are likely multitude, including inappropriate dosing, trial design, and the impact of a strong placebo effect. However, these issues are now becoming the focus of investigation (for instance, the work lead by Dr Goetz, see Chap. 2). With this emerging understanding, and new validated clinical rating scales, we see clear hope that the success of preclinical science and early-stage clinical development can be capitalized upon in the coming decade (Chap. 17).

Two features of LID research described above, strong predictive value of animal models, and increasing understanding of clinical translation should de-risk the process of investing in developing therapies for LID and are beginning to make LID an attractive indication for pharmaceutical companies. This is particularly the case with respect to indication switching of compounds, where the mechanism of action of such agents overlaps with one of the validated targets for LID. In such instances, LID can represent an attractive opportunistic route for a rapid transition to demonstration of efficacy, firstly in a nonhuman primate model and subsequently clinical proof of concept, for a compound for which the pharmacokinetic, metabolism, and safety properties are already well understood.

It is clear from the above discussion, and indeed the spectrum of transmitter systems covered in individual chapters herein, that LID is not a simple problem of enhanced dopaminergic signalling. Complex cascades of compensation, for loss of dopaminergic transmission, and plasticity, driven by pulsatile dopaminergic stimulation, impact on multiple transmitter systems and contribute to the development and expression of LID. It is clear, in both nonhuman primates and in Phase II clinical studies, where available, that the actions of agents acting on any single pharmacological target are likely to have a range of efficacy across a patient population. Thus, in any individual, the contribution of different transmitter systems to the mechanisms underlying their LID is likely idiosyncratic. A corollary of this is that no single antidyskinetic agent is likely to be able to completely suppress the expression of LID once established. It is thus, perhaps, surprising that, hitherto, therapeutic approaches modulating multiple targets have been little studied. Indeed, for most of the potential therapeutic agents/targets discussed in the preceding chapters, combination with standard clinical care, amantadine has not been rigorously investigated. As we move forward, an understanding of such interactions could prove extremely valuable. Firstly, on a purely logistical level, it could define whether clinical trials to assess efficacy of an approach should/could include patients already receiving benefit from amantadine therapy and indeed should exclude those who have not received such amantadine benefit. Such an understanding could dramatically empower our ability to demonstrate efficacy in clinical studies, for instance, optimizing power calculations of study sample size and recruitment. Secondly, and more importantly, by overlooking potential interactions between different targets, we may be missing significant opportunities for synergy and improved efficacy. The idea of developing therapies that combine actions at multiple targets is beginning to gain traction in the serotonergic space. Thus, there is some perception that combination of 5-HT1A and 5-HT1B agonists might, by allowing the use of lower doses of both, be able to deliver benefits of both targets while minimizing any adverse effects of one or the other. One approach to this is to develop compounds that are multifunctional, acting on more than one receptor. The problem we envisage with this approach is that it seems unlikely that a single molecule can capture the relative combination of multiple receptor blockade/stimulation that would provide optimal efficacy. This is compounded by our impression that there is likely no single optimal dose for any compound across a population. Certainly, in nonhuman primates, we find that the lowest effective dose of drug, in terms of antidyskinetic action, varies by a factor of tenfold or more even within a study. For a combination of two more targets, such variability would be compounded. A more attractive approach to modulating multiple targets is, to our minds, polypharmacy where the dose of each agent can be tailored/titrated to an individual's response. This might be achieved with a therapy combining multiple active molecules, though this is associated with multiple development challenges and, as with the single multifunctional molecule, a combination therapy may be limited by being only available in one, or at least invariable, combinations. We therefore propose that the therapeutic landscape for LID will/should evolve in a way in which multiple drugs/targets are developed in parallel and that the armamentarium should be built organically. As compounds are developed, studies in nonhuman primates should be expanded to focus on synergy, additivity, or lack of, between targets and also on defining whether certain populations of patients might befit more than others from drugs for a particular target. With respect to the latter, it is already clear that some classes of drug act preferentially to reduce LID of a choreic, rather than dystonic, phenotype, and vice versa, while others reduce dyskinesia elicited by levodopa but not dopamine receptor agonists. At present, such considerations are rarely taken into account when transitioning a development project from nonhuman primate to clinical development but could become even more important in defining how to employ therapies once they reach the market. Moreover, as agents other than amantadine become available, neurologists will learn in an empirical manner which patients respond best to which combinations, and best practice for combining the multiple pharmacologies will become defined.

The discussion above, and indeed the vast majority of the chapters preceding, has focussed primarily upon the issue of understanding and managing LID once it has become established. A major opportunity to reduce the impact of LID on patients with PD exists if we can prevent, *de novo*, the development of LID once dopamine replacement therapy is initiated. Over the last decade, and more, the issue of continuous dopaminergic stimulation to prevent the development of dyskinesia has gained much attention. However, to date, it has proved impossible to deliver antiparkinsonian benefits equivalent to levodopa while avoiding the development of LID. Alternative strategies should be investigated, and we have been attracted to the potential of adjunctive therapies that leverage the antiparkinsonian benefit of levodopa but combine levodopa with an agent that reduces its propensity to lead to the development of LID. Compounds with potential to achieve this goal, as indicated by nonhuman primate studies, include those acting as A2A adenosine receptor, D3 dopamine receptor, and NR2B NMDA receptor antagonists. Indeed, we have long espoused that such an indication represents the biggest opportunity for A2A adenosine antagonists in PD. It should be noted that of these three potential therapeutic targets for reducing the development of LID, none would be considered, by us at least, as having significant potential in reducing LID once it has been established. Thus, we note that the pharmacology of the development of LID is very different than that of its expression. Moreover, this leads us to believe that separate paths of drug development are needed to prevent rather than diminish previously established dyskinesia. One attraction of developing agents to prevent the development of LID is that, unlike approaches to suppress established LID, which as discussed above, have yet to succeed at Phase III, a path through Phase III to market has already been demonstrated for de novo therapy that leads to reduced development of LID, for dopamine agonists. However, none of the three targets for preventing development of LID proposed above has been investigated for such potential in clinical studies. The reasons for this are likely not solely scientific. The magnitude of a clinical proof-of-concept study, likely several hundred patients over 3–5 years, is perceived as too large to justify the potential investment for anticipated reward. We feel this undervalues the impact of LID, both from a clinical perspective and also from a commercial market perspective. In the absence of a true disease-modifying agent in PD, an agent that was able to prevent development of LID while allowing the antiparkinsonian benefit of levodopa would form the basis of product with potential for annual sales in excess of $1bn. A major challenge that faces us today is to convince our partners and colleagues in the pharmaceutical sector that a de novo therapy to slow or prevent LID development to represent an unmet need with potential impact equivalent to a disease-modifying therapy or a symptomatic therapy in a disorder with greater incidence than PD. The translatability of our animal models and their value in de-risking investment should help in this respect.

In conclusion, the reviews presented through this book illustrate the significant advances that have been made in LID over recent years. They highlight the rapidly changing face of this important disease area. The discussion presented in this last chapter is given to encourage thought and debate and highlight the opportunities that remain ahead of us.

Index

S.H. Fox, J.M. Brotchie (eds.), *Levodopa-Induced Dyskinesia in Parkinson's Disease*, DOI 10.1007/978-1-4471-6503-3

CPSIA information can be obtained
at www.ICGtesting.com
Printed in the USA
LVHW05*1952220818
587776LV00001B/45/P